FEEDING SYSTEMS AND
FEED EVALUATION MODELS

FEEDING SYSTEMS AND FEED EVALUATION MODELS

Edited by

M.K. Theodorou

*Department of Animal Science and Microbiology
Institute of Grassland and Environmental Research
Aberystwyth
UK*

and

J. France

*Department of Agriculture
University of Reading
UK*

CABI *Publishing*

CABI *Publishing* is a division of CAB *International*

CABI Publishing
CAB International
Wallingford
Oxon OX10 8DE
UK

CABI Publishing
10 E 40th Street
Suite 3203
New York, NY 10016
USA

Tel: +44 (0)1491 832111
Fax: +44 (0)1491 833508
Email: cabi@cabi.org

Tel: +1 212 481 7018
Fax: +1 212 686 7993
Email: cabi-nao@cabi.org

A catalogue record for this book is available from the British Library, London, UK.

Library of Congress Cataloging-in-Publication Data
Feeding systems and feed evaluation models / edited by M.K. Theodorou and J.
  France.
     p.   cm.
  Includes bibliographical references.
  ISBN 0-85199-346-X (alk. paper)
    1. Feeds—Evaluation.   I. Theodorou, M.K. (Michael K.)   II. France, J. (Jim).
  SF97.F38 1999
  636.08'4—dc21   99-35147
                                                                              CIP

ISBN 0 85199 346 X

Typeset by AMA DataSet Ltd, UK
Printed and bound in the UK by Biddles Ltd., Guildford and King's Lynn

Contents

Contributors

G. Alderman, *The University of Reading, Department of Agriculture, Earley Gate, Reading, Berkshire RG6 6AT, UK*

R.L. Baldwin, *Department of Animal Science, University of California, Davis, CA 95616-8521, USA*

A. Bannink, *DLO-Institute for Animal Science and Health (DL-DLO), Department of Ruminant Nutrition, PO Box 65, 8200 AB Lelystad, The Netherlands*

D.E. Beever, *The University of Reading, Department of Agriculture, Earley Gate, Reading, Berkshire RG6 6AT, UK*

J.L. Black, *John L. Black Consulting, Locked Bag 21, Warrimoo, NSW 2774, Australia*

G.A. Broderick, *US Dairy Forage Research Center, Agricultural Research Service, USDA, Madison, WI 53706, USA*

J.G. Buchanan-Smith, *Department of Animal and Poultry Science, University of Guelph, Guelph, Ontario, Canada, N1G 2W1*

I.H. Burger, *Waltham Centre for Pet Nutrition, Waltham-on-the-Wolds, Melton Mowbray, Leicestershire LE14 4RT, UK*

A. Chesson, *Rowett Research Institute, Bucksburn, Aberdeen AB21 9SB, UK*

L.I. Chiba, *Department of Animal and Dairy Sciences, Animal Science Building, Auburn University, AL 36849-5415, USA*

R.C. Cochran, *Department of Animal Sciences and Industry, Kansas State University, Manhattan, KA 66506, USA*

D. Cuddeford, *Department of Veterinary Clinical Studies, Royal (Dick) School of Veterinary Studies, The University of Edinburgh, Veterinary Field Station, Easter Bush, Roslin, Midlothian EH25 9RG, UK*

Y. Cui, *Institute of Hydrobiology, The Chinese Academy of Sciences, Wuhan, Hubei 430072, China*

J. Dijkstra, *Animal Nutrition Group, Wageningen Institute of Animal Sciences (WIAS), Wageningen Agricultural University, Marijkeweg 40, 6709 PG Wageningen, The Netherlands*

K.C. Donovan, *Department of Animal Science, University of California, Davis, CA 95616-8521, USA*

D.G. Fox, *Department of Animal Science, Cornell University, Ithaca, NY 14853, USA*

J. France, *The University of Reading, Department of Agriculture, Earley Gate, Reading, Berkshire RG6 6AT, UK*

W.J.J. Gerrits, *Animal Nutrition Group, Wageningen Institute of Animal Sciences (WIAS), Wageningen Agricultural University, PO Box 338, 6700 AH Wageningen, The Netherlands*

G. Hof, *Animal Nutrition Group, Wageningen Institute of Animal Sciences, Wageningen Agricultural University, ZODIAC, Marijkeweg 40, Wageningen, The Netherlands*

S. Leeson, *Department of Animal and Poultry Science, University of Guelph, Guelph, Ontario, Canada, N1G 2W1*

A.C. Longland, *Institute of Grassland and Environmental Research, Plas Gogerddan, Aberystwyth, Ceredigion SY23 3EB, UK*

R.S. Lowman, *Brackenhurst College, Southwell, Nottinghamshire N25 0QF, UK*

M.G. MacLeod, *Roslin Institute (Edinburgh), Roslin, Midlothian EH25 9PS, UK*

S.R. McLennan, *Queensland Beef Industry Institute, Animal Research Institute, Yeerongilly 4105, Brisbane, Australia*

D.P. Poppi, *School of Land and Food, University of Queensland 4072, Brisbane, Australia*

C.K. Reynolds, *Department of Agriculture, The University of Reading, Earley Gate, PO Box 236, Reading, Berkshire RG6 6AT, UK*

L.A. Sinclair, *Harper Adams University College, Edgmond, Newport, Shropshire TF10 8NB, UK*

J.D. Summers, *Department of Animal and Poultry Science, University of Guelph, Guelph, Ontario, Canada, N1G 2W1*

S. Tamminga, *Animal Nutrition Group, Wageningen Institute of Animal Sciences, Wageningen Agricultural University, ZODIAC, Marijkeweg 40, Wageningen, The Netherlands*

M.K. Theodorou, *Institute of Grassland and Environmental Research, Plas Gogerddan, Aberystwyth, Ceredigion SY23 3EB, UK*

R.G. Wilkinson, *Harper Adams University College, Edgmond, Newport, Shropshire TF10 8NB, UK*

S. Xie, *Institute of Hydrobiology, The Chinese Academy of Sciences, Wuhan, Hubei 430072, China*

Preface

As we look back over the millennium, it is difficult to imagine man's evolution in the absence of domesticated livestock. Likewise, domesticated animals are so dependent upon man that in his absence their very existence would be jeopardized to the point where they would not thrive and some would fail to survive. Livestock husbandry developed because it was possible to demonstrate a clear and beneficial relationship between man's intervention and productivity. Thus, it was recognized that an animal's performance could be manipulated and that certain animal feeds or combinations of feeds were nutritious and enhanced performance, whereas others were of less benefit, some being detrimental or even toxic.

As the nutritional sciences developed in modern times, and in an attempt to formalize traditional knowledge, methodologies were elaborated to characterize feedstuffs in order to predict the performance of livestock. Today, we can describe the biological, chemical and physical properties of feedstuffs with an array of sophisticated instrumentation and to an ever-increasing degree of accuracy. However, such exacting science is pointless unless these feedstuff measurements have practical value. By the middle of the 19th century, tables of feeds ranked by nitrogen content were available. The turn of the 20th century saw the beginning of research culminating in feed formulation strategies for ruminants based upon, for example, total digestible nutrients, and starch or corn equivalents. Although these systems of feed evaluation underpin the present-day concepts of digestible and metabolizable energy, their value was lost by the latter part of the 20th century because of the introduction of more intensive livestock production systems. In modern times, the practical goal of any feeding system is to optimize the efficiency of feed utilization, animal output and ultimately financial return to the producer. Broadly speaking, the science underpinning this goal can be broken down into three components and these are melded together in the current net energy and net protein feeding systems. At a scientific level, the components of our modern-day feeding systems are therefore concerned with (i) methods to describe the animal feedstuffs, (ii) evaluation of the effect on animal response of the ingested nutrients and (iii) the development of suitable predictive routines (normally based upon empirical equations, but as we look to the future, augmented or replaced altogether by mechanistic models) to determine how a desired level of performance can be achieved using various diets or dietary constituents.

As they move into the new millennium, livestock farmers are faced with new challenges. Animals of higher genetic potential have been produced, although in many parts of the world punitive legislation has been introduced to limit the pollution of land and water courses caused by their intensive production. Public concern over animal welfare and the use of genetically modified crops has meant that intensive livestock farming is now less acceptable in certain parts of the world and we are uncertain if this trend is set to continue. Fish and crustacean agriculture has increased more than threefold over the past decade, to a point where farmed fish consume 3 to 4% of total world feed production, bringing with it many problems of resource allocation and pollution control. Ownership of horses and companion animals has also seen unprecedented growth, largely due to the increasing affluence and leisure time of people in the more developed regions of the world.

Although our current feeding systems were developed on the back of a profusion of measurements, they are in fact based upon relatively few nutritional concepts. In the future it will be necessary to define feeds and feed ingredients more tightly to allow predictions to be made with much smaller margins of error whilst taking account of the environmental conflict of production and the quality of the resultant produce. Future rationing systems will therefore benefit from a greater insight into the effects of nutrition on the utilization of energy and protein within the body. Perhaps more importantly, the ability to predict responses and the partitioning of absorbed nutrients will only be achieved by a discriminating view of the biological processes which determine the animal's productive response to its feed. By assembling the feed evaluation methods, feeding systems and feeding models available for livestock production today, we believe that we will be assisting in the development of the new feeding systems of tomorrow.

The scope of the book is best seen in the Contents, but for ease of exposition it is arranged in three sections: methods, systems and models. The chapters stand fairly well alone and can be read out of sequence. While we have tried to cover the areas we judged most relevant to the agricultural scientist, inevitably our own experience and competence and those of our contributors have greatly influenced the choice of topics and their depth of treatment. None the less, all concerned with researching and teaching animal agriculture should find this book of value, including those involved in extension work. The book should also be essential reading for graduate students, and undergraduates of the nutritional sciences could derive much benefit from its study.

ACKNOWLEDGEMENTS

We would like to acknowledge the proof reading contributions made by Geoff Alderman and Lori Jones, and the significant organizational and word processing skills of Christine March.

M.K. Theodorou and J. France

1 Feed Evaluation for Animal Production

J. France[1], M.K. Theodorou[2], R.S. Lowman[3]
and D.E. Beever

[1]The University of Reading, Department of Agriculture, Reading, Berkshire;
[2]Institute of Grassland and Environmental Research, Aberystwyth, Ceredigion;
[3]Brackenhurst College, Southwell, Nottinghamshire, UK

INTRODUCTION

This chapter outlines the role of feed evaluation in animal production, providing an overview of feed evaluation methods, current feeding systems and empirical and mechanistic modelling, and attempts to link these methods, systems and models.

THE NEED FOR FEED EVALUATION

Animal production is concerned with providing food (and clothing) of animal origin for man. Animal production science, which underpins this goal, provides the rational basis for livestock management practices. Feed evaluation concerns the use of methods to describe animal feedstuffs with respect to their ability to sustain different types and levels of animal performance. In feed evaluation, emphasis is placed on determining specific chemical entities, although the physical characteristics of the feed are also important. Subsequently, the acquired data are used, with appropriate animal indices, in feeding systems comprising suitable predictive routines (normally based on empirical equations) to determine whether a desired level of animal performance can be achieved from various diets.

The practical goal of feed evaluation is to optimize the efficiency of feed utilization, animal output and, ultimately, financial return to the producer. In this context, it is important to establish the potential of major feedstuffs and the need for appropriate supplements in order to overcome nutritional deficiencies and raise the level of performance. With respect to the animal, the level of performance will be dictated by the amount of feed voluntarily consumed and the efficiency of utilization of the major nutrients, namely energy and protein. Furthermore, the composition of the animal products (e.g. fat and protein in meat and milk) is important, as energy retention *per se* is no longer an adequate index of the performance of the animal or of the nutritive value of the feed.

To meet the requirements, recourse to the feedstuff and characterization of the appropriate indices that are thought likely to influence the level and type of animal performance is necessary. In this chapter, consideration is given to identifying the most important indices, and how these can be used to improve the prediction of animal performance. This is done with reference to the ruminant, as ruminant species have a central role in animal production; other chapters in this volume are concerned with the performance of the monogastric animal.

DIGESTION AND METABOLISM

The first stage in the identification of the most important indices is a consideration of the biology and chemistry of the feed and the links between feed composition on the one hand and output of animal products on the other. It is therefore essential to understand how feed is digested and the main metabolic pathways occurring within the animal.

For most diets, extensive degradation and fermentation of the digestible nutrients occurs within the rumen. Plant cell walls contribute the bulk of the available carbon and energy in the rumen, although soluble sugars, including fructans in grasses, can make a significant contribution, particularly during the initial fermentation period, due to their rapid degradation. Degraded carbohydrates are either used directly by the microbial population in the synthesis of microbial biomass or are fermented with the concomitant production of ATP, which is essential to meet the energy requirements of the population. The fermentation process is associated with the production of volatile fatty acids (VFAs), the extent of VFA production (i.e. moles produced per gram of carbohydrate fermented) being related to the maintenance requirements and growth rates of the microorganisms. The molar proportions of individual VFAs reflect the metabolic activities of the dominant microbial species, whose presence is in part regulated by the type of feed consumed.

Unless significant quantities of undegradable protein are included in the diet, a large proportion of the ingested protein is degraded to amino acids, with further dissimilation by microbial deamination to ammonia. Microorganisms incorporate both amino acids and ammonia into microbial protein, but ammonia is the major precursor. Efficiency of capture of degraded nitrogen (N) is known to vary, and depends upon the availability of N substrates (e.g. ammonia or amino acids) and energy (ATP). When ruminally degradable N is supplied in excess, the efficiency of microbial N capture falls, absorption of ammonia from the rumen increases and microbial protein flow to the small intestine is reduced (in relation to N supply).

The supply and absorption of starch from the small intestine is low on most diets, the most notable exception being diets containing ground maize, where up to 30% of the ingested starch can escape rumen degradation and be absorbed as glucose from the small intestine. Protein and lipid digestion and absorption are the other major activities occurring in the small intestine, whilst processes in the caecum and colon are confined largely to the fermentation of residual carbohydrates, with the resultant production of some VFAs to augment the major contribution from the rumen.

Thus, degradation and fermentation in the rumen and digestion in the hind-gut influences the quantity and quality of nutrients absorbed, and these processes are affected by nutrient interactions especially with respect to energy and protein.

At the tissue level, nutrient utilization depends upon the nature and quantity of the absorbed nutrients and the physiological and hormonal status of the animal. A priority of the animal is maintenance of body function, which requires the supply of ATP via oxidation, with acetate being a major precursor. Equally, obligatory protein turnover requires the diversion of some absorbed amino acids to refurbish existing tissues. These costs can be considered as essential maintenance costs but, due to other inefficiencies within the tissues, they do not reflect the total loss of energy and protein at the tissue level.

The mechanisms controlling protein synthesis and protein degradation have yet to be fully elucidated. There have, however, been many attempts to manipulate protein metabolism by the use of exogenous hormones to widen the gap between synthesis and degradation, so permitting an increased deposition of tissue protein. Protein synthesis *per se* is influenced by the quantity of amino acids arriving at the productive tissues and the composition of these amino acids in relation to the amino acid composition of the tissue or milk being synthesized. Inefficiencies in the transfer of absorbed amino acids to tissue (or milk) protein occur, and this has led to the development of specific amino acids (methionine and lysine) as feed additives protected from degradation in the rumen.

The synthesis of milk fat and body fat from acetate requires glucose to supply glycerol-6-phosphate and NADPH, as an essential cofactor. Propionate augments the supply of glucose absorbed directly from the gut whilst, in situations of acute glucose insufficiency, gluconeogenesis of amino acids occurs. In excess of requirements, the efficiency of acetate disposal declines, excessive oxidation of acetate occurs via futile cycles, and heat production ensues. This in turn acts as a negative feedback on voluntary feed intake, along with plasma acetate levels, which increase under such conditions.

METHODS USED IN FEED EVALUATION

Once the nutrient requirements of the animal have been established, a diet that provides the correct balance of nutrients can be formulated if accurate information on the feedstuffs is available. The vast majority of feedstuffs consist of plants and plant products, though products of animal origin such as fish meal, meat and bone meal, and milk are also fed.

Ruminants are fed predominantly on forages, either in the form of grazed pasture or after conservation, usually as silage. Most silage is made from grass but, increasingly, alternative crops such as forage legumes, forage maize, whole crop cereals, kale, etc. are being used as a source of conserved feed. Concentrate feeds, high in carbohydrates, e.g. cereal grains, or in proteins, e.g. soybean, may also be fed, depending on the level of animal performance required.

Due to their very nature, concentrate feeds generally show little variation in composition. Forages, however, are extremely variable, their composition being highly dependent upon the stage of growth at harvest (or at grazing), the plant species, the proportion of leaf to stem, the fertilizer treatment, etc. Young grass, especially if heavily fertilized, is high in protein and non-protein nitrogen compounds but low in soluble and cell wall structural carbohydrates; the cell wall is relatively unlignified and is therefore highly degradable. At the other extreme, mature grass is high in structural carbohydrates but the cell walls are highly lignified and of low digestibility. Moreover, mature grass, although generally high in available carbohydrates, such as fructans, is low in protein.

The composition of silage tends to reflect that of the crop at the time of ensilage, at least with regard to the structural plant cell wall components. Important differences include (i) the low pH of the silage brought about by fermentation of the soluble carbohydrates to form organic acids, mainly lactic acid, by the ensilage bacteria (lactic acid bacteria) and (ii) the fact that proteolysis (partially mediated by plant enzymes) has increased the proportion of non-protein nitrogen in the resultant silage. Similar considerations apply to alternative crops such as forage legumes and forage maize for example, except protein and starch, respectively in these crops survive the ensilage process and are important available protein and energy constituents of the resultant silages.

Chemical Analysis

The analysis of ruminant feeds generally involves determining the dry matter (DM), organic matter (OM), structural carbohydrate (fibre or non-starch polysaccharide, NSP), soluble carbohydrate, starch (where applicable) and crude protein (CP) content of the feedstuff. Silages require further analysis, notably for their pH, ammonia N and organic acid contents; recent research suggests that their true protein content should also be characterized.

The DM of a feedstuff is usually determined by oven drying at 60 or 100°C, whilst silages require special treatment (e.g. toluene DM determinations) due to their high content of volatile organic acids; thus DM is usually determined by distillation. OM is determined by dry ashing (at 500°C until all the carbon has been removed); the residue or ash can be used to determine the content of individual mineral elements in the feedstuff.

The most widely used methods for analysing the structural constituents, or fibre, are the detergent extraction methods of Van Soest. These methods involve extraction of plant biomass with neutral detergent to leave a fibrous residue of predominantly cellulose, hemicellulose and lignin (i.e. the neutral detergent fibre or NDF of plant cell walls) or with acid detergent to leave a residue of cellulose and lignin (i.e. the acid detergent fibre or ADF of plant cell walls). As these are gravimetric procedures, the exact composition of the NDF and ADF residues is not known. The fibre content of a feedstuff may be described more accurately by NSP analysis, whereby alditol acetate derivatives of carbohydrate monomers derived from acid hydrolysis of washed, polymeric, de-starched samples are quantified by gas chromatography. With NSP analysis in addition to obtaining details of the chemical composition of the fibre, the values measured are independent of food processing and storage, and hence the amounts present can be quantified more accurately.

Crude protein content is calculated from the nitrogen (N) content, determined by the Kjeldahl procedure involving acid digestion and distillation. More recently, Dumas methods, involving combustion and determination of released gaseous N, are being used. Ammonia N in fresh silage is determined on water extracts by either distillation or use of specific ion-sensitive electrodes. These methods measure N rather than protein; the quantity of N is therefore multiplied by 6.25 (assuming the N is derived from protein containing 16% nitrogen) to obtain an approximate protein value.

In recent years, near infrared reflectance spectroscopy (NIRS) has also been adopted for determining the composition of feedstuffs. In terms of accuracy, precision, speed and unit cost of analysis, the NIRS technique, provided it is calibrated correctly, is preferable to traditional laboratory methods. However, the technique ultimately relies on a set of stan ard samples whose composition has been determined by traditional methods.

Digestibility

In addition to chemical composition, several methods have been developed to characterize feedstuffs in terms of their digestibility. These comprise *in vivo*, *in situ* and *in vitro* methods. *In vivo* measurements provide the standard measure of digestibility as they represent the actual animal response to a dietary treatment. However, such trials cannot be considered routine in most laboratories, and cannot be carried out for all the possible feeding situations found in practice. Therefore, a number of *in vitro* and *in situ* methods (e.g. batch culture digestibility, enzyme digestibility, gas production, polyester bag) have been developed to estimate digestibility and the extent of ruminal degradation of feedstuffs, and to study their variation in response to changes in rumen conditions. Thus *in vitro* and *in situ* techniques may be used to study individual processes, providing information about their nature and sensitivity to various changes. This information is of great importance in the development of mechanistic models.

SYSTEMS OF FEED EVALUATION

Despite the apparent complexity of nutrient metabolism, current systems (see, for example, AFRC, 1993) to predict energy and protein utilization and voluntary intake are relatively simple. This *per se* does not suggest that they are inadequate, as the value of such must be judged against their ability to predict animal performance accurately in practice.

Energy Concepts

Metabolizable energy (ME) is the currently accepted unit of energy, representing an approximation of the total amount of energy available for metabolism, without characterization of that energy with respect to specific nutrients. In practice, few direct estimates of ME contents are available, and many are derived from digestibility and urine output measurements, conducted at maintenance levels of feeding with mature sheep, and adjusted for estimates of methane production. Alternatively, for forages and compound feeds, frequently ME contents are predicted from laboratory assessments (e.g. *in vitro* digestibility) and previously derived relationships.

It is difficult to assess to what extent these procedures bias the estimate of ME content. In general, however, current approaches to determine ME content and hence ME intake are considered satisfactory, but estimates of the efficiency of utilization of ME for maintenance (k_m), growth (k_f) and lactation (k_l) are subject to doubt. Currently, estimates of diet metabolizability (ME gross energy^{-1}, or q) are used in predictive routines to determine efficiencies of energy utilization, with some recognition of different dietary classes (e.g. primary and re-growths of forage; mixed diets) being reflected in the equations used to predict k_f. This is probably an oversimplification of the processes involved. From the data on growing cattle in Table 1.1, it can be seen that energy retention predicted by the AFRC system is substantially greater than that obtained from comparative slaughter, suggesting that the metabolism of growth is represented insufficiently. A major criticism of current energy

Table 1.1.   Comparison of energy retention (MJ day^{-1}) obtained by comparative slaughter (CS) and predicted by AFRC.

Diet		Energy retention (MJ day^{-1}) estimated by	
Silage	Barley (g kg^{-1} total DM)	CS	AFRC
Late cut	0	5.5	9.5
Late cut	230	9.2	13.3
Late cut	560	14.6	15.7
Early cut	0	12.2	18.0

Source: Beever *et al.* (1988) and AFRC (1993).

systems is that they are unable to predict the composition of gain, which is now of primary importance to producers.

Similar problems exist with current systems regarding lactation. With the AFRC system, for example, an underestimate of 5% in ME content can lead to an overestimate of feed allowance by as much as 1.5 kg of DM day^{-1} for a 600 kg cow producing 30 kg of milk, due to compounding errors in estimating q. The system tends to under-predict milk energy output in early lactation, whilst prediction of milk output over short periods of lactation appears to be satisfactory. A major criticism of all energy systems is that they are unable to predict milk constituent yield. Equally, the current systems cannot be used to predict how performance will change in response to deliberate changes in feeding strategy.

Protein Concepts

In general, the concepts included in the AFRC protein system are scientifically defensible, and recognition of energy–protein interactions, albeit only at the ruminal level, represents a significant advance. However, reservations exist about the values ascribed to several of the factors. Microbial protein synthetic efficiency is not constant in reality. Equally, values obtained from the nylon bag technique regarding likely yields of ruminally degradable dietary N (RDN) and undegradable dietary N (UDN) should be treated with caution, especially as they offer no qualification of the form of the RDN. Furthermore, the application of a constant biological value to absorbed protein is an oversimplification. The extent of microbial synthesis, the supply of undegraded dietary protein to the small intestine, protein availability within the small intestine, and gut and liver metabolism of specific amino acids each can markedly influence the composition of the amino acids which ultimately arrive at the productive tissues. Not surprisingly, direct evaluation of the AFRC protein system has revealed a number of discrepancies.

Intake Prediction

A number of empirical equations have been derived to predict voluntary feed intake by ruminants, especially dairy cows. The seven equations compared by Neal *et al.* (1984) are typical.

The accuracy of the predictions of DM intake was examined using live weights recorded weekly (W), at calving (C) and using estimated values based on breed weight (B) coupled with a notional pattern of live weight change (Table 1.2). Mean square prediction errors (MSPE) were calculated for each week using data obtained from cows and heifers offered silage and concentrate, and summarized over the whole experimental period. Whilst there was considerable variation between the equations, the basis used for live weight had little effect on the overall accuracy of the predictions. The greatest

Table 1.2. Comparison of the mean square predicted error (MSPE) values for seven intake equations compared.

Equation	MSPE			
	W	C	B	Average
ARC (1980)	4.0	5.1	3.4	4.2
Bines *et al.* (1977)	10.8	5.4	6.4	7.5
Lewis (1981)	2.5	3.1	2.8	2.8
MAFF (1975)	3.3	4.7	3.0	3.7
Modified MAFF	3.8	5.0	3.3	4.0
Vadiveloo and Holmes (1979) (Eqn 1)	2.1	2.8	2.4	2.4
Vadiveloo and Holmes (1979) (Eqn 2)	2.8	3.4	3.9	3.4
Average	4.2	4.2	3.6	

MSPE values for equations were based on live weights recorded weekly (W), at calving (C) and using estimated values based on breed weight (B). Source: Neal *et al.* (1984).

accuracy was with the three equations that included some dietary characteristics as variables.

Assuming an efficiency of utilization of ME for milk production of 0.62 and energy values of feed and milk of 11.0 (ME) and 2.9 (gross energy) MJ kg^{-1}, respectively, an MSPE of 3.0 (kg DM)^2 is equivalent to an average error of ± 1.7 kg in daily DM intake which could, at zero body energy change, lead to an error of ± 4.0 kg day^{-1} in milk output. This level of error is large. As voluntary feed consumption is the net result of a variety of interactions occurring at both the rumen and tissue levels, more detailed representations are needed than can be provided by simple regression equations. This suggests a role for mechanistic modelling.

POTENTIAL OF MECHANISTIC MODELLING

New feeding systems need to be based on the mechanisms that govern the response of animals to nutrients, dealing with quantitative aspects of the digestion and metabolism in the ruminant animal. From the drawbacks of current systems previously described, new mechanistic systems of feed evaluation need to represent the individual nutrients and substrates in organs and tissues, including the major interactions between nutrients and substrates and between organs and tissues. Here is where mechanistic mathematical modelling can be applied to represent quantitatively concepts and mechanisms (France and Thornley, 1984; Baldwin, 1995). These models are constructed by looking at the structure of the system, dividing it into its key components and analysing the behaviour of the whole system in terms of its individual components and their interactions with one another.

The accuracy of prediction of animal response using mechanistic models currently may be lower than that achieved by the empirical methods used in practical application; hence, the use of the former models has been constrained mainly to research. However, these research models are very useful in evaluating the adequacy of current knowledge and data, identifying those areas where research efforts should be focused.

In contrast to extant models for feed evaluation, new research findings can easily be incorporated and integrated into models constructed mechanistically, which will improve their ability to predict the availability of nutrients and their subsequent utilization within the digestive tract. Therefore, many mechanistic models have been built on the achievements of previously reported models, introducing additional knowledge and detail based upon newly emerging experimental information and specific failures of previous models to simulate reality. With this approach, it can be expected that in the future, mechanistic models will yield superior predictions of animal performance and will be more generally applicable than empirical models. As a consequence of this evolution, the need to re-examine the appropriateness of specific analyses associated with feed evaluation will arise.

REFERENCES

Agricultural and Food Research Council (1993) *Energy and Protein Requirements of Ruminants*. An advisory manual prepared by the AFRC Technical Committee on Responses to Nutrients. CAB International, Wallingford, UK.

Agricultural Research Council (1980) *The Nutrient Requirements of Ruminant Livestock*. Commonwealth Agricultural Bureaux, Farnham Royal, Slough, UK.

Baldwin, R.L. (1995) *Modeling Ruminant Digestion and Metabolism*. Chapman and Hall, London.

Beever, D.E., Cammell, S.B., Thomas, C., Spooner, M.C., Haines, M.J. and Gale, D.L. (1988) The effect of date of cut and barley substitution on gain, and on the efficiency of utilisation of grass silage by growing cattle. II. Nutrient supply and energy partition. *British Journal of Nutrition* 60, 307–319.

Bines, J.A., Napper, D.J. and Johnson, V.W. (1977) Long term effects of level of intake and diet composition on the performance of lactating dairy cows. 2. Voluntary intake and ration digestibility in heifers. *Proceedings of the Nutrition Society* 36, 146A.

France, J. and Thornley, J.H.M. (1984) *Mathematical Models in Agriculture: A Quantitative Approach to Problems in Agriculture and Related Sciences*. Butterworths, London.

Lewis, M. (1981) Equations for predicting silage intake by beef and dairy cattle. In: *Proceedings of the Sixth Silage Conference*. Edinburgh, pp. 35–36.

Ministry of Agriculture, Fisheries and Food (1975) Energy allowances and feeding systems for ruminants. *Technical Bulletin, No 33*. HMSO, London.

Neal, H.D. St C., Thomas, C. and Cobby, J.M. (1984) Comparison of equations for predicting voluntary intake by dairy cows. *Journal of Agricultural Science, Cambridge* 103, 1–10.

Vadiveloo, J. and Holmes, W. (1979) The prediction of the voluntary feed intake of dairy cows. *Journal of Agricultural Science, Cambridge* 93, 553–563.

2 Feed Characterization

A. Chesson

Rowett Research Institute, Bucksburn, Aberdeen, UK

INTRODUCTION

Farmers throughout the world have developed intuitive and often quite elaborate methods for assessing the feeding value of available feed resources. Such knowledge became established within a community because a relationship with production could be demonstrated and used to predict and manipulate performance. Interestingly, subsequent attempts to relate a farmer's historical understanding of nutritive value to current laboratory-based methods of proximate analysis have often proved quite unsuccessful (Thappa *et al.*, 1997). The development of laboratory methods to characterize feedstuffs was an attempt to formalize this traditional knowledge, allowing feed evaluation to be put on the numerical basis required by a nascent feed industry. Tables of feeds ranked by nitrogen content were available by the middle of the 19th century (Bossingault, 1843) with total digestible nutrients following (Wolf, 1874). However, it was not until the turn of the 20th century that work underpinning the present concepts of digestible and metabolizable energy and the net energy of feeds was published (Armsby, 1903).

Although feeds can be described with increasing sophistication, there is little point in so doing unless any new measure introduced has practical value. For most of the 20th century, methods of feed evaluation have developed in parallel with the needs of farmers and the feed industry, aided by an increasing understanding of the digestive processes which fuel animal growth and production. During this period, pressure for the development of improved methodology was limited, and change was by evolution rather than revolution. However, the latter part of the century has seen a far greater urgency to refine the methods used to characterize feed resources. Changes to production methods and feed formulation, the introduction of new feed resources and animals with greater genetic potential for growth, coupled with falling margins for the livestock producer, have shown the need to define feeds and their ingredients more tightly in order to allow predictions to be made with much smaller margins of error than previously. As a result, antinutritional factors, whose effects previously might have gone unnoticed or been ignored, now assume greater importance, and bioavailability is no longer considered synonymous with concentration. Above all, there is now a recognition that the composition of ingredients can vary significantly with variety, geographical origin, agronomic practice and year of harvest. Consequently, a single description applied to all samples of an ingredient is proving inadequate for some least-cost formulation purposes. Growing concerns about the environmental pollution of land and water courses by intensive production systems and the punitive legislation introduced in some countries are added reasons to avoid overformulation and ensure maximum capture of nutrients by the animal. The Netherlands were the first to introduce such legislation in 1984; since then the allowed allocation for application of phosphorus and nitrogen to fields has been steadily reduced, and farmers are required to pay a levy for the disposal of any surplus produced. This has led to a significant fall in phosphorus excretion in Dutch pig herds, from a mean of 1.62 kg per pig in 1973 to 0.67 kg per pig in 1996 (Jongbloed and Lenis, 1999).

Feed manufactures have to respond to global trends, and at present the world's fastest growing livestock industry is for chicken meat, which in 1997 exceeded that of beef cattle; conservative estimates of continuing growth are put at 8% per annum for the next 10 years. In general, the demand for feed resources in most developing countries is growing at a rate approximately twice that of the more industrialized nations. The intensive systems needed to meet this consumer-led demand for quality white meat are well suited to tight feed formulation and the use of additives. Aquafeed for commercial fish farming, although small in volume compared with feed for terrestrial species, is also a rapidly growing outlet for feed manufacturers and a relatively new area for nutritionists concerned with formulation. Again, the focus for growth is with the developing economies of South America, China and the other countries of the Pacific rim. In the more developed economies, the pet food market is perhaps the healthiest of all outlets, not because of the volume of sales but because of the greater added value carried by the finished product compared with other manufactured feeds. Both pet food and aquafeed have the same requirements for characterization as the more traditional outlets for manufactured feedstuffs.

Most countries in the world now publish tables of feed composition based on local measurements or taken from work done elsewhere. In addition, databases containing feed characteristics are held by all feed compounders and used to help formulate their own products. These databases hold a wealth of information not readily obtainable from other sources, representing the cumulative experience of the company nutritionists. They are often used to set upper limits for inclusion levels for particular ingredients for reasons which may not be fully understood but which experience has shown to be desirable in terms of nutritional value or the physical attributes of the finished feed. An official 'taxonomy' of feedstuffs was established (International Feed Nomenclature and Number, see Harris *et al.*, 1980). Unfortunately, this became too unwieldy and is no longer widely supported although it still forms the basis of a number of other schemes. One reason why the original International Feed Institute scheme failed was because there remains a plethora of feed ingredients available to the compounder, particularly by-products of the extraction and fermentation industries, whose local description may not equate readily to a recognized formal description. Even when an ingredient can be matched to a formal description, terms such as bran and offal or processing descriptions such as 'cooked' can encompass material of widely varying composition and nutritive value. However, as Table 2.1 shows, the volume of the more exotic ingredients incorporated into manufactured feeds remains small in comparison with the cereal grains and soybean.

Existing tables of feed composition are still based on the crude nutrient fractions of the proximate analysis scheme and the cell wall fractions of the detergent fibre system (Van Soest, 1967). Additional information on fatty acid, amino acid, mineral and vitamin contents may also be included.

The proximate analysis of feeds, an arbitrary and empirical series of tests, which allowed some prediction of animal performance, was in essence first described by Henneberg and Stohmann (1864) in the 19th century. Under this scheme, five fractions; moisture, ash, ether extract, crude protein ($N \times 6.25$) and crude fibre are determined directly and the sum of all five expressed as $g\ kg^{-1}$ subtracted from 1000 to generate a sixth fraction, the nitrogen-free extract (NFE). Severely criticized for its imprecision over the intervening years,

Table 2.1. Use of raw materials in the production of compounds and blends[a] for livestock in the UK in 1997 (data from the Ministry of Agriculture and Fisheries).

Raw material	Usage [thousand tonnes] (%)	Raw material	Usage [thousand tonnes] (%)
Wheat	3267.6 (30.8)	Other seed meals	448.8 (4.2)
Barley	720.2 (6.8)	Field beans	34.3 (0.3)
Oats	44.8 (0.4)	Field peas	117.0 (1.1)
Whole/flaked maize	79.8 (0.8)	Sugar beet pulp	269.2 (2.5)
Other grains	5.5 (0.1)	Molasses	344.4 (3.2)
Rice bran extractions	81.4 (0.8)	Citrus/other fruit pulp	100.9 (0.9)
Maize gluten feed	443.1 (4.2)	Animal products[b]	19.2 (0.2)
Cereal by-products	1111.1 (10.5)	Fish meal	218.7 (2.1)
Whole oilseeds	86.3 (0.8)	Oils and fats	262.2 (2.5)
Rapeseed meal/cake	581.0 (5.5)	Protein concentrates	30.0 (0.3)
Soybean meal/cake	1170.2 (11.0)	Other ingredients[c]	639.6 (6.0)
Sunflower meal/cake	538.2 (5.1)		
		Total	10,613.5

[a]Total production for cattle was 3585.6 thousand tonnes of which 2279.6 was used for dairy cows. Over the same period, total manufactured feed for pigs was 2451.3 thousand tonnes, for poultry 3617.3 thousand tonnes and for sheep and horses, 617.8 and 161.5 thousand tonnes, respectively.
[b]The use of mammalian meat and bone meal was banned in 1996. Use of poultry products (meat and bone meal, feather meal) was not affected.
[c]Includes all raw materials not listed elsewhere.

proximate analysis has been modified or replaced by other methods in the majority of laboratories. Crude fibre and NFE were found particularly wanting because of their lack of any consistent relationship with recognizable components of crop plants. Crude fibre contained some, but not all, of the polymers which constitute the plant cell wall, while NFE could encompass water-soluble carbohydrate, starch, organic acids and much of the pectic fraction of some cell walls. These measures have now been superseded by the direct determination of the water-soluble components and by neutral detergent fibre (NDF). NDF can be equated with the cell wall content of grasses and cereals and, if preceded by a starch extraction, the cell wall content of many other feed ingredients. It does, for reasons detailed later, underestimate the cell wall content of legumes. Acid detergent fibre (ADF), which provides a measure of just the cellulose and lignin content of the cell wall, has found value in the description of forages because of its statistical relationship with degradability.

The limitation of these chemical analytical methods of feed description applied to predicting nutritive value are well recognized and documented, and will not be considered further here. Biological approaches to estimating organic matter digestibility and energy content based on *in vitro* digestibility measurements made with rumen microorganisms (Tilley and Terry, 1963; Menke *et al.*, 1979) or cell wall-degrading enzymes (Dowman and Collins, 1982) have often proved more successful, particularly for ruminants (Aiple *et al.*, 1996). These methods are the subject of Chapter 4. Directly determined available energy and available protein values are not a routine option for the compounder who ideally needs to know whether a particular delivery to the mill falls within the specifications used in formulation. The compounder thus has the conflicting need of more rapid, preferably on-line, methods of feed characterization and an increasing need for a more detailed assessment of the raw material. The immediate interest in and application of near infrared reflectance spectroscopy (NIRS) to predict moisture and protein in grain and oil in oilseeds, despite the financial outlay and difficulties associated with early systems, points to the value placed by the feed industry on the ability to characterize all incoming batches of feed ingredients.

The requirements for descriptions of feed, which allow a good prediction of responses, clearly differ between livestock classes and particularly between ruminants and non-ruminants. These needs are addressed separately below, together with those of the compounder for more rapid and robust methods of analysis.

CHARACTERIZATION OF FEED FOR NON-RUMINANT LIVESTOCK

Production diets for pigs and poultry throughout the world are heavily dependent on cereal grains, typically wheat or barley in countries with a temperate climate and maize in warmer regions. The grain is supplemented with an additional source of protein, usually soybean. Pigs and broiler chicks have a high capacity for starch utilization and, provided that the grain is disrupted sufficiently and the endosperm is accessible, ileal digestibilities of 97–98% and overall digestibilities close to 100% are recorded. In contrast, grain protein is far less accessible, and 20–30% of the nitrogen may be excreted and lost to the pig or bird. Thus, while an estimate of total starch provides a close approximation to what is available to the non-ruminant, estimating total nitrogen cannot provide a direct measure of amino acid supply or the quality of the protein fed.

Starch

Proximate analysis coupled with an estimate of starch content provides a description of most cereal grains which can be equated readily with the amount of cell wall and cell contents present. This is shown for barley grain in Fig. 2.1A. As starch represents approximately two-thirds of barley dry matter,

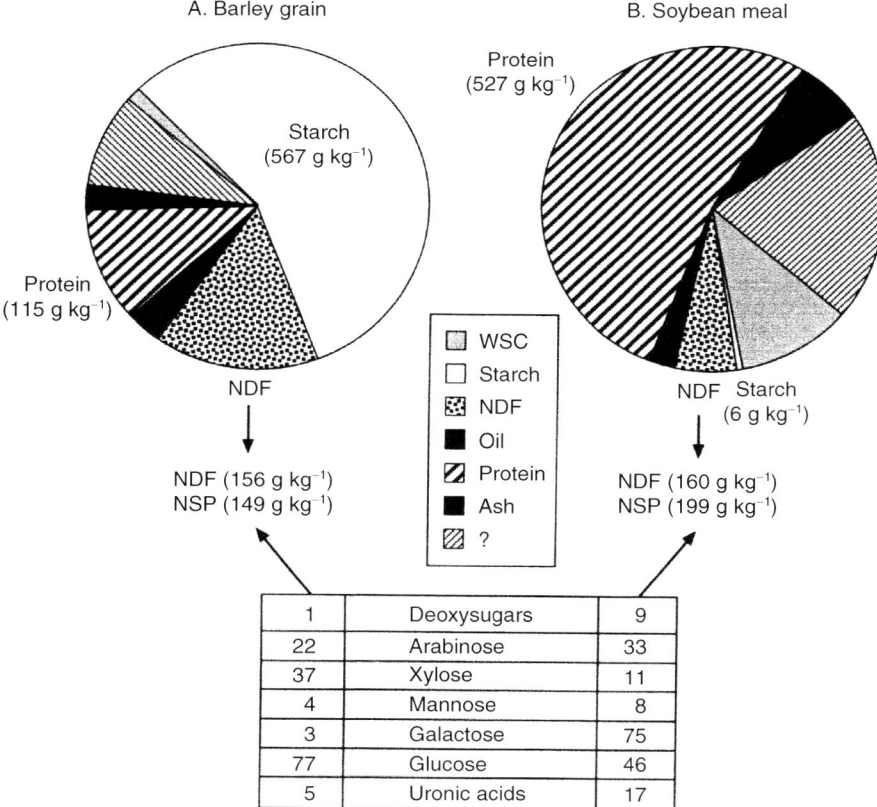

Fig. 2.1. A comparison between the composition of a starch-rich cereal grain and a protein-rich legume seed meal. Non-starch polysaccharide (NSP), measured as the sum of the component monosaccharides, and neutral detergent fibre (NDF) are similar in the cereal, but NDF significantly underestimates the cell wall content of the legume.

evidently it is the prime determinant of metabolizable energy (ME) in non-ruminant feeds. Because of this, it is relatively straightforward to take account of those fluctuations in the feed value of grains that are a product of differences in gross composition. A comparison of three methods of analysis used to calculate the ME value of cereal grain for pigs, the first requiring determination of ash, crude protein, crude fibre, fat, starch and water-soluble sugars by wet chemical methods, the second the same parameters derived from NIRS, and the third only starch and crude protein, found no significant differences between methods (Lengerken *et al.*, 1997). The implication of this and other studies (Gerlach, 1990) is that an estimate of starch and crude protein alone, whether by wet chemistry or NIRS, provides sufficient inform-ation to calculate ME. The earlier suggestion by Carré and Brillouet (1989) to base regression equations for poultry on the water-insoluble cell wall (WICW) content of feeds would support this view. Since the contribution to ME provided by the fermentation of cell walls is negligible in most poultry species, WICW measures the inert fraction of the feed, providing a reproducible indirect measure of the digestible components, notably starch, protein and any oil present.

Surprisingly little attention has been applied to the direct estimation of starch content by NIRS in the raw materials most commonly found in pig and poultry diets (Gerlach, 1990), particularly when it is realized that NIRS works best when applied to specific chemical entities (Givens *et al.*, 1997). The commercial network for raw material control established in Spain seems to be the only systematic move to substitute NIRS for proximate analysis methods (Moya *et al.*, 1995). At present, the feed industry remains tied to polarimetric methods because of legal requirements, while most research laboratories use an enzymatic method based on that described by Aman and Hesselman (1984). The advantage of this method over previous enzyme-based assays is the use of a thermostable bacterial α-amylase (Termamyl™ Novo Nordisk) which allows digestion at temperatures up to 90°C ensuring easy gelatinization of the starch granule.

Because of the high degradability of cereal starch and the relatively limited use of other sources of starch in pig and poultry diets, there have been few attempts to establish a simple *in vitro* assay able to predict bioavailability. Those that have been developed lack the discrimination necessary to detect small differences in the digestibility of wheat and barley starch (Drake *et al.*, 1991). Such methods may have more value when applied to less digestible potato or legume starch or when applied to the prediction of digestible energy in feedstuffs and complete diets (Boisen and Fernández, 1997). Methods commonly used in the food industry to describe the physical state of food, in particular the response of starch to various forms of processing such as extrusion cooking, have yet to find value in the characterization of feeds. The 'Rapid Visco Analyzer' developed for the baking industry is possibly the method with the greatest potential (Eerlingen *et al.*, 1997). This produces so-called pasting curves which describe the viscosity of a stirred suspension of the starch-rich material when subjected to a heating/cooling regime (Fig. 2.2). The initial rise in viscosity, caused by the disruption of the granules and

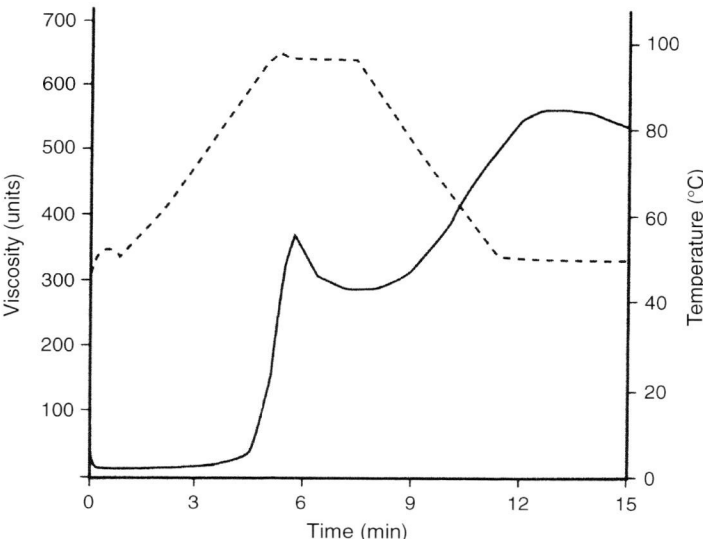

Fig. 2.2. Pasting curve for milled barley produced using a Rapid Visco Analyzer. The temperature changes applied are shown by the dotted line and the changes in viscosity of the stirred suspension by the solid line.

exudation of the starch, provides a measure of the integrity of the starch granule. In extruded feed, for example, this peak can be much reduced, reflecting the extent to which processing has led to gelatinization. The second peak produced as the suspension cools is due to gelation effects and provides a measure of the integrity of the amylose and amylopectin polymers. This can be an important factor in pellet binding and pellet quality (Thomas *et al.*, 1998). Well-established methods for the description of starch based on differential scanning calorimetry also exist and have the added advantage of unambiguously demonstrating the presence of retrograded starch (Gruchala and Pomeranz, 1993). However, it has yet to be demonstrated that this is a problem of any significance in prepared feeds.

Non-starch Polysaccharide

Since the contribution of starch to the energy value of most production diets for pigs can be readily assessed, the remaining major source of variation is the extent to which the plant cell wall (fibre) fraction is fermented in the hind-gut and the organic acids produced are absorbed and utilized (Whittemore, 1997). NDF provides an accurate measure of the cell wall content of members of the *Poaceae* (formally *Gramineae* – grasses, cereals) but, as has long been recognized by ruminant nutritionists, substantially underestimates the cell wall content of other plants, notably the legumes (Fig. 2.1B). The pectic and associated polysaccharides that are solubilized by neutral detergent and discounted by this gravimetric procedure tend to be fermented more readily and completely by rumen microorganisms (Chesson and Monro, 1982; Hall *et al.*, 1998), and, presumably, by those inhabiting the pig hind-gut, than the neutral detergent residue and so are likely to contribute more to the overall energy content of the diet. Any measure of total cell wall content minus NDF provides a measure of neutral detergent-soluble polysaccharides (Hall *et al.*, 1997) but not their contribution to the energy value of the diet. *In vitro* methods for estimating fibre degradability in ruminants can and have been applied to this fraction (Hall *et al.*, 1998) but may not be necessary. It may be sufficient to assume that this fraction is fully digestible and to calculate the energy supplied. This may depend on the relative proportions of uronic acids and hexose and pentose sugars present, since the efficiency of conversion of different sugars to volatile fatty acids (VFAs) by the gut flora differs. As a guide, this will be approximately two-thirds of the energy supplied by the same sugar when metabolized directly by the host.

Variation in the ME value of cereals does not always depend on those differences in nutrient density which can be detected by classical proximate analysis. It has long been recognized that the ME of barley and oats fed to poultry can be improved by addition of enzymes. Initially, since the addition of crude amylase preparations produced benefits, this was thought to be due to an impaired utilization of starch. Subsequently, it was demonstrated that the substrates for added enzymes were soluble mixed-linked glucans leached from the endosperm cell wall and that the early amylases were effective because they also contained β-glucanase activity (Ricks *et al.*, 1962). In time, it was recognized further that the increase in ME accompanying enzyme treatment was not the result of a net improvement in digestibility *per se* but primarily a restoration of ME to the level predicted by composition. Soluble mixed-linked glucan acts as an antinutritional compound, slowing uptake of nutrients in the digestive tract by increasing the viscosity of digesta and the thickness of the unstirred layer coating the mucosal surface, and so reducing ME (Bedford and Classen, 1992; Bedford, 1995; Pasquier *et al.*, 1996). This was the effect reversed by enzyme addition.

Wheat, rye and triticale contain substantially lower concentrations of mixed-liked glucans than barley and, it was assumed, were free from this problem. However, as the proportion of wheat in broiler diets was increased, similar problems began to be noted, firstly in Australia (Rogel *et al.*, 1987; Annison, 1991) and latterly in Europe (Wiseman and Inborr, 1990). Wheat samples were identified in which the observed ME value was substantially lower than that which would have been predicted on the basis of composition and where ME could be (partially) restored to the expected value by addition of enzymes. Unlike barley and oats, however, the principal soluble polysaccharide increasing the viscosity of the digesta was the more structurally complex arabinoxylan, and the enzyme activity required was a xylanase.

Compounders may make allowance for a small improvement in ME which accompanies enzyme treatment of barley and most wheat by assigning a higher ME value to the grain or otherwise altering the specifications of the mixed feed. However, enzyme addition, which is an added cost to the feed producer, is not always of value. Some wheat, when incorporated into poultry feed at more than 60% or fed to young pigs, can produce a marked depression of ME and require enzyme addition, while other wheat samples behave as expected and do not greatly benefit. The ability rapidly to characterize or at least identify a 'problem' grain would be of considerable value. *In vitro* methods based on the extraction of soluble polysaccharide and measurement of its viscosity in solution can be related to bird performance (Bedford, 1995). This provides a relatively rapid method of distinguishing wheat with a high soluble arabinoxylan content likely to increase the viscosity of digesta from wheat with a lower soluble arabinoxylan content with a limited effect on viscosity (Bedford and Classen, 1993). However, such measures are often poorly representative of viscosity measurements made *in vivo* and are unable to explain the often considerable differences in bird to bird response to the same diet. In the case of wheat arabinoxylan, this is perhaps not surprising because polymer structure is at least as important a determinant of antinutritional effects as polymer content (Austin *et al.*, 1999). Direct determination of apparent metabolizable energy (AME) remains the only certain way of identifying 'problem' wheat and measuring the response to enzyme (Scott, 1996). There does, however, appear to be a genetic element governing the amount of soluble polysaccharide present in wheat and barley, and this may allow variety to be used as the basis on which to incorporate enzyme (Pérez-Vendrell *et al.*, 1996; Bedford, 1997). However, knowledge of the varieties grown in the area from which the mill receives its wheat alone will not be sufficient, and some routine testing will be required. The obvious choice of NIRS has not been widely used for this purpose although its value has been demonstrated in relation to the measurement of β-glucans in human food (Givens *et al.*, 1997) and feed barley (Pérez-Vendrell *et al.*, 1997).

Protein and Amino Acid Supply

The availability of protein and essential amino acids in feedstuffs is normally established *in vivo* in pigs cannulated at the terminal ileum or in caecectomized birds, and the values obtained are corrected for endogenous nitrogen. These corrected values can then be applied to a directly determined protein or amino acid content to establish availability or to the calculated content of individual amino acids (Henry *et al.*, 1988). Various attempts have been made to short-circuit this laborious process by use of *in vitro* enzyme-based systems which seek to mimic digestion and which could be used on a more regular basis to predict energy and protein values *in vivo*. Invariably these are based on porcine pepsin and pancreatin (Savoie, 1994) and, for pigs, a third stage digestion with cell wall-degrading enzymes may be included to take account of hind-gut fermentation (Boisen and Fernández, 1997). *In vitro* digestions can never hope to simulate the *in vivo* system in all respects, and optimizing the system for one ingredient or group of ingredients is unlikely to provide a universal solution for all feedstuffs. Although reasonable predictions can be made of protein digestibility (Table 2.2) and the

Table 2.2. Comparison of results from an *in vitro* enzyme-based system for estimating crude protein digestibility with the apparent ileal digestibility values determined *in vivo* in pigs (from Drake *et al.*, 1991).

Feedstuff	Fraction of total protein digested	
	In vitro	*In vivo*
Soybean meal	0.76	0.74
Sunflower meal	0.66	0.70
Rapeseed meal	0.60	0.59
Wheat	0.64	0.69
Maize	0.60	0.59

bioavailability of limiting amino acids, the errors associated with the methods are no better than those resulting from the use of table values. As a result, present *in vitro* characterization of dry matter or protein digestibility does not offer a routine method sufficiently able to discriminate between the natural variation found between samples of the same feed ingredient. This inherent variability means that most feed producers overformulate for digestible amino acids by about 6–8%, depending on ingredients, to ensure that at least 80% of the finished feed falls at or above specification (Fawcett and Webster, 1996; Simmins and van Kempen, 1999).

NIRS has been used extensively to estimate protein content, particularly of cereal grains, and is one of the more robust applications of this method. Less attention has been paid to establishing whether NIRS can be used for the more specific purpose of quantifying the limiting amino acids. Some success has been reported for lysine, threonine, tryptophan and methionine in wheat (Williams *et al.*, 1984) and for a wider range of amino acids in maize and soybean (Shenk, 1992). Two preliminary reports of applications of NIRS to predicting digestible amino acids in poultry (Van Kempen and Jackson, 1996) and in poultry and pigs (Harrison *et al.*, 1991) indicated the feasibility of the approach but recognized the need for far larger sample sets to generate calibrations with application to the feed industry.

Variation in protein digestibility and that of the constituent amino acids can arise for a number of reasons, including damage through processing, the presence of inhibitors in the seed, differences in seed composition or simply structural differences in seed anatomy. Ideally, methods to characterize ingredients should take account of these factors. A dilemma faced by all commercial nutritionists is the inability very rapidly and easily to detect batches of ingredients which have reduced amino acid availability because of overheating. Relatively simple but sensitive tests such as protein solubility in potassium hydroxide (Parsons, 1996) or the specific detection of the lysine Maillard product(s) (Hurrel, 1990) exist, but none are in routine use.

Data on antinutritional factors have to be used with care since they may only partly explain the observed variation in protein digestibility or may conceal underlying and more important influences. Thus, while a highly significant overall negative relationship between the presence of condensed tannins, measured as catechin equivalents, and the digestibility of individual amino acids in cultivars of sorghum grain could be demonstrated, this concealed the fact that several cultivars had identical tannin contents but markedly different nutrient digestibilities (Elkin *et al.*, 1996). In a similar vein, the digestibility of pea proteins, an important source of amino acids for monogastrics because of their high lysine content, exhibits considerable variation; no simple relationship could, however, be found with the concentration of protease inhibitors present in the seeds (Carré and Conan, 1989). Only when the behaviour of the individual seed proteins in the digestive tract of poultry was studied was it recognized that much of the observed variation was due to the presence in seeds of different proportions of albumin and globulins with different degradability characteristics (Crévieu *et al.*, 1997). In both of these cases, the means to quantify the antinutritional factors were available, but in neither case would this have aided the prediction of nutritive value.

CHARACTERIZATION OF FEEDS FOR RUMINANTS

Freshly grazed or conserved forages represent approximately 75% of the diet consumed by domestic ruminants. It is perfectly possible to obtain up to 20 kg day^{-1} milk production on good pasture alone, and more in the presence of somatotropin (Hoogendoorn *et al.*, 1990). Only in highly intensive production systems, such as those involving high-yielding dairy cows or rapidly growing steers, is the protein and energy content of a forage-only diet inadequate. In these situations, more protein and energy-dense ingredients have to be introduced in the form of concentrates. The characterization of forages and other feed ingredients for ruminants is far from straightforward because the nature of the microbial conversion process ensures that the rate of nutrient release is at least as important a factor as the total amount of nutrient present or its maximum bioavailability. This is readily demonstrated by comparison of the performance of ruminants on grass or clover swards; although of very similar composition, clover supports a higher level of performance because nutrients are released at a faster rate than from grass (Ulyatt, 1971; Beever *et al.*, 1986). Although, for legislative purposes in Europe, a relatively simple regression equation based only on a cellulase digestibility (neutral detergent–cellulase organic matter digestibility, NCD) value and fat content is used to derive the energy content of compounded feeds, this is not applicable to fresh forage and silage. The lipid content of grass is very low (~ 40 g kg^{-1} of dry matter), while NCD does not take account of the important contribution of water-soluble carbohydrate (WSC) in forage and organic acids in silage to the supply of energy. Various forms of compromise have evolved which allow a measure of characterization in the laboratory but which fall short of providing data on the rate of nutrient supply. These describe feedstuffs in terms of analytical fractions of carbohydrate and protein, which reflect how readily they are fermented in the rumen and, in some cases, marry these values to actual *in vitro* degradability measurements. Thus WSC and protein are considered to be degraded immediately and completely, to be followed by a more slowly digested cell wall and (rumen-degradable) protein fraction, leaving a residue of undigested protein and carbohydrate to escape the rumen. Although these fractions perform better than crude nutrients (proximate analysis) in predicting the energy content of raw ingredients and compound feeds, they are still less accurate than predictions based on *in vitro* digestion (Aiple *et al.*, 1996). In part at least, this poorer performance is a consequence of the heterogeneity of cell wall material forming the detergent fibre fractions and an inability to take account of their architecture (Chesson, 1993; Jung and Allen, 1995).

The Plant Cell Wall (Fibre) Fraction

Neutral detergent fibre (NDF) has proven of value in ruminant nutrition, providing a robust measure of the cell wall content of grasses and cereals, able to distinguish between warm and cool season grasses, forages and concentrates, and roughage and energy feeds (Mertens, 1997). While NDF equates with the cell wall content in grasses, this is not the case with most other crop plants. Legumes and other non-grass species contain relatively high concentrations of pectic polysaccharides that are extracted by neutral detergent and not included in their NDF fraction. However, neutral detergent solubles, which are readily measured (Hall *et al.*, 1997), tend to be readily degraded and contribute little to the non-degradable fraction of most cell walls. Thus there is a broad negative relationship between NDF as a fraction of dry matter, regardless of botanical origin, and the degradation rate of potentially digestible NDF (Sauvant *et al.*, 1995), and between NDF content and intake (Sudweeks *et al.*, 1981; Mertens, 1992). However, the amount of cell wall present is a poor index of other factors important to the digestive process of herbivores, such as salivation and pH control in the rumen or particle comminution. Forages have physical properties that may refine compositional data or even replace them in importance. In an attempt to integrate a physical concept into feed evaluation, a measure of 'diet fibrosity' was proposed (Sudweeks *et al.*, 1981) and was shown to relate to VFA production (Beauchemin, 1991), nutrient utilization (Grant *et al.*, 1990) and the rate of dry matter intake (Sauvant *et al.*, 1995). Unfortunately, fibrosity, measured as time spent eating/ruminating kg^{-1} dry matter intake, is not a readily measured parameter. The energy input required to grind a feed to a known particle size, proposed earlier (Paul and Mika, 1981), is perhaps an easier measure, although used in isolation it is as limited in value as a single chemical analysis. The physical effectiveness or 'roughage value unit' proposed by Mertens (1992) and defined as the dry matter NDF content in particles less than 1.2 mm diameter also acknowledges the physical attributes of forages and is more readily determined than the fibrosity index. This has been used practically to 'adjust' the value for NDF (eNDF) to improve recommendations for more accurately meeting the fibre requirements of dairy cows (Mertens, 1997).

Although there is implicit acknowledgement of the importance of the three-dimensional nature of feed particles in measures such as fibrosity, the spatial distribution of the cell wall and other nutrient fractions has been largely ignored in feed evaluation and its importance underrated. In a comparison of measurements relating to the anatomy of the plant obtained by image analysis and various chemical characteristics including proximate analysis (Travis *et al.*, 1996), anatomical features were able to account for virtually as much of the variation in degradability as was analytical chemistry (Table 2.3).

Considering the inherent limitations of chemical parameters for describing and predicting digestion characteristics, it is somewhat surprising that NIRS has proved relatively successful in predicting both the chemical composition and organic matter digestibility of forages (Murray, 1993). NIRS is best when measuring a definable chemical entity or a property which relates directly to a chemical entity, and most vulnerable when attempting to measure a multi-faceted parameter such as digestibility, which involves an interaction between the feed and the host animal (Givens *et al.*, 1997). Nonetheless, the robustness of the method has improved as a result of improved instrumentation, standardization across laboratories and, more importantly, because of the ability to relate spectral regions to relevant chemical characteristics. As a result, NIRS is now the method of choice for many extension workers. The increasingly widespread use of NIRS applied to forages, and particularly to silage, has revealed some problems of accuracy largely relating to moisture and particle size (Baker and Barnes, 1990). However, account can be taken of these problems in calibration equations (Baker *et al.*, 1994).

Voluntary dry matter intake can be a more significant factor than digestibility in many forage-based production systems, and NIRS has also been applied to the prediction of intake. Silage is notoriously difficult to characterize fully using conventional wet chemical analysis, and any directly determined and rapid prediction of intake would be of considerable value to the industry. Results to date look promising. Steen *et al.* (1995) were able to demonstrate that NIRS was able to predict intake of grass silage by cattle. However, the extensive drying and fine milling of the dried samples required to achieve precision was thought to lead to the loss of volatiles and, potentially, to affect the accuracy of prediction of some samples. Subsequently it was demonstrated that prediction of intake could be made with wet silage with similar accuracy to that based on the use of dried samples (Gordon *et al.*, 1998).

Table 2.3. Value of chemical and anatomical measures in regression models to predict the degradability of wheat and barley straw (from Travis *et al.*, 1996).

Term	Percentage of variance explained
Chemical features only	66.7
Anatomical features only	58.2
Combined five-term model[a]	77.8

[a]Five-term model: thickness of sclerenchyma and epidermis cell walls, density of epidermis cell walls, NDF content and lignin content.

Protein Supply

While NIRS will measure the total or crude protein content of feeds, it has proved of limited value in differentiating between the various forms of protein available to the ruminant animal. The present UK metabolizable protein system for the estimation of protein requirements of ruminants (AFRC, 1992) recognizes that protein, like the carbohydrate present in feed, can be considered as a number of fractions with different degradability characteristics. As indicated previously, immediately soluble protein, akin to WSC, is assumed to be readily utilized by the rumen microflora (QDN, quickly degraded nitrogen), while a second insoluble fraction is degraded more slowly (SDN, slowly degraded nitrogen) leaving a rumen-undegradable fraction (UDN, undegradable nitrogen). This undegradable fraction may be degraded post-rumen, providing a non-microbial source of protein to the animal. Assessment of protein quality ideally requires a measure of these fractions coupled with the rate at which the SDN fraction is degraded. Various laboratory methods can be used to measure the immediately soluble fraction, and the amount of nitrogen remaining associated with ADF (ADIN, acid detergent-insoluble nitrogen) used to indicate the size of the non-degradable fraction. Subtraction of these values from the total protein will provide an estimate of the amount of SDN present. However, this is a crude approximation, and the various regression equations developed to predict needs for growth, lactation or wool production require kinetic data. Typically, values for QDN, SDN, UDN and the necessary rate data are still obtained by the nylon (Dacron) bag method described by Orskov and MacDonald (1979). As a result, diet formulation for ruminants may utilize actual crude protein values and, occasionally, AIDN, but otherwise is based wholly on table values and is unable to take account of variation in raw ingredients.

CHARACTERIZATION OF FEEDS FOR FISH

Fin fish and crustacean aquaculture production increased more than threefold during the period 1986–1996 and now approaches 20 million tonnes per annum worldwide. The need for compound aquafeed has grown in parallel, although the actual amount manufactured is difficult to estimate. Higher estimates suggest that fish consume 3–4% of total world feed production, representing about 18 million tonnes of finished product (Gill, 1997). More conservative estimates would put manufactured levels at around 8–9 million tonnes (Tacon, 1998).

There are several trends in aquafeed production that will require improved characterization of raw materials and finished feeds. Much of the world's captured fish production occurs in non-industrialized countries with a low per capita income, and there is a recognized need to utilize non-food grade, locally available resources whenever possible; these are likely to be highly variable in composition and nutritional value. This conflict of interest is particularly true for fishery products that are the ingredients of choice for fish feeding. These invariably represent the only locally available source of high-quality protein for the human population, and their diversion to support fish production often has severe consequences. As a result, there is a need to replace fishery products in aquafeeds with the more variable protein-rich products of terrestrial agriculture (Allen, 1997). Some indication of the range of raw ingredients already utilized in the Mediterranean region is given in Table 2.4. The chemical composition and nutritional value of many of these products are poorly characterized, particularly with respect to the protein, available amino acid, lipid and mineral content. In addition, many raw materials are by-products of industries with old facilities which often damage the quality of the by-product entering the food chain. Meat meals, in particular, are subject to heat and other forms of damage, and are more variable in nutritional value than plant protein sources such as soybean meal (Tacon, 1998).

Knowledge of the digestibility of nutrients, particularly protein, in the target species is essential. This alone presents a major problem, since the potential number of target species is far larger than that of domestic livestock and ranges from freshwater fishes such as carp and tilapia, to crustaceans (prawns, crayfish, etc.). Compared with raw materials of marine origin, other feed ingredients are usually deficient in essential amino acids, particularly lysine and methionine, and therefore data on the essential amino acid content and availability of potential ingredients are needed. No absolute requirement for carbohydrate has been demonstrated for fish (NRC, 1993), but starch and WSC are utilized and reduce the need for using expensive protein as an energy source. However, the extent to which carbohydrate is utilized is species dependent and is least in carnivorous species. Starch utilization is greatly improved by gelatinization, and the integrity of the starch granule is often as important a characteristic of the finished feed as the amount present. The 'Rapid Visco Analyzer' offers one method of monitoring the disruption of starch granule structure (Fig. 2.2).

Antinutritional factors may assume far greater importance in fish feeding than in other farmed species. Even meat meals may pose difficulties because of the presence of bone fragments that damage the digestive tract. Depending on the raw ingredients incorporated, it may prove necessary to monitor for the

Table 2.4. Typical raw materials used in the manufacture of aquafeeds in the Mediterranean region.

Ingredients of animal origin	Source	Ingredients of plant origin	Source
Blood meal	Local	Cottonseed meal	Local
Fish meal	Local/imported	Maize	Imported
Gelatine	Local	Maize gluten	Local
Meat meal	Imported	Maize starch	Local
Meat and bone meal	Local	Rice bran	Local
Poultry by-products	Local	Soybean meal	Local/imported
Shrimp meal	Local	Wheat bran	Local
Fish oils	Local/imported	Maize oil	Local

various enzyme inhibitors and phytate present in seed meals and for the presence of mycotoxins because of the greater susceptibility of fish to such factors. Provided essential fatty acid requirements are met, saturated animal fat has no adverse effects on fish (Reinitz, 1980). However, inclusion of saturated fats does affect body composition and may reduce the unsaturated acid content of the product, with possible consequences for the 'healthy' image of fish. For this reason, some characterization of fat in the completed diet may be of value.

CHARACTERIZATION OF FEED FOR COMPANION ANIMALS

A staggering $11 billion are spent annually in the USA on feeding dogs alone, with Japan following at $7 billion. The UK, the traditional home of dog lovers, falls well behind at $2.6 billion, although the annual expenditure per dog ($363) is considerably greater and is second only to Japan and Germany. Since the bulk of this expenditure is the purchase of prepared dog food, a market of this size cannot be ignored. The prime aim of pet nutrition is to maintain a relatively constant live weight for periods up to 14 years in the case of dogs, including well into old age. Thus feed formulation for dogs, and to a lesser extent cats, owes as much to information on long-lived species such as the human as to nutritional studies of the production animal. In every other respect, pet foods are formulated as for feeds for other animal species and are based on a set of nutrient requirement tables, an appreciation of the bioavailability of nutrients present in raw materials (Morris and Rogers, 1994) and the chemical composition of the raw materials. The most commonly used calculation of the energy content (ME) of diets for dogs is based simply on the content of crude protein. crude fat and carbohydrate (excluding fibre) (AAFCO, 1997). The major problem in meeting the objective of maintaining a stable weight is the lack of control the industry has in the supply of food to individual animals. Owners may not take sufficient account of the very wide range of adult body weights found in different breeds (1–100 kg for dogs) or the declining need for nutrients as the animal ages.

Commercial pet foods are either sold dry (6–10% moisture), usually as extruded products based on cereals and animal protein by-products with some fat to aid palatability and sufficient antioxidant to avoid fat rancidity, or as wet products (78–82% moisture) in cans. Wet dog foods typically are made from offal meat and textured vegetable protein, with added guar or carrageenan gum to provide a gel.

Food for companion animals is selected by the owner, and the choice of diet is more likely to reflect the owners perception of good nutrition than any information provided by the manufacturer. As a result, feed characterization may have to respond to factors not normally considered important in production animals. Foremost is stool quality and consistency, used by owners as a marker of animal health, and as much consideration is given to dietary fibre in cat and dog foods as in human nutrition (Sunvold, 1996). Diets must also be formulated to avoid clinical conditions such as various skin diseases. For this reason, attention may also be given to the ratio of *n*-3 and *n*-6 fatty acids. As a general rule, any nutritional topic highlighted in terms of human nutrition is likely at some stage to impact on foods for companion animals.

FUTURE NEEDS OF THE FEED INDUSTRY

It is evident from the foregoing account that the feed industry needs a better assessment of the range of ingredients available for incorporation into finished feeds and the variation in composition and nutritional value that occur between batches of the same raw ingredient. The former is a more acute problem for producers of fish feed, particularly when circumstances require that local sources of ingredients form the bulk of the raw materials. Application of existing methods of feed evaluation can substantially alleviate this problem, allowing a more complete and relevant database on composition and digestibility to be constructed. Experience will show which ingredients are most subject to variation and how the extent of such variation can be monitored.

The feed industry in the more industrialized societies has already completed this exercise and manufacturers have databases of proven value. Tightening formulation in such situations requires that every batch of ingredients is tested separately and results obtained within the time frame allowed by the flow of materials through the mill. In a modern mill, this will certainly be less than 48 h for the bulk ingredients; to date, only NIRS has met this need. There are other spectrophotometric methods that are able to provide rapid characterization of complex materials. However, far more information is provided about chemical structure from the mid-IR region, in which absorption bands arise from the vibrational and rotational changes of specific chemical bonds and from the overtones and combinations of these fundamental vibrations which appear in the near IR, than from the visual or UV regions of the spectrum. While there exists a rational basis for the application of NIRS to specific chemical entities within feeds, it is questionable whether this also applies to dynamic properties. The assumption underpinning the application of NIRS to digestibility or intake is that the factors controlling these processes can be described adequately in chemical terms. This seems unlikely since it is evident that the three-dimensional structure of the feed is also a factor in both digestion and intake. The more that such physical or biological parameters intervene, the more spectrophotometric methods will be prone to error.

Other rapid analytical methods exist but have yet to be applied to any extent to feed characterization. Mass spectrometry, which also generates data in seconds, was developed to provide information on the relative abundance of molecular fragments and is difficult to quantify in terms of the parent molecules. One form of mass spectrometry, pyrolysis–mass spectrometry (PY–MS), involving the rapid thermal dissociation of a material and the separation and detection of the mass fragments produced, can provide fingerprint pyrograms able to distinguish between complex structures on a semi-quantitative basis (Boon, 1989). Methods for improving the quantitation of mass spectrometry, such as isotope ratioing, do exist and could be developed further. PY–MS and the many other new ionization techniques linked to mass spectrometry could well have a future in feed characterization. The third basic tool used in determining chemical structure, nuclear magnetic resonance (NMR) spectroscopy, is most sensitive when samples are in solution. Solid-state NMR lacks resolution, is relatively slow in acquiring a signal and is difficult to quantify. At its present stage of development, it is difficult to see what advantages it offers over existing methods.

Method development is a continuous process in the feed laboratory, particularly as applied to the detection of specific ingredients or contaminants. Developments range from better extraction technologies (microwave, super-critical fluid) to faster and more sensitive detection methods (ELISA, bio-sensors). Feed laboratories can only come under increasing pressure as quality control standards tighten, as legislation requires more information to be declared and as permitted levels of contamination are reduced further.

One additional factor which may increase the pressure for batch analysis but which seems to have been overlooked is the continuing introduction of genetically modified crop plants. As demonstrated by the proliferation of new rapeseed varieties being developed for non-feed use around the world (Murphy, 1996), the number of different by-products available for incorporation into compound feeds could increase dramatically. Feed producers may well find that the parameters for by-products currently used in least-cost formulation are no longer applicable, and certainly a greater awareness of, and control over, the sourcing of ingredients will become essential. The assumption of substantial equivalence cannot be assumed.

ACKNOWLEDGEMENTS

My thanks to Dr A. Alldrick (Campden and Chorleywood Food Research Association) for the RVA data. The Scottish Office Department of Agriculture, Environment and Fisheries provided financial support for this work.

REFERENCES

AAFCO (1997) In: *Official Publication of the Association of American Feed Control Officers*. AAFCO Inc., Atlanta, Georgia, pp. 137–140.

Agricultural and Food Research Council (1992) Technical Committee on Responses to Nutrients Report 9. *Nutrition Abstracts and Reviews Series B* 62, 789–825.

Aiple, K.-P., Steingass, H. and Drochner, W. (1996) Prediction of the net energy content of raw materials and compound feeds for ruminants by different laboratory methods. *Archives of Animal Nutrition* 49, 213–220.

Allan, G. (1997) Alternative feed ingredients for intensive aquaculture. In: Corbett, J.L., Choct, M., Nolan, J.V. and Rowe, J.B. (eds) *Recent Advances in Animal Nutrition in Australia 1997*. University of New England Publishing Unit, Armidale, Australia, pp. 98–109.

Åman, P. and Hesselman, K. (1984) Analysis of starch and other main constituents of cereal grains. *Swedish Journal of Agricultural Research* 14, 135–139.

Annison, G. (1991) Relationship between levels of soluble nonstarch polysaccharides and the apparent metabolizable energy of wheats assayed in broiler chickens. *Journal of Agricultural and Food Chemistry* 39, 1252–1256.

Armsby, H.P. (1903) *The Principles of Animal Nutrition*, 1st edn. John Wiley and Sons, New York.

Austin, S.C., Wiseman, J. and Chesson, A. (1999) Influence of non-starch polysaccharide structure on the metabolisable energy of UK wheat fed to poultry. *Journal of Cereal Science* 29, 77–88.

Baker, C.W. and Barnes, R. (1990) The application of near infra-red spectroscopy to forage evaluation in the agricultural development and advisory service. In: Wiseman, J. and Cole, D.J.A. (eds), *Feedstuffs Evaluation*. Butterworths, London, pp. 337–354.

Baker, C.W., Givens, D.I. and Deaville, E.R. (1994) Prediction of organic matter digestibility *in vivo* of grass silage by near infrared spectroscopy: effect of calibration method, residual moisture and particle size. *Animal Feed Science and Technology* 50, 17–26.

Beauchemin, K.A. (1991) Effects of neutral detergent fiber concentration and alfalfa hay quality on chewing, rumen function and milk production of dairy cows. *Journal of Dairy Science* 74, 3140–3148.

Bedford, M.R. (1995) The optimum dose of a xylanase-based enzyme offered to broilers fed a wheat based diet increases as the bird ages. *Poultry Science* 74 (Suppl. 1), 18.

Bedford, M.R. (1997) Factors affecting response of wheat based diets to enzyme supplementation. In: Corbett, J.L., Choct, M., Nolan, J.V. and Rowe, J.B. (eds), *Recent Advances in Animal Nutrition in Australia 1997*. University of New England Publishing Unit, Armidale, Australia, pp. 1–7.

Bedford, M.R. and Classen, H.L. (1992) Reduction of intestinal viscosity through manipulation of dietary rye and pentosanase concentration is effected through changes in the carbohydrate composition of the intestinal aqueous phase and results in improved growth and feed conversion. *Journal of Nutrition* 122, 560–569.

Bedford, M.R. and Classen, H.L. (1993) An *in vitro* assay for prediction of broiler intestinal viscosity and growth when fed rye-based diets in the presence of exogenous enzymes. *Poultry Science* 72, 137–143.

Beever, D.E., Dhanoa, M.S., Losada, H.R., Evans, R.T., Cammell, S.B. and France, J. (1986) The effect of forage species and stage of harvest on the processes of digestion occurring in the rumen of cattle. *British Journal of Nutrition* 56, 439–454.

Boisen, S. and Fernández, J.A. (1997) Prediction of the total tract digestibility of energy in feedstuffs and pig diets by *in vitro* analyses. *Animal Feed Science and Technology* 68, 277–286.

Boon, J.J. (1989) An introduction to pyrolysis mass spectrometry of lignocellulosic material. In: Chesson, A. and Orskov, E.R. (eds), *Physico-chemical Characterisation of Plant Residues for Industrial and Feed Use*. Elsevier Applied Science, London, pp. 25–49.

Bossingault, J.B. (1843) *Economic Rurale Consideree dans ses Reports avec la Chemie, la Physique et la Metiorologie*. Bechet Heunne, Paris.

Carré, B. and Brillouet, J.M. (1989) Determination of water-insoluble cell walls in feeds: interlaboratory study. *Journal of the Association of Official Analytical Chemists* 72, 463–481.

Carré, B and Conan, L. (1989) Relationship between trypsin-inhibitor content of pea seeds and pea protein digestibility in poultry. In: Huisman, J., Van der Poel, A.F.B. and Liener, I.E. (eds), *Recent Advances of Research in Antinutritional Factors in Legume Seeds*. Pudoc, Wageningen, The Netherlands. pp. 103–106.

Chesson, A. (1993) Mechanistic models of forage cell wall degradation. In: Jung, H.G., Buxton, D.R., Hatfield, R.D. and Ralph, J. (eds) *Forage Cell Wall Structure and Digestibility*. American Society of Agronomy, Inc., Crop Science Society of America, Inc., Soil Science Society of America, Inc., Madison, Wisconsin, pp. 347–376.

Chesson, A. and Monro, J.A. (1982) Legume pectic substances and their degradation in the ovine rumen. *Journal of the Science of Food and Agriculture* 33, 852–859.

Crévieu, I., Carré, B., Chagneau, A.-M., Quillien, L., Guéguen, J. and Bérot, S. (1997) Identification of resistant pea (*Pisum sativum* L.) protein in the digestive tract of chickens. *Journal of Agricultural and Food Chemistry* 45, 1295–1300.

Dowman, M.G. and Collins, F.C. (1982) The use of enzymes to predict the digestibility of animal feeds. *Journal of the Science of Food and Agriculture* 33, 689–696.

Drake, A.P., Fuller, M.F. and Chesson, A. (1991) Simultaneous estimations of precaecal protein and carbohydrate digestion in the pig. In: Fuller, M.F. (ed.), *In Vitro Digestion for Pigs and Poultry*. CAB International, Wallingford, UK, pp. 162–176.

Elkin, R.G., Freed, M.B., Hamaker, B.R., Zhang, Y. and Parsons, C.M. (1996) Condensed tannins are only partially responsible for variations in nutrient digestibilities of sorghum grains. *Journal of Agricultural and Food Chemistry* 44, 848–853.

Eerlingen, R.C., Jacobs, H., Block, K. and Delcour, J.A. (1997) Effects of hydrothermal treatments on the rheological properties of potato starch. *Carbohydrate Research* 297, 347–356.

Fawcett, R. and Webster, M. (1996) Valuing variance reduction. *Proceedings of the Australian Poultry Science Symposium* 8, 53–64.

Gerlach, M. (1990) NIR measuring technology for quality evaluation of feeds. *Kraftfutter* 2, 67–74.

Gill, C. (1997) World feed panorama: high cost of feedstuffs: global impact, response. *Feed International* 18, 6–16.

Givens, D.I., De Boever, J.L. and Deaville, E.R. (1997) The principles, practices and some future applications of near infrared spectroscopy for predicting the nutritive value of foods for animals and humans. *Nutrition Research Reviews* 10, 83–114.

Gordon, F.J., Cooper, K.M., Park, R.S. and Steen, R.W.J. (1998) The prediction of intake potential and organic matter digestibility of grass silages by near infrared spectroscopy analysis of undried samples. *Animal Feed Science and Technology* 70, 339–351.

Grant, R.J., Colenbrader, V.F. and Mertens, D. (1990) Milk fat depression in dairy cows: role of particle size of alfalfa hay. *Journal of Dairy Science* 73, 1823–1835.

Gruchala, L. and Pomeranz, Y. (1993) Enzyme-resistant starch: studies using differential scanning calorimetry. *Cereal Chemistry* 70, 163–170.

Hall, M.B., Lewes, B.A., Van Soest, P.J. and Chase, L.E. (1997) A simple method for estimation of neutral detergent-soluble fibre. *Journal of the Science of Food and Agriculture* 74, 441–449.

Hall, M.B., Pell, A.N. and Chase, L.E. (1998) Characteristics of neutral detergent-soluble fiber fermentation by mixed ruminal microbes. *Animal Feed Science and Technology* 70, 23–39.

Harris, L.E., Haendler, H., Riviere, R. and Rechaussat, L. (1980) International feed data-bank system; an introduction to the system with instructions for describing feeds and recording data. Publication 2 International Feedstuffs Unit, Utah State University, Logan, Utah.

Harrison, M.D., Ballard, M.R.M., Barclay, R.A., Jackson, M.E. and Stilborn, H.L. (1991) A comparison of true digestibility for poultry and apparent ileal digestibility for swine. A classical *in vitro* method and NIR spectrophotometry for determining amino acid digestibility. In: Verstegen, M.W.A., Huisman, J. and den Hartog, L.A. (eds), *Digestive Physiology in the Pig (5th International Symposium)*. Purdoc, Wageningen, The Netherlands, pp. 254–259.

Henneberg, W. and Stohmann, W. (1864) *Begründung einer Rationellen Fütterung der Wiederkauer*. Schwetschke und Söhne, Braunschweig, Germany.

Henry, Y., Vogt, H. and Zoiopolus, P.E. (1988) Feed evaluation and nutritional requirements III. 4. Pigs and poultry. *Livestock Production Science* 19, 299–354.

Hoogendoorn, C.J., McCutcheon, S.N., Lynch, G.A., Wickham, B.W. and MacGibbon, A.K.H. (1990) Production responses of New Zealand Friesian cows at pasture to exogenous recombinately derived bovine somatotropin. *Animal Production* 51, 431–439.

Hurrell, R.F. (1990) Influence of the Maillard reaction on the nutritional value of foods. In: Finot, P.A., Aeschbacher, H.U., Hurrell, R.F. and Liardon, R. (eds), *The Maillard Reaction in Food Processing, Human Nutrition, and Physiology*. Birkhauser Veerlag, Basel, Switzerland, pp. 245–258.

Jongbloed, A.W. and Lenis, N.P. (1999) Nutrition as a tool to reduce the impact on the environment. *Options Méditerranéennes* 37, 229–240.

Jung, H.G. and Allen, M.S. (1995) Characteristics of plant cell walls affecting intake and digestibility of forages by ruminants. *Journal of Animal Science* 73, 2774–2790.

Lengerken, J.V., Peterhänsel, M., Hartmann, G. and Halle, H.J. (1997) How reliable are estimating methods for grain? *Kraftfutter* 4, 183–185.

Menke, K.H., Raab, L., Salewski, A., Steingass, H., Fritz, D. and Schneider, W. (1979) The estimation of the digestibility and metabolizable energy content of ruminant feedingstuffs from gas production when they are incubated with rumen liquor *in vitro*. *Journal of Agricultural Science, Cambridge* 93, 217–222.

Mertens, D.R. (1992) Non structural and structural carbohydrates. In: Van Horn, H.H. and Wilcox, C.J. (eds), *Large Dairy Herd Management*. American Dairy Science Association, Champaign, USA, pp. 219–235.

Mertens, D.R. (1997) Creating a system for meeting the fibre requirements of dairy cows. *Journal of Dairy Science* 80, 1463–1481.

Morris, J.G. and Rogers, Q.R. (1994) Assessment of the nutritional adequacy of pet foods through the life cycle. *Journal of Nutrition* 124, 2520S–2534S.

Moya, L., Garrido, A., Guerrero, J.E., Lizaso, J. and Gomez, A. (1995) Quality control of raw materials in the compound feed industry. In: Batten, G., Flinn, P.C., Welsh, L.A. and Blakeney, A.B. (eds), *Leaping Ahead with Near Infrared Spectroscopy*. Royal Australian Chemical Institute, Melbourne, Australia, pp. 111–116.

Murphy, J.M. (1996) Engineering oil production in rapeseed and other oil crops. *Trends in Biotechnology* 14, 206–213.

Murray, I. (1993) Forage analysis by near infra-red spectroscopy. In: Davies, A., Baker, R.D., Grant, S.A. and Laidlaw, A.S. (eds), *Sward Management Handbook*, 2nd edn. British Grassland Society, Reading, UK, pp. 285–312.

National Research Council (1993) *Nutrient Requirements of Fish*. National Academy Press, New York.

Orskov, E.R. and McDonald, I. (1979) The estimation of protein degradability in the rumen from incubation measurements weighted according to rate of passage. *Journal of Agricultural Science, Cambridge* 92, 499–503.

Parsons, C.M. (1996) Digestible amino acids for poultry and swine. *Animal Feed Science and Technology* 59, 147–153.

Pasquier, B., Armand, M., Guillon, F., Castelain, C., Borel, P., Barry, J.L., Pieroni, G. and Lairon, D. (1996) Viscous soluble dietary fibre alter emulsification and lipolysis of triacylglycerols in duodenal medium *in vitro*. *Journal of Nutritional Biochemistry* 7, 293–302.

Paul, C. and Mika, V. (1981) Mahlwiderstandsmessungen an rauhfutter II. Beziehungen zwischen mahlwiderstand und futterwertmerkmalen. *Landbauforschung Volkenrode* 31, 163–169.

Pérez-Vendrell, A.M., Brufau, J., Molino-Cano, J.L., Francesch, M. and Guasch, J. (1996) Effect of cultivar and environment on β-(1,3)–(1,4)-D-glucan content and acid extract viscosity of Spanish barleys. *Journal of Cereal Science* 23, 285–292.

Reinitz, G. (1980) Acceptability of animal fat in diets for rainbow trout at two environmental temperatures. *Progressive Fish Culturalist* 42, 218–222.

Rickes, E.L., Ham, E.A., Morcatelli, E.A. and Ott, W.H. (1962) The isolation and biological properties of a β-glucanase from *B. subtilis. Archives of Biochemistry and Biophysics* 96, 371–375.

Rogel, A.M., Annison, E.F., Bryden, W.L. and Balnave, D. (1987) The digestion of wheat starch in broiler chickens. *Australian Journal of Agricultural Science* 38, 639–649.

Sauvant, D., Dijkstra, J. and Mertens, D. (1995) Optimisation of ruminal digestion: a modelling approach. In: Journet, M., Grenet, E., Farce, M.-H., Theriez, M. and Demarquilly, C. (eds), *Recent Developments in the Nutrition of Herbivores*. INRA Editions, Paris, pp. 143–165.

Savoie, L. (1994) Digestion and absorption of food – usefulness and limitations of *in vitro* models. *Canadian Journal of Physiology and Pharmacology* 72, 407–414.

Scott, T.A. (1996) Assessment of energy levels in feedstuffs for poultry. *Animal Feed Science and Technology* 62, 15–19.

Shenk, J.S. (1992) NIRS analysis of natural agricultural products. In: Hildrum, K.I., Isaksson, T., Naes, T. and Tanberg, A. (eds), *Near Infra-red Spectroscopy. Bridging the Gap Between Data Analysis and NIR Applications*. Ellis Horwood, London, pp. 235–240.

Shenk, J.S. (1995) NIRS technology for the feed industry. *Seminar for the Feed and Agricultural Industry*, Gent, Belgium.

Simmins, P.H. and van Kempen, T.A.T.G. (1998) Reduction of nitrogen in monogastrics. Improvement of precision in feed manufacturing. *Options Méditerranéennes* 37, 275–283.

Steen, R.W.J., Gordon, F.J., Mayne, C.S., Proots, R.E., Kilpatrick, D.J., Unsworth, E.F., Barnes, R.J., Porter, M.G. and Pippard, C.J. (1995) Prediction of intake of grass silage by cattle. In: Gainsworthy, P.C. and Cole, D.J.A. (eds), *Recent Advances in Animal Nutrition 1995*. Butterworth, London, pp. 67–89.

Sudweeks, E.M., Ely, L.O., Mertens, D.R. and Sisk, L.R. (1981) Assessing amounts and form of roughages in ruminant diets: roughage value index system. *Journal of Animal Science* 53, 1406–1411.

Sunvold, G.D. (1996) Dietary fiber for dogs and cats: a historical, perspective. In: Carey, D.P., Norton, S.A. and Bolse, S.M. (eds), *Recent Advances in Canine and Feline Nutritional Research*. Orange Frazer Press, Ohio, pp. 3–14.

Tacon, A.G.J. (1998) Global trends in aquaculture and aquafeed production. *FAO/GLOBEFISH Research Programme Report*, FAO, Rome.

Thapa, B., Walker, D.H., and Sinclair, F.L. (1997) Indigenous knowledge of the feeding value of tree fodder. *Animal Feed Science and Technology* 67, 97–114.

Thomas, M., van Vliet, T. and van der Poel, A.F.B. (1998) Physical quality of pelleted animal feed. 3. Contribution of feedstuff components. *Animal Feed Science and Technology* 70, 59–78.

Tilley, J.M.A. and Terry, R.A. (1963) A two stage technique for the *in vitro* digestion of forage crops. *Journal of the British Grasslands Society* 18, 104–111.

Travis, A.J., Murison, S.D., Hirst, D.J., Walker, K.C. and Chesson, A. (1996) Comparison of the anatomy and degradability of straw from varieties of wheat and barley that differ in susceptibility to lodging. *Journal of Agricultural Science, Cambridge* 127, 1–10.

Ulyatt, M.J. (1971) Studies on the causes of differences in pasture quality between perennial ryegrass, short-rotation ryegrass and white clover. *New Zealand Journal of Agricultural Research* 14, 352–367.

Van Kempen, T. and Jackson, D. (1996) NIRS may provide rapid evaluation of amino acids. *Feedstuffs* 68, 12–15.

Van Soest, P.J. (1967) Development of a comprehensive system of feed analysis and its application to forages. *Journal of Animal Science* 26, 119–128.

Whittemore, C.T. (1997) An analysis of methods for the utilisation of net energy concepts to improve the accuracy of feed evaluation in diets for pigs. *Animal Feed Science and Technology* 68, 89–99.

Williams, P.C., Preston, K.R., Norris, K.H. and Starkey, P.M. (1984) Determination of amino acids in wheat and barley by near infrared reflectance spectroscopy. *Journal of Food Science* 49, 17–20.

Wiseman, J. and Inborr, B. (1990) The nutritive value of wheat and its effects on broiler performance. In: Haresign, W. and Cole, D.J.A. (eds), *Recent Advances in Animal Nutrition 1990*. Butterworth, London, pp. 79–102.

Wolf, E. (1874) *Rationellen Füttering der Landwirtschaftlichen Nutztiere*. Paul Parey, Berlin.

3

Intake, Passage and Digestibility

D.P. Poppi[1], J. France[2] and S.R. McLennan[3]

[1]School of Land and Food, University of Queensland, Brisbane, Australia; [2]The University of Reading, Department of Agriculture, Reading, Berkshire, UK; [3]Queensland Beef Industry Institute, Animal Research Institute, Brisbane, Australia

INTRODUCTION

The most important factor influencing the production response of an animal is the total quantity of nutrients absorbed. Thus intake and digestibility are key parameters in any feed evaluation system, and of these intake is the most important as it accounts for most differences between feed types. The measurement of intake and digestibility is simple with housed animals but it is a requirement in any feed evaluation system to be able to predict intake and to do so quickly. Most animals graze, and it is in seeking to explain differences in animal performance in the field that quantification of intake becomes important. This is difficult in the field.

INTERDEPENDENCE OF INTAKE, PASSAGE AND DIGESTIBILITY

The original model of Waldo *et al.* (1972) has been the basis of the interaction of intake, passage and digestibility. This model involved the concept of potential digestibility, fractional digestion rate and fractional passage rate (Fig. 3.1).

Whilst the detail of this simple model has been modified, it has withstood the test of time. Potential digestibility (PD) is defined as that fraction which disappears after a long incubation period, and the indigestible fraction is that part which is not available to the microorganisms. Since both fractions are part of the same plant particle, then the fractional passage rate k_p applies to both fractions. It follows that the final digestibility value is a function of these two rate parameters. Hence:

$$\text{Digestibility} = \text{function } k_d/(k_d + k_p)$$

$$= \text{PD} \times [k_d/(k_d + k_p)] \tag{3.1}$$

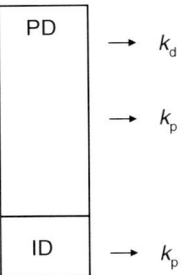

Fig. 3.1.   A schematic representation of the disappearance of potentially digestible (PD) and indigestible (ID) material from the rumen by fractional digestion rate (k_d) and fractional passage rate (k_p) (from Waldo *et al.*, 1972).

Since in the physical model of intake regulation, the pool size (rumen pool) is assumed to be constant, then:

$$\text{Intake} = \text{Rumen pool size} / [(\text{PD}/(k_d + k_p)) + (\text{ID}/k_p)] \qquad (3.2)$$

In this simple model, the interaction of intake and digestibility is explained and reasons may be proposed for variation in the intake/digestibility relationship and for variation in digestibility with feeding conditions, e.g. level of intake. Thus intake and digestibility are both functions of the key parameters of potential digestibility, k_d and k_p, and it is their relative quantitative values which determine the final values for intake and digestibility. It is these latter two values which drive the feeding standards and in turn determine the level of nutrients absorbed and final animal productivity.

An assumption used in this model to predict intake is that pool size as in rumen fill is a constant. Whilst this does occur for a number of situations, it does not hold for a number of important situations, e.g. grass versus legume diets, lactating versus dry animals. Sutherland (1988) has proposed another concept in that particles do not escape from the rumen pool until they have been digested to a certain extent. The same basic principles apply but the end result is a more constant proportion of the PD fraction of the cell wall digested, and a variable rumen fill is a possible outcome (Poppi *et al.*, 1994).

These basic principles have been expanded with various degrees of complexity (Baldwin *et al.*, 1976; Mertens and Ely, 1979; Poppi *et al.*, 1981a; Mertens, 1987; Pond *et al.*, 1988; Allen, 1996) and while these have led to a better understanding of the factors involved, the underlying hypothesis is still that proposed by Waldo *et al.* (1972).

There are alternative theories on intake regulation whereby metabolic and physical factors interact to result in a final intake value. Some have even proposed that the preoccupation with physical regulation as explained in the Waldo *et al.* (1972) model has overshadowed the real metabolic effects limiting intake (Tolkamp and Ketelaars, 1992; Ketelaars and Tolkamp, 1996).

THE COMPETING MODELS OF INTAKE REGULATION

A brief outline of the various models of intake regulation will be given. This will serve to focus attention on the parameters that need to be determined in any feed evaluation system. Any unifying theory of intake regulation must seek to explain some observations. These may be listed as:

1. For many forage-based diets, intake may vary considerably whilst rumen fill is relatively constant.
2. Rumen fill may vary between diets.
3. A nutrient deficiency leads to a depression in intake or, more correctly, the addition of a nutrient in a circumstance can lead to a rapid increase in intake from which it is assumed the nutrient was limiting.
4. Different requirements, as set by the physiological state of the animal, enable animals to increase intake when previously intake was thought to be limited by some characteristic of the diet, usually its digestibility, e.g. a lactating cow has a much higher intake of grass than her dry twin cow even though digestibility of the grass is the same for both animals.

These features can be defined as nutritional factors that influence intake. Poppi *et al.* (1987) have defined two components which influence intake of grazing animals as distinct from housed animals, namely harvesting and nutritional components. The nutritional component relates to the more traditional area of research, and the physical and metabolic factors which contribute to this will be outlined in more detail later. Factors associated with harvesting material during grazing are both behavioural and nutritional in origin. It is not proposed to examine these in detail here as they have been examined extensively elsewhere (Spedding *et al.*, 1966; Stobbs, 1973; Chacon and Stobbs, 1976; Poppi *et al.*, 1987; Gordon and Lascano, 1993; Demment and Laca, 1994; Newman *et al.*, 1994, 1995; Edwards *et al.*, 1995).

The grazing animal needs to harvest material (mainly leaf) from swards which can be very heterogeneous in distribution of leaf both horizontally (spatial distribution of plants) and vertically (sward structure). The drive to consume nutrients is set metabolically but is constrained by behavioural interactions associated with the difficulty in locating (site selection) and prehending leaf (selection within a site) (Gordon and Lascano, 1993; Demment and Laca, 1994; Newman *et al.*, 1994).

The ease or difficulty of harvesting leaf interacts with the desire to consume nutrients to set a final intake which may be less than that required metabolically or less than is physically possible from considerations of rumen fill and retention time. Chacon and Stobbs (1976) showed this elegantly when cows grazed a sward at a high stocking rate or low leaf allowance (kg green leaf DM head^{-1} day^{-1}). They removed some rumen contents at the end of the grazing period but the cows did not return to graze, demonstrating that physical rumen factors did not constrain their intake and that a behavioural factor associated with the difficulty of harvesting leaf was predominant. A similar group of cows grazing the same sward at a low stocking rate (high leaf allowance) returned to grazing for a substantial period when some rumen contents were removed, suggesting that physical factors were important for this group of cows.

This was elaborated on further by a simulation exercise by Cruickshank (1986). He used measured retention times in the rumen of white clover and ryegrass, and by varying rate of intake was able to determine the rate at which the rumen would fill or accumulate digesta over a grazing period (Fig. 3.2). Rate of intake of grazing animals varies with sward structure and leaf allowance (Stobbs,1973; Chacon and Stobbs, 1976; Demment and Laca, 1994; Newman *et al.*, 1994). Animals on white clover would only reach the maximum level of rumen fill possible (potential rumen fill) with very high rates of intake (6 g kg^{-1} W day^{-1}), whilst those on ryegrass could do so even with low rates of intake (3 g kg^{-1} W day^{-1}) providing the grazing period was long enough (Fig. 3.2). This illustrates the interaction between rate of intake in the grazing animal as set by sward conditions and the animal's physiological drive to eat, and nutritional factors associated with the rate of nutrient release and rate of disappearance from the rumen. The following discussion will concentrate on factors important for hand-fed animals, but these factors also interact with behavioural factors associated with prehension in grazing animals.

The nutritional models of intake regulation and the parameters that need to be examined are those associated with physical regulation of intake and those associated with metabolic regulation of intake.

The physical regulation of intake has been largely applied to forages. Blaxter *et al.* (1956, 1961) used this concept to derive a relationship between intake and digestibility, and used the rate of passage concept to explain this relationship. Central to this model is that intake is regulated by the rate at which digesta leaves the rumen by digestion and passage and that rumen fill is relatively constant. It was part of an expansion of work in that period seeking to explain firstly the relationship between intake and digestibility and secondly the large variability in the relationship which became apparent as more studies were done (Minson, 1990). Nevertheless the intake/digestibility relationship is still the basis for the prediction of feeding and nutritional value of feed types by most feeding standards (Jarrige *et al.*, 1986; NRC, 1987; AFRC, 1991; Poppi, 1996; see also Chapter 7).

The idea of the rate of disappearance of digesta from the rumen focused attention on the parameters which contribute to that, i.e. digestion and passage, rather than the end result of their quantitative interaction, i.e. digestibility. Relationships between intake and rate of passage (Poppi *et al.*, 1981b), and intake and rate of digestion and potential digestibility (Gill *et al.*, 1969; Hovell *et al.*, 1986) were examined. Some good relationships were found, just as some good relationships between intake and digestibility were found (Minson, 1990); there was, however. no overall relationship. Further parameterization resulted in models incorporating particle size, differential passage rate and digestion rate with particle size or with chemically defined components with digestion and passage characteristics (Baldwin *et al.*, 1976; Mertens and Ely, 1979; Poppi *et al.*, 1981a; Kennedy and Murphy, 1988; Ellis *et al.*, 1991; Fisher, 1996).

The major outcome of this approach was the realization that the Waldo concept provided a useful model by which to examine intake regulation and that particle dynamics, in particular the rate of large particle breakdown and small particle disappearance from the rumen, were key parameters. In a simulation using a rumen model, Poppi *et al.* (1981a) showed that intake was

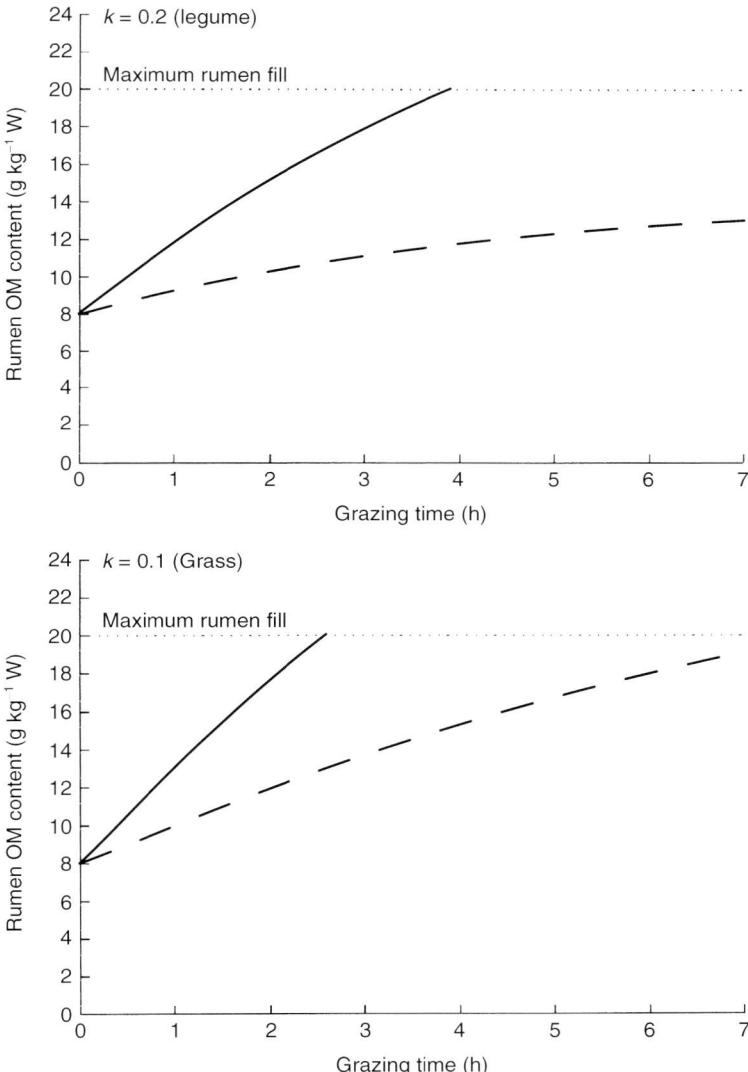

Fig. 3.2. Predicted increase in organic matter (OM) content in the rumen over a 7 h continuous grazing period, at two rates of OM intake, namely 3 (dashed line) and 6 (solid line) g kg⁻¹ W h⁻¹, when fractional disappearance rate from the rumen (k) is 0.2 (legume) or 0.1 (grass). The maximum rumen fill ever observed in lambs of 20 g OM kg⁻¹ W is indicated (adapted from Cruickshank, 1986).

sensitive to variation in the parameters in the order k_p > PD > k_{ds} > k_b > l_p; where k_p is the fractional passage rate of small particles, PD is potential digestibility, k_{ds} is the fractional digestion rate of small particles, k_b is the fractional breakdown of large particles to small particles, and l_p is the proportion of large particles entering the rumen.

Whilst this approach was useful in explaining outcomes, it did little to develop simple new parameters by which to evaluate feed or predict intake and digestibility. It did, however, focus on attributes that needed to be determined. The rate of digestion and potential digestibility have long been examined by *in vitro* and *in situ* methods, and these models provided a framework by which they could be incorporated mathematically and mechanistically to predict intake and digestibility rather than using an empirical relationship (e.g. Hovell *et al.*, 1986). The rate and extent of digestion have been examined by various mathematical approaches (France *et al.*, 1990; Dhanoa *et al.*, 1995). Similarly, with the rate of passage, attention had focused on particle size and particle breakdown, intuitively following on from the measurements of passage rate (Campling and Freer, 1962; Freer and Campling, 1963; Poppi *et al.*, 1981a; Mosely and Jones, 1984) and the large increase in intake when forages were ground and pelleted (Minson, 1963).

An important outcome of the early simulation work (Poppi *et al.*, 1981a) was that the passage of small particles had by far the major influence on retention time and intake. This observation challenged the preoccupation with large particle dynamics and was another indicator that physical factors may not be the major regulators of intake with forage-based diets. Most of the DM in the rumen (63–77%) was in the form of small particles (Poppi *et al.*, 1981a; Thiago *et al.*, 1992), and the question was what was constraining their escape. Beever *et al.* (1980–1981) showed that water outflow and fractional passage were largely a function of the osmotic interplay within the rumen as a result of osmotically active compounds (VFAs, etc.) released on digestion and saliva flow as a direct function of eating and ruminating times. Particles are carried out suspended in water, so factors influencing water outflow became important. Sutherland (1988) provided perhaps the most elegant treatise in this area when he showed that particles needed to be an appropriate specific gravity before they passed out through the omasal orifice and that this was related to the extent of digestion of the particle and its lack of buoyancy through entrapment of gas. This followed on from the extensive studies of Welch (1967, 1982) who showed that both particle size and specific gravity were important. Sutherland's model was, however, a significant conceptual departure from the particle size model.

The separation of the physical and metabolic components of intake was presented by Conrad *et al.* (1964) and Dinius and Baumgardt (1970). This proposed that, at high net energy intakes, intake was regulated by metabolic factors, but that, at some statistically defined digestibility, physical factors became dominant in regulating intake. The metabolic factors which control intake are many (Baile and Forbes, 1974).

Forbes (1977) outlined a model of intake regulation based on physical or metabolic aspects. Predicted intake was the lower of the two calculated intakes set by physical or metabolic parameters. The intake set by physical mechanisms was based on rumen fill and digestibility. The intake set by metabolic mechanisms was based on a metabolizable energy (ME) intake requirement. This was the first model to incorporate physical and metabolic parameters. This approach was expanded in a later book (Forbes, 1995).

Weston (1985, 1996) proposed an interaction between physical and metabolic regulation when he sought to explain why rumen fill declined as *ad libitum* net energy (NE) intake increased. This went against the concept of a physical regulation where rumen fill was supposed to be a constant. He proposed a concept that had rumen fill declining as NE intake approached NE requirement. Perhaps the converse should be invoked where the animal will push rumen fill and its associated parameters to higher and higher levels in an attempt to satisfy requirement when the NE value of the diet decreases. Presumably there is an upper limit to rumen fill, set by physical dimensions and pressure on internal organs, within which the animal will keep. Hutton (1963) and Hutton *et al.* (1964), in observing the higher rumen fill of lactating cows and their much higher intake over their dry twin pair, dismissed physical regulation as a controlling factor and invoked metabolic demand as the main regulator.

These observations cast doubt on the universal application of physical regulation and highlighted metabolic features. This became apparent with diets of high quality, both forage- and formulated concentrate-based feeds, and naturally raised the question of whether these features were applicable to low-quality diets. The most common example, that of a protein deficiency, was described by Egan (1970). Here the situation is one of low intake when protein content is low, and also low rumen fill. The features of a physical regulation are not apparent and are quickly modified on addition of the limiting nutrient to the diet such that intake increases markedly. Balance of nutrients becomes an important concept in nutrient metabolic regulation, e.g. the protein/energy ratio or protein content of diet and mineral content of the diet, e.g. phosphorus (Oldham, 1984). Thus, in feed evaluation systems, nutrient content in relation to DM or ME became important in defining adequacy. The underlying concept is the supply of ATP relative to the supply of building blocks. The supply of amino acids for protein synthesis of storage proteins and transient enzymes drives the life of the cell and the organism, thus in any metabolic regulation it is this balance which is important.

There is a problem with metabolic regulation. It seems logical that if a nutrient is defined as a key nutrient, then an animal seeks to consume that nutrient to the level required for genes to express themselves. This was the underlying concept in the early dilution experiments of Dinius and Baumgardt (1970) where it was proposed that animals ate for a set net energy. Yet in both free choice and infusion experiments, animals rarely do this (Cropper, 1987; Gherardi and Black, 1989; Weston, 1996), and a more usual observation is a trade-off of physically regulating pathways and various metabolic pathways for some nutrient intake below that required. Tolkamp and Ketelaars (1992) and Ketelaars and Tolkamp (1996) have proposed an alternative metabolic pathway to the interplay of protein synthesis, homeostatic mechanisms and ATP use which is related to the production of free radicals and oxygen efficiency based on an evolutionary model of how cells work. In this, they also proposed an optimization procedure in intake regulation whereby the benefits and costs of feeding are assessed to achieve a balance. Illius and Jessop (1996), in trying to integrate a number of regulatory pathways, proposed that intake was regulated imprecisely within the boundaries of these pathways. Poppi *et al.* (1994) earlier had proposed a simpler, more expedient approach. They identified six pathways with physical and metabolic characteristics (rate of intake, rumen turnover, faecal output, genetic limit to protein deposition, heat dissipation and ATP degradation) and proposed upper limits to these. Animals may eat to maximize a specific nutrient intake but upper limits in one or more of these pathways may stop the animal achieving that intake. Another important observation was that in any experimental approach to examining intake regulation, the proposed parameter is varied, e.g. protein content, rate of digestion, etc. If there is more than one pathway close to a maximum, then the other pathways may provide a hidden limit to intake and so intake may not be changed by that particular experimental approach and the hypothesis would be rejected incorrectly, e.g. the minimum protein content of a diet for a cold environment is lower than that required for a hot environment because of the limitations of heat dissipation in a hot environment and the need for heat to maintain body temperature in a cold environment and so greater use of substrate cycles.

The use of the metabolism models of France *et al.* (1987) and Gill *et al.* (1984) in this exercise (Poppi *et al.*, 1994) illustrated their usefulness in integrating a variety of potential metabolic signals. Whilst one feature, that of ATP degradation, has been criticized (Illius and Jessop, 1996), the use of a model in this way provides a means to examine nutrient balance as a regulator of intake and, by inference, relates back to nutrient composition in the ingested feed and the balance and rate of release of nutrients on digestion (see Chapters 4, 13, 14 and 15). The approach of using metabolism models in feed evaluation, especially in relation to intake, is limited by the concepts and data which drive these models. However, they are attractive, and Poppi *et al.* (1989) used this concept successfully to examine substitution of forage intake when a young animal consumes varying levels of milk.

Ketelaars and Tolkamp (1996) recently have invoked oxygen efficiency as a means of intake regulation, still using the oxygen free radical idea and evolutionary theory for the optimization of intake to maximize oxygen/NE. Recently, Ellis *et al.* (1999) suggested that another way to examine this is to use the instantaneous rate of heat production as an integrating mechanism (Fig. 3.3). This relies on the curvilinear nature of heat production increasing at a faster rate as ME intake is increased. It is proposed that this represents the increase in metabolic pathways involved in metabolizing nutrients for protein and fat synthesis as nutrient absorption increases, and that there are consequences in this, e.g. acid–base balance, which need to be controlled for homeostatic reasons. Nicolaides and Even (1985) suggested that the cue to start eating was a rapid decline in heat production, and so both initiation and cessation signals for eating may be related to instantaneous heat production which is indicative of underlying instantaneous metabolism and any homeostatic consequences. In

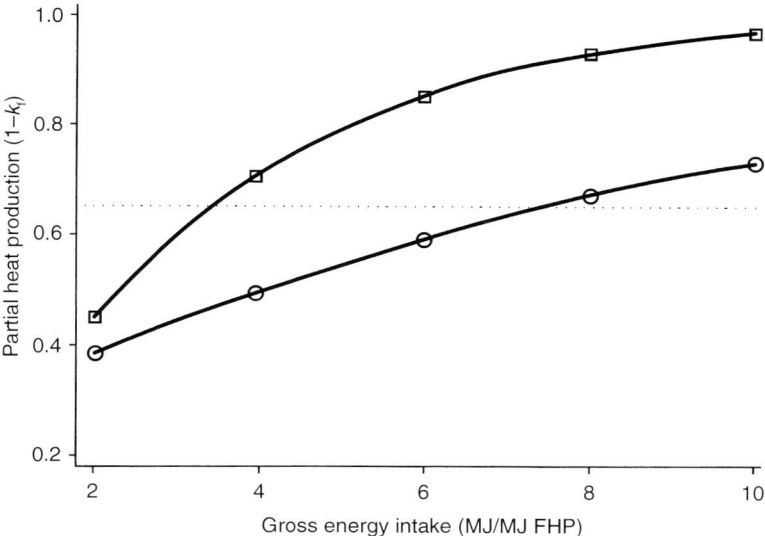

Fig. 3.3. The increase in heat production as gross energy intake increases as multiples of fasting heat production (FHP) for diets varying in metabolizability (q): ($q = 0.4$, □; $q = 0.7$, ○). The dotted line represents the maximum partial heat production where intake will be maximum for that diet quality (Blaxter and Boyne, 1978; Ellis *et al.*, 1999).

all of these (Weston, 1985, 1996; Poppi *et al.*, 1994; Illius and Jessop, 1996; Ketelaars and Tolkamp, 1996; Ellis *et al.*, 1999), there is an inherent lack of precision for each pathway or integrative procedure.

Williams (1985) points out the lack of precision in meal intake but the precision in long-term intake regulation in pigs, an idea in agreement with the above models. The underlying concept in all of these models is that there are negative consequences for under- or overshooting the metabolic pathways. Whether this leads to precision or chaos in the long term has not been resolved, but it is no doubt involved in the variability of intake between days within an animal and also between animals, giving population variability.

This approach has identified the important features for feed evaluation, that of key nutrients, nutrient balance and the level of a particular balance of nutrients, which can be accommodated for a particular physiological state. Thus a diet manifestly deficient in protein is still eaten but only to a low level, which often is not adequate for maintenance. The same diet fed to a cold-exposed animal which has increased or initiated the activity of certain metabolic pathways to utilize substrates other than protein to generate heat will be eaten at higher levels, enabling the animal to survive adequately.

RATE OF DIGESTION

This is covered by Broderick and Cochran (Chapter 4) but the interaction of rate of digestion, rate of passage and potential digestibility and their influence on intake and digestibility were outlined earlier. Poppi *et al.* (1994) outlined another feature of relevance to rumen fill. In this, they used the concept of Sutherland (1988) that a particle must be digested extensively so that it loses its buoyancy before it can pass from the rumen. It follows that the rate of digestion becomes an important variable influencing the rate of passage, in that particles become available for passage in a time-dependent fashion rather than in a pathway which operates simultaneously on a pool of identical particles. The conceptual difference is important; if the rate and extent of digestion are the limiting features for escape of particles, then the only way in which the animal can increase intake is for the animal to increase the pool size, i.e. the amount of digesta in the rumen. This is observed in lactating animals (Hutton, 1963; Hutton *et al.*, 1964) or those with balloons inserted to test the physical regulation of intake (Cruickshank *et al.*, 1987). It would also explain the lower level of rumen fill of animals consuming a legume diet compared with those consuming a grass diet (Poppi *et al.*, 1994).

Another important feature of the rate of digestion in feed evaluation is its role in microbial growth and hence the yield and balance of nutrients released on digestion (Fox *et al.*, 1992; Illius and Jessop 1996; Bannink *et al.*, 1997). A more speculative role is on intake regulation where the rate of release of nutrients and their synchrony over the day might affect the pattern and level of intake.

RATE OF PASSAGE

The importance of the rate of passage in influencing intake and digestibility was outlined clearly by Blaxter *et al.* (1956) and Waldo *et al.* (1972). Classical marker kinetic studies were done by Campling and Freer (1962) and Grovum and Williams (1973); and a critique of marker kinetics and the lack of rigour was outlined by Warner and Stacy (1968) and Faichney (1993). It is clear that the rate of disappearance of digesta from the rumen is a key factor integrating with metabolic signals to determine the final intake of an animal (Weston, 1996). The rate of disappearance is markedly influenced by the rate of passage, and the practical observation of the large increase in intake achieved by grinding and pelleting roughages suggested that particle size or rate of particle disintegration was important. Various models and experimental approaches sought to determine this, and although large particle breakdown was a key feature, it was also found that the rate of passage of small particles was the more important variable (Poppi *et al.*, 1981a). However, the emphasis in feed evaluation programmes has been parameters relating to particle disintegration, particularly of large particles. Thus systems to simulate chewing (Troelsen and Bigsby, 1964; Kennedy *et al.*, 1997), grinding energy (Chenost, 1966; Laredo and Minson, 1973; Weston, 1985; McLeod *et al.*, 1990; Henry *et al.*, 1996b), leaf shear strength (Mackinnon *et al.*, 1989; Inoue *et al.*, 1994; Henry *et al.*, 1996c) and compression energy (Baker *et al.*, 1993; Henry *et al.*, 1996a,b,c; Ru and Fortune, 1997) have been developed and correlated with intake with variable success.

Another approach has concentrated on what drives the passage of small particles generated on disintegration of the ingested particle. Particles pass out of the rumen suspended in water, so the factors influencing water outflow become important (Warner and Stacy, 1968; Beever *et al.*, 1980–1981). The main factors are the osmotic conditions created by the release of osmotically active compounds on digestion, e.g. VFAs and the input of saliva through eating and rumination. This provides an interaction between digestion rate and passage rate.

Passage rate is one of the most important parameters influencing disappearance of digesta from the rumen and intake, and various methods have been developed for its estimation, some based on the plant (see above) and some within the animal. If, however, the physical model is rejected as having any influence and intake is controlled only by a metabolic mechanism, then estimates of the rate of passage might simply reflect the effect of a level of intake rather than the cause of a level of intake (Ketelaars and Tolkamp, 1996). Measuring the rate of passage has depended on the kinetics of marker disappearance from the rumen or whole tract, where the markers are either internal, e.g. lignin, or externally applied particles, e.g. rare earth markers (Ellis and Huston, 1968; Hartnell and Satter, 1979; Eliman and Orskov, 1981, 1982; Colucci *et al.*, 1982; Amaning-Kwarteng, 1986a,b). These have been used to delineate sections of the gastrointestinal tract (Grovum and Williams, 1973; France *et al.*, 1985; Cruickshank *et al.*, 1989; Van Soest *et al.*, 1992), delineate particle size (Poppi *et al.*, 1981a; Pond *et al.*, 1988; Faichney, 1993), to identify pools without any defined anatomical features (Dhanoa *et al.*, 1985; Pond *et al.*, 1988; Ellis *et al.*, 1991) and for estimating faecal output (France *et al.*, 1988). The kinetics have been integral to the development of feed evaluation and formulation models, e.g. CNCPS (Sniffen *et al.*, 1992). They are often misused, with scant regard for the mathematical rigour needed (Warner and Stacy, 1968; France *et al.*, 1985; Matis *et al.*, 1989).

Two approaches have been taken in examining marker kinetics relating to passage; the deterministic approach of France *et al.* (1985) and the stochastic approach of Matis *et al.* (1989). Both seek to determine the retention time of defined particles in defined pools, but the marker concentration data are analysed differently to estimate the same parameters. In the deterministic approach of France *et al.* (1985), pools are characterized, rate constants for passage determined and assumptions made on the biological mechanism. In the stochastic approach of Matis *et al.* (1989), the data are fitted statistically to estimate retention time parameters of pools which may or may not have a defined anatomical compartment.

Both methods can give the same answer to a data set for the parameter, retention time, but they differ in the mathematical characteristics and the assumed mechanism which gives rise to the estimate of retention time of a marker in a pool. Choice of approach is as much a philosophical outlook on parameter estimation as it is on the availability of packages and procedures to use.

Two questions arise with marker kinetic data. Firstly, do they just reflect the level of intake set by some other metabolic mechanism? Secondly, do they just reflect the rumen conditions set by other characteristics of the diet such as rumen fill in relation to nutrient demand (Weston, 1996), particle shape (Thornton and Minson, 1973) and the rate and extent of digestion (Sutherland, 1988). The first question cannot be answered satisfactorily until the extent of metabolic control has been better defined. Vega and Poppi (1997) addressed some aspects of the second question. They marked particles in the size range 0.5–1.0 mm which had been either digested extensively (faecal particles) or not digested (ground feed particles) and inserted them into the rumen of animals on four different diets (Table 3.1). Furthermore, both grass and legume particles were used as they differed in shape: grass was needle-like while legume was rounded in shape. The extent of digestion and particle type had no influence on retention time, but the diet and hence the rumen conditions markedly influenced the retention time. This did not support Sutherland's concept (Sutherland, 1988), at least for small particles, and did show the marked influence of rumen conditions and type as influenced by diet. Thus it strikes a note of caution in examining passage rate in isolation from other features that might influence intake and rumen conditions, such as metabolic regulation, rate of digestion, osmotic factors and mixing characteristics of the pool.

CONCLUSIONS

The procedures and hypotheses outlined for intake, passage and digestibility need to lead to simple methods by which to evaluate a feed.

The simple methodology aims to predict empirical values for intake, metabolizable energy and metabolizable protein content of feedstuffs to be used in feeding standards. These are outlined in other chapters in this book, and Poppi (1996) collated a number of predictive equations and procedures for the prediction of intake. Essentially, regression relationships of the parameter (intake, digestibility) are established against chemical composition or *in vitro* values determined by NIRS, wet chemistry or biological means.

The complex methodology aims to use models for intake, digestibility and passage to interlink with digestion and metabolism models to predict animal

Table 3.1. The fractional outflow rate (FOR) from the rumen of CrEDTA- or Yb-labelled small particles (0.5–1.5 mm) of lucerne or pangola hay inserted into the rumen of sheep consuming diets of concentrate, pangola hay, lucerne chaff or lucerne pellets (from Vega and Poppi, 1997).

FOR (h^{-1})	Concentrate	Pangola hay	Lucerne chaff	Lucerne pellets
CrEDTA	0.049	0.042	0.064	0.099
Yb	0.029	0.031	0.047	0.054

performance. The vast array of methods and approaches previously discussed in this chapter has enabled these models to be constructed but, more importantly, has focused attention on the key features that need to be considered. The newer ideas on intake regulation have also focused on metabolic factors that need to be considered. Thus, four key features need to be characterized by any evaluation method such that it can incorporate the new ideas on intake regulation and nutrient use. They are: (i) nutrient balance released on digestion, (ii) rate of release and absorption of nutrients, (iii) rate of disintegration of particles and (iv) rate and potential extent of digestion. These features will enable the integration of physical and metabolic parameters in a modelling process.

REFERENCES

Agricultural and Food Research Centre (1991) Voluntary intake of cattle. Technical Committee on Responses to Nutrients, Report No. 8. *Nutrition Abstracts and Reviews* Series B 61, 815–823.

Allen, M.S. (1996) Physical constraints on voluntary intake of forages by ruminants. *Journal of Animal Science* 74, 3063–3075.

Amaning-Kwarteng, K., Kellaway, R.C., Leibholtz, J. and Kirby, A.C. (1986a) Rumen degradation and fractional outflow rates of nitrogen supplements given to cattle eating sodium hydroxide-treated straw. *British Journal of Nutrition* 55, 387–398.

Amaning-Kwarteng, K., Kellaway, R.C. and Kirby, A.C. (1986b) Supplemental protein degradation, bacterial protein synthesis and nitrogen retention in sheep eating sodium hydroxide-treated straw. *British Journal of Nutrition* 55, 557–569.

Baile, C.A. and Forbes, J.M. (1974) Control of feed intake and regulation of energy balance in ruminants. *Physiological Reviews* 54, 160–214.

Baker, S.K., Klein, L., De Boer, E.S. and Purser, D.B. (1993) Genotypes of dry, mature subterranean clover differ in shear energy. In: *Proceedings of the XVII International Grassland Congress*, Palmerston North, New Zealand, pp. 592–593.

Baldwin, R.L., Koong, L.J., Ulyatt, M.J. and Smith, N. (1976) Towards a synthesis. In: Sutherland, T.M., McWilliam, J.R. and Leng, R.A. (eds), *Reviews in Rural Science. 2. From Plant to Animal Protein.* University of New England Publishing Unit; Armidale, Australia, pp. 175–181.

Bannink, A., Visser, H. De, and Van Vuuren, A.M. (1997) Comparison and evaluation of mechanistic rumen models. *British Journal of Nutrition* 78, 563–581.

Beever, D.E., Black, J.L. and Faichney, G.J. (1980–1981) Simulation of the effects of rumen function on the flow of nutrients from the stomach of sheep, part 2 – assessment of computer predictions. *Agricultural Systems* 6, 221–241.

Blaxter, K.L. and Boyne, A.W. (1978) The estimation of the nutritive value of feeds as energy sources for ruminants and the derivation of feeding systems. *Journal of Agricultural Science, Cambridge* 90, 47–68.

Blaxter, K.L., Graham, N.McC. and Wainman, F.W. (1956) Some observations on the digestibility of food by sheep, and on related problems. *British Journal of Nutrition* 10, 69–91.

Blaxter, K.L., Wainman, F.W. and Wilson, R.S. (1961) The regulation of food intake by sheep. *Animal Production* 3, 51–61.

Campling, R.C. and Freer, M. (1962) The effect of specific gravity and size on the mean time of retention of inert particles in the alimentary tract of the cow. *British Journal of Nutrition* 16, 507–518.

Chacon, E. and Stobbs, T.H. (1976) Influence of progressive defoliation of a grass sward on the eating behaviour of cattle. *Australian Journal of Agricultural Research* 27, 709–727.

Chenost, M. (1966) Fibrousness of forages: its determination and its relation to feeding value. *Proceedings of the X International Grassland Congress*, Helsinki, Finland, pp. 406–411.

Colucci, P.E., Chase, L.E. and Van Soest, P.J. (1982) Feed intake, apparent digestibility and rate of particulate passage in dairy cows. *Journal of Dairy Science* 65, 1445–1456.

Conrad, H.R., Pratt, A.D. and Hibbs, J.W. (1964) Regulation of feed intake in dairy cows. 1. Change in importance of physical and physiological factors with increasing digestibility. *Journal of Dairy Science* 47, 54–62.

Cropper, M.R. (1987) Growth and development of sheep in relation to feeding strategy. PhD Thesis, University of Edinburgh.

Cruickshank, G.J. (1986) Nutritional constraints to lamb growth at pasture. PhD Thesis, Lincoln University, New Zealand.

Cruickshank, G.J., Poppi, D.P. and Sykes, A.R. (1987) The influence of intraruminal water-filled balloons on pasture intake by grazing sheep. In: Rose, M. (ed.), *Herbivore Nutrition Research, Occasional Publication of the Australian Society of Animal Production*. Australian Society of Animal Production, Brisbane, Australia, pp. 97–98.

Cruickshank, G.J., Poppi, D.P. and Sykes, A.R. (1989) Theoretical considerations in the estimation of rumen fractional outflow rate from various sampling sites in the digestive tract. *British Journal of Nutrition* 62, 229–239.

Demment, M.W. and Laca, E.A. (1994) Reductionism and synthesis in the grazing sciences: models and experiments. *Proceedings of the Australian Society of Animal Production* 20, 6–16.

Dhanoa, M.S., Siddons, R.C., France, J. and Gale, D.L. (1985) A multicompartmental model to describe marker excretion patterns in ruminant faeces. *British Journal of Nutrition* 53, 663–671.

Dhanoa, M.S., France, J. and Siddons, R.C. (1995) A non-linear compartmental model to describe forage degradation kinetics during incubation in polyester bags in the rumen. *British Journal of Nutrition* 73, 3–15.

Dinius, D.A. and Baumgardt, B.R. (1970) Regulation of food intake in ruminants. 6. Influence of caloric density of pelleted rations. *Journal of Dairy Science* 53, 311–316.

Edwards, G.R., Parsons, A.J., Penning, P.D. and Newman, J.A. (1995) Relationship between vegetation state and bite dimensions of sheep grazing contrasting plant species and its implication for intake rate and diet selection. *Grass and Forage Science* 50, 378–388.

Egan, A.R. (1970) Nutritional status and intake regulation in sheep. VI. Evidence for variation in setting of an intake regulatory mechanism relating to the digesta content of the reticulo-rumen. *Australian Journal of Agricultural Research* 21, 735–746.

Eliman, M.E. and Orskov, E.R. (1981) Determination of rate of outflow of protein supplements from the rumen by measuring the rate of excretion of chromium-treated protein supplements and polyethylene glycol in the faeces. *Animal Production* 32, 386.

Eliman, M.E. and Orskov, E.R. (1982) Effect of the feeding level of grass on the rate of outflow of protein supplements from the rumen of sheep. *Proceedings of the Nutrition Society* 41, 29A.

Ellis, W.C. and Huston, J.E. (1998) [144]Ce–[144]Pr as a particulate digesta flow marker in ruminants. *Journal of Nutrition* 95, 67–78.

Ellis, W.C., Matis, J.H. and Kennedy, P.M. (1991) Passage and digestion of plant tissues in herbivores. In: Ho, Y.W., Wong, W.K., Abdullahand, N. and Tajuddin, Z.A. (eds), *Recent Advances on the Nutrition of Herbivores*, Malaysian Society of Animal Production, pp. 227–236.

Ellis, W.C., Poppi, D.P., Matis, J.H., Lippke, H., Hill, T.M. and Rouquette, F.M. (1999) Dietary–digestive–metabolic interactions determining the nutritive potential of ruminant diets. In: *Proceedings of the V International Herbivore Symposium*, San Antonio, Texas, USA, pp. 423–481.

Faichney, G.J. (1993) Digesta flow. In: Forbes, J.M. and France, J. (eds), *Quantitative Aspects of Ruminant Digestion and Metabolism*, CAB International, Wallingford, UK, pp. 53–85.

Fisher, D.S. (1996) Modeling ruminant feed intake with protein, chemostatic and distension feedbacks. *Journal of Animal Science* 74, 3076–3081.

Forbes, J.M. (1977) Interrelationships between physical and metabolic control of voluntary food intake in fattening, pregnant and lactating sheep: a model. *Animal Production* 24, 91–101.

Forbes, J.M. (1995) *Voluntary Food Intake and Diet Selection in Farm Animals*. CAB International, Wallingford, UK.

Fox, D.G., Sniffen, C.J., O'Connor, J.D., Russell, J.B. and Van Soest, P.J. (1992) A net carbohydrate and protein system for evaluating cattle diets. III. Cattle requirements and diet adequacy. *Journal of Animal Science* 70, 3578–3596.

France, J., Thornley, J.H.M., Dhanoa, M.S. and Siddons, R.C. (1985) On the mathematics of digesta flow kinetics. *Journal of Theoretical Biology* 113, 743–758.

France, J., Gill, M., Thornley, J.H.M. and England, P. (1987) A model of nutrient utilization and body composition in beef cattle. *Animal Production* 44, 371–385.

France, J., Dhanoa, M.S., Siddons, R.C., Thornley, J.H.M. and Poppi, D.P. (1988) Estimating the production of faeces by ruminants from faecal marker concentration curves. *Journal of Theoretical Biology* 135, 383–391.

France, J., Thornley, J.H.M., Lopez, S., Siddons, R.C., Dhanoa, M.S., Van Soest, P.J. and Gill, M. (1990) On the two-compartment model for estimating the rate and extent of feed degradation in the rumen. *Journal of Theoretical Biology* 146, 269–287.

Freer, M. and Campling, R.C. (1963) Factors affecting the voluntary intake of food by cows. 5. The relationship between the voluntary intake of food, the amount of digesta in the reticulo-rumen and the rate of disappearance of digesta from the alimentary tract with diets of hay, dried grass or concentrates. *British Journal of Nutrition* 17, 79–88.

Gherardi, S.G. and Black, J.L. (1989) Influence of post-rumen supply of nutrients on rumen digesta load and voluntary intake of a roughage by sheep. *British Journal of Nutrition* 62, 589–599.

Gill, M., Thornley, J.H.M., Black, J.L., Oldham, J.D. and Beever, D.E. (1984) Simulation of the metabolism of absorbed energy-yielding nutrients in young sheep. *British Journal of Nutrition* 52, 621–649.

Gill, S.S., Conrad, H.R. and Hibbs, J.W. (1969) Relative rate of *in vitro* cellulose disappearance as a possible estimator of digestible dry matter intake. *Journal of Dairy Science* 52, 1687–1690.

Gordon, I.J. and Lascano, C. (1993) Foraging strategies of ruminant livestock on intensively managed grasslands: potential and constraints. In: *Proceedings of the XVII International Grassland Congress*. Palmerston North, New Zealand, pp. 681–690.

Grovum, W.L. and Williams, V.J. (1973) Rate of passage of digesta in sheep. 3. Differential rates of passage of water and dry matter from the reticulo-rumen, abomasum and caecum and proximal colon. *British Journal of Nutrition* 30, 231–240.

Hartnell, G.F. and Satter, L.D. (1979) Determination of rumen fill, retention time and ruminal turnover rates of ingesta at different stages of lactation in dairy cows. *Journal of Animal Science* 48, 381–392.

Henry, D.A., Baker, S.K. and Purser, D.B. (1996a) Measuring the compression and shear energies of green plant material. *Proceedings of the Australian Society of Animal Production* 21, 497.

Henry, D.A., Simpson, R.J. and Hosking, B.J. (1996b) Physical and chemical characteristics of forages as predictors of rate of feed intake by sheep. *Proceedings of the Australian Society of Animal Production* 21, 494.

Henry, D.A., MacMillan, R.H. and Simpson, R.J. (1996c) Measurement of the shear and tensile fracture properties of leaves of pasture grasses. *Australian Journal of Agricultural Research* 47, 587–603.

Hovell, F.D.DeB, Ngambi, J.W.W., Barber, W.P. and Kyle, D.J. (1986) The voluntary intake of hay by sheep in relation to its degradability in the rumen as measured in nylon bags. *Animal Production* 42, 111–118.

Hutton, J.B. (1963) The effect of lactation on intake in the dairy cow. *Proceedings of the New Zealand Society of Animal Production* 23, 39–52.

Hutton, J.B., Hughes, J.W., Newth, R.P. and Watanabe, K. (1964) The voluntary intake of the lactating dairy cow and its relation to digestion. *Proceedings of the New Zealand Society of Animal Production* 24, 29–42.

Illius, A.W. and Jessop, N. (1996). Metabolic constraints on voluntary intake in ruminants. *Journal of Animal Science* 74, 3052–3062.

Inoue, T., Brookes, I.M., John, A., Hunt, W.F. and Barry, T.N. (1994) Effects of leaf shear breaking load on the feeding value of perennial ryegrass (*Lolium perenne*) for sheep. I. Effects on leaf anatomy and morphology. *Journal of Agricultural Science, Cambridge* 123, 129–136.

Jarrige, R., Demarquilly, C., Dulphy, J.P., Hoden, A., Robelin, J., Beranger, C., Geay, Y., Journet, M., Malterre, C., Micol, D. and Petit, M. (1986) The INRA 'Fill Unit' system for predicting the voluntary intake of forage-basal diets in ruminants: a review. *Journal of Animal Science* 63, 1737–1758.

Kennedy, P.M. and Murphy, M.R. (1988) The nutritional implications of differential passage of particles through the ruminant alimentary tract. *Nutrition Research Reviews* 1, 189–208.

Kennedy, P.M., Toscas, P.J., Faddy, M.J. and Minson, D.J. (1997) Use of a multi-exponential model to assess the effect of fermentation in the reticulorumen on particle fermentability as simulated from artificially macerating leaf and stem fractions of two tropical grasses. *Animal Feed Science and Technology* 66, 111–128.

Ketelaars, J.J.M.H. and Tolkamp, B.J. (1996) Oxygen efficiency and the control of energy flow in animals and humans. *Journal of Animal Science* 74, 3036–3051.

Laredo, M.A. and Minson, D.J. (1973) The voluntary intake, digestibility and retention time by sheep of leaf and stem fractions of five grasses. *Australian Journal of Agricultural Research* 24, 875–888.

Matis, J.H., Ellis, W.C. and Allen, D.M. (1989) On a stochastic approach to modelling ruminant digestion and metabolism. In: Robson, A.B. and Poppi, D.P. (eds),

Proceedings of the Third International Workshop on Modelling Digestion and Metabolism in Farm Animals Lincoln University, New Zealand, pp. 139–158.

Mackinnon, B.W., Easton, H.S., Barry, T.N. and Sedcole, J.R. (1989) The effect of reduced leaf shear strength on the nutritive value of perennial ryegrass. *Journal of Agricultural Science, Cambridge* 111, 469–474.

McLeod, M.N., Kennedy, P.M. and Minson, D.J. (1990) Resistance of leaf and stem fractions of tropical forage to chewing and passage in cattle. *British Journal of Nutrition* 63, 105–119.

Mertens, D.R. (1987) Predicting intake and digestibility using mathematical models of rumen function. *Journal of Animal Science* 64, 1548–1558.

Mertens, D.R. and Ely, L.O. (1979) A dynamic model of fiber digestion and passage in the ruminant for evaluating forage quality. *Journal of Animal Science* 49, 1085–1095.

Minson, D.J. (1963) The effect of pelleting and wafering on the feeding value of roughage – a review. *Journal of the British Grassland Society* 18, 39–44.

Minson, D.J. (1990) *Forage in Ruminant Nutrition*. Academic Press, London.

Mosely, G. and Jones, D.I.H. (1984) The physical digestion of perennial ryegrass (*Lolium perenne*) and white clover (*Trifolium repens*) in the foregut of sheep. *British Journal of Nutrition* 52, 381–390.

National Research Council (1987) *Predicting Feed Intake of Food Producing Animals*. National Academy of Sciences, Washington, DC.

Newman, J.A., Parsons, A.J. and Penning, P.D. (1994) A note on the behavioural strategies used by grazing animals to alter their intake rates. *Grass and Forage Science* 49, 502–505.

Newman, J.A., Parsons, A.J., Thornley, J.H.M., Penning, P.D. and Krebs, J.E. (1995) Optimal diet selection by a generalist grazing herbivore. *Functional Ecology* 9, 255–268.

Nicolaidis, S. and Even, P. (1985) Physiological determinant of hunger, satiation, and satiety. *American Journal of Clinical Nutrition* 42, 1083–1092.

Oldham, J.D. (1984) Protein–energy interrelationships in dairy cows. *Journal of Dairy Science* 67, 1090–1114.

Pond, K.R., Ellis, W.C., Matis, J.H., Ferreiro, H.M. and Sutton, J.D. (1988) Compartment models for estimating attributes of digesta flow in cattle. *British Journal of Nutrition* 60, 571–595.

Poppi, D.P. (1996) Prediction of food intake in ruminants from analyses of food composition. *Australian Journal of Agricultural Research* 47, 489–504.

Poppi, D.P., Minson, D.J. and Ternouth, J.H. (1981a) Studies of cattle and sheep eating leaf and stem fractions of grasses. III. The retention time in the rumen of large feed particles. *Australian Journal of Agricultural Research* 32, 123–137.

Poppi, D.P., Minson, D.J. and Ternouth, J.H. (1981b) Studies of cattle and sheep eating leaf and stem fractions of grasses. II. Factors controlling the retention of feed in the reticulo-rumen. *Australian Journal of Agricultural Research* 32, 109–121.

Poppi, D.P., Hughes, T.P. and L'Huillier, P.J. (1987) Intake of pasture by grazing ruminants. In: Nicol, A.M. (ed.), *Livestock Feeding on Pasture*. New Zealand Society of Animal Production Occasional Publication No. 10, pp. 55–63.

Poppi, D.P., Gill, M., France, J. and Dynes, R.A. (1989) Additivity in intake models. In: Robson, A.B. and Poppi, D.P. (eds), *Proceedings of the Third International Workshop on Modelling Digestion and Metabolism in Farm Animals*. Lincoln University, New Zealand, pp. 29–46.

Poppi, D.P., Gill, M. and France, J. (1994) Integration of theories of intake regulation in growing ruminants. *Journal of Theoretical Biology* 167, 129–145.

Ru, Y.J. and Fortune, J.A. (1997) Effect of cultivar and grazing intensity in spring on shear and compression energies of dry mature subterranean clover (*Trifolium subterranean* L.). *Australian Journal of Agricultural Research* 48, 1199–1206.

Sniffen, C.J., O'Connor, J.D., Van Soest, P.J., Fox, D.G. and Russell, J.B. (1992) A net carbohydrate and protein system for evaluating cattle diets II. Carbohydrate and protein availability. *Journal of Animal Science* 70, 3562–3577.

Spedding, C.R.W., Large, R.U. and Kydd, D.D. (1966) The evaluation of herbage species by grazing animals. In: *Proceedings of the X International Grassland Congress*. Helsinki, Finland, pp. 479–483.

Stobbs, T.H. (1973) The effect of plant structure on the intake of tropical pastures. 1. Variation in the bite size of grazing cattle. *Australian Journal of Agricultural Research* 24, 808–819.

Sutherland, T.M. (1988) Particle separation in the fore stomachs of sheep. In: Dobson, A. and Dobson, M. (eds), *Aspects of Digestive Physiology in Ruminants*. Cornell University Press, Ithaca, New York, pp. 43–73.

Thiago, L.R.L., Gill, M. and Sissons, J.W. (1992) Studies of method of conserving grass herbage and frequency of feeding in cattle. *British Journal of Nutrition* 67, 319–336.

Thornton, R.F. and Minson, D.J. (1973) The relationship between apparent retention time in the rumen, voluntary intake and apparent digestibility of legume and grass diets in sheep. *Australian Journal of Agricultural Research* 24, 889–898.

Tolkamp, B.J. and Ketelaars, J.J.M.H. (1992) Toward a new theory of feed intake regulation in ruminants. 2. Costs and benefits of feed consumption: an optimization approach. *Livestock Production Science* 30, 297–317.

Troelsen, J.E. and Bigsby, F.W. (1964) Artificial mastication – or new approach for predicting voluntary forage consumption by ruminants. *Journal of Animal Science* 23, 1139–1142.

Van Soest, P.J., France, J. and Siddons, R.C. (1992) On the steady-state turnover of compartments in the ruminant gastrointestinal tract. *Journal of Theoretical Biology* 159, 135–145.

Vega, A. De and Poppi, D.P. (1997) Extent of digestion and rumen condition as factors affecting passage of liquid and digesta particles in sheep. *Journal of Agricultural Science, Cambridge* 128, 207–215.

Waldo, D.R., Smith, L.W. and Cox, E.L. (1972) Model of cellulose disappearance from the rumen. *Journal of Dairy Science* 55, 125–129.

Warner, A.C.I. and Stacy, B.D. (1968) The fate of water in the rumen. 1. A critical appraisal of the use of scluble markers. *British Journal of Nutrition* 22, 369–387.

Welch, J.G. (1967) Appetite control in sheep by indigestible fibres. *Journal of Animal Science* 26, 849–854.

Welch, J.G. (1982) Rumination, particle size and passage from the rumen. *Journal of Animal Science* 54, 885–894.

Weston, R.H. (1985) The regulation of feed intake in herbage-fed ruminants. *Proceedings of the Nutrition Society of Australia* 10, 55–62.

Weston, R.H. (1996) Some aspects of constraint to forage consumption by ruminants. *Australian Journal of Agricultural Research* 47, 175–197.

Williams, I.H. (1985) Food intake and reproduction in pigs. *Proceedings of the Nutrition Society of Australia* 10, 63–69.

4

In Vitro and *In Situ* Methods for Estimating Digestibility with Reference to Protein Degradability

G.A. Broderick[1] and R.C. Cochran[2]

[1]US Dairy Forage Research Centre, Agricultural Research Service, USDA, Madison, Wisconsin; [2]Department of Animal Sciences and Industry, Kansas State University, Manhattan, Kansas, USA

INTRODUCTION

In theory, *in vivo* methods should be ideal for measuring nutrient digestibility in the animal. However, *in vivo* techniques require large amounts of feed and suffer from considerable variation due to the animal and to other factors; this variation necessitates use of sufficient experimental replication to obtain reliable results. However, the expense of obtaining adequate replication, when added to the costs of maintaining and sampling numbers of large animals, can make *in vivo* studies prohibitively expensive. Moreover, animal welfare concerns are likely to contribute to further reductions in *in vivo* experimentation. This has led to increased interest in using *in vitro* and *in situ* methods for estimating digestibility in the gastrointestinal tract.

The microbiology of nutrient digestion in the rumen is very complex. A multiplicity of organisms and enzymes carry out ruminal digestion of cell walls (Weimer, 1993; White *et al.*, 1993) and of starch and other non-structural carbohydrates (McAllister *et al.*, 1993). Protein degradation in the rumen also depends on numerous organisms elaborating an array of protein-, peptide- and amino acid-degrading activities (Jouany, 1996; Wallace, 1996; Wallace and Cotta, 1988; Paster *et al.*, 1993; Attwood and Reilly, 1995). The animal secretes many digestive enzymes into the abomasum and small intestine to effect nutrient release prior to absorption into the body. To yield useful data, *in vitro* and *in situ* techniques must somehow mimic *in vivo* digestion processes. Ideally, *in vitro* and *in situ* estimates of rate and extent of digestion should be quantitatively similar to those obtained *in vivo*. Estimates that are only correlated with *in vivo* values are also useful, indicating that important, perhaps limiting, characteristics of *in vivo* digestion had been simulated by the experimental system.

IN VITRO TECHNIQUES

Estimating Ruminal Digestion *In Vitro*

The two basic approaches to making *in vitro* estimates of ruminal digestion involve incubations with: (i) ruminal microorganisms or (ii) cell-free enzymes. The former methodology employs ruminal digesta generally obtained from cannulated donor animals. The latter approach is based on using enzymes, most often from commercial chemical houses, that are intended to have activities similar to mixed ruminal microorganisms. Use of cell-fee enzymes has the clear advantage of obviating the need for cannulated donor ruminants.

Perhaps the most widely applied ruminal *in vitro* method is that of Tilley and Terry (1963) for predicting total tract digestion of forage organic matter (OM). This procedure mimics the two-stage digestion that occurs in the ruminant *in vivo*: forage samples are prepared using standard protocols and incubated in buffered ruminal inoculum for 48 h, followed by a 24-h digestion in pepsin at acid pH. This method is very robust, partly because *in vivo* OM digestibility is estimated from standard curves developed by regressing the extent of *in vivo* OM digestion on the extent of OM digestion in the two-stage *in vitro* system. Ruminal inocula can vary substantially in microbial numbers from day to day, even when obtained from the same animal fed the same diet. Thus the microbial numbers in starting inocula for Tilley and Terry incubations also would be expected to vary. However, because of the rapid doubling times for most ruminal bacteria, microbial numbers probably converge to within narrow ranges in only a few hours after beginning an incubation, and will be stable for most of the 48 h of the ruminal phase. Thus, so long as care is taken to maintain anaerobiosis and prevent excessive losses of bacteria, end-point measurements of *in vitro* OM are not influenced greatly by variation in microbial numbers in the original inoculum. Initial microbial numbers are likely to influence the early phases of the progress curve of *in vitro* digestion, but neither the final extent of *in vitro* OM digestion, nor the *in vivo* OM digestibility predicted from that result, would be greatly altered. The regression in the original report of Tilley and Terry (1963) was developed using data from more than 100 individual forage samples, thus adding to the robustness of the method as originally applied. As always, successful application of a bio-assay such as that of Tilley and Terry requires that the unknown forages being assayed are represented appropriately among the standards relating *in vitro* and *in vivo* digestibility.

The procedure of Goering and Van Soest (1970) has also gained wide acceptance for estimating *in vivo* OM and dry matter (DM) digestibility from *in vitro* neutral detergent fibre (NDF) digestion. In this method, the extent of NDF digestion is determined after a 48-h incubation with ruminal inoculum; neutral detergent solubles are assumed to have a constant, high *in vivo* digestibility (of 98%). The Goering and Van Soest (1970) technique is the basis of a number of approaches to estimating ruminal and total tract digestibility, including application in forage evaluation programmes (e.g. Holt *et al.*, 1978). Weiss (1994) reviewed *in vitro* ruminal procedures for quantifying digestion, particularly those patterned after these two methods. More recently, there has been extensive work on using *in vitro* gas production to estimate ruminal OM and DM digestibility (Cone, 1997; Pell *et al.*, 1997; Theodorou *et al.*, 1997); these techniques, and continuous culture *in vitro* methods, are not discussed in this chapter.

Use of cellulases to estimate the extent of ruminal NDF and total tract OM and DM digestion has considerable promise. Overall, the most effective cellulase for solubilizing DM and for predicting the extent of *in vivo* digestibility of forages has been the mixture of fibrolytic enzymes elaborated by *Trichoderma* spp. (Jones and Hayward, 1975; Dowman and Collins, 1982; De Boever *et al.*, 1988). European work has shown that pre-treatment of feedstuffs, either digesting for 24 h with acid–pepsin (Jones and Hayward, 1975) or extracting with neutral detergent solution (Dowman and Collins, 1982), followed by a 24 h incubation with *Trichoderma* cellulase, substantially improved estimates of OM digestibility compared with application of the same treatments after the cellulase. In a number of cases, better predictions of OM digestibility were obtained using pepsin–cellulase pre-treatment (Moss and Givens, 1990; Givens *et al.*, 1993a), while the neutral detergent pre-treatment was more effective in other experiments that studied dried herbage (Bughrara and Sleper, 1986), silages (Givens *et al.*, 1989, 1993b) and spring-grown herbage (Givens *et al.*, 1990a). Often, similar proportions of the variation in ruminal *in vivo* (Stakelum *et al.*, 1988; Givens *et al.*, 1990b) and *in vitro* digestibility (Bughrara *et al.*, 1989) have been explained using either acid–pepsin or neutral detergent pre-treatments. Applying an amylase-neutral detergent pre-treatment was effective for maize silages (Givens *et al.*, 1995). Alderman (1985) proposed this approach for evaluating compound feedstuffs and described routine methods for its application. Several recent papers have addressed the use of blends of carbohydrases to assess the extent of digestion of starch and non-structural carbohydrates (Cone and Vlot, 1990; Cone, 1991; Richards *et al.*, 1995).

Emphasis must now be placed on conducting *in vitro* incubations with ruminal microorganisms or cell-free enzymes to predict kinetic parameters of digestion, including lag times, which are critically important in determining the extent of ruminal *in vivo* NDF digestion (Mertens, 1993; Wilman *et al.*, 1996). However, *in vitro* methods for estimating rates and other kinetic parameters of digestion probably will not be as robust as the Tilley and Terry (1963) or Goering and Van Soest (1970) techniques. The importance of determining kinetic parameters of digestion was discussed in a recent review on use of cell-free enzymes to quantify ruminal carbohydrate and protein degradation in the animal (Broderick, 1997).

In Vitro Prediction of Ruminal Protein Degradability

Ruminal in vitro *systems*
Quantifying the rate and extent of protein degradation using *in vitro* incubations with mixed ruminal organisms is complicated by microbial uptake of the

products of protein degradation as well as degradation of microbial and residual feed proteins present in the inoculum. Microbial catabolism of inoculum protein can be dealt with, at least theoretically, using blanks that contain all components of the incubation except the test protein. However, microbial incorporation of the breakdown products from the test protein cannot be 'blanked out' in the same way, and degradation is not, therefore, a simple function of accumulation of nitrogenous end-products. It is necessary either to distinguish newly formed microbial protein from feed protein remaining undegraded or somehow to account for this underestimation of end-product. An example of this potential flaw is illustrated by an early report on use of *in vitro* ammonia (NH_3) production to assess protein degradation. Results showing an inverse relationship between starch content of feed and apparent NH_3 release led to speculation that starch may inhibit amino acid deamination (Warner, 1956). Problems related to use of ruminal *in vitro* NH_3 production to estimate protein degradability have been reviewed elsewhere (Broderick, 1982).

Borchers (1967) attempted to circumvent microbial uptake of protein degradation products by adding toluene, an inhibitor of amino acid deamination, to *in vitro* inocula and then quantifying degradation from amino acid release. Later, hydrazine was found to be more effective than toluene in inhibiting microbial amino acid disappearance; hydrazine also inhibited microbial NH_3 uptake *in vitro* (Broderick and Balthrop, 1979). Broderick (1978) added 1.0 mM hydrazine to ruminal *in vitro* incubations containing limited amounts of feed protein; protein degradation was quantified from N accumulation as amino acid plus NH_3, after correction for amino acid and NH_3 production in blanks without added protein. Evidence suggesting that direct utilization of amino acids was not prevented totally by hydrazine alone led to the inclusion of chloramphenicol with hydrazine in the inhibitor *in vitro* (IIV) system to completely shut off microbial uptake of protein degradation products (Broderick, 1987). Neither compound depressed proteolytic activity in short-term incubations of up to 6 h.

The IIV method was used to estimate the protein degradation rate (k_d) by incubating limited substrate concentrations (i.e. under first-order conditions) with the inhibited ruminal inoculum; the extent of degradation was computed using the observed rate and an assumed ruminal passage rate (k_p), typically 0.06 h^{-1}, from the ratio: $k_d/(k_d + k_p)$ (Ørskov and McDonald, 1979). The IIV procedure successfully predicted relative differences in lactation performance of cows fed solvent and expeller soybean meal (SBM) (Broderick *et al.*, 1990), identified the optimal extent of heating required for protecting protein in soybeans (Faldet *et al.*, 1991) and lucerne hay (Broderick *et al.*, 1993) and served as the basis of a solubility test (Hsu and Satter, 1995) and a near infrared spectrometric (NIRS) calibration to estimate protein degradability in roasted soybeans (Tremblay *et al*, 1996). However, the IIV procedure has several significant limitations: (i) accumulation of NH_3, amino acids and small peptides may make the system subject to end-product inhibition, particularly in the presence of rapidly degraded proteins; (ii) degradation rates determined for feeds containing high levels of NH_3 and free amino acids (e.g. grass and legume forage silages) are less accurate because breakdown of more slowly degraded residual protein must be quantified from the appearance of additional NH_3 and amino acids in the presence of high backgrounds; and (iii) estimation of degradation rates from the gentle slopes obtained with slowly degraded proteins appears to be less accurate. These latter two problems also plague *in situ* methodology. Veresegyházy *et al.* (1993) reported that more than 19 mM of added NH_3 or 12 mM of added free amino acids inhibited the rate of casein degradation by ruminal inoculum *in vitro*.

Regardless of these limitations, degradation rates determined with the IIV method compared favourably with those determined *in vitro* using uninhibited ruminal inoculum in which degradation rates were computed from net appearance of NH_3 plus net synthesis of microbial crude protein (CP) estimated from the ^{15}N enrichment method of Hristov and Broderick (1994) (described below). Generally, degradation rates and ruminal escapes determined for seven proteins were of a similar magnitude to those obtained for the same proteins using the IIV method (Table 4.1). Mean degradation rates estimated by IIV averaged 86% of those determined using the ^{15}N method. Casein was the most degradable of the seven proteins and the one most likely to show reduced degradation because of end-product inhibition in the IIV system. Regressing the ^{15}N rates on the IIV rates yielded the equations: $Y = -0.034 + 1.77 \ X$ ($r^2 = 0.889$; including casein data); and $Y = 0.037 + 0.996 \ X$; ($r^2 = 0.872$; excluding casein data). That the slope of the regression with casein was substantially greater than 1.0, but approximated to 1.0 without

Table 4.1. Comparison of ruminal *in vitro* degradation rates and estimated ruminal escapes[a] for seven proteins determined from net microbial growth plus NH_3 release (Hristov and Broderick, 1994) or using the limited substrate inhibitor *in vitro* procedure.

Protein source	Net microbial growth plus NH_3[b]		Inhibitor *in vitro*[c]	
	Rate, k_d (h^{-1})	Escape (% of total N)	Rate, k_d (h^{-1})	Escape (% of total N)
Casein	0.569	10	0.307	17
Solvent soybean meal	0.148	29	0.137	30
Expeller soybean meal	0.036	63	0.030	65
Low-solubles fish meal	0.026	70	0.034	63
High-solubles fish meal	0.063	49	0.066	46
Maize gluten meal	0.034	64	0.017	75
Roasted soybeans	0.050	55	0.045	56

[a]Estimated escape, % of total N = $[k_p/(k_d + k_p)]$, where $k_p = 0.06 \ h^{-1}$.
[b]Degradation rate and escape computed from the extent of CP degradation at 6 h estimated from amount of N detected in net growth of microbial CP plus NH_3.
[c]Degradation rates and estimated escapes obtained previously for the same proteins using 4-h inhibitor *in vitro* incubations (Broderick, 1987). Means are from Broderick and Clayton (1992) (low- and high-solubles fish meal), Broderick *et al.* (1990) (maize gluten meal) and Faldet *et al.* (1991) (roasted soybeans); means for casein and for solvent and expeller soybean meal are from all three trials.

casein, suggested that the IIV method yielded similar degradation rates as the uninhibited ^{15}N system for all except the most rapidly degraded proteins.

A 'Michaelis–Menten' approach based on the IIV system was developed partly to correct for the underestimation of degradation rates for rapidly degraded proteins (Broderick and Clayton, 1992). Rather than following the time course of degradation in incubations using a single, limited amount of protein substrate (~ 2.4 mg of CP ml^{-1} of ruminal fluid equivalent in the inoculum), several protein levels, from very small to very large (1.9–45 mg of CP ml^{-1} of ruminal fluid equivalent), are incubated for a single, relatively short time period in the Michaelis–Menten approach. Fractional degradation rate, k_d, is estimated as the tangent through the origin of the velocity versus substrate concentration curve; the slope of this line is the ratio of the maximum velocity to the Michaelis constant (i.e. $k_d = V_{max}/K_m$). Non-linear regression analysis using the integrated Michaelis–Menten equation of data from 2-h incubations yields direct estimates of this ratio (Broderick and Clayton, 1992). A comparison of how the degradation rate was derived for the same sample of freeze-dried lucerne using the limited substrate and Michaelis–Menten approaches is illustrated in Fig. 4.1.

Degradation rates estimated by the Michaelis–Menten IIV method for certain proteins were more rapid, and more consistent with *in vivo* estimates of ruminal protein escape, than were those obtained using the limited substrate IIV approach. For example, mean estimates of rate and escape for protein in casein and lucerne hay were, respectively, about 0.30 h^{-1} and 15% and 0.07 h^{-1} and 40% by the limited substrate method; and 0.75 h^{-1} and 8%, and 0.20 h^{-1} and 25% by the Michaelis–Menten procedure. Literature values of ruminal escape for casein (10%; McDonald and Hall, 1954; Broderick, 1978) and lucerne hay (28%; NRC, 1989) were more similar to the Michaelis–Menten results. Degradation rates and ruminal escapes were also determined with the limited substrate and Michaelis–Menten IIV procedures for a number of legume forages (Broderick and Albrecht, 1997). Linear regression of Michaelis–Menten rates on those estimated using limited substrate methodology (Fig. 4.2) showed a high correlation between methods ($r^2 = 0.914$); the slope of the regression was 1.06 and not different from unity ($P > 0.17$). However, the intercept was 0.047 h^{-1}, suggesting that rates determined using the limited substrate approach that approximated to zero for several of the tannin-containing forages were in fact more rapid, corresponding to ruminal escapes of about 56% $\{100 \times 0.06/[0.06 + 0.047]\}$. Nevertheless, both approaches ranked these forages the same for protein degradability. It should be noted that determination of degradation rates with either IIV method requires use of rapid NH_3 and total amino acid analyses. This may be an important practical limitation to widespread application of the technique.

Neutze *et al.* (1993) applied the limited substrate IIV system described above, except they measured protein degradation from the net release of trichloroacetic acid-soluble N. The advantage of their approach is that the extent of degradation can be measured by simple Kjeldahl or Dumas N assays rather than by a dual NH_3 plus total amino acid analyses. That they obtained more complete recovery of degraded CP present as small peptides was

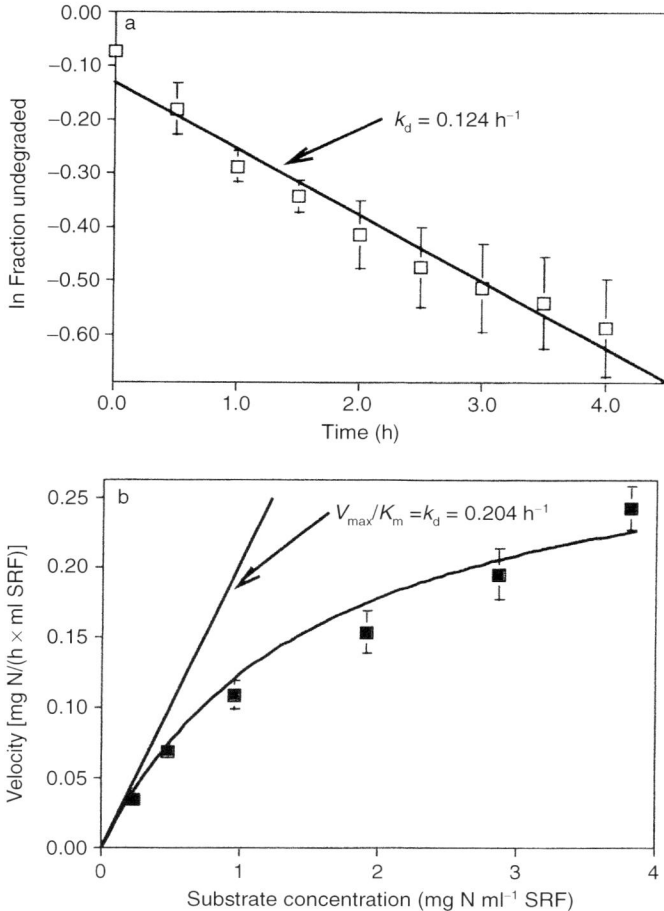

Fig. 4.1. Protein degradation rates (h^{-1}) estimated for one sample of freeze-dried lucerne foliage using (a) the limited-substrate inhibitor *in vitro* (IIV) method as usually applied (Broderick, 1987), or (b) the Michaelis–Menten IIV method (Broderick and Clayton, 1992). Vertical lines represent ± 1 SEM.

indicated by the faster rate observed for casein. Degradation rates reported for a number of SBM samples (Neutze *et al.*, 1993) were of similar magnitude to rates determined from NH_3 and total amino acid release for several other samples of SBM (Broderick, 1987; Broderick and Clayton, 1992).

The alternative approach of quantifying microbial CP formation plus N release was used by Raab *et al.* (1983) in a ruminal *in vitro* system to which starch was added in graded amounts to incubations containing the test protein. Net (i.e. blank-corrected) gas production (Menke *et al.*, 1979) and NH_3 concentration were measured for up to 24 h. The negative slope from regressing NH_3 concentration on gas production resulted from greater microbial

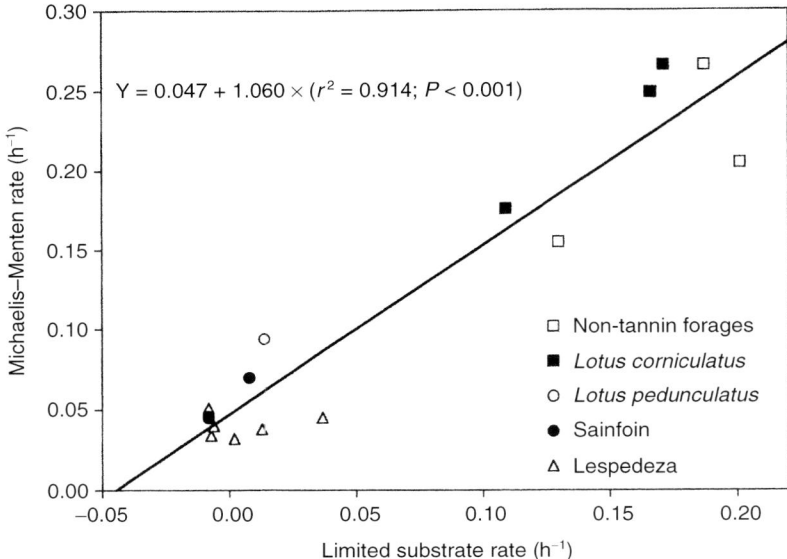

Fig. 4.2. Regression of protein degradation rates obtained for 16 forages using the Michaelis–Menten inhibitor *in vitro* (IIV) method (Broderick and Clayton, 1992) on protein degradation rates for the same forages obtained using the limited-substrate IIV method (Broderick, 1987). Data are from three forages that do not contain tannins (non-tannin forages) plus 13 that contain varying amounts of condensed tannins (Broderick and Albrecht, 1997).

uptake for growth at the higher starch levels. Regressions of NH_3 concentration on gas production were extrapolated to zero gas production; the intercept, NH_3 concentration at zero gas production, was assumed to represent the NH_3 concentration where there was no N incorporated into microbial protein. In this way, correction was made for microbial uptake of N as NH_3 (and amino acids). Later, these workers found virtually all of the N end-product present as NH_3, with very little released N remaining as free amino acids and peptides after 6 or 24 of incubation (Krishnamoorthy *et al.*, 1990). In both reports, the extent of protein degradation most often was determined after 24 h incubations, however, proportionately less protein was degraded in the time period between 6 and 24 h of incubation. This suggested that the system may not exhibit linear microbial growth and protein breakdown at times later than 6 h. Computations of degradation rates from the extents of degradation at 6 and 24 h indicated more satisfactory results using 6-h data. The method of Raab *et al.* (1983) appears promising and may also be valuable in analyses of the dynamics of protein degradation. However, this procedure requires that the relationship between microbial N uptake and gas production be linear over the whole range of incubation conditions. An alternative method, such as ^{15}N incorporation, should be used to confirm that fermentation and gas production do not become uncoupled from microbial growth in this system (Blümmel *et al.*, 1997). The recent upsurge in work on gas production methodology, and the wider availability of equipment to measure gas production, makes re-evaluation of this method very timely.

An *in vitro* procedure was developed to estimate the rate and extent of protein degradation in uninhibited ruminal inoculum by using $^{15}NH_3$ to quantify microbial uptake of protein breakdown products (Hristov and Broderick, 1994). Incubations were conducted for 6 h using buffered ruminal inoculum containing soluble carbohydrates and $(^{15}NH_4)_2SO_4$. The degradation rate was computed from net (i.e. blank-corrected) release of NH_3 plus net synthesis of microbial CP (estimated from ^{15}N enrichment of isolated microbial cells and total solids); ruminal escapes were computed assuming a passage rate of $0.06\ h^{-1}$. Comparison of degradation rates and ruminal escapes for seven proteins determined using this system with those obtained with the limited substrate IIV method were discussed earlier (Table 4.1). This method is promising in that uninhibited ruminal organisms are used in incubations with 'normal' microbial growth and without significant end-product accumulation. The procedure may prove useful with problematic protein sources such as legume and grass silages containing large amounts of background non-protein N. Alternative approaches might estimate microbial N formation from incorporation of ^3H- or ^{14}C-labelled amino acids, since the radioisotopes are less costly and more easily measured than ^{15}N.

Other techniques developed using mixed ruminal organisms *in vitro* have included the application of dyes (Mahadevan *et al.*, 1979) or radioisotopes (Wallace, 1983) to label the test proteins to quantify their breakdown in incubations with mixed ruminal organisms or proteases. These methods worked well for soluble proteins, but proved unreliable with insoluble proteins due to non-homogeneous labelling (Broderick *et al.*, 1988).

Commercial proteases

Use of commercial cell-free proteases to estimate ruminal protein degradation followed closely after the widespread application of N solubility as an index of protein degradability, and an extensive literature has developed. Ideally, proteases would yield estimates of degradation fractions and rates that are numerically similar to those generated by mixed ruminal organisms. However, there has been little research effort to predict, for example, the rate of N disappearance from *in situ* bags using cell-free proteases. Generally, laboratory estimates of the extent of ruminal degradation obtained with proteases have only been correlated to *in situ* estimates of the extent of degradation. Theoretical concerns about *in situ* methodology are discussed later in this chapter. Finding appropriate standard methods for assessing the rate and extent of ruminal protein degradation has been a major limitation to progress in identifying the best *in vitro* technique to adopt. Therefore, researchers cannot always be faulted for relying heavily on 'boot-strapping' approaches such as using *in situ* results to evaluate *in vitro* methods. This has often restricted the application of *in vitro* procedures, such as protease methods, to predicting *in situ* extents of degradation, rather than attempting to predict rates and extents of ruminal protein degradation *in vivo*.

Table 4.2 summarizes results from a wide range of cell-free proteases tested in different studies. Incubation conditions varied greatly in these trials, even when the same protease was used. For example, *Streptomyces griseus* protease, the protease that has been tested most extensively, was used at activities ranging from 0.0002 to 9.4 units mg^{-1} of substrate CP. Caution is also necessary when comparing results among these disparate reports because of the many different feed proteins that were tested. Nevertheless, several generalizations can be made. Correlations tended to be higher using proteases other than that of *S. griseus* (Table 4.2); r^2 ranged from 0.39 to 0.91 for *S. griseus*, from 0.66 to 0.95 for ficin, from 0.92 to 0.94 for two neutral proteases, from 0.59 to 0.88 for papain, 0.88 for pancreatin and (except for an r^2 of 0.07 for grains without amylase addition) from 0.65 to 0.90 for bromelain. Poos-Floyd *et al.* (1985) compared five proteases and found that, *S. griseus*, the protease with the lowest correlation to *in vivo* results, actually yielded the greatest extents of *in vitro* digestion. Assoumani *et al.* (1992) observed, using an *S. griseus* system, that the large positive intercept associated with the best fit indicated that the protease system digested proportionately lower amounts of more slowly degraded feeds. Aufrere *et al.* (1991) reported an r^2 of 0.88 with *S. griseus* at pH 8.0 using a large number of feeds. Antoniewicz and Kosmala (1998) had poor results with lucerne forages using the *S. griseus* method of Aufrere, but others have had more success with other versions of the *S. griseus* assay. Licitra *et al.* (1993) modified the *S. griseus* method by using a constant ratio of enzyme activity to substrate true protein and found improved predictions of *in situ* degradability. Chaudhry (1998), also using *S. griseus*, found rates of degradation for casein, wheat gluten and maize gluten that approximated to those reported for the IIV system; *S. griseus* proved more effective than papain in these experiments.

Including enzymes to remove the carbohydrates that are normally digested *in vivo*, but which occlude protein attack *in vitro*, shows promise for improving *in vitro* assays of protein degradation. Added carbohydrases

Table 4.2. Summary of results obtained from comparing *in vitro* protein degradation, estimated using various proteases, with *in situ* extents of protein degradation.

Protease source	Protein feeds tested (refs)[a] (mg protein degraded h^{-1})	Correlation coefficient (r^2)[b]
Streptomyces griseus	266 (10)	0.39–0.91
Ficin	227[c] (5)	0.66–0.95
Bromelain	44 (2)	0.53–0.90
Papain	25 (2)	0.59–0.88
Porcine pancreatin	74 (2)	0.71–0.88
Aspergillus oryzae	12 (1)	0.92–0.94
Bacillus subtilis	8 (1)	0.92

[a]Number of protein samples tested and number of papers referenced.
[b]Correlation coefficients obtained when the extent of *in vitro* protein degradation was regressed on the *in situ* extent of protein degradation.
[c]Includes 135 replicate samples from eight feeds.

improved the predictions of *in situ* protein degradation for cereal grains in systems using *S. griseus* and a neutral protease (Assoumani *et al.*, 1992), ficin (Kosmala *et al.*, 1996) and bromelain (Tomankova and Kopecny, 1995). For example, Tomankova and Kopecny (1995) found an r^2 of 0.07 without amylase and of 0.65 with amylase addition to bromelain incubations with 19 grains. Cone *et al.* (1996), however, found a slight reduction in correlations to *in situ* results with a number of concentrates and forages when they added two carbohydrases to *S. griseus* incubations. Abdelgadir *et al.* (1997) observed that addition of either cellulase or Driselase® (a broad-spectrum carbohydrase) to *S. griseus* protease resulted in improved estimates of *in vivo* protein degradability for lucerne and prairie hays. The data of Abdelgadir *et al.* (1997) also suggested that this improvement resulted partly from proteolytic activity present in the carbohydrase preparations. Thus, while use of carbohydrases appears to be valuable for feeds with high starch content, the picture seems less clear for forages. Often, there has been substantial numerical disparity between the extents of digestion observed with the *in vitro* and *in situ* systems, despite regression equations that apparently explain most of the *in situ* variation. Antoniewicz and Kosmala (1998) observed high correlations between the extent of protein digestion using ficin or pancreatin and the extent of *in situ* digestion of 24 lucerne forages.

Ideal performance of a cell-free protease system *in vitro* would predict the rate as well as the extent of protein degradation. However, there has been relatively little effort along this line. In their early work, Krishnamoorthy *et al.* (1983) used incubations with *S. griseus* protease at high activity (6.6 U ml^{-1}) to estimate kinetically the sizes of pools corresponding to protein fractions B2, B3 and C in the Cornell net carbohydrate protein system; they also used *S. griseus* protease at low activities (0.066 and 0.022 U ml^{-1}) to estimate degradation rates of SBM. Although Krishnamoorthy *et al.* (1983) observed reproducible degradation patterns and rates for SBM at 0.022 U ml^{-1} of *S. griseus* protease, no *in situ* or *in vivo* degradation rates were reported for the same protein source. Roe *et al.* (1991) were unable to predict *in situ* degradation rates of four SBM and two distiller's grains using *S. griseus*, ficin or neutral fungal protease plus amylase. Inspection of the degradation patterns for all six proteins revealed considerable deviation between *in situ* patterns and those obtained using proteases (Roe *et al.*, 1991). Mahadevan *et al.* (1987) observed substantially different degradation rates with proteases extracted from mixed ruminal organisms and those from either *S. griseus* or *Aspergillus oryzae*. For example, *S. griseus* protease yielded digestion rates for blood meal and fish meal, two proteins known to be very resistant to ruminal degradation, that were more rapid than the rate obtained for SBM, while proteases from mixed ruminal microbes yielded digestion rates for blood meal and fish meal that were only about half as rapid as that for SBM (Table 4.3). Kohn and Allen (1995a) developed an assay for protein degradation based on proteases extracted from mixed ruminal microbes using butanol and acetone. These 'cell-free' proteases were used to assess degradability of the protein in several forages (Kohn and Allen, 1995b). Although largely successful, the difficulty of

Table 4.3. Rates of degradation of different proteins by protease isolated from mixed ruminal organisms and by *Streptomyces griseus* protease (Mahadevan *et al.*, 1987)[a].

	Protein degraded (mg h^{-1})	
Protein source	Ruminal protease	*S. griseus* protease
Soybean meal	1.06	0.70
Maize gluten meal	0.37	0.35
Blood meal	0.53	0.96
Fish meal	0.62	1.65

[a]Incubations conducted with 25 mg of CP ml^{-1} medium with protease activities added to levels to hydrolyse 5.0 mg of azo-casein h^{-1}. Rates were determined from linear regression of the amount degraded against time from 0 to 6 h.

isolating even small amounts of protease from mixed ruminal microbes reduces the chance that this approach will be widely applied.

Because ruminal microbial proteases have specificities that differ substantially from those of commercial proteases, simple use of individual proteolytic enzymes is unlikely to yield reliable estimates of the kinetic parameters of ruminal degradation. Luchini *et al.* (1996) characterized the proteolytic activity of mixed ruminal microbes using 13 artificial substrates and then elaborated blends of commercial proteases to try to match this proteolytic activity. Although three different blends had hydrolytic activities toward the artificial substrates that were similar to those of mixed ruminal microbes, none of the blends degraded standard feed proteins at similar rates. For example, degradation rates obtained for 15 proteins using one of these protease blends ranged from only 0.014 to 0.097 h^{-1}, while degradation rates obtained with ruminal inoculum ranged from 0.01 to 0.23 h^{-1}; rates deviated most with the most rapidly degraded proteins (Luchini *et al.*, 1996; Fig. 4.3). All of the protease blends were particularly inaccurate for assessing degradability for heat-treated soybean products. Carbohydrate-digesting enzymes were not tested in this system. Despite the success of using cell-free proteases to predict the extents of *in situ* degradability, clearly these enzymes, as applied thus far, have not captured many of the important characteristics of the activities of ruminal organisms that determine the rates and extents of ruminal protein degradation.

ESTIMATING DIGESTION IN THE ABOMASUM AND SMALL INTESTINE

A number of *in vitro* procedures have been developed to assess protein digestion in the abomasum and small intestine based on the enzymatic procedure of Akeson and Stahmann (1964). With this method, proteins are first incubated with acid–pepsin, the sample is then neutralized to about pH 7.5, and incubated with pancreatin. Akeson and Stahmann (1964) found a high

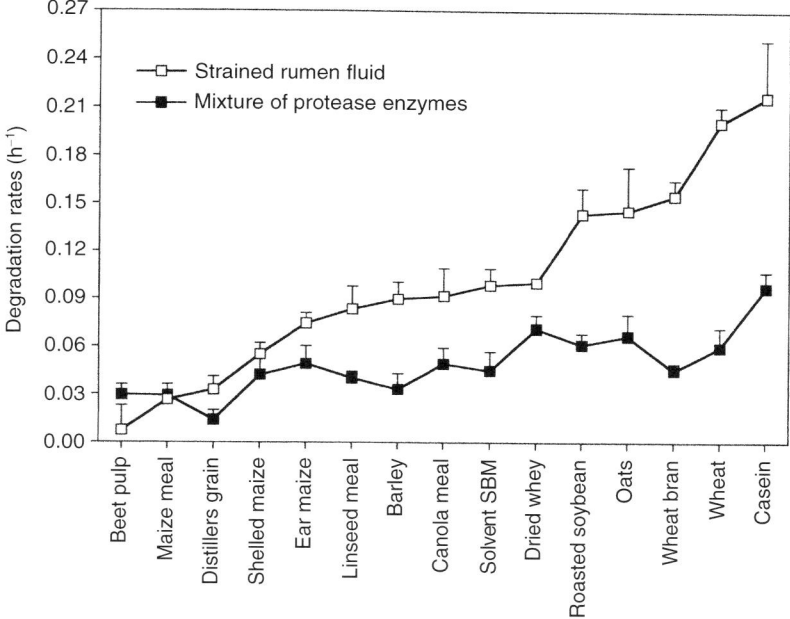

Fig. 4.3. Degradation rates (h^{-1}) for 15 protein sources obtained using a blend of four cell-free proteases or a ruminal *in vitro* inoculum (Luchini *et al.*, 1996). The protease blend was a mixture of trypsin, carboxypeptidase B, chymotrypsin and carboxypeptidase A at, respectively, 116.6, 0.5, 2.5 and 0.5 enzyme U ml^{-1}; the ruminal *in vitro* method was based on the limited-substrate inhibitor *in vitro* technique (Broderick, 1987). Vertical lines represent ± 1 SD.

correlation (r = 0.98) between the amount of N solubilized by this method and the extent of protein digestion in the rat. Calsamiglia and Stern (1995) tested a similar *in vitro* enzymatic technique using a 1-h acid–pepsin step, followed by a 24-h pancreatin step, and found high correlations (r = 0.91) to *in vivo* protein digestion for 34 feed proteins. Antoniewicz *et al.* (1998) observed even higher correlations (r = 0.98) between N solubilized using a pepsin–pancreatin digestion method and N disappearance from intestinal mobile bags for the 12-h ruminal *in situ* residues from 79 samples of forages and concentrates. Calsamiglia and Stern (1995) used their pepsin–pancreatin technique to obtain post-ruminal CP digestibilities on the 16-h *in situ* residues from ten protein concentrates; the values recorded were: 90% (solvent SBM), 89% (maize gluten meal), 81% (dried distiller's grains), 81% (ring-dried blood meal), 80% (fish meal), 80% (cottonseed meal), 77% (dried brewer's grains), 67% (hydrolysed feather meal), 63% (batch-dried blood meal) and 55% (meat and bone meal). The 16-h *in situ* incubation was thought to approximate the mean retention time for a ruminal passage rate of 0.06 h^{-1} (1/0.06 = 16.7 h). However, it should be noted that the incubation time required to yield a CP residue equivalent to the amount that would escape from the rumen at a given passage rate is a function of the degradation rate as well as the passage rate (Broderick, 1994). Proteins with degradation rates of 0.15 (e.g. solvent SBM) and 0.02 h^{-1} (e.g. ring-dried blood meal) would require incubation times of, respectively, 8.4 and 14.4 h, not 16 h, to yield *in situ* residues corresponding to the proportions that escape the rumen at a passage rate of 0.06 h^{-1}.

Several trials have confirmed the value of using pepsin–pancreatin, or even acid–pepsin alone (Menden and Cremer, 1970), for assessing post-ruminal protein digestion. Craig and Broderick (1981) observed, for heat-treated cottonseed meals, a high correlation ($r^2 = 0.996$) between CP true digestibility in the rat and total amino acid release in an *in vitro* pepsin–pancreatin assay. Low post-ruminal digestibility was found to explain the poor effectiveness of hydrolysed feather meal as a source of escaped protein, despite substantial resistance to ruminal degradation. Mansfield *et al.* (1994) observed that a mixture of animal by-product proteins, in which hydrolysed feather meal provided 46% of CP, gave rise to lower NH$_3$ and greater flow of non-ammonia N in continuous culture ruminal fermentors. However, compared with SBM, this dietary mixture lowered the milk protein content and yield when fed to lactating cows. Waltz *et al.* (1989) found greater ruminal escape, but lower intestinal amino acid absorption, for hydrolysed feather meal protein than for SBM or blood meal. Two samples of hydrolysed feather meal had digestibilities in acid–pepsin of 9 and 33%, compared with digestibilities of 77, 93, 95 and 98% for meat and bone meal, solvent SBM, fish meal and blood meal, respectively (England *et al.*, 1997).

ESTIMATING DIGESTIBILITY USING *IN SITU* TECHNIQUES

Attempts to evaluate the nutritive content of feedstuffs by measuring the disappearance of feed material from small bags exposed to digestive processes began as early as the 1700s (Sauer *et al.*, 1983). Current incarnations of *in situ* procedures trace back to the work of Quin *et al.* (1938) for the measurement of ruminal digestion and Sauer *et al.* (1983) for the measurement of post-stomach digestion. The usefulness of the *in situ* approach lies in the ability to expose feedstuffs to digestive conditions thought to be similar to those existing *in vivo*. Thus, insight is gained as to the potential extent of digestibility of different dietary constituents in one or more segments of the digestive tract. Moreover, if residence time in a particular segment of the digestive tract is known, and if repeated access to *in situ* bags within that segment is possible, the rate of digestion of various dietary constituents can be estimated.

In Situ Procedures for Describing Ruminal Digestion

Probably the most common application of the *in situ* approach in the latter half of the 20th century has been for describing the rate and/or extent of ruminal digestion in various ruminant species. The dietary constituents most commonly evaluated with this procedure include dietary DM (or OM), fibre and protein. In addition, Emanuele *et al.* (1991) suggested that the *in situ* approach might be useful for describing the ruminal (and post-ruminal) disappearance of various minerals from forages.

Several thorough reviews of *in situ* procedures for estimating ruminal degradability have been published in the last 15 years (Ørskov, 1982; Lindberg, 1985; Nocek, 1988; AFRC, 1992; Michalet-Doreau and Ould-Bah, 1992; Madsen and Hvelplund, 1994; Huntington and Givens, 1995; Wilkerson *et al.*, 1995). These reviews clarified major factors that can affect *in situ* degradability and often proposed standardized procedures that would be helpful in minimizing variation. The major factors influencing variation were identified as basal diet, bag characteristics, sample and animal characteristics, replication, incubation conditions, rinsing technique, microbial correction and data analysis; these are discussed below.

Although the *in situ* methods have been widely applied, procedures used in different laboratories have been highly variable. The importance of procedural differences on estimates of the rate and extent of ruminal digestion have been assessed in several 'ring tests' conducted in recent years (AFRC, 1992; Madsen and Hvelplund, 1994; Wilkerson *et al.*, 1995). In general, these efforts (which were devoted primarily to the study of protein degradability) showed considerable variation among laboratories in their estimates of ruminal degradability; the magnitude of this variation was considered to be unacceptable (AFRC, 1992; Madsen and Hvelplund, 1994). As a result, a strong consensus exists among those that have reviewed *in situ* literature or conducted ring tests that a significant degree of standardization is called for in this area (Lindberg, 1985; Nocek, 1988; AFRC, 1992; Michalet-Doreau and Ould-Bah, 1992; Madsen and Hvelplund, 1994; Wilkerson *et al.*, 1995). Indeed, such standardization is crucial to be able to have confidence in results from *in situ* assays and to be able to compare results reliably among laboratories.

Standardized Procedures for *In Situ* Methodology

Basal diet

Under some circumstances, basal diet has had limited impact on *in situ* N or DM disappearance (de Faria and Huber, 1984; Grummer *et al.*, 1984). In contrast, most research supports the view that the basal diet significantly alters estimates of the rate or extent of degradation (Van Keuren and Heinemann, 1962; Hopson *et al.*, 1963; Figroid *et al.*, 1972; Weakley *et al.*, 1983; Marinucci *et al.*, 1992; Vanzant *et al.*, 1996). Most researchers recommend a standardized basal diet as part of their protocol (Table 4.4) but the choice of the basal diet and the recommended level of feeding have varied. Ruminal degradability is a function of pool size, degradation rate and passage rate; degradability results will, therefore, have greatest applicability to production conditions where diets are fed that are similar to those used when the data were collected. On the other hand, standard diets are desirable if one intends to develop degradability values for inclusion in standard feed tables. Clearly, the shortcoming of using a

Table 4.4. Comparison of standardized procedures or recommendations for measuring ruminal degradability (emphasis on protein degradability).

Item	Ørskov (1982)	Lindberg (1985)	Nocek (1988)	AFRC (1992)	Michalet-Doreau and Ould-Bah (1992)	Madsen and Hvelplund (1994)	Wilkerson et al. (1995)
Basal diet							
Type	Similar to diet of application	50:50 hay to concentrate	Similar to diet of application	60:40 hay to concentrate	Similar to diet of application	66.7:33.3 hay to concentrate	All forage
Level of feeding	Unspecified	Maintenance	Ad libitum	Maintenance	Unspecified	Maintenance	Unspecified
Bag							
Material	Polyester	Polyester	Polyester	Polyester	Polyester	Polyester	Polyester
Pore size	20–40 µm	35–40 µm	40–60 µm	40–50 µm	40–60 µm	30–50 µm	53 µm
Sample size:surface area	12–20 mg cm^{-2}	10–15 mg cm^{-2}	10–20 mg cm^{-2}	12 mg cm^{-2}	15 mg cm^{-2}	10–15 mg cm^{-2}	12.5 mg cm^{-2}
Sample grind							
Concentrates	2.5–3 mm	1.5 mm	2 mm	2.5 mm	1.5–3 mm	1.5–2.5 mm	2 mm
Roughages	2.5–5 mm	1.5 mm	5 mm	4 mm	1.5–3 mm	1.5–2.5 mm	2 mm
Animal species	Sheep or cattle	Cattle	Animal of application	Unspecified	Unspecified	Cattle, sheep, goats	Cattle
Replication							
No. of animals	2–4	3–4	Unspecified	3	Unspecified	3	1–4
No. of days	1–2	Unspecified	Unspecified	1	Unspecified	1	1–4
No. of bags	1–4	Unspecified	Unspecified	1	Unspecified	2	1–8
Pre-incubation	Unspecified	Unspecified	Pre-soak	None	None	Unspecified	
Incubation conditions							
Position in rumen	Contact with liquid and solids	Ventral sac (free movement)	Unspecified	Liquid (free movement)	Unspecified	Unspecified	Ventral sac (free movement)
Entry/removal order	Enter at same time	Unspecified	Retrieve at same time	Either approach	Enter at same time	Enter at same time	Retrieve at same time
Times	2, 6, 12, 24, 36 h (? for roughages)	2, 4, 8, 16, 24 h (48 h for roughages)	6–12 points up to 24 h; others > 25 h	2, 6, 8, 24, 48 h (72 h for roughages)	Adequate to describe curve	0, 2, 4, 8, 16, 24, 48 h	16 h for test, times for rate unspecified
Rinsing	Hand	Hand	Hand	Machine	Unspecified	Machine	Hand
Microbial correction	Corrected	Unspecified	Correct	Unspecified	Unspecified	Unspecified	Unspecified
Standard feed	Unspecified	Unspecified	Use	Unspecified	Use	Use	Unspecified

single standard for dietary conditions is the limitation in applicability to a wide range of production situations. A range of diets (e.g. with different forage to concentrate ratios) could be used in developing tabular values, although the additional work this would require may make this unfeasible. A reasonable compromise seems to be to offer a mixed diet (e.g. 60% roughage and 40% concentrate) that meets the animal's nutrient requirements and to feed this diet at maintenance (Table 4.5). Accurate application to production situations of tabular values derived in this manner would require adaptation of results based on relationships established for diet type and level of feeding to degradability. Of course, degradability values measured under conditions that most closely mimic the production situation will still probably be the most accurate. Thus, the logical choice of basal diet and level of feeding might vary depending on the intended use of the values obtained.

Bag characteristics

The choice of material for *in situ* bags represents a balance between preventing excessive loss of small particles from the bag while permitting sufficient

Table 4.5. Recommended procedures for standard ruminal *in situ* trials.

Item	Recommendation
Basal diet	
Type	60:40 hay to concentrate
Level of feeding	Maintenance
Bag	
Material	Polyester
Pore size	40–60 μm
Sample size:surface area	10–15 mg cm^{-2}
Sample grind size	
Concentrates	2 mm
Roughages	2 mm
Animal species	Cattle or sheep (or species of application if for use in other than cattle or sheep)
Replication	
No. of animals	2
No. of days	2
No. of bags	1
Pre-incubation	None
Incubation conditions	
Position in rumen	Ventral sac (free movement)
Entry/removal order	If standardized mechanical rinsing is used, sequential removal seems viable. Otherwise, sequential entry followed by removal and rinsing at the same time seems advisable
Times	Adequate to describe curve
Rinsing	Machine: five 1-min rinses and 2-min spins at the low water setting (~45 l)
Microbial correction	Yes, when residue detectable
Standard feed	Yes, if available

influx of fluid and microorganisms and efflux of end-products (Ørskov, 1982). Considerable research documents the fact that bag material significantly affects degradability estimates (Figroid *et al.*, 1972; Van Hellen and Ellis, 1977; Weakley *et al.*, 1983; Udén and Van Soest, 1984; Nocek, 1985; Marinucci *et al.*, 1992). In general, it appears desirable to use an intermediate pore size. Small pore sizes (5–10 μm), while minimizing influx of particles, are prone to inhibit fermentation due to gas build up (Udén and Van Soest, 1984; Marinucci *et al.*, 1992); thus, most standardized protocols, therefore recommend use of bags with a pore size of between 30 and 60 μm. Varvikko and Vanhatalo (1990) suggested that the free surface area (the proportion of the material comprised of holes) may also be an important consideration in the selection of bag material. However, at present, there is insufficient research available upon which to judge the importance of this attribute.

The size of each *in situ* bag depends on the amount of material to be incubated. The choice of sample size will be dictated largely by the quantity of residue needed for analysis. The AFRC (1992) suggested that a sample size of approximately 5 g is adequate, regardless of feed type. To ensure adequate sample fermentation, most standardized protocols recommend a sample:surface area ratio of 10 to 15 mg of DM cm^{-2}. Virtually all such standardized protocols recommend the use of bags constructed from synthetic polyester material.

Sample characteristics

Sample preparation is conducted largely to facilitate handling, analysis and simulation of the comminution normally occurring via mastication and rumination. Udén and Van Soest (1984) suggested that the potential for small particle loss would be too great to recommend use of grinder screens with apertures smaller than 2 mm. In contrast, very large particle size might limit bacterial attachment and has been suggested to increase variability in degradability estimates (Michalet-Doreau and Ould-Bah, 1992). As a result, intermediate screen apertures (1.5–3 mm) are most commonly chosen for *in situ* analyses. For roughages, some authors recommend using larger apertures (4–5 mm). In such cases, the residues frequently are reground after incubation before further chemical analysis.

Caution is warranted when using *in situ* procedures with some feeds. For example, maize gluten meal has been observed to form a gelatinous mass inside *in situ* bags upon exposure to ruminal fluid (Cozzi *et al.*, 1993), a property that depressed degradability. Cozzi *et al.* (1993) were able to obtain *in situ* degradability estimates closer to *in vivo* values by incubating maize gluten meal in a mixture with other fibrous material that was poorly degraded (e.g. sawdust or straw washed with neutral detergent). These materials inhibited the formation of the gelatinous mass and thereby facilitated microbial access to the maize gluten meal.

Animal characteristics

While it is ideal to use the animal species for which the information is intended (Nocek, 1988), the limited differences in ruminal digestion and microbial efficiency between sheep and cattle fed the same diet (ARC, 1980) suggest that there is flexibility of choice with regard to species. One of the most significant limitations concerning using sheep or other small ruminants is the reduced capacity for incubating large numbers of bags at one time (AFRC, 1992). Thus, while choice of species is flexible in most standardized protocols, there is a preference for use of cattle.

Replication

Differences exist in the literature as to the relative importance of these sources of variability. Ørskov (1982) identified the principal sources of variation for *in situ* trials conducted within a given laboratory as those due to animal, day and bag, in the order: animal > day > bag. Figroid *et al.* (1972) also noted that the animal was a significant source of variation within *in situ* trials. In contrast, other reports indicated that animal effects were either not significant (Weakley *et al.*, 1983; Nocek, 1985) or were less important than day effects (Wilkerson *et al.*, 1995; Vanzant *et al.*, 1996). Although recommendations for replication differed among standardized protocols, most suggested using multiple animals (2–4) with less emphasis on multiple replications of day or bag.

As noted previously, results from various 'ring tests' indicated that variation among laboratories also can be very significant (AFRC, 1992; Madsen and Hvelplund, 1994; Wilkerson *et al.*, 1995). Clearly, this calls for as much standardization as possible among laboratories when conducting *in situ* procedures. Interestingly, a significant source of variation among laboratories in estimating protein degradation in one ring test (Madsen and Hvelplund, 1994) was variation in CP determinations. This highlights that great care must be taken even in the execution of standardized protocols. One method by which reliability of standard procedures can be evaluated is through the use of standard feeds. Several researchers (Nocek, 1988; Michalet-Doreau and Ould-Bah, 1992; Madsen and Hvelplund, 1994) suggested that a standard feed should be included with each *in situ* experiment.

Incubation conditions

The position of individual *in situ* bags in the rumen can have a significant impact on nutrient disappearance (Figroid *et al.*, 1972; Hawley, 1981). Furthermore, the physical contact afforded by immersion of *in situ* bags in ruminal contents appears to be an important mechanism for ensuring adequate movement of fluid and gas into and out of *in situ* bags (Marinucci *et al.*, 1992). Thus, use of a system that allows the bags to remain immersed and yet freely movable within the ruminal contents appears desirable. One method that has been used to accommodate these conditions is the insertion of individual polyester bags within a larger mesh bag (Wilkerson *et al.*, 1995). In their work, Wilkerson *et al.* (1995) placed a maximum of 24 individual *in situ* bags into loose-knit polyester mesh bags with dimensions of approximately 36 × 42 cm; these mesh bags were then inserted into the ventral rumen for the incubation period. This system can be adapted to permit easy retrieval of the mesh bag via a cord attached to the ruminal cannula while still allowing for free ruminal movement of the mesh bag. If the mesh bag tends to move to the top of the fibre mat (where immersion in ruminal fluid is likely to be inconsistent), it is probably desirable to add a small weight to the mesh bag to help to keep it immersed. This probably was unnecessary in the work of Wilkerson *et al.* (1995) because their *in situ* bags were weighted via the rubber stoppers (wrapped with rubber bands) used in sealing their *in situ* bags.

An additional item of interest with regard to incubation conditions is the order in which *in situ* bags are added to or removed from the rumen. The dominant approaches have been to add all bags at one time and then remove them sequentially or to add them sequentially and then remove them all at the same time. The former has been criticized on the basis of variation introduced from batch to batch by differences in rinsing technique (Nocek, 1988). Use of a standardized mechanical rinsing procedure may minimize this concern. The objection to sequential addition and removal at a single time relates to the incubation conditions to which the *in situ* bags are exposed. Michalet-Doreau and Ould-Bah (1992) note that, because ruminal conditions are not static, introducing bags at different times will subject them to different incubation conditions. Ideally, exposure to similar incubation conditions is preferable if potential variability in rinsing can be reduced (e.g. via mechanical rinsing). An alternative approach for minimizing differences in incubation conditions is to insert bags with different incubation times on different days, but at the same time relative to feeding (Paine *et al.*, 1982). However, as with sequential removal, this procedure seems most suitable when using standardized mechanical rinsing.

Rinsing methodology

Rinsing *in situ* bags after ruminal incubation is used to arrest microbial activity and to minimize the presence of contaminants (Lindberg, 1985). A variety of procedures have been used for rinsing *in situ* bags, the most common of which is hand rinsing. Typically, hand-rinsing procedures entail exposing the bags to cold tap water in conjunction with mild manipulation of the bag (Nocek, 1988); this is usually done until the rinse water is clear. In some instances, a limited stream of water has been run through the open bag subsequent to thorough rinsing of the closed bag (Wilkerson *et al.*, 1995). Hand-rinsing techniques appear to be a significant source of variation in the determination of ruminal degradability (Nocek, 1988; Wilkerson *et al.*, 1995), and mechanical rinsing frequently is considered to be a desirable alternative (Cherney *et al.*, 1990; AFRC, 1992; Madsen and Hvelplund, 1994). Almost all mechanical rinsing techniques have involved commercial washing machines; but the rinsing methods have been variable. Madsen and Hvelplund (1994) and AFRC (1992) recommended use of a single, long-term rinse (from 10 to 60 min). In contrast, Cherney *et al.* (1990) and Coblentz *et al.* (1997) suggested that multiple, short-term rinses were preferable. Coblentz *et al.* (1997) monitored the change in N in rinse water following rinses that entailed 1 min of agitation and 2 min spin (~ 45 litre capacity). They noted that the concentration of N in the rinse water declined dramatically after the first rinse but did not plateau until after four rinse–spin cycles. No further drop in rinse-water N was observed with as many as ten rinse–spin cycles. Given the desirability of working toward a consistent end-point, they suggested that this information could be useful in standardizing machine rinsing procedures. Using multiple (≥ 5) short-term rinses to achieve a consistent end-point seems preferable to a single, long-term rinse.

Microbial correction

Early research conducted by Mathers and Aitchison (1981) verified that *in situ* residues may contain significant quantities of microbial matter. Although under some conditions microbial contamination has been reported to be negligible (Nocek, 1985; Coblentz *et al.*, 1997), many other studies (Nocek and Grant, 1987; Olubobokun and Craig, 1990; Olubobokun *et al.*, 1990; Wanderly *et al.*, 1993) corroborate the work of Mathers and Aitchison (1981). The potential for microbial contamination to affect results was particularly problematic for the determination of protein degradation in low-protein forages (Wanderly *et al.*, 1993). When corrections for microbial contamination were used in the calculation of *in situ* protein degradation, lag times typically were reduced and estimates of rate and extent of digestion were increased (Olubobokun *et al.*, 1990; Wanderly *et al.*, 1993). However, the effects of this correction can be variable (Nocek and Grant, 1987). In general, it appears that corrections for microbial contamination are necessary under those conditions where significant microbial residues are detected and where corrections will have a large impact on results (e.g. the determination of protein degradation in low-quality roughages).

Data analysis

There are numerous alternatives for mathematical description of *in situ* degradation curves. The most common approaches generally entail either linear regression of log-transformed values or the use of non-linear regression (Nocek and English, 1986). In the former procedure, the natural logarithm of the fraction remaining at each time point (after subtraction of undegradable pool C) is calculated, and these values are regressed against incubation time (Waldo *et al.*, 1972). Pool sizes typically are determined via solubility (degradable pool A), extended incubation in the rumen (C) and by difference (pool B = 100 − A − C). If an obvious delay exists in the initiation of degradation of B, a lag component should be included in the model (McDonald, 1981; Mertens, 1993). Effective degradability in the rumen is a function of the sizes of pools A and B, the rate of degradation (k_d) of pool B and the rate of passage (k_p) of pool B from the rumen (Ørskov and McDonald, 1979). The passage rates used most often are those derived for ruminal solids. Effective degradability is calculated as: A + {B [$k_d/(k_d + k_p)$]}. In some cases, when the plot of log-transformed values against time has deviated significantly from linearity, researchers have used curve-peeling techniques to partition pool B and to determine the associated rates (Nocek and English, 1986). Alternatively, untransformed values can be subjected to non-linear regression analysis using commercially available statistical packages (Van Milgen and Baumont, 1995). When using this approach, both the form of the model and initial parameter estimates need to be provided as inputs for the analysis; effective degradability is calculated as described in the log transformation approach. The number of parameters that can be reasonably determined by this approach will depend on the number of points and the placement of those points. Broderick (1994) suggested that as many five points may be required for each regression parameter estimated. Optimal incubation times (i.e. placement of points) will depend on the shape of the curve (Michalet-Doreau and Ould-Bah, 1992); however, one generally should give particular attention to adequate description of the early, rapid phases of degradation, to points of significant change in rate of degradation and to the asymptote. Several researchers (Nocek and English, 1986; Messman *et al.*, 1991; Van Milgen and Baumont, 1995) have compared various mathematical approaches (including those noted above) for estimating rates, pool sizes and degradability. In general, these researchers were unable to identify a single, ideal mathematical approach based on common statistical criteria (e.g. goodness of fit, residual patterns, etc.). Indeed, Nocek and English (1986) evaluated four typical methods and concluded that no single approach would be appropriate under all conditions. Thus, choice of mathematical model should be made after consideration of the data being evaluated (i.e. number of points, pattern of degradation, etc.) and of standard indicators of model appropriateness.

CRITICISMS OF RUMINAL *IN SITU* METHODS

In situ methods may be criticized on theoretical grounds for at least four reasons: (i) microbial contamination of the residue leads to underestimation of degradation of both DM and protein; (ii) disappearance from the bag of undegraded small particles leads to overestimation of degradation; (iii) disappearance from the bag of soluble undegraded nutrients, particularly protein N that becomes classed with degraded fraction A, leads to overestimation of extent of degradation; and (iv) physical separation of the bag contents from ruminal digesta as a whole leads to underestimation of degradation. *In situ* methodology can be modified to address the first two of these sources of error by using, for example, nucleic acid analyses to estimate microbial N and DM contamination of bag residues (e.g. Olubobokun and Craig, 1990) and by adjusting the N and DM effluxing rapidly from *in situ* bags (fraction A) for the proportion that is present in small particles (Lopez *et al.*, 1994; Madsen *et al.*, 1995). However, the other two problems are not dealt with so easily. Soluble proteins vary greatly in degradability and cannot be assumed to be broken down completely in the rumen. Mahadevan *et al.* (1980) reported that some soluble proteins extracted from solvent SBM were less degradable than other insoluble proteins from the same source. Broderick *et al.* (1988) observed that *in vitro* degradabilities for soluble proteins ranged from 0.10 to 0.80 h^{-1}. Messman *et al.* (1994) found that 16–80% of various soluble forage proteins identified by electrophoresis remained undegraded after ruminal *in vitro* incubations of 5 h. Physical restriction of ruminal microorganisms from access to contents of *in situ* bags and restriction of digestion product removal probably reduce *in situ* digestion rates. Meyer and Mackie

(1986) reported that microbial counts in contents within nylon bags with 53 μm pore size were only 60% of those of the surrounding ruminal contents. Gas retention, which would be expected to make bags more buoyant and reduce contact with ruminal liquids, was observed in *in situ* bags (Udén *et al.*, 1974). Udén and Van Soest (1984) found substantial influx of small particles into *in situ* bags even with small (20 μm) pore sizes. Concerns about *in situ* methodology have been discussed elsewhere in more detail (Broderick, 1994).

IN SITU PROCEDURES FOR DESCRIBING POST-RUMINAL DIGESTION

Because of the time, expense and amount of test material required for determining nutrient digestibility in the gastrointestinal tract, Sauer *et al.* (1983) proposed introducing small *in situ* bags into the duodenum and collecting them in the faeces for this purpose. This initial work was with swine, although the procedure was soon adapted for use in ruminants (Hvelplund, 1985). Most of the work with the mobile bag technique has evaluated protein digestion, although Emanuele *et al.* (1991) also used it to describe mineral digestion. Initial evaluations of protein disappearance using the mobile bag technique indicated good agreement with conventional methods (Sauer *et al.*, 1983; Hvelplund, 1985; Cherian *et al.*, 1988; Sauer *et al.*, 1989). However, subsequent studies have shown that the mobile bag technique is subject to many of the same methodological concerns that plague ruminal *in situ* measurements (Varvikko and Vanhatalo, 1990; Vanhatalo and Ketoja; 1995; Vanhatalo *et al.*, 1995).

In general, current application of the mobile bag procedure entails placing a small sample (1.0–1.5 g for concentrates or ruminally digested material, and 0.5 to 1.2 g for forages) into small polyester bags; typical dimensions being 25 × 40 mm for swine and 35 × 50 mm for cattle (Cherian *et al.*, 1988; Sauer *et al.*, 1989; Varvikko and Vanhatalo, 1990; Frydrych, 1992; Vanhatalo and Ketoja, 1995; Vanhatalo and Varvikko, 1995; Vanhatalo *et al.*, 1995). The material used typically has a pore size of approximately 40–50 μm; Varvikko and Vanhatalo (1990) suggested that the percentage of free surface space is also an important consideration. For monogastrics, samples commonly are subjected to an acid–pepsin incubation before placement in the duodenum. For ruminants, sample preparation frequently has included a period of incubation in the rumen in polyester bags, in some cases followed by incubation in acid–pepsin. However, Vanhatalo *et al.* (1995) suggested that it may be unnecessary to expose ruminally incubated samples to a subsequent acid–pepsin treatment before placement in the duodenum. Sauer *et al.* (1989) placed four bags per day into the duodenum of barrows (two bags inserted at each of two feeding times). Larger quantities (frequently 12 per day) were used when mature cows were the subjects, although some time was allowed to elapse between the insertion of individual bags to ensure unimpeded movement within the duodenum. Bags commonly are collected from the faeces; some researchers have placed an upper time limit for bag appearance in the faeces in order for an observation to be considered viable (Frydrych, 1992; Masoero *et al.*, 1994). Rinsing procedures have been similar to those used for ruminal *in situ* procedures. The impact of non-feed N contamination of mobile bag contents on estimates of protein digestibility appears to be variable. Vanhatalo and Varvikko (1995) noted a significant influence on estimates for low-N, high-fibre feeds but little effect for a high-protein concentrate. This may require correction for microbial contamination under conditions where significant quantities of microbial matter are detected.

SUMMARY AND CONCLUSIONS

The microbiology and enzymology of carbohydrate and protein digestion in the gastrointestinal tract are very complex and difficult to model using simple *in vitro* and *in situ* methods. *In vitro* systems should yield accurate predictions of digestion fractions, rates and lag times. However, there has been little effort to quantify these kinetic parameters using *in vitro* methods. Ruminal *in vitro* assays have proven to be robust for predicting the extent of OM and DM digestion in the total tract; *in vitro* assays using cellulases, when sample pre-treatment was included, have been successful for predicting the extent of *in situ* and total tract OM and DM digestibility. Results estimating protein degradation in the rumen have not been as satisfactory. Ruminal *in vitro* methods in which microbial growth is inhibited, or in which microbial N incorporation is quantified, show promise for estimating the rate and extent of ruminal protein degradation. The inhibitor *in vitro* (IIV) procedure has proven to be reliable in a number of applications but is limited to short-term incubations of 4–6 h. Also, the IIV technique has not been adopted widely beyond the laboratory where it was developed, possibly because it requires rapid analysis of many samples for NH_3 and total amino acids. The extents of protein hydrolysis obtained using a number of cell-free proteases have been well correlated to the extents of *in situ* degradation, but, thus far, cell-free proteases have not proven as reliable for predicting rates of ruminal protein degradation; pepsin–pancreatin assays have, however, yielded reliable estimates of post-ruminal protein digestibility. Much more emphasis must be placed in the future on using *in vitro* assays to predict the rates of ruminal carbohydrate and protein degradation that are required in the current models of ruminant nutrient utilization.

In situ methodologies have the potential to contribute useful information regarding effective ruminal degradability and post-ruminal digestion. However, in those areas where significant limitations have been demonstrated (e.g. microbial N contamination of *in situ* residues), appropriate correction techniques must be applied to ensure the most accurate results. The *in situ* method is not suitable in some specific situations, notably the description of degradation rates of soluble dietary constituents with variable degradabilities. The magnitude of inter- and intra-laboratory variation observed with *in situ* procedures highlights the need for increased standardization of *in situ* methodology. Increased standardization of these techniques is crucial to our ability to have confidence in the results from such assays and to be able reliably to compare findings across laboratories.

DISCLAIMER

Mention of a trademark or proprietary product in this chapter does not constitute a guarantee or warranty of the product by the USDA or the Agricultural Research Service and does not imply its approval to the exclusion of other products that also may be suitable.

REFERENCES

Abdelgadir, I.E.O., Cochran, R.C., Titgemeyer, E.C. and Vanzant, E.S. (1997) *In vitro* determination of ruminal protein degradability of alfalfa and prairie hay via a commercial protease in the presence or absence of cellulase or driselase. *Journal of Animal Science* 75, 700–713.

Agricultural and Food Research Council (1992) Nutrient requirements of ruminant animals: protein. AFRC Technical Committee on Responses to Nutrients. Report No. 9. *Nutrition Abstracts and Reviews Series B* 62, 787–835.

Agricultural Research Council (1980) *The Nutrient Requirements of Ruminant Livestock*, 1st edn. CAB International, Wallingford, UK, pp. 126–127.

Akeson, W.R. and Stahmann, M.A (1964) A pepsin pancreatin digest index of protein quality evaluation. *Journal of Nutrition* 83, 257–264.

Alderman, G. (1985) Prediction of the energy value of compound feeds. In: Haresign, W.H. (ed.), *Recent Advances in Animal Nutrition* – 1985. Butterworths, London, pp. 3–52.

Antoniewicz, A.M. and Kosmala, I. (1998) Estimation of ruminal effective degradability of dried lucerne forage based on *in vitro* digestion by protease from *Streptomyces griseus*, ficin or pancreatin. In: *Proceedings of the International Symposium on In Vitro Techniques for Measuring Nutrient Supply to Ruminants*. British Society of Animal Science, Reading, UK, pp. 118–119.

Antoniewicz, A.M., Kosmala, I., Hvelplund, T. and van Vuuren, A. (1998) Use of pepsin–pancreatin system to estimate intestinal digestibility of protein in ruminant feeds and undegraded residues. In: *Proceedings of the International Symposium on In Vitro Techniques for Measuring Nutrient Supply to Ruminants*. British Society of Animal Science, Reading, UK, pp. 115–117.

Assoumani, M.B., Vedeau, F., Jacquot, L. and Sniffen, C.J. (1992) Refinement of an enzymatic method for estimating the theoretical degradability of proteins in feedstuffs for ruminants. *Animal Feed Science and Technology* 39, 357–368.

Attwood, G.T. and Reilly, K. (1995) Identification of proteolytic rumen bacteria isolated from New Zealand cattle. *Journal of Applied Bacteriology* 79, 22–29.

Aufrere, J., Graviou, D., Demarquilly, C., Verite, R., Michalet-Doreau, B. and Chapoutot, P. (1991) Predicting *in situ* degradability of feed proteins in the rumen by two laboratory methods (solubility and enzymatic degradation). *Animal Feed Science and Technology* 33, 97–116.

Blümmel, M., Makkar, H.P.S. and Becker, K. (1997) *In vitro* gas production: a technique revisited. *Journal of Animal Physiology and Animal Nutrition* 77, 24–34.

Borchers, R. (1967) Proteolytic activity of rumen fluid *in vitro. Journal of Animal Science* 24, 1033–1038.

Broderick, G.A. (1978) *In vitro* procedures for estimating rates of ruminal protein degradation and proportions of protein escaping the rumen undegraded. *Journal of Nutrition* 108, 181–190.

Broderick, G.A. (1982) Estimation of protein degradation using *in situ* and *in vitro* methods. In: Owens, F.N. (ed.), *Protein Requirements for Cattle.* Oklahoma State University Press, Stillwater, Oklahoma, pp. 72–80.

Broderick, G.A. (1987) Determination of protein degradation rates using a rumen *in vitro* system containing inhibitors of microbial nitrogen metabolism. *British Journal of Nutrition* 58, 463–476.

Broderick, G.A. (1994) Quantifying forage protein quality. In: Fahey, G.C., Jr, Collins, M.D., Mertens, D.R. and Moser, L..E. (eds), *Forage Quality, Evaluation, and Utilization.* American Society of Agronomy, Madison, Wisconsin, pp. 200–228.

Broderick, G.A. (1998) Can cell-free enzymes replace rumen microorganisms to model energy and protein supply? In: *Proceedings of the International Symposium on In Vitro Techniques for Measuring Nutrient Supply to Ruminants.* British Society of Animal Science, Reading, UK, pp. 99–114.

Broderick, G.A. and Albrecht, K.A. (1997) Ruminal *in vitro* degradation of protein in tannin-free and tannin-containing forage legume species. *Crop Science* 37, 1884–1891.

Broderick, G.A. and Balthrop, J.E., Jr (1979) Chemical inhibition of amino acid deamination by ruminal microbes *in vitro. Journal of Animal Science* 49, 1101–1111.

Broderick, G.A. and Clayton, M.K. (1992) Rumen protein degradation rates estimated by nonlinear regression analysis of Michaelis–Menten *in vitro* data. *British Journal of Nutrition* 67, 27–42.

Broderick, G.A., Wallace, R.J., Ørskov, E.R. and Hansen, L. (1988) Comparison of estimates of ruminal protein degradation by *in vitro* and *in situ* methods. *Journal of Animal Science* 66, 1739–1745.

Broderick, G.A., Ricker, D.B. and Driver, L.S. (1990) Expeller soybean meal and corn by-products versus solvent soybean meal for lactating dairy cows. *Journal of Dairy Science* 73, 453–462.

Broderick, G.A., Yang, J.H. and Koegel, R.G. (1993) Effect of steam heating alfalfa hay on utilization by lactating dairy cows. *Journal of Dairy Science* 76, 165–174.

Bughrara, S.S. and Sleper, D.A. (1986) Digestion of several temperate forage species by a prepared cellulase solution. *Agronomy Journal* 78, 94–98.

Bughrara, S.S., Sleper, D.A., Belyea, R.L. and Marten, G.C. (1989) Quality of alfalfa herbage estimated by a prepared cellulase solution and near IR reflectance spectroscopy. *Canadian Journal of Plant Science* 69, 833–840.

Calsamiglia, S. and Stern, M.D. (1995) A three-step *in vitro* procedure for estimating intestinal digestion of protein in ruminants. *Journal of Animal Science* 73, 1459–1465.

Chaudhry, A.S. (1998) Suitability of pure enzymes or centrifuged rumen fluid to estimate protein degradation. In: *Proceedings of the International Symposium on In Vitro Techniques for Measuring Nutrient Supply to Ruminants.* British Society of Animal Science, Reading, UK, pp. 120–122.

Cherian, G., Sauer, W.C. and Thacker, P.A. (1988) Effect of predigestion factors on the apparent digestibility of protein for swine determined by the mobile nylon bag technique. *Journal of Animal Science* 66, 1963–1968.

Cherney, D.J.R., Patterson, J.A. and Lemenager, R.P. (1990) Influence of *in situ* bag rinsing technique on determination of dry matter disappearance. *Journal of Dairy Science* 73, 391–397.

Coblentz, W.K., Fritz, J.O., Cochran, R.C., Rooney, W.L. and Bolsen, K.K. (1997) Protein degradation in response to spontaneous heating in alfalfa hay by *in situ* and ficin methods. *Journal of Dairy Science* 80, 700–713.

Cone, J.W. (1991) Degradation of starch in feed concentrates by enzymes, rumen fluid and rumen enzymes. *Journal of the Science of Food and Agriculture* 54, 23–34.

Cone, J.W. (1998) The development, use and application of the gas production technique at ID-DLO. In: *Proceedings of the International Symposium on In Vitro Techniques for Measuring Nutrient Supply to Ruminants.* British Society of Animal Science, Reading, UK, pp. 65–78.

Cone, J.W. and Vlot, M. (1990) Comparison of degradability of starch in concentrates by enzymes and rumen fluid. *Journal of Animal Physiology and Animal Nutrition* 63, 142–148.

Cone, J.W., Van Gelder, A.H., Steg, A. and Van Vuuren, A.M. (1996). Prediction of *in situ* rumen escape protein from *in vitro* incubation with protease from *Streptomyces griseus. Journal of the Science of Food and Agriculture* 72, 120–126.

Cozzi, G., Bittante, G. and Polan, C.E. (1993) Comparison of fibrous materials as modifiers of *in situ* ruminal degradation of corn gluten meal. *Journal of Dairy Science* 76, 1106–1113.

Craig, W.M. and Broderick, G.A. (1981) Effect of heat treatment on true digestibility in the rat, *in vitro* proteolysis and available lysine content of cottonseed meal protein. *Journal of Animal Science* 52, 292–301.

De Boever, J.L., Cottyn, B.G., Andries, J.I., Buysse, F.X. and Vanacker, J.M. (1988) The use of a cellulase technique to predict digestibility, metabolizable and net energy of forages. *Animal Feed Science and Technology* 19, 247–260.

de Faria, V.P. and Huber, J.T. (1984) Influence of dietary protein and energy on disappearance of dry matter from different forage types of dacron bags suspended in the rumen. *Journal of Animal Science* 59, 246–252.

Dowman, M.G. and Collins, F.C. (1982) The use of enzymes to predict the digestibility of animal feeds. *Journal of the Science of Food and Agriculture* 33, 689–696.

England, M.L., Broderick, G.A., Shaver, R.D. and Combs, D.K. (1997) Comparison of *in situ* and *in vitro* techniques for measuring ruminal degradation of animal by-product proteins. *Journal of Dairy Science* 80, 2925–2931.

Emanuele, S.M., Staples, C.R. and Wilcox, C.J. (1991) Extent and site of mineral release from six forage species incubated in mobile dacron bags. *Journal of Animal Science* 69, 801–810.

Faldet, M.A., Voss, V.L. Broderick, G.A. and Satter. L.D. (1991) Chemical, *in vitro* and *in situ* evaluation of heat treated soybean proteins. *Journal of Dairy Science* 74, 2548–2554.

Figroid, W., Hale, W.H. and Theurer, B. (1972) An evaluation of the nylon bag technique for estimating rumen utilization of grains. *Journal of Animal Science* 35, 113–120.

Frydrych, Z. (1992) Intestinal digestibility of rumen undegraded protein of various feeds as estimated by the mobile bag technique. *Animal Feed Science and Technology* 37, 161–172.

Givens, D.I., Everington, J.M. and Adamson, A.H. (1989) The digestibility and metabolizable energy content of grass silage and their prediction from laboratory measurements. *Animal Feed Science and Technology* 24, 27–44.

Givens, D.I., Everington, J.M. and Adamson, A.H. (1990a) The nutritive value of spring-grown herbage produced on farms throughout England and Wales (UK) over 4 years. II. The prediction of apparent digestibility *in vivo* from various laboratory measurements. *Animal Feed Science and Technology* 27, 173–184.

Givens, D.I., Everington, J.M. and Adamson, A.H. (1990b) The nutritive value of spring-grown herbage produced on farms throughout England and Wales (UK) over 4 years. III. The prediction of energy values from various laboratory measurements. *Animal Feed Science and Technology* 27, 185–196.

Givens, D.I., Moss, A.R. and Adamson, A.H. (1993a) Influence of growth stage and season on the energy value of fresh herbage. 2. Relationships between digestibility and metabolizable energy content and various laboratory measurements. *Grass and Forage Science* 48, 175–180.

Givens, D.I., Moss, A.R. and Adamson, A.H. (1993b) Prediction of the digestibility and energy value of grass silage conserved in big bales. *Animal Feed Science and Technology* 41, 297–312.

Givens, D.I., Cottyn, B.G., Dewey, P.J.S. and Steg, A. (1995) A comparison of the neutral detergent-cellulase method with other laboratory methods for predicting the digestibility *in vivo* of maize silages from three European countries. *Animal Feed Science and Technology* 54, 55–64.

Goering, H.K. and Van Soest, P.J. (1970) Forage fiber analyses (apparatus, reagents, procedures and some applications). *Agricultural Handbook Number 379*, ARS-USDA, Washington, DC.

Grummer, R.R., Clark, J.H., Davis, C.L. and Murphy, M.R. (1984). Effect of ruminal ammonia-nitrogen concentration on protein degradation *in situ*. *Journal of Dairy Science* 67, 2294–2301.

Hawley, A.W.L. (1981) Effect of bag location along a suspension line on nylon bag digestibility estimates in bison and cattle. *Journal of Range Management* 34, 265–266.

Holt, E.C., Conrad, B.E., Ellis, W.C., Jones, R.M., Lovelace, D.A. and Norris, M.J. (1978) *Bermudagrass (Cynodon dactylon) Research in Central and North Texas (Varieties, Yields). Progress Report no. 3482.* Texas Agricultural Experiment Station.

Hopson, J.D., Johnson, R.R. and Dehority, B.A. (1963) Evaluation of the dacron bag technique as a method for measuring cellulose digestibility and rate of forage digestion. *Journal of Animal Science* 22, 448–453.

Hristov, A. and Broderick, G.A. (1994) *In vitro* determination of ruminal protein degradability using [^{15}N]-ammonia to correct for microbial nitrogen uptake. *Journal of Animal Science* 72, 1344–1353.

Hsu, J.T. and Satter, L.D. (1995) Procedures for measuring the quality of heat-treated soybeans. *Journal of Dairy Science* 78, 1353–1361.

Huntington, J.A. and Givens, D.I. (1995) The *in situ* technique for studying the rumen degradation of feeds: a review of the procedure. *Nutrition Abstracts and Reviews Series B* 65, 63–93.

Hvelplund, T. (1985) Digestibility of rumen microbial protein and undegraded dietary protein estimated in the small intestine of sheep and by *in sacco* procedure. *Acta Agriculturae Scandinavica, Supplement* 25, 132–144.

Jones, D.I.H. and Hayward, M.V. (1975) The effect of pepsin pretreatment of herbage on the prediction of dry matter digestibility from solubility in fungal cellulase solutions. *Journal of the Science of Food and Agriculture* 26, 711–718.

Jouany, J.P. (1996) Effect of rumen protozoa on nitrogen utilization by ruminants. *Journal of Nutrition* 126, 1335S–1346S.

Kohn, R.A. and Allen, M.S. (1995a) *In vitro* protein degradation of feeds using concentrated enzymes extracted from rumen contents. *Animal Feed Science and Technology* 52, 15–20.

Kohn, R.A. and Allen, M.S. (1995b) Prediction of protein degradation of forages from solubility fractions. *Journal of Dairy Science* 78, 1774–1788.

Kosmala, I., Antoniewicz, A., De Boever, J., Hvelplund, T. and Kowalczyk, J. (1996) Use of enzymatic solubility with ficin (EC 3.4.22.3) to predict *in situ* feed protein degradability. *Animal Feed Science and Technology* 59, 245–254.

Krishnamoorthy, U., Sniffen, C.J., Stern, M.D. and Van Soest, P.J. (1983) Evaluation of a mathematical model of rumen digestion and an *in vitro* simulation of rumen proteolysis to estimate the rumen-undegraded nitrogen content of feedstuffs. *British Journal of Nutrition* 50, 555–568.

Krishnamoorthy, U., Steingass, H. and Menke, K.H. (1990) The contribution of ammonia, amino acids and short peptides to estimates of protein degradability *in vitro*. *Journal of Animal Physiology and Animal Nutrition* 63, 135–141.

Licitra, G., Carpino, S., Van Soest, P.J. and Sniffen, C.J. (1993) Improvement of the *Streptomyces griseus* method for degradable protein in ruminant feeds. *Journal of Dairy Science* 76 (Suppl. 1), 175 (abstract).

Lindberg, J.E. (1985) Estimation of rumen degradability of feed proteins with the *in sacco* technique and various *in vitro* methods: a review. *Acta Agriculturae Scandinavica, Supplement* 25, 64–97.

Lopez, S., France, J. and Dhanoa, M.S. (1994) A correction for particulate matter loss when applying the poly-ester bag method. *British Journal of Nutrition* 71, 135–137.

Luchini, N.D., Broderick, G.A. and Combs, D.K. (1996) Characterization of the proteolytic activity of commercial proteases and strained ruminal fluid. *Journal of Animal Science* 74, 685–692.

Madsen, J. and Hvelplund, T. (1994) Prediction of *in situ* protein degradability in the rumen: results of a European ringtest. *Livestock Production Science* 39, 201–212.

Madsen, J., Hvelplund, T., Weisbjerg, M.R., Bertilsson, J., Olsson, I., Sporndly, R., Harstad, O.M., Volden, H., Tuori, M., Varvikko, T., Huhtanen, P. and Olafsson, B.L. (1995) The AAT/PBV protein evaluation system for ruminants, a revision. *Norwegian Journal of Agricultural Sciences* (Suppl. 19), 1–37.

Mahadevan, S., Erfle, J.D. and Sauer, F.D. (1979) A colorimetric method for the determination of proteolytic degradation of feed proteins by rumen microorganisms. *Journal of Animal Science* 48, 947–953.

Mahadevan, S., Erfle, J.D. and Sauer, F.D. (1980) Degradation of soluble and insoluble proteins by *Bacteroides amylophilus* protease and by rumen microorganisms. *Journal of Animal Science* 50, 723–728.

Mahadevan, S., Erfle, J.D. and Sauer, F.D. (1987) Preparation of protease from mixed rumen microorganisms and its use for the *in vitro* determination of the degradability of true protein in feedstuffs. *Canadian Journal of Animal Science* 67, 55–64.

Mansfield, H.R., Stern, M.D. and Otterby, D.E. (1994) Effects of beet pulp and animal by-products on milk yield and *in vitro* fermentation by rumen microorganisms. *Journal of Dairy Science* 77, 205–216.

Marinucci, M.T., Dehority, B.A. and Loerch, S.C. (1992) *In vitro* and *in vivo* studies of factors affecting digestion of feeds in synthetic fiber bags. *Journal of Animal Science* 70, 296–307.

Masoero, F., Fiorentini, L., Rossi, F. and Piva, A. (1994) Determination of nitrogen intestinal digestibility in ruminants. *Animal Feed Science and Technology* 48, 253–263.

Mathers, J.C. and Aitchison, E.M. (1981) Direct estimation of the extent of contamination of food residues by microbial matter after incubation within synthetic fibre bags in the rumen. *Journal of Agricultural Science, Cambridge* 96, 691–693.

McAllister, T.A., Phillippe, R.C., Rode, L.M. and Cheng, K.J. (1993) Effect of the protein matrix on the digestion of cereal grains by ruminal microorganisms. *Journal of Animal Science* 71, 205–212.

McDonald, I. (1981) A revised model for the estimation of protein degradability in the rumen. *Journal of Agricultural Science, Cambridge* 96, 251–252.

McDonald, I.W. and Hall, R.J. (1954) The conversion of casein into microbial proteins in the rumen. *Biochemical Journal* 67, 400–405.

Menden, E. and Cremer, H.D. (1970) Laboratory methods for the evaluation of changes in protein quality. In: Albanese, A.A. (ed.), *Newer Methods of Nutritional Biochemistry*, Vol. IV. Academic Press, New York, pp. 123–127.

Menke, K.H., Raab, L., Salewski, A., Steingass, H., Fritz, D. and Schneider, W. (1979) The estimation of the digestibility and metabolizable energy content of ruminant feedingstuffs from the gas production when they are incubated with rumen liquor *in vitro. Journal of Agricultural Science, Cambridge* 93, 217–222.

Mertens, D.R. (1993) Kinetics of cell wall digestion and passage in ruminants. In: Jung, H.G., Buxton, D.R., Hatfield, R.D. and Ralph, J. (eds), *Forage Cell Wall Structure and Digestibility*. American Society of Agronomy, Crop Science Society of American and Soil Science Society of America, Madison, Wisconsin, pp. 485–498.

Messman, M.A., Weiss, W.P. and Erickson, D.O. (1991) Effects of nitrogen fertilization and maturity of bromegrass on *in situ* ruminal digestion kinetics of fiber. *Journal of Animal Science* 69, 1151–1161.

Messman, M.A., Weiss, W.P. and Koch, M.E. (1994) Changes in total and individual proteins during drying, ensiling and ruminal fermentation of forages. *Journal of Dairy Science* 77, 492–500.

Meyer, J.H.F. and Mackie, R.I. (1986). Microbiological evaluation of the intraruminal *in sacculus* digestion technique. *Applied and Environmental Microbiology* 51, 622–629.

Michalet-Doreau, B. and Ould-Bah, M.Y. (1992) *In vitro* and *in sacco* methods for the estimation of dietary nitrogen degradability in the rumen: a review. *Animal Feed Science and Technology* 40, 57–86.

Moss, A.R. and Givens, D.I. (1990) Chemical composition and *in vitro* digestion to predict digestibility of field-cured and barn-cured grass hays. *Animal Feed Science and Technology* 31, 125–138.

National Research Council (1989) *Nutrient Requirements of Dairy Cattle*, 6th edn (Update, 1989). National Academy Press, Washington, DC.

Neutze, S.A., Smith, R.L. and Forbes, W.A. (1993) Application of an inhibitor *in vitro* method for estimating rumen degradation of feed protein. *Animal Feed Science and Technology* 40, 251–265.

Nocek, J.E. (1985) Evaluation of specific variables affecting *in situ* estimates of ruminal dry matter and protein digestion. *Journal of Animal Science* 60, 1347–1358.

Nocek, J.E. (1988). *In situ* and other methods to estimate ruminal protein and energy digestibility: a review. *Journal of Dairy Science* 71, 2051–2069.

Nocek, J.E. and English, J.E. (1986). *In situ* degradation kinetics: evaluation of rate determination procedure. *Journal of Dairy Science* 69, 77–87.

Nocek, J.E. and Grant, A.L. (1987) Characterization of *in situ* nitrogen and fiber digestion and bacterial nitrogen contamination of hay crop forages preserved at different dry matter percentages. *Journal of Animal Science* 64, 552–564.

Olubobokun, J.A. and Craig, W.M. (1990) Quantity and characteristics of micro-organisms associated with ruminal fluid or particles. *Journal of Animal Science*, 68, 3360–3370.

Olubobokun, J.A., Craig, W.M. and Pond, K.R. (1990) Effects of mastication and microbial contamination on ruminal *in situ* forage disappearance. *Journal of Animal Science* 68, 3371–3381.

Ørskov, E.R. (1982) *Protein Nutrition in Ruminants*, 1st edn. Academic Press, Inc., London.

Ørskov, E.R. and McDonald, I. (1979) The estimation of protein degradability in the rumen from incubation measurements weighted according to rate of passage. *Journal of Agricultural Science, Cambridge* 92, 499–503.

Paine, C.A., Crawshaw, R. and Barber, W.P. (1982) A complete exchange method for the *in sacco* estimation of rumen degradability on a routine basis. In: Thomson, D.J., Beever, D.E. and Gunn, R.G. (eds), *Forage Protein in Animal Production*. Occasional Publication No. 6, British Society of Animal Production, pp. 177–178.

Paster, B.J., Russell, J.B., Yang, C.M.J., Chow, J.M., Woese, C.R. and Tanner, R. (1993) Phylogeny of the ammonia-producing ruminal bacterial *Peptostreptococcus anaerobius*, *Clostridium sticklandii*, and *Clostridium aminophilum* spp. nov. *International Journal of Systematic Bacteriology* 43, 107–110.

Pell, A.N., Pitt, R.E., Doane, P.H. and Schofield, P. (1998) The development, use and application of the gas production technique at Cornell University. In: *Proceedings of the International Symposium on In Vitro Techniques for Measuring Nutrient Supply to Ruminants*. British Society of Animal Science, Reading, UK, pp. 45–54.

Poos-Floyd, M., Klopfenstein, T. and Britton, R.A. (1985) Evaluation of laboratory techniques for predicting ruminal protein degradation. *Journal of Dairy Science* 68, 829–839.

Quin, J.I., van der Wath, J.G. and Myburgh, S. (1938) Studies on the alimentary tract of merino sheep in South Africa. IV. Description of experimental technique. *Onderstepoort Journal of Veterinary Science and Animal Industry* 11, 341–360.

Raab, L., Cafantaris, B., Jilg, T. and Menke, K.H. (1983). Rumen protein degradation and biosynthesis. 1. A new method for determination of protein degradation in rumen fluid *in vitro*. *British Journal of Nutrition* 50, 569–582.

Richards, C.J., Pedersen, J.F., Britton, R.A., Stock, R.A. and Krehbiel, C.R. (1995) *In vitro* starch disappearance procedure modifications. *Animal Feed Science and Technology* 55, 35–45.

Roe, M.B., Chase, L.E. and Sniffen, C.J. (1991) Comparison of *in vitro* techniques to the *in situ* technique for estimation of ruminal degradation of protein. *Journal of Dairy Science* 74, 1632–1640.

Sauer, W.C., den Hartog, L.A., Huisman, J., van Leeuwen, P. and de Lange, C.F.M. (1989) The evaluation of the mobile nylon bag technique for determining the apparent protein digestibility in a wide variety of feedstuffs for pigs. *Journal of Animal Science* 67, 432–440.

Sauer, W.C., Jørgensen, H. and Berzins, R. (1983) A modified nylon bag technique for determining apparent digestibilities of protein in feedstuffs for pigs. *Canadian Journal of Animal Science* 63, 233–237.

Stakelum, G., Morgan, D. and Dillon, P. (1988) A comparison of *in vitro* procedures for estimating herbage digestibility. *Irish Journal of Agricultural Research* 27, 104–105.

Theodorou, M.K., Lowman, R.S., Davies, Z.S., Cuddeford, D. and Owen, E. (1998) The physical and chemical principles of feed evaluation techniques in ruminant nutrition based on gas measurement. In: *Proceedings of the International Symposium on In Vitro Techniques for Measuring Nutrient Supply to Ruminants.* British Society of Animal Science, Reading, UK, pp. 55–63.

Tilley, J.M.A. and Terry, R.A. (1963) A two stage technique for the *in vitro* digestion of forage crops. *Journal of the British Grassland Society* 18, 104–111.

Tomankova, O. and Kopecny, J. (1995) Prediction of feed protein degradation in the rumen with bromelain. *Animal Feed Science and Technology* 53, 71–80.

Tremblay, G.F., Broderick, G.A. and Abrams, S.M. (1996) Estimating ruminal protein degradation of roasted soybeans using near infrared reflectance spectroscopy. *Journal of Dairy Science* 79, 276–282.

Udén, P. and Van Soest, P.J. (1984). Investigations of the *in situ* bag technique and a comparison of the fermentation in heifers, sheep, ponies and rabbits. *Journal of Animal Science* 58, 213–221.

Udén, P., Parra, R. and Van Soest, P.J. (1974). Factors influencing reliability of the nylon bag technique. *Journal of Dairy Science* 57, 622–629.

Vanhatalo, A. and Ketoja, E. (1995) The role of the large intestine in post-ruminal digestion of feeds as measured by the mobile-bag method in cattle. *British Journal of Nutrition* 73, 491–505.

Vanhatalo, A. and Varvikko, T. (1995) Effect of rumen degradation on intestinal digestion of nitrogen of ^{15}N-labelled rapeseed meal and straw measured by the mobile-bag method in cows. *Journal of Agricultural Science, Cambridge* 125, 253–261.

Vanhatalo, A., Aronen, I. and Varvikko, T. (1995) Intestinal nitrogen digestibility of heat–moisture treated rapeseed meals as assessed by the mobile-bag method in cows. *Animal Feed Science and Technology* 55, 139–152.

Van Hellen, R.W. and Ellis, W.C. (1977). Sample container porosities for rumen *in situ* studies. *Journal of Animal Science* 44, 141–146.

Van Keuren, R.W. and Heinemann, W.W. (1962). Study of a nylon bag technique for *in vivo* estimation of forage digestibility. *Journal of Animal Science* 21, 340–345.

Van Milgen, J. and Baumont, R. (1995) Models based on variable fractional digestion rates to describe ruminal *in situ* digestion. *British Journal of Nutrition* 73, 793–807

Vanzant, E.S., Cochran, R.C., Titgemeyer, E.C., Stafford, S.D., Olson, K.C., Johnson, D.E. and St Jean, G. (1996) *In vivo* and *in situ* measurements of forage protein degradation in beef cattle. *Journal of Animal Science* 74, 2773–2784.

Varvikko, T. and Vanhatalo, A. (1990) The effect of differing types of cloth and of contamination by non-feed nitrogen on intestinal digestion estimates using porous synthetic-fibre bags in a cow. *British Journal of Nutrition* 63, 221–229.

Veresegyházy, T., Fekete, S. and Hegedûs, M. (1993) Factors influencing ruminal bacterial activity. I. Effect of endproduct concentration on proteolytic activity. *Journal of Animal Physiology and Animal Nutrition* 70, 1–5.

Waldo, D.R., Smith, L.W. and Cox, E.L. (1972) Model of cellulose disappearance from the rumen. *Journal of Dairy Science* 55, 125–129.

Wallace, R.J. (1983). Hydrolysis of ^{14}C-labelled proteins by rumen microorganisms and by proteolytic enzymes prepared from rumen bacteria. *British Journal of Nutrition* 50, 345–355.

Wallace, R.J. (1996) Ruminal microbial metabolism of peptides and amino acids. *Journal of Nutrition* 126, 1326S–1334S.

Wallace, R.J. and Cotta, M.A. (1988). Metabolism of nitrogen-containing compounds. In: Hobson, P.N. (ed.), *The Rumen Microbial Ecosystem*. Elsevier Applied Science, London, pp. 217–249.

Waltz, D.M., Stern, M.D. and Illg, D.J. (1989) Effect of ruminal protein degradation of blood meal and feather meal on the intestinal amino acid supply to lactating cows. *Journal of Dairy Science* 72, 1509–1518.

Wanderley, R.C., Huber, J.T., Wu, Z., Pessarakli, M. and Fontes, C., Jr (1993) Influence of microbial colonization of feed particles on determination of nitrogen degradability by *in situ* incubation. *Journal of Animal Science* 71, 3073–3077.

Warner, A.C.I. (1956) Proteolysis by rumen microorganisms. *Journal of General Microbiology* 14, 749–762.

Weakley, D.C., Stern, M.D. and Satter, L.D. (1983) Factors affecting disappearance of feedstuffs from bags suspended in the rumen. *Journal of Animal Science* 56, 493–507.

Weimer, P.J. (1993) Microbial and molecular mechanisms of cell wall degradation – session synopsis. In: Jung, H.G., Buxton, D.R., Hatfield, R.D. and Ralph, J. (eds), *Forage Cell Wall Structure and Digestibility*. American Society of Agronomy, Crop Science Society of American and Soil Science Society of America, Madison, Wisconsin, pp. 485–498.

Weiss, W.P. (1994) Estimation of digestibility of forages by laboratory methods. In: Fahey, G.C. Jr, Collins, M., Mertens, D.R. and Moser, L.E. (eds), *Forage Quality, Evaluation, and Utilization*. American Society of Agronomy, Crop Science Society of American and Soil Science Society of America, Madison, Wisconsin, pp. 644–680.

White, B.A., Mackie, R.I. and Doerner, K.C. (1993) Enzymatic hydrolysis of forage cell walls. Forage cell wall structure and digestibility In: Jung, H.G., Buxton, D.R., Hatfield, R.D. and Ralph, J. (eds), *Forage Cell Wall Structure and Digestibility*. American Society of Agronomy, Crop Science Society of American and Soil Science Society of America, Madison, Wisconsin, pp. 455–484.

Wilkerson, V.A., Klopfenstein T.J. and Stroup, W.W. (1995) A collaborative study of *in situ* forage protein degradation. *Journal of Animal Science* 73, 583–588.

Wilman, D., Foulkes, G.R. and Givens, D.I. (1996) The rate and extent of cell-wall degradation *in vitro* for 40 silages varying in composition and digestibility. *Animal Feed Science and Technology* 63, 111–122

5 Measurement of Energy Metabolism

C.K. Reynolds

Department of Agriculture, The University of Reading, Earley Gate, Reading, UK

INTRODUCTION

There is a vast number of publications describing the measurement of energy metabolism in farm animals and the interpretation of the measurements obtained. A few are referenced within the following chapter, which is intended as an overview of the subject for those who do not work actively in the area. The result is far from complete, but will hopefully provide a starting point for further investigation. Measurements of energy metabolism are the basis of current rationing systems for feeding energy to livestock, which many believe to be inadequate. This is not unlike the situation when the current systems were developed some 30 years ago. Energy is the first limiting nutrient in most animal production systems, thus the need for an accurate and precise feed-rationing system which can budget energy balances, and predict responses, will remain. There are many who feel that the classical techniques for the study of energy metabolism have reached the end of their useful life, to borrow the words of Sir Kenneth Blaxter (Blaxter and Graham,1955), that the approach and the information obtained have been 'sucked dry'. Time will tell. In the meantime, an enlightened understanding of the current energy feeding systems and their basis is needed for them to be improved, or replaced.

HISTORICAL PERSPECTIVES

Limitations of existing systems for feed evaluation and ration formulation in the 'post-war' boom of agricultural research of the 1950s and 1960s led to the development of a number of 'new' feeding systems based on measurements of energy metabolism using balance trials or comparative body composition methodology. Some of the systems in place, such as the total digestible nutrient (TDN) approach, tended to overestimate the feeding value of forages (e.g. lucerne hay) compared to concentrates (e.g. maize meal) by not accounting for differences in their net energy value. Debate arising over this specific comparison was a driving force behind the establishment of six respiration calorimeters for dairy cows at the US Department of Agriculture's Research Center in Beltsville, Maryland (Flatt *et al.*, 1958; Van Soest, 1992). A sustained programme of research on the energy metabolism of lactating dairy cattle at this facility, and others in Europe, lead to the development of a Net Energy for Lactation (NE$_l$) system for describing the nutrient requirements of dairy cattle and the energy value of feeds (Moe *et al.*, 1972; NRC, 1989). In Europe, similar concerns about the Starch Equivalent systems in use (Blaxter, 1986) lead to the simultaneous and conjoint development of various metabolizable energy (ARC, 1965) and net energy (Vermorel and Coulon, 1998) systems for ruminant livestock. Similar systems were also developed for pigs and poultry, although in these species the effects of forages on the suitability of digestible energy (DE) as a basis for rationing energy are of less concern. Current systems for rationing energy for pigs are effectively based on DE (Chapter 9), whilst a metabolizable energy (ME) system is widely used for poultry as faeces and urine are not separated in the measurement of digestion (Chapter 10; McDonald *et al.*, 1995).

Although the current systems were developed on the back of a profusion of new measurements of energy metabolism, the concepts on which they are based and the techniques used for measurements of energy metabolism and feeding value of ration components have a long history. The work of Kellner in Germany and Armsby in the US led to the development of net energy (NE) approaches for describing feed energy utilization by ruminants at the turn of the century. However, the NE values published by Armsby (Armsby, 1917) were expressed as therms, whilst Kellner's system expressed the energy value of feeds in terms of starch equivalents (Kellner, 1926). Due to the difficulty of obtaining measurements of the NE value of individual feeds, both approaches used values for a large proportion of feeds which were estimated from their digestible nutrient content (Morrison, 1954; McDonald *et al.*, 1995). The Starch Equivalent system was used widely in Europe until replaced by more refined NE systems. In the US, Armsby's NE system tended to be used as a supplement to the TDN system, which later incorporated Morrison's own estimated NE values, so called 'corn equivalents', which were derived from feeding trials (Morrison, 1954). The period between the ground breaking experimentation of Kellner, Zuntz, Rubner, Armsby and others and the larger scale studies 60 years later was not devoid of research in energy metabolism of farm animals. Work continued at a number of locations (NRC, 1935; van Es, 1994) and substantial contributions were made, but a working NE system was never achieved, largely due to the limited number of measurements of feed utilisation which the technology of the day could produce.

One concern with the number of facilities operating independently in the area of energy metabolism was a lack of standardization of approach and terminology. In 1935 many of the US scientists working in the area of energy metabolism (S. Brody, Missouri; E.B. Forbes, Pennsylvania; M. Kleiber, California; T.S. Hamilton and H.H. Mitchell, Illinois; E.G. Ritzman, New

Hampshire) met at State College, Pennsylvania, site of the Armsby calorimeter, to discuss and debate current findings and approaches. The meeting was held under the auspices of the Committee on Animal Nutrition of the National Research Council, the organization which coordinates the publication of current feeding standards in the USA. This conference on energy metabolism provided an early opportunity for standardization of terminology, methodology and conceptual frameworks such as maintenance energy requirements and metabolic body size (NRC, 1935). Similarly, the European Association of Animal Production held a symposium on Energy Metabolism in Farm Animals in 1958 to provide a forum for comparison and discussion of energy metabolism methodology and results. This symposium, held every 3 years (Table 5.1), has also contributed to the standardization of terminology and approaches used, such as the formulae for calculating heat production from respiratory exchange (Brouwer, 1965). In spite of the success of this symposium in achieving these objectives, many of the feeding standards developed in individual countries have used varied approaches, assumptions and terminology. This has lead to confusion amongst those seeking to compare the relative merits and weaknesses of the individual systems. In addition, the lack of understanding of the historical basis for the assumptions and approaches used in studies of energy metabolism has led to misinterpretations of results and comparisons of data obtained using differing methodologies.

TERMINOLOGY AND ASSUMPTIONS

There are a variety of units for the expression of the energy value of feeds and animal products. The calorie has been the term of choice in the US, but the internationally accepted term is now the Joule. In the US, the NRC published (Harris, 1966) and later revised (NRC, 1981) a glossary of terms for describing the components of energy metabolism. This glossary gives a comprehensive breakdown of the various components of mammalian energetics, along with a suggested acronym for each component. However, these terms and acronyms have not been universally adopted. Terminology differs between energy feeding systems used in different countries and between systems used for different species as well, and as in the present chapter, is subject to the interpretation of individual authors. This is especially true for the various efficiency constants, which can be particularly confusing for the uninitiated. Differences in terminology between countries using different language are to be expected. However, in comparing energy feeding standards, the problems arising from the use of different assumptions and calculations for individual terms is made worse by the lack of universally accepted terminology and/or abbreviations. Editorial boards for scientific journals can establish 'acceptable' terminology for measurements of energy metabolism, but acceptable terminology varies between journals. In reviewing the literature, one should be aware that the terminology used to report measurement of energy metabolism has evolved with time as well.

Table 5.1. Symposia on energy metabolism of farm animals sponsored by the European Association of Animal Production (Moe, 1981). For a brief history see van Es (1994).

No.	Year	Site	Reference
*	1935	USA	Report of the Conference on Energy Metabolism held at State College, Pennsylvania (1935). NRC, Washington, DC.
1	1958	Denmark	Symposium on Energy Metabolism. Principles, Methods, and General Aspects (1958) Thorbek, G. and Aersoe, H. (eds) EAAP Publ. No. 8. Statens Husdyrugsudvalg, Copenhagen.
2	1961	Netherlands	Symposium on Energy Metabolism. Mehods and Results of Experiments with Animals (1961) Brouwer, E. and van Es, A.J.H. (eds) EAAP Publ. No. 10. EAAP, Wageningen.
3	1964	Scotland	Energy Metabolism (1965) Blaxter, K.L. (ed.) EAAP Publ. No. 11. Academic Press, London.
4	1967	Poland	Energy Metabolism of Farm Animals (1969) Blaxter, K.L., Kielanowski, J. and Thorbek, G. (eds) EAAP Publ. No. 12. Oriel Press, Newcastle upon Tyne.
5	1970	Switzerland	Energy Metabolism (1970) Schurch, A. and Wenk, C. (eds) EAAP Publ. No. 13. Juris Verlag, Zurich.
6	1973	West Germany	Energy Metabolism of Farm Animals (1974) Menke, K.H., Lantzsch, H.J. and Reichl, J.R. (eds) EAAP Publ. No. 14, Universitat Hohenheim Dokumentationsstelle, B.D.R.
7	1976	France	Energy Metabolism of Farm Animals (1976) Vermorel, M. (ed) EAAP Publ. No. 19. G. de Bussac, Clermont-Ferrand, France.
8	1979	England	Energy Metabolism (1979) Mount, L.E. (ed.) EAAP Publ. No. 26. Butterworths, London.
9	1982	Norway	Energy Metabolism of Farm Animals (1982) Ekern, A. and Sundstol, R. (eds) EAAP Publ. No. 29. Agricultural University of Norway.
10	1985	USA	Energy Metabolism of Farm Animals (1986) Moe, P.W., Tyrrell, H.F. and Reynolds, P.J. (eds) EAAP Publication No. 32, Rowman and Littlefield, New Jersey.
11	1988	Netherlands	Energy Metabolism of Farm Animals (1989) Close, W.H. and van der Honing, Y. (ed) EAAP Publication No. 43. Pudoc, Wageningen, Netherlands.
12	1991	Switzerland	Energy Metabolism of Farm Animals (1991) Wenk, C. and Boessinger, M. (eds) EAAP Publication No. 58. ETH-Zentrum, Zurich.
13	1994	Spain	Energy Metabolism of Farm Animals (1994) Aguilera, J.F. (ed) EAAP Publication No. 76. CSIC, Madrid.
14	1997	Northern Ireland	Energy Metabolism of Farm Animals (1998) McCracken, K.J., Unsworth, E.F. and Wylie, A.R.G. (eds) CAB International, Wallingford.

*Early conference sponsored by the National Research Council, Committee on Animal Nutrition, USA.

Calculations and Abbreviations

The idealized flow of energy through animals suggested by the NRC (1981) is shown in Fig. 5.1. All energy-feeding systems begin with gross energy, the total energy in food provided to the animal. Gross energy intake, or intake energy (IE), is the total amount of energy consumed. Apparently digested energy (DE) is measured by subtracting faecal energy (FE) from IE. This is distinguished from true digested energy (TDE), which accounts for metabolic faecal energy (F_mE) and heat of fermentation (H_fE). Subtracting urine and gaseous energy (mainly methane) gives metabolizable energy (ME), which is in a sense a net measurement as the energy in urine is partly a consequence of

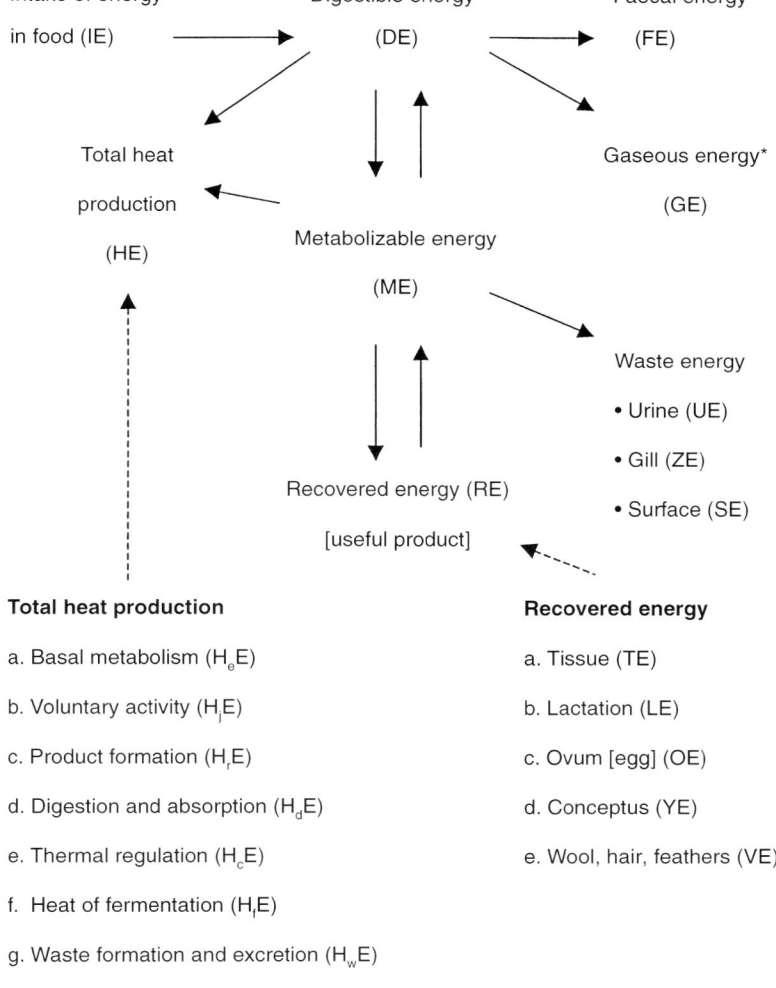

Total heat production

a. Basal metabolism (H_eE)

b. Voluntary activity (H_jE)

c. Product formation (H_rE)

d. Digestion and absorption (H_dE)

e. Thermal regulation (H_cE)

f. Heat of fermentation (H_fE)

g. Waste formation and excretion (H_wE)

Recovered energy

a. Tissue (TE)

b. Lactation (LE)

c. Ovum [egg] (OE)

d. Conceptus (YE)

e. Wool, hair, feathers (VE)

Fig. 5.1. Energy flow in animals and suggested terms (NRC, 1981). *May be considered useful.

metabolism. True metabolizable energy (TME) then is the energy truly available for metabolism, which is obtained by accounting for endogenous urine energy (U_eE) not of food origin. TME is used in evaluating poultry feeds by comparing the energy voided in fasted (or glucose fed) birds and birds fed a test meal (McDonald *et al.*, 1995). In other species the effects of body protein loss (or gain) on U_eE losses can be obtained by adjusting to zero nitrogen retention, giving nitrogen-corrected ME (M_nE).

On a net basis, ME can be lost as heat energy (HE) or recovered as energy in specific products (RE), such as lactation energy (LE) or body tissue energy (TE), the sum of recovered energy representing 'energy balance' in many publications. This subdivision requires the measurement of either HE or RE. In growing animals, energy is retained as TE, whilst in dairy cows milk energy production must also be measured. Historically NE was a term applied to total energy balance at a given level of food intake (Armsby, 1917; Morrison, 1954), or ME minus HE. However, in the glossary proposed by the NRC (1981) the term NE is specifically applied to the change in recovered energy relative to a change in IE, which represents an incremental efficiency. Incremental efficiencies are often represented by 'k' values, or 'partial' efficiency constants, which can be calculated for a variety of products and inputs. For example, k_m, k_g and k_l represent the partial efficiencies of ME use for maintenance, TE gain and lactation energy, respectively, which are calculated as the increase in energy recovered in these products with increasing ME. These efficiency constants are typically derived using linear regression, although one should never assume absolute linearity for any biological response. Certainly, the relationship between retained energy (or conversely HE) and ME is different above and below maintenance. This relationship is often described by two straight lines, intersecting at the point where ME = HE, although there is evidence that the response is curvilinear (Blaxter, 1989; NRC, 1996).

The Concept of Maintenance

The manner in which maintenance energy requirements are described and accounted for varies between feeding systems (Moe and Tyrrell, 1973; Moe, 1981). In energetic terms, maintenance represents the state of retained energy (energy balance) being zero, thus the point where ME = HE. The total NE requirement for maintenance is theoretically equal to the heat production at zero food intake, or fasting metabolism, whilst NE for maintenance (NE_m) represents the relationship between energy balance (which is negative) and ME below maintenance. Thus k_m is equal to fasting HE divided by maintenance ME (ME_m). Alternatively, NE for production (recovered energy) represents the relationship between energy balance (which is positive) and ME above maintenance. The partial efficiency of ME use for energy recovered in a given product is ideally calculated with corrections for ME_m (i.e. ME − ME_m). If NE_m and NE for production are expressed as a combined, linear function which theoretically intersects fasting HE, then at higher intakes (as occur in lactating dairy cows at production intakes), the slope is more similar to the one for NE for production than NE_m (Moe and Tyrrell, 1973). Alternatively, at lower intakes the slope becomes more similar to the regression for NE_m. This is one reason that systems used for describing energy requirements and feed values for ruminants fed at lower intakes tend to use separate terms for NE_m and NE for production, whilst systems in use for lactating dairy cows use a single combined NE term (Moe and Tyrrell, 1973). This is also justified on the basis that measurements of energy metabolism of lactating cows at lower intakes or fasting are difficult to obtain and considered irrelevant to the lactating cow consuming 3–4 times her ME_m (Moe *et al.*, 1972).

Whilst older systems assumed a constant efficiency of DE or ME use for production, virtually all the major energy rationing systems in use today assume a variable efficiency of ME use. This results in the use of different efficiency constants for calculating RE from ME, or the assignment of different NE values for feeds. An alternative approach, used in the Australian energy feeding system for ruminants (Australian Standing Committee on Agriculture, 1990), includes a variable maintenance cost that increases with higher intakes, rather than varying k_m or k_g as done in the current UK system. In reality, these systems differ more in terms of application rather than underlying scientific principle, but problems of application are an important consideration in the development of any effective feeding system (Moe and Tyrrell, 1973).

Net Energy for Lactation

In the US system for feeding energy to dairy cows (NRC, 1989), nutrient requirements and feed energy value are expressed in terms of NE_l, which is LE with corrections for tissue energy loss or gain, energy costs of gestation and energy cost of excess protein intake (Moe *et al.*, 1972). The corrections for tissue energy loss or gain are not simply energy balance (milk plus tissue energy), but include adjustments based on the efficiency of conversion of tissue energy into milk energy obtained using multiple regression (Moe *et al.*, 1971). As the system is based on describing the net energy value of feeds for milk energy production, this correction is also applied to retained tissue energy to reflect the energetic value of body tissue for milk synthesis. This logic applies even though the tissue energy may not be converted into milk energy until the rising phase of a subsequent lactation. The adjustments for gestation energy (Moe and Tyrrell, 1972) and the cost of excess protein (Tyrrell *et al.*, 1970) were also based on relationships derived from regression procedures. These adjustments are applied to account for effects of physiological state or biochemical responses to ration imbalances on the energetic value of individual feeds. The calculation of NE_l then is as follows:

NE_l (MJ) = milk energy + adjusted TE + 0.0301(excess N) + 0.1841(fetal tissues) where

- adjusted TE = TE/1.14 if TE > 0 or
               = TE(0.84) if TE < 0,
- fetal tissues = (cow live weight/600)exp$^{0.0174t}$, where t = days pregnant and
  the exponential equation is assumed to be for a 600 kg cow and vary
  linearly with cow live weight,
- excess N (g) = digested N − milk N − fetal N − 0.456(cow live weight$^{0.75}$),
  where fetal N = 0.3259(fetal tissues) and cow tissue N requirements are
  assumed to be 0.456 g kg^{-1} live weight$^{0.75}$.

As for earlier NE systems, in the absence of measured NE$_l$ values for many
feeds, tabular values were obtained from TDN values using regression (Moe
et al., 1972). In this regard, corrections are applied to data obtained at lower
levels of intake to account for depressions in digestibility or metabolizability
(NRC, 1989). Thus in using tables of feed composition, uncorrected ME values
should not be directly compared to corrected NE$_l$ values. In this regard,
extreme care must be exercised in using tabular values of the digestible
nutrient content of feeds obtained using sheep or non-lactating animals to
estimate ME or NE content of feeds fed to lactating cows (Moe and Tyrrell,
1975; Tyrrell and Moe, 1975; Sutton *et al.*, 1997).

MEASUREMENT OF ENERGY METABOLISM

There are numerous reviews, detailed descriptions and interpretations
published of the techniques used for the measurement of energy metabolism
in farm animals and humans (e.g. Blaxter, 1967, 1971, 1989; Flatt, 1969;
McLean and Tobin, 1987; McDonald *et al.*, 1995). The series of symposia on
energy metabolism in farm animals (Table 5.1) is also an excellent source of
information on the subject. It is beyond the scope of this chapter to provide a
detailed shopping list for the initiation of a programme of energy metabolism
research using classical methodology. The approaches used are indeed
relatively simple, but complex in the number of measurements required, and
have changed little since the 'Determination of the Source of Animal Heat' by
Despretz and Dulong in 1822 (see McLean and Tobin, 1987). What has
changed is the technology available for obtaining these measurements, making
measurements of respiratory exchange and heat production more precise and
less labour intensive. Indeed, advancements in technology for obtaining and
recording measurements and calculation of results enabled the onslaught of
energy metabolism studies coinciding with the initiation of the EAAP Energy
Metabolism Symposia. Technology for calorimetric measurements of energy
metabolism continues to evolve and result in modifications to the particular
approaches used, but those undertaking research in this area should not
ignore the volumes of archived material describing the work of their
predecessors.

Measurement of Heat Energy

Measurements of heat production can be obtained either directly or indirectly using calorimeters. As the name implies, direct calorimeters measure heat (both non-evaporative and evaporative) produced by an animal within them directly, using thermocouples or changes in the amount of heat produced in cooling the chamber. Types of direct calorimeters include iso-thermal, heat-sink, convection and differential, which are all described in the exhaustive book on calorimetry by McLean and Tobin (1987). The calorimeter used by Armsby was a heat-sink calorimeter that was accurate, as well as responsive to acute changes in heat production, but very complex and labour intensive to operate. Although modern gradient layer calorimeters are now highly automated, their complexity makes them expensive. In order to measure accurately the HE produced, the animal is typically contained within a closed chamber under environmental control. This allows the measurement of respiratory exchange, which can also be used to estimate HE indirectly. Comparison of estimates of HE based on respiratory exchange with direct measurements in the Armsby calorimeter were important in establishing the validity of the respiratory exchange approach (McLean and Tobin, 1987). Owing to the historical complexity of direct calorimeters, the majority of measurements of energy metabolism in farm animals in the last 40 years have been obtained using indirect calorimeters. This is especially true for lactating dairy cows.

There have generally been two basic approaches used for indirect calorimetry. In the first, HE is estimated from respiratory exchange based on established relationships between oxygen (O_2) consumption and HE, with adjustments for the proportions of fat and carbohydrate oxidized based on CO_2 production (respiratory quotient), and for amino acid oxidation based on urinary nitrogen excretion (UN). In addition, the incomplete oxidation of nutrients lost as methane (CH_4) is accounted for in ruminants. These calcula-tions were condensed into a single equation using simultaneous equations and effectively standardized by a sub-committee established by the EAAP Energy Symposium and chaired by E. Brouwer. In 1957 Brouwer published a revision of a similar formula developed by Zuntz in 1897 and revised by Forbes and others in the intervening years (Brouwer, 1957). A modification for methane losses was added (Brouwer, 1958) and the equation adopted by the sub-committee in 1965 (Brouwer, 1965) has been universally accepted and used without question by many:

$$\text{HE (MJ)} = 16.18(O_2) + 5.16(CO_2) - 5.90(UN) - 2.42(CH_4)$$

In the equation, gases are expressed in litres and UN is expressed in grams. The measurement of respiratory exchange is usually obtained whilst animals are housed in a chamber, and two approaches can be used. In the closed-circuit respiration chamber, an airtight system is maintained with chamber air circulated through scrubbers for removing CO_2 and H_2O, with O_2 introduced into the system. In this system, O_2 use is based on the required input, whilst CO_2 production is obtained from the change in weight of absorbent. These systems are not convenient for measurements of methane production or for use with large animals, thus most measurements of respiratory exchange for large animals are obtained using open-circuit respiration chambers. In open-circuit systems respiratory exchange is based on the difference in the concentration of gases entering the chamber in outside air and leaving in chamber exhaust, which is then multiplied by flow rate through the chambers after correction for temperature, humidity and pressure. Chambers are operated under negative pressure, thus do not have to be absolutely air tight, although changes in the gas concentration of air in the room housing the chamber are a concern for a leaky chamber. Modern infra-red analysis of CO_2 and CH_4 and paramagnetic analysis of O_2 has greatly reduced the labour required for measuring gas concentrations, and the 'flow-through' open-circuit respiration chamber is used widely for measurements of energy metabolism in farm animals. Respiratory exchange can also be measured using head chambers, face masks, tracheal cannulas or mouthpieces (McLean and Tobin, 1987). Brody (1945) used face masks extensively for measurements of respiratory exchange in farm animals.

Measurement of Recovered Energy

The second indirect approach is to estimate RE based on measurements of total carbon and nitrogen balance (Blaxter, 1967). This approach is based on the assumption that energy is stored in the body as fat or protein, with minimal change in body carbohydrate (glycogen) stores in the long term. Protein deposition is estimated from body nitrogen retention, whilst fat storage is estimated from body carbon retention after correction for carbon storage as protein. The amount of energy retained as fat and protein is then estimated using factors derived primarily from muscle analysis, although other approaches have been used (Flatt, 1969). As for estimates of HE by respiratory exchange, formulae for calculating energy balance from measurements of carbon and nitrogen balance were recommended by the Energy Symposium subcommittee on constants chaired by Brouwer (1965). This approach requires the measurement of CO_2 and CH_4 production, but not O_2 consumption, and separates energy retention into fat and protein components. The approach was used widely in the past, but as carbon analysis is required on all inputs and outputs for the animal, adds to the analytical burden of estimating energy balance. Difficulties and cost of carbon analysis compared to measurement of gross energy content using a bomb calorimeter have made the estimation of heat production the preferred indirect approach in recent years.

Another indirect approach used widely in recent years, especially in humans, is to use isotope dilution procedures to estimate respiratory exchange. One approach is to estimate body CO_2 production using dilution of labelled CO_2 in blood. In the other approach, known as the 'doubly labelled water' technique, the turnover of H_2 and O_2 is estimated by following the concentration of $^2H_2$ and ^{18}O in urine after an injection of $^2H_2^{18}O$. The difference in their rate of turnover is proportional to the rate of CO_2 production, as H_2 is eliminated as H_2O, whilst O_2 is eliminated as both H_2O and CO_2. The merits and limitations of these approaches have been widely discussed in the literature (McLean and Tobin, 1987). The major advantage of dilution approaches is that subjects do not have to be confined to a stationary or portable respiration apparatus, thus effects of normal activity on energy metabolism can be included. They also require fewer measurements, but are inherently less accurate than direct measurements of respiratory exchange.

Comparative Slaughter Balance

Another approach for estimating energy retention, which has been applied primarily to growing animals, is to compare the total energy content of groups of animals before and after a sufficient experimental period. The composition of the initial slaughter group is then assumed to be equal to the initial composition of animals slaughtered at the end of the experiment, and energy retention is calculated as the difference in total body energy content between the two groups (Flatt, 1969; Blaxter, 1989). As the name implies, this requires slaughter and analysis of the energy content of a representative sample of the total carcass, thus within animal comparisons are impossible and larger numbers of animals are required to account for animal variation. The approach is very precise for smaller animals where the entire carcass can be processed, but carcass processing can be difficult in larger ruminants. Companion digestion trials are also conducted to determine DE and UE, but for ruminants, methane losses (and thus ME) are often estimated from DE and not measured directly.

The difficulty of conducting slaughter balance studies in larger animals, the terminal nature of the approach, and the inability to apply the technique in humans, has led to the development of a plethora of indirect methods for estimating body composition which do not require the slaughter of experimental subjects. The basis of many of these approaches is that there is an inverse relationship between body fat and water and that within species the proportions of water, protein and ash in the fat-free empty body can be predicted based on the results of large scale slaughter trials (Reid *et al.*, 1968). Thus if body water or fat content can be estimated, the proportions of the other components can be predicted. The energy content of the body can then be estimated as for the carbon and nitrogen balance technique. A major stumbling block for these approaches is the contributions of gut fill and water to empty body weight, which can be extremely large in ruminants. This makes estimation of empty body weight difficult, but also compromises the use of dilution techniques to estimate body water content (Flatt, 1969). Approaches used to estimate body water content generally involve measurement of the dilution of injected substances that are rapidly and uniformly distributed in body water, wherein lies the problem with gut water contents for ruminants. A variety of compounds have been used, but of those listed by Flatt (1969), deuterium and urea have seen the most attention from animal scientists in recent years (e.g. Andrew *et al.*, 1995). Attempts have been made to address the problem of gut water content by using multiple pool models to relate the dilution of the marker to measured body water in validation studies. However, the resulting equations have not proved accurate in practice (Crooker *et al.*, 1998) and there has not been widespread adoption of these approaches. Although measurements of the body composition of dairy cows at various stages of lactation are needed, and are extremely costly to obtain directly, the flux of water through the gut and mammary gland of a high yielding cow makes the application of these approaches especially difficult. As already mentioned for estimates of CO_2 production, dilution procedures for estimating body composition are inherently less accurate and precise than direct approaches.

Another approach for estimation of carcass water content is the estimation of specific gravity by underwater weighing procedures (Flatt, 1969). Specific gravity can then be used to estimate body water content, which is then used to predict fat, protein and ash content and thus their energy value. This approach was used extensively in the development of the California Net Energy system for rationing beef cattle (Lofgreen and Garret, 1968; NRC, 1996). A major advantage of the approach used is that the system was based on measurements from animals fed under normal industry conditions, rather than the artificial and restrained environment of a respiration chamber.

Other approaches have involved the prediction of body fat from absorption of marker compounds or the prediction of body protein content from estimated body K content. The list of procedures used is long, but there are a number of procedures currently in use for the estimation of body composition based on recently developed technologies, such as nuclear magnetic resonance, CAT scans, ultrasound scanning or, more recently, dual energy X-ray absorption (Geers *et al.*, 1998). The cost of many of these procedures, the need for subjects to remain perfectly still and the design of the systems for use in humans makes the application of these technologies to larger, less cooperative farm animals difficult. But they have been used to predict the composition of smaller (often anaesthetized) animals or their carcasses. In addition, ultrasound scanning is used widely in animal agriculture and now used to estimate changes in body fat content based on measurements of subcutaneous fat depth, as an adjunct to visual condition scoring approaches.

COMPARISON OF APPROACHES

Comparisons of results from simultaneous measurements of HE and RE obtained using 'balance trials' (direct or indirect calorimetric measurements of HE or RE from carbon and nitrogen balance) were extensive in the early part of this century. At the time, many calorimeters were constructed for the simultaneous use of more than one of these techniques. On the whole, these approaches yielded very similar results when the techniques were rigorously applied and experimental errors were minimized (Blaxter, 1967). Indeed, differences between measured and calculated energy balance from 129 measurements in the Armsby calorimeter resemble a bell-shaped curve

(Blaxter, 1967). Blaxter concluded that 'there is no reason to suppose' there was any significant inaccuracy or bias in the balance trial approach to measuring energy metabolism. At the time, there were few comparisons of results from energy balance measurements with those obtained using comparative slaughter trials. Comparisons made in chickens have shown good agreement between the two approaches (Blaxter, 1967; McDonald *et al.*, 1995). In contrast, direct comparisons of the effects of specific feeds on energy metabolism in ruminants have found that measurements obtained using respiration calorimetry have yielded consistently higher estimates of energy retention and k_g than those obtained using slaughter balance (e.g. Waldo *et al.*, 1990; Webster, 1989). This is similar to the bias observed for measurements of nitrogen retention obtained using short-term balance trials, which are frequently higher than direct measures of nitrogen retention by 20% or more (Johnson, 1986). Indeed, for this reason some workers use a correction factor to account for this bias in short-term measurements of nitrogen balance. There are a number of reasons for these discrepancies. First, errors of measurement in digestion trials are cumulative and thus all errors are included in the variable calculated by difference, which is tissue energy or protein retention (Johnson, 1986). Therefore, if any feed, faeces or urine is unaccounted for, the loss is assumed to be included in body tissue. Urine is acidified to prevent ammonia volatilization, but some losses from urine or faeces are unavoidable. Frequent scraping and attention to faecal collections will greatly reduce this error. In the Beltsville respiration chambers the residual faeces accumulating on collection equipment is accounted for by measuring the energy and nitrogen content of an initial wash (a wet scraping) of the chamber and faecal collection equipment. In addition, hair, scurf and spilled feed are collected from the floor of the chamber and analysed. Other losses can occur after sampling, and great care must be taken to avoid losses of volatile ammonia and energy during sample storage, processing and analysis.

In addition to the accumulation of errors of measurement in balance trials, another consideration when comparing them to comparative slaughter trials is the fact that the animals are restrained in respiration chambers or digestion stalls, thus they have limited activity other than standing and changing position. In addition, the environment of the chamber is controlled. Therefore the energy lost in activity and, to a lesser extent, temperature regulation is reduced. This is one reason energy retention tends to be lower in slaughter balance trials conducted under practical conditions. In addition, there may be interactions between intake level and activity which differ for the two techniques (Webster, 1989).

ADDITIONAL CONSIDERATIONS

Potential sources of error in measurements of energy metabolism are well documented (e.g. Blaxter, 1967, 1971; Johnson, 1986; McLean and Tobin, 1987). The original publication of what has come to be known as the Brouwer equation contained a number of disclaimers (Brouwer, 1957, 1958), which should not be ignored by the users. First, the statement that measurements were obtained 'in a not too short experimental period' preceded each list of equations. The primary concern was that short-term changes in body temperature or blood and tissue CO_2 concentrations would compromise the validity of the equation. In most cases the measurements obtained represent daily rates of exchange, typically averaged over a number of days. Certainly this is 'not too short an experimental period'. Adjustments were also included for H_2 production and hippuric acid excretion in urine, but they are seldom used. In addition, the disclaimer specifies that under physiological states leading to incomplete combustion (specifically ketosis), or when specific nutrients such as sucrose or ethanol are oxidized, the equation should be revised (McLean and Tobin, 1987). An additional provision was that the equation should not be used if the respiratory quotient (RQ) was outside the range of 0.707 and 1.00 (NRC, 1935). Another consideration is that the equation was developed using UN as an indicator of the amount of protein oxidized, whilst in ruminants fed excessive amounts of rumen degradable protein a large portion of urinary nitrogen is derived directly from ammonia absorbed into the portal vein. However, errors of UN have a relatively minor effect on HE compared to errors in airflow and O_2 concentration measurement (Johnson, 1986). In spite of these concerns, the use of alternative equations, derived using other approaches or reference compounds, do not have a dramatic effect on calculated HE (Blaxter, 1967).

Another potential source of error in digestion trials is the use of bladder catheters to collect urine in females. In males, a soft collection funnel suspended from the belly and evacuated by a vacuum is an excellent approach (Varga *et al.*, 1990) and preferred over the use of a metabolism crate. No matter how carefully and aseptically they are established, bladder catheters are a potential source of irritation to the urinary tract which can in many animals increase the volume of urine produced and nitrogen excreted, in addition to the general effects of distress and immune response. The use of urine collection devices attached to the genital region (e.g. Fellner *et al.*, 1988), in combination with plastic chutes for distribution of faeces into collection vessels (see Morrison, 1954), can with experience provide an excellent separation and collection of urine and faeces in females (Sutton *et al.*, 1997). While not completely irritation free for the animal, the system is much preferred to the use of bladder catheters, which in the author's opinion should be avoided vigorously. Although acetone can be used for loosening the cement and removing the collection device without hair loss, the major drawback of this approach is that it should not be used at intervals too frequent to allow adequate hair growth for attaching the device to the genital area.

Other considerations for the use of respiration calorimetry include the adaptation of animals to facilities prior to experimentation, the separation of respiratory exchange and digestion trials, and environmental control. Adaptation of animals to calorimeters is critical to avoid depressions in intake, milk yield and nervous behaviour, all of which can dramatically alter energy metabolism. Nervous behaviour may be obvious in animals that do not settle and are constantly bawling or refuse to lie down, or may be less apparent in some individuals. Regardless, heat energy can be elevated in animals that are not adapted. Having more than one chamber and windows allowing animals in adjacent chambers to see each other can reduce the stress of confinement for many animals. The subject of adaptation and stress of confinement has been the subject of much discussion at the EAAP Energy Symposia (Table 5.1).

Depending on the construction of respiration chambers, it may not be possible to obtain measurements of respiratory exchange, digestion and urine output simultaneously. If these measurements are obtained separately, then care should be taken to insure that conditions under which the measurements are obtained are as similar as possible. Ideally the animals will be housed under environmental conditions which are similar to those of the chambers. Alternatively, to avoid dramatic changes the temperature of the chambers may be adjusted to that of the housing in which experimental subjects are maintained when they are not in the chambers. Environmental control can be a particular problem in ruminants, and particularly lactating dairy cows that lose large amounts of water through respiration.

Variations in intake can have immediate effects on respiratory exchange, thus if intake is reduced when animals enter the respiration chambers then measurements of heat production will not be quantitatively comparable to measurements of feed digestion and urine nitrogen output. This has serious consequences for the calculation of HE and energy balance. For this reason intakes may be set below *ad libitum* for a period of time prior to measurements, but this will change the physiological state of the animal. This is especially a problem in the conduct of energy metabolism studies with lactating dairy cows, where restriction of intake below *ad libitum* can influence the remainder of the lactation curve and the response to dietary perturbations (Blaxter, 1956). Ideally, measurements of DE and HE should be obtained simultaneously to avoid disparities in intake and other conditions during the measurements.

TISSUE 'CALORIMETRY'

The combination of multiple techniques for measuring nutrient metabolism was suggested as an approach which would provide important insights into the mechanisms underlying production responses to variation in diet composition, intake and physiological state in ruminants (Annison, 1965). This view has been echoed repeatedly (Moe, 1981; Webster, 1989). The combination of measurements of HE and RE using calorimetry with measurements of the metabolism of specific nutrients (e.g. using isotopic labelling) or specific tissues has provided important insights into the processes underlying energetic responses to nutrition or changes in physiological state. Measurements of the contribution of individual tissues to body O_2 consumption can be obtained *in vivo* by the use of multicatheterization procedures (Huntington *et al.*, 1989). Surgical placement of chronic, in-dwelling catheters enable the measurement of blood flow and venous–arterial concentration difference for O_2, CO_2 and other nutrients and metabolites across specific tissues. The net removal of O_2

from blood or CO_2 release into blood can then be calculated. Combination of these measurements with measurements of body respiratory exchange have shown that the tissues drained by the hepatic portal vein, the portal-drained viscera (PDV) and liver each account for roughly 20 to 25% of body O_2 consumption, whilst accounting for less than 13% of body mass (Reynolds, 1994). Although the prediction of body HE from O_2 consumption is based on measurements for the whole body (McLean, 1972), the relationship has been used to estimate HE by body tissues. Alternatively, thermocouples have been used to measure transfer of heat into the portal vein directly, accounting for the contribution of H_fE. These measurements agreed reasonably well with measurements of PDV HE based on O_2 consumption (Webster *et al.*, 1975). The high rate of O_2 consumption by these tissues highlights their importance to the maintenance requirement and energy balance of the animal, as well as the energetic response to changes in diet composition and intake (Reynolds *et al.*, 1991). Interpretation of CO_2 production rates by these tissues is compromised by the fact that metabolic processes in the liver use CO_2, whilst CO_2 absorbed into the portal vein can be a product of fermentation or arise from salivary bicarbonate. In addition, CO_2 can also be transferred from blood to the lumen of the gut, and vice versa (Hoernicke *et al.*, 1965). For these reasons measurements of CO_2 production by the PDV and liver vary considerably and CO_2 removal is sometimes measured when very rapid sampling is employed. Thus measurements of tissue RQ, especially for the PDV, must be interpreted with extreme caution.

FUTURE DIRECTIONS

It has been suggested that measurements of fasting metabolism and k_m are irrelevant to the energy metabolism of animals at production intakes and their importance in estimating maintenance requirements over emphasised (Webster *et al.*, 1974; Webster, 1989). Similarly, it has been suggested that there has been too much emphasis on obtaining measurements of ME_m and k_g in growing animals, as these terms have no absolute meaning, but are simply components of the linear regression of recovered energy on ME (Webster, 1989; Table 5.1). In comparing effects of diet or physiological state on energy metabolism, if at equal ME one treatment results in a higher HE than another, then the increase in HE may be the result of a higher maintenance requirement, a reduced efficiency of ME use for production of recovered energy, or both. In many cases, the reduction in efficiency may be due to an increase in the mass of metabolically active tissues such as the gut or liver. Does this increase represent a maintenance cost or a production cost? For measurements of whole body HE, the design of the trial and the mathematical description of the results have in the past determined the answer. Approaches other than linear regression have been used to resolve energy balance measurements in the past, and in the future more emphasis on the use of alternative models and more enlightened approaches are needed (Moe, 1981).

There are many, and many of them with considerable experience in the field of energy metabolism, who believe measurements of energy metabolism using classical approaches are nearing (or well past) the end of their useful life. Certainly a limitation, but also a strength, of current feeding systems based on measurements of energy metabolism is that DE or ME is not a nutrient *per se*, but the sum of a number of processes resulting in the assimilation of specific energy yielding nutrients. For the past 30 years those involved in the development of the feeding systems in use today have recognized the need for feeding systems based on a clearer understanding of the role of specific absorbed nutrients and their metabolism in determining productive responses of farm animals (Moe, 1981; Webster, 1989). Forty years ago animal nutritionists were criticizing the day's feeding standards, and worked hard to improve them. Today the current energy feeding standards, which were built on the back of 200 years of energy metabolism research, are being challenged and criticized, largely for their inability to predict productive responses. For dairy cattle, a major concern is the ability to predict the partition of ME use between milk and body tissue. Newer systems based on models of digestion are now in use and being refined, but they are being used to predict ME, not specific energy yielding substrates (Sniffen *et al.*, 1992). In practice, the ability to predict the absorption of specific nutrients and their metabolism may be limited by the ability to obtain adequate measurements of the food character-istics needed to 'feed' predictive models. More mechanistic models of nutrient absorption and metabolism are also in use, but need more refinement to achieve practical application (e.g. Chapter 14). Any new feeding system must be flexible and adaptable to the circumstances confronting the user. Today's nutritionist has access to a variety of rationing systems, and may use different systems depending on the application. In addition, many develop their own 'customized' feeding system using components of individual systems with modifications based on experience. Depending on the end user, new rationing systems should allow that flexibility, but in today's research environment copy-right restrictions may limit this versatility.

FINALLY

In calling the Conference on Energy Metabolism (NRC, 1935), P.E. Howe, the chairman of the Committee on Animal Nutrition, presented the following statement of the problem, which is reproduced verbatim:

> Studies of energy, energy metabolism, and efficiency of feed utilization have a relationship to agriculture in establishing:
>
> (a) Fundamental concepts of the energy requirements of animals of different ages, sexes, and conditions of production, including work
> (b) Fundamentals of the utilization of feed, the nutritive elements in feed and the interrelation of the various feed stuffs.
> (c) The characteristics of animals.

The last 65 years has seen considerable progress in addressing these issues, and improvements in feeding standards. However, future rationing systems will benefit from a greater insight into the effects of nutrition on the utilization of specific energy yielding nutrients within the body. Perhaps more importantly, the ability to predict responses and the partition of absorbed nutrients will only be achieved by an enlightened representation of the characteristics of animals which determine their productive response to feeds, and applicable measurements of the components of feed that determine those responses.

REFERENCES

Andrew, S.M., Erdman, R.A. and Waldo, D.R. (1995) Prediction of body composition of dairy cows at three physiological stages from deuterium oxide and urea dilution. *Journal of Dairy Science* 78, 1083–1095.

Annison, E.F. (1965) Absorption from the ruminant stomach. In: Dougherty, R.W. (ed.), *Physiology of Digestion in the Ruminant*. Butterworths, London, pp. 185–197.

ARC (1965) *The Nutrient Requirements of Farm Livestock. No.2. Ruminants*. Agricultural Research Council, London, UK.

Armsby, H.P. (1917) *The Nutrition of Farm Animals*. Macmillan, New York.

Australian Standing Committee on Agriculture (1990) *Feeding Standards for Australian Livestock: Ruminants*. CSIRO, Melbourne.

Blaxter, K.L. (1956) Starch equivalents, ration standards and milk production. *Proceedings of the British Society of Animal Production* 1956, 3–31.

Blaxter, K.L. (1967) Techniques in energy metabolism studies and their limitations. *Proceedings of the Nutrition Society* 26, 86–96.

Blaxter, K.L. (1971) Methods of measuring the energy metabolism of animals and interpretation of results obtained. *Federation Proceedings* 30, 1436–1443.

Blaxter, K.L. (1986) An historical perspective: the development of methods for assessing nutrient requirements. *Proceedings of the Nutrition Society* 45, 177–183.

Blaxter, K.L. (1989) *Energy Metabolism in Animals and Man*. Cambridge University Press, Cambridge, UK.

Blaxter, K.L. and Graham, N.McC. (1955) Methods of assessing the energy values of foods for ruminant animals. *Proceedings of the Nutrition Society* 14, 131–139.

Brody, S.A. (1945) *Bioenergetics and Growth*. Hafner Publ., New York.

Brouwer, E. (1957) On simple formulae for calculating the heat expenditure and the quantities of carbohydrate and fat oxidized in metabolism of men and animals, from gaseous exchange (oxygen intake and carbonic acid output) and urine-N. *Acta Physiology and Pharmacology, Neerlandica* 6, 795–802.

Brouwer, E. (1958) On simple formulae for calculating the heat expenditure and the quantities of carbohydrate and fat metabolized in ruminants, from data on gaseous exchange and urine-N. In: Thorbeck, G. and Aersoe, H. (eds) *Energy Metabolism of Farm Animals*. EAAP Publ. No. 8. Statens Husdyrugsudvalg, Copenhagen, pp. 182–194.

Brouwer, E. (1965) Report of the sub-committee on constants and factors. In: Blaxter, K.L. (ed.) *Energy Metabolism of Farm Animals*. EAAP Publ. No. 11. Academic Press, London, pp. 441–443.

Crooker, B.A., Weber, W.J. and Andrew, S.M (1998) Development and use of deuterium oxide dilution equations to predict body composition of Holstein cows. In: McCracken, K.J., Unsworth, E.F. and Wylie, A.R.G. (eds), *Energy Metabolism of Farm Animals*. CAB International, Wallingford, Oxon, UK, pp. 177–180.

Fellner, V., Weiss, M.F., Belo, A.T., Belyea, R.L., Martz, F.A., and Orma, H. (1988) Urine cup for collection of urine from cows. *Journal of Dairy Science* 71, 2250–2255.

Flatt, W.P. (1969) Methods of calorimetry (B) Indirect. In: Cuthbertson, D. (ed.) *The International Encyclopedia of Food and Nutrition. Volume 17: Nutrition of Animals of Agricultural Importance. Part 2*. Pergamon Press, Oxford, pp. 1–30.

Flatt, W.P., Van Soest, P.J., Sykes, J.F. and Moore, L.A. (1958) A description of the Energy Metabolism Laboratory at the U.S. Department of Agriculture, Agricultural Research Centre in Beltsville, Maryland. In: Thorbeck, G. and Aersoe, H. (eds), *Energy Metabolism of Farm Animals*. EAAP Publ. No. 8. Statens Husdyrugsudvalg, Copenhagen, pp. 53–64.

Geers, R., De Pauw, B., Spincemaille, G., Ville, H., Vits, J., Rombouts, G., Duchateau, W. and Perremans, S. (1998) The integration of dexa, ultrasound scans and growth modelling technology to estimate body composition in growing pigs. In: McCracken, K.J., Unsworth, E.F. and Wylie, A.R.G. (eds), *Energy Metabolism of Farm Animals*. CAB International, Wallingford, Oxon, UK, pp. 201–204.

Harris, L.E. (1966) *Biological Energy Interrelationships and Glossary of Energy Terms*. Publ. 1411, National Academy of Sciences, Washington, DC.

Hoernicke, H., Williams, W.F., Waldo, D.R. and Flatt, W.P. (1965) Composition and absorption of rumen gases and their importance for the accuracy of respiration trials with tracheostomized ruminants. In: Blaxter, K.L. (ed.), *Energy Metabolism of Farm Animals*. EAAP Publ. No. 11. Academic Press, London, pp. 165–178.

Huntington, G.B., Reynolds, C.K. and Stroud, B. (1989) Techniques for measuring blood flow in the splanchnic tissues of cattle. *Journal of Dairy Science* 72, 1583–1595.

Johnson, D.E. (1986) Fundamentals of whole animal calorimetry: use in monitoring body tissue deposition. *Journal of Animal Science* 63 (Suppl. 2), 111–114.

Kellner, O. (1926) *The Scientific Feeding of Animals*. Duckworth, London.

Lofgreen, G.P. and Garrett, W.N. (1968) A system for expressing net energy requirements and feed values for growing and finishing cattle. *Journal of Animal Science* 27, 793–806.

McDonald, P., Edwards, R.A., Greenhalgh, J.F.D. and Morgan, C.A. (1995) *Animal Nutrition*, 5th edn. Longman Scientific & Technical, Harlow, Essex, UK.

McLean, J.A. (1972) On the calculation of heat production from open-circuit calorimetric measurements. *British Journal of Nutrition* 27, 597–600.

McLean, J.A. and Tobin, G. (1987) *Animal and Human Calorimetry*. Cambridge University Press, Cambridge, UK.

Moe, P.W. (1981) Energy metabolism of dairy cattle. *Journal of Dairy Science* 64, 1120–1139.

Moe, P.W. and Tyrrell, H.F. (1972) Metabolizable energy requirements of pregnant dairy cows. *Journal of Dairy Science* 55, 480–483.

Moe, P.W. and Tyrrell, H.F. (1973) The rationale of various energy systems for ruminants. *Journal of Dairy Science* 37, 183–189.

Moe, P.W. and Tyrrell, H.F. (1975) Efficiency of conversion of digested energy to milk. *Journal of Dairy Science* 58, 602–610.

Moe, P.W., Tyrrell, H.F. and Flatt, W.P. (1971) Energetics of body tissue mobilisation. *Journal of Dairy Science* 54, 548–553.

Moe, P.W., Flatt, W.P. and Tyrrell, H.F. (1972) Net energy value of feeds for lactation. *Journal of Dairy Science* 55, 945–958.

Morrison, F.B. (1954) *Feeds and Feeding. A Handbook for the Student and Stockman,* 21st edn. Morrison Publ. Co., Ithaca, New York.

NRC (1935) *Report of the Conference on Energy Metabolism, State College, Pennsylvania, 14–15 June 1935, under the Auspices of the Committee on Animal Nutrition.* National Research Council, Washington, DC.

NRC (1981) *Nutritional Energetics of Domestic Animals and Glossary of Energy Terms,* 2nd edn. National Academy Press, Washington, DC.

NRC (1989) *Nutrient Requirements of Dairy Cattle,* 6th edn. National Academy Press, Washington, DC.

NRC (1996) *Nutrient Requirements of Beef Cattle,* 7th edn. National Academy Press, Washington, DC.

Reid, J.T., Bensadoum, A., Bull, L.S., Burton, J.H., Gleeson, P.A., Han, I.K., Joo, Y.D., Johnson, D.E., McManus, W.R., Paladines, O.L., Stroud, J.W., Tyrrell, H.F., Van Niekerk, B.D.H., and Wellington, G.W. (1968) Some peculiarities in the body composition of animals. In: Reid, J.T., Breidenstein, B.C., Hansard, S.L., Stonaker, H.H. and Zobrisky, S.E. (eds), *Body Composition in Animals and Man.* Publication 1598, National Academy of Sciences, Washington, DC, pp. 19–44.

Reynolds, C.K., Tyrrell, H.F. and Reynolds, P. J. (1991) Effects of forage-to-concentrate ratio and intake on energy metabolism in growing beef heifers: whole body energy and nitrogen balance and visceral heat production. *Journal of Nutrition* 121, 994–1003.

Reynolds, C.K. (1994) Quantitative aspects of liver metabolism in ruminants. In: Engelhardt, W.V., Leonhard-Marek, S., Breves, G. and Giesecke, D. (eds), *Ruminant Physiology: Digestion, Metabolism, Growth and Reproduction: Proceedings of the Eighth International Symposium on Ruminant Physiology.* Ferdinand Enke Verlag, Stuttgart, Germany.

Sniffen, C.J., O'Connor, J.D., Van Soest, P.J., Fox, D.G. and Russell, J.B. (1992) A net carbohydrate and protein system for evaluating cattle diets: II. Carbohydrate and protein availability. *Journal of Animal Science* 70, 3562–3577.

Sutton, J.D., Abdalla, A.L., Phipps, R.H., Cammell, S.B. and Humphries, D.J. (1997) The effect of the replacement of grass silage by increasing proportions of urea-treated whole-crop wheat on food intake and apparent digestibility and milk production by dairy cows. *Animal Science* 65, 343–351.

Tyrrell, H.F. and Moe, P.W. (1975) Effect of intake on digestive efficiency. *Journal of Dairy Science* 58, 1151–1163.

Tyrrell, H.F., Moe, P.W. and Flatt, W.P. (1970) Influence of excess protein intake on energy metabolism of the dairy cow. In: Schurch, A. and Wenks, C. (eds), *Energy Metabolism of Farm Animals.* EAAP Publ. No. 13. Juris Verlag, Zurich, Switzerland, pp. 69–72.

Van Es, A.J.H. (1994) The symposia on energy metabolism of farm animals of the EAAP. In: *Energy Metabolism of Farm Animals.* EAAP Publ. No. 76. CSIC Publishing Service, Madrid, Spain, pp. 409–418.

Van Soest, P.J. (1982) *Nutritional Ecology of the Ruminant.* O & B Books, Inc., Corvallis, Oregon, USA.

Varga, G.A., Tyrrell, H.F., Huntington, G.B., Waldo, D.R. and Glenn, B.P. (1990) Utilization of nitrogen and energy by Holstein steers fed formaldehyde- and formic acid–treated alfalfa or orchardgrass silage at two intakes. *Journal of Animal Science* 68, 3780–3791.

Vermorel, M. and Coulon, J.B. (1998) Comparison of the national research council energy system for lactating cows with four European systems. *Journal of Dairy Science* 81, 846–855.

Waldo, D.R., Varga, G.A., Huntington, G.B., Glenn, B.P. and Tyrrell, H.F. (1990) Energy components of growth in Holstein steers fed formaldehyde- and formic acid treated alfalfa or orchardgrass silages at equalized intakes of dry matters. *Journal of Animal Science* 68, 3792–3804.

Webster, A.J.F. (1989) Energy utilisation during growth and reproduction–discussion. In: *Energy Metabolism of Farm Animals*. EAAP Publ. No. 43. Pudoc, Wageningen, Netherlands, pp. 85–88.

Webster, A.J.F., Brockway, J.M. and Smith, J.S. (1974) Prediction of the energy requirements for growth in beef cattle. 1. The irrelevance of fasting metabolism. *Animal Production* 19, 127–139.

Webster, A.J.F., Osuji, P.O., White, F. and Ingram, J.F. (1975) The influence of food intake on portal blood flow and heat production in the digestive tract of sheep. *British Journal of Nutrition* 34, 125–139.

6 Feeding Systems for Dairy Cows

S. Tamminga and G. Hof

Animal Nutrition Group, Wageningen Institute of Animal Sciences, Wageningen Agricultural University, Wageningen, The Netherlands

INTRODUCTION

Production of milk by dairy cows in Western Europe and North America has reached extremely high levels, and annual yields of 12,000 kg per animal are no longer exceptional. With a normal lactation curve, this means a peak production of 55–60 kg at between 30 and 50 days post-partum. Normal levels of feed intake are highly inadequate to meet nutrient requirements for such levels of production. Hence, sophisticated feeding systems and strategies are required to maintain a normal lactation curve with a good persistency after peak production in which many aspects have to be taken into account. There is also a growing interest in ways to manipulate milk composition and for which special feeding strategies are needed.

Lactating dairy cows are characterized by a continuously changing stage of lactation. In practice, this means that the dairy farmer has to deal with trying to exploit the genetic potential of the cows by feeding the right amount of the right feeds at each lactation stage. In order to adapt the feeding strategy to these changes, a dairy farmer has to comply with numerous constraints, which relate to the animal as well as the feed. Examples of such constraints are the feed intake capacity of the animal, the energy and protein content and composition of the feed, and the required forage to concentrate ratio. A combination of all such constraints can be called a feeding system, and such a system may comprise well-defined as well as poorly defined elements. Despite their limited value as predictors of responses to changes in dietary inputs, feed evaluation for energy and protein can be called well-defined elements. Less well-defined elements are factors such as feed intake or the optimal ratio between structural and non-structural compounds needed for optimal rumen fermentation.

Feed evaluation for dairy cows has a long history. Important milestones in this history were the introduction of the concept of hay equivalents by

Albrecht Thaer (1809), followed by the introduction of the Weende analysis by Henneberg and Stohmann (1864), the introduction of the starch equivalent in Europe (Kellner, 1912), the Scandinavian fodder unit (Mølgaard, 1929) and the concept of total digestible nutrients in the USA (Morrison, 1936). Between 1970 and 1980, various European countries introduced systems for energy based on the results of energy balance trials in dairy cows instead of fat deposition in growing animals. Most European and North American countries have now opted for the net energy of lactation (NEL) or metabolizable energy (ME) (Van der Honing and Alderman, 1988).

Evaluation of protein has a similar history. Initially, crude protein, based on N × 6.25, was used, followed, about a century ago, by digestible crude protein and, more recently, the concepts of metabolizable protein or absorbed protein), i.e. protein absorbed from the small intestine, have been introduced.

Apart from feed intake and feed evaluation, diets for dairy cows are formulated on the basis of additional guidelines and constraints. Examples are substitution of forages by concentrates, type and physical characterization of carbohydrates, rumen fill values and interactions between energy, proteins and carbohydrates. In this chapter, feeding systems for dairy cattle presently in use and their limitations will be described and discussed.

FEED INTAKE

Many attempts have been made to predict feed and dry matter intake (DMI) in ruminants, including dairy cows. In a recent review, Ingvartsen (1994) lists about 40 prediction equations for dairy cattle and some 20 for fattening cattle. The more complex equations contain animal factors, such as milk yield and body weight, food (quality) factors such as dry matter digestibility, or a factor representing rumen fill, as variables in their prediction of DMI.

Minson (1990) presented a large number of general factors influencing forage intake by housed as well as grazing ruminants. Factors considered important for dairy cows were pregnancy, lactation and body condition as animal factors and heat or cold stress as an environmental factor. Forage factors for housed animals were forage species, cultivar, stage of growth, soil fertility, climatic condition and conservation method. In grazing animals, essentially the same factors were important and, in addition, selection by the animals and availability (sward density) of the forage. Van Soest (1994) suggests that the regulation of intake is an integration of various factors in the metabolic system of the animal, a view largely supported by Forbes (1995), but challenged by others (Ketelaars and Tolkamp, 1992a,b; Tolkamp and Ketelaars, 1992). The latter authors propose and give evidence that feed intake in ruminants is under metabolic control, involving an optimal ratio of oxygen consumed to available net energy.

More specific factors which have been observed to limit feed intake in ruminants are fill and distension of the rumen (Van Soest, 1994; Forbes, 1995); the amount of rumen inert bulk present in the rumen (Dado and Allen, 1995);

concentrations of acetate in the rumen, with receptors in the rumen wall (Forbes, 1995); osmolality in the rumen (Grovum, 1995); and concentrations of propionate in the liver (Forbes, 1995). Forbes (1995) also presents evidence that many of these factors are additive. Daily feed intake is the sum of the amount eaten in individual meals, and one can therefore argue that feed intake is determined by intake behaviour (Gill and Romney, 1994). Bite size, bite rate and eating time may also become factors limiting daily feed intake, particularly under grazing conditions (Chilibroste *et al.*, 1997).

According to Dulphy and Demarquilly (1994), intake in ruminants is dominated by two factors, the ingestibility of the fed forage and the intake capacity of the animal. In order to account for differences in voluntary intake caused by differences in ingestibility 'fill units' were defined in France (Jarrige *et al.*, 1986) and in Denmark (Kristensen and Ingvartson, 1986). To predict intake, INRA in France uses fill units for sheep, cattle and dairy cows derived from intakes of a standard feed of 75, 95 and 140 g of DM kg^{-1} $W^{0.75}$, respectively (Dulphy and Demarquilly, 1994). Some typical fill values are given in Table 6.1, the variation between the ratios indicates that measuring the ingestibility of feeds is difficult because of the numerous factors involved in its regulation.

An important and intriguing question is whether feed intake controls, i.e. 'pushes', milk production or whether milk production controls, i.e. 'pulls', feed intake. The lack of an adequate answer to this question is probably one of the main reasons why the regulation of feed intake in dairy cows is still poorly understood (Gill and Romney, 1994). In early lactation, milk production seems to develop independently of feed intake; under these circumstances milk production is pulling feed intake rather than feed intake pushing milk production (De Visser, 1993).

Interesting in this respect are the observations that when bovine somatotropin is administered to dairy cows, milk production increases almost immediately whereas feed intake, if affected at all, follows after 2–3 weeks, sometimes even after an initial slight decline. A similar situation occurs when the frequency of milking is increased from two to three times per day. The milk production response is quite rapid but the increased feed intake lags several weeks behind, which suggests that under such conditions milk production is also pulling feed intake, probably mediated by a diverted blood flow towards the mammary gland (Bequette and Backwell, 1997).

Table 6.1. 'Fill values' for dairy cows as used in France (FVF) and Denmark (FVD).

Feed	FVF	FVD	Ratio
Untreated barley straw	1.60	0.85	0.53
NH_3-treated barley straw	1.15	0.70	0.60
Grass silage	1.20	0.45	0.38
Grass silage	1.25	0.50	0.40
Grass silage	1.35	0.55	0.41
Dried beet pulp	1.05	0.20	0.19
Fodder beets	0.60	0.25	0.42

After having reached peak production, feed intake seems to push milk production. This can be concluded from the findings of Broster (1972) who observed a residual effect on milk production in later lactation of a high peak yield, resulting from high concentrate allowances. Observations made by Østergard (1979) showed, however, that flat feeding for a longer period around peak yield might result in the same total milk production.

SUBSTITUTION

Forages are important ingredients in dairy diets but, because of their relatively low energy density and high fill value resulting in limited intake, forages alone are usually insufficient to maintain high milk yields. Supplementation with concentrates is, therefore, needed. Unless the compound feed compensates for a deficiency of nutrients in the forage, essential to maintain an optimal microbial activity in the rumen, supplementation with concentrates will increase total energy intake but decrease forage intake.

The exact mechanism of substitution is not entirely clear but, following the theories of feed intake regulation, one may assume that replacing effective rumen fill, influencing rumen osmolality and the animal's genetic potential to utilize nutrients for milk production are most likely to be the underlying mechanisms.

The substitution rate, defined as the decrease in forage DMI per kg of supplemented concentrates, decreases with the level of milk production and increases with the amount of concentrates and with the quality of the forage (Forbes, 1995). Some typical substitution rates are presented in Table 6.2. The results show an increased substitution with a better forage quality and also with the amount of concentrates fed; an increased milk yield potential, on the other hand, decreases the substitution rate.

Table 6.2. Substitution rates of forage (kg DM) by concentrates (kg).

Forage quality (MJ NEL)	Milk yield (kg year^{-1})	kg concentrates					
		2	4	6	8	10	12
5.5	6000	0.25	0.25	0.28	0.32	0.35	0.38
	7000	0.20	0.25	0.27	0.30	0.33	0.36
	8000	0.20	0.22	0.25	0.28	0.31	0.33
6.2	6000	0.25	0.30	0.32	0.36	0.39	0.43
	7000	0.25	0.28	0.30	0.34	0.37	0.40
	8000	0.20	0.25	0.28	0.31	0.35	0.38
6.9	6000	0.30	0.33	0.37	0.40	0.44	0.48
	7000	0.30	0.30	0.34	0.38	0.41	0.44
	8000	0.25	0.28	0.32	0.35	0.39	0.41

NEL = net energy of lactation.
Source: IKC (1993).

ENERGY

Energy in animal feeds is stored in organic compounds, broadly known as carbohydrates, proteins and lipids. Their heats of combustion are 17.4, 24.0 and 39.8 MJ kg^{-1}, respectively. In biological terms, the unit MJ, in which heat of combustion is expressed, has only limited significance. Differences in heat of combustion between nutrients are caused mainly by the ratio in which hydrogen and oxygen (H:O ratio) are present. On a molar basis, H:O ratios approximate to 2.0, 4.4 and 11.6 in carbohydrates, proteins and lipids, respectively. In order to provide 'fuel', needed for metabolic processes, the energy has to be present in ATP. A high H:O ratio requires a large amount of external O_2 to oxidize the component, with a concomitant formation of ATP formed mainly in a process known as oxidative phosphorylation.

Before metabolically active organs and tissues can extract energy from the digested organic compounds, carbohydrates, proteins and lipids need to be converted into simpler forms. This conversion starts in the digestive tract with the hydrolytic cleavage of the chemical bonds, which keep the monomers (monosaccharides, amino acids and fatty acids) together in their complicated polymeric structures. Although some energy is liberated, hydrolytic cleavage of bonds in carbohydrates, proteins or lipids does not result in the formation of ATP, because the amount of energy is insufficient to link phosphate to ADP and form ATP. In dairy cows, like all ruminants, microbial enzymes in the reticulo-rumen bring about hydrolytic cleavage. The hydrolysis does not release sufficient energy to result in the formation of ATP, and the rumen microorganisms continue their action and degrade the monomers further. Because of lack of O_2, part of the substrate and other compounds as well are used as electron acceptors to enable the oxidation of the substrate. The further degradation of carbohydrates and proteins yields carbon dioxide (CO_2), which is used subsequently as a sink for the surplus hydrogen [H], resulting in the formation of methane. As a consequence, before yielding suitable precursors for further utilization by organs and tissues of the dairy cow, carbohydrates, proteins and lipids lose varying proportions of their energy as heat (Table 6.3). From the H:O ratios mentioned above, it is easy to see why carbohydrates are the best possible energy source for microorganisms.

Table 6.3. Heat of combustion (MJ kg^{-1}), ATP yield and required energy input for ATP formation (Blaxter, 1989).

	H/O	MJ kg^{-1}	ATP kg^{-1}	kJ ATP^{-1}
Carbohydrates	2.0	17.7	219	81
Protein	4.4	23.7	227	104
Lipids	11.6	39.3	505	78
Acetic acid	2.0	14.6	167 (124)[a]	87 (143)[a]
Propionic acid	3.0	20.6	243 (222)[a]	85 (80)[a]
Butyric acid	4.0	24.8	307 (167)[a]	81 (106)[a]

[a]Calculated back to the precursor carbohydrates.

The energy input required per mole of ATP also differs between the different nutrients, particularly when one takes the original ingested or digested feed as the energy source. In the case of the volatile fatty acids (VFAs), acetic acid (HAc), propionic acid (HPr) and butyric acid (HBu), losses occur in the rumen due to energy losses in CH_4 and fermentation heat. With a good quality dairy diet, HAc, HPr and HBu are formed in a molar ratio 60:25:15, and some 12% of digestible energy (DE) is lost in CH_4, 6% in the urine and an additional 7% in fermentation heat. The losses change with the fermentation pattern, and a higher HPr results in lower losses, but the ratio of the energy lost in CH_4 and fermentation heat remains almost constant (Blaxter, 1989).

In dairy cow nutrition, energy is partitioned into gross energy (GE), digestible energy (DE), metabolizable energy (ME) and net energy (NE). GE expresses the heat of combustion of the organic part of a feed and can be divided into faecal energy (FE) and DE. DE is the difference between heat of combustion of the feed and the heat of combustion of what is apparently left of the feed after passing through the digestive tract. Passage of feed through the digestive tract requires labour, which ultimately is also lost as heat. It also causes the release of endogenous compounds, mainly protein and lipids, which are partly reabsorbed and partly lost in the faeces. The re-synthesis of these compounds also requires energy, which again ultimately is lost as heat. Supply and requirements of energy in dairy production are usually expressed as ME or NE. Both terms are somewhat misleading. ME represents energy not lost in faeces, urine or CH_4 and its name suggests that ME is available for further processing in metabolic transactions. As already explained above, this is not entirely correct; part of it is lost as fermentation heat; part of the energy excreted in the urine has already undergone metabolic processing and a significant part of the energy lost in urine is lost as urea. If this urea results from an excessive breakdown of feed (crude) protein in the rumen, it does not contribute to ME. If, on the other hand, it results from amino acid oxidation or metabolic protein turnover, it should theoretically be considered as part of ME.

For dairy cows, as for other farm animals, energy requirements have been defined for maintenance, for milk production and for reproduction. Requirements for maintenance are considered constant and have been defined as the energy losses occurring when a non-producing animal is in energetic equilibrium. If ME is used for maintenance, the term NE expresses the fasting metabolic rate (i.e. energy lost as heat under fasting). If, on the other hand, energy is used for production, NE represents the energy deposited in desired animal products such as meat, milk, body reserves, the fetus or labour. The energy required to maintain the integrity of producing organs and tissues is considered as inefficiency, but no clear distinction is possible between these losses and energy lost in maintenance. The difference is only dictated by the definitions. Feed values and requirements are expressed in these units (Blaxter, 1989; Van der Honing and Alderman, 1988). In the case of milk production, this is usually standardized to a fixed fat content (FCM, milk with 4% fat) and sometimes also a fixed protein (FPCM, fat 4%, protein 3.3%)

content. To calculate FCM or FPCM, the following equations can be used (CVB, 1990):

$$kg\ FCM = (0.4 + 0.15\%\ fat)\ (kg\ milk) \tag{6.1}$$

$$kg\ FPCM = (0.337 + 0.116\%\ fat + 0.06\%\ protein)\ (kg\ milk) \tag{6.2}$$

From the foregoing discussion, it will be obvious that, in its present form, energy evaluation in dairy cows gives a rather poor description of supply and demand of nutrients for milk production. When energy-yielding compounds are not used as fuel after oxidation, but are used as precursors for animal products, three different groups can be distinguished, i.e. lipogenic, glucogenic and aminogenic. Products are animal proteins (aminogenic) in meat and milk, fatty acids in animal fats (lipogenic) in meat and milk, and lactose in milk and glycerol in fats (glucogenic). All three groups can be used as fuel and also as precursors for the synthesis of fatty acids. The synthesis of lactose, glycerol and protein requires glucose or glucogenic precursors and (aminogenic) amino acids, respectively. Because proteins have a genetically determined amino acid sequence, the synthesis of protein not only requires sufficient amino acids, but they also have to be supplied in the correct ratio.

Nutrients which are available for dairy cows after absorption from the digestive tract are the VFAs (HAc, HPr and HBu) from the rumen, glucose originating from starch digested in the small intestine (Nocek and Tamminga, 1991), long chain fatty acids from lipids digested in the small intestine either of feed or of microbial origin, and amino acids resulting from the intestinal digestion of feed protein escaping rumen degradation or from microbial protein. In fact, during milk production, nutrients from turnover and recycling additionally are available; these are not treated separately but are adjusted for by efficiency of utilization. Nucleic acids, mainly originating from microbial biomass synthesized in the rumen, are also absorbed from the small intestine (Armstrong and Hutton, 1975) but they have little nutritive value as they are almost quantitatively excreted in the urine (Chen and Gomez, 1992).

Of the absorbed nutrients, HPr and glucose are glucogenic; HAc, HBu and long chain fatty acids are lipogenic, and amino acids are potentially aminogenic but can also be either glucogenic or lipogenic. Some scope exists to vary the degradation behaviour through the fermentation pattern in the rumen (Tamminga *et al.*, 1990). Varying the ration composition and feeding strategy may alter the ratio in which lipogenic, glucogenic and aminogenic nutrients become available. Shifting the site of digestion of protein and starch from the rumen to the intestine results in more protein and starch reaching the small intestine and increases the proportion of aminogenic and glucogenic energy, respectively. The fermentation pattern in the rumen can also be altered; rumen-degradable starch usually results in a shift towards more HPr (Nocek and Tamminga, 1991), whereas more cell walls result in more HAc (Tamminga, 1993); the inclusion of soluble sugars often results in an increase of HBu.

The production of biomass in dairy cows, regardless of whether this is in microbial cells in the rumen or in the carcass or in milk, always requires energy as fuel and energy in precursors. In energy for fuel, the different forms of energy can substitute for each other, but for the energy in precursors the possibilities of substitution are limited. Because of this lack of substitution, one may expect that one of the three forms of energy (lipogenic, glucogenic or aminogenic) is limiting and, as a result, will interfere with the utilization of the other forms.

PROTEIN

With the recognition of the significance of rumen fermentation for nutrient supply in ruminants, it became obvious that crude protein or digestible crude protein did not give an adequate description of the amount and nature of the amino acids which are provided to organs and tissues after absorption from the digestive tract. Most countries have therefore developed and introduced protein evaluation systems, which take into account the transactions in the digestive tract. Examples are the metabolizable protein system (USA, UK), the PDI system (France, Italy), the AAT/PDV system (Denmark) and the DVE/OEB system (The Netherlands, Belgium). The systems are generally based on similar basic principles. In a review, Oldham (1996) lists ten factors for which the systems introduced in different countries use common principles. These are:

- rumen-degraded (crude) protein
- rate of degradation in the rumen and digesta passage
- microbial protein yield as a function of energy
- the proportion of true protein in microbial crude protein
- an allowance for recycling of urea N to the rumen
- true absorption from the small intestine of microbial true protein
- absorption from the small intestine of undegraded dietary crude protein
- endogenous losses
- efficiency of conversion of absorbed protein to animal protein
- protein demands for the processes of maintenance, growth, pregnancy, milk or (muscle) fibre production.

Although the systems have the same basic principles in common, numerically there are differences, both with regard to the characterization of the feeds and with regard to requirements. Examples are digesta passage rates for which Scandinavian countries use 0.08, the Dutch systems uses 0.045 for roughages and 0.06 for concentrates and the systems in use in the USA, the UK and Australia use variable values. Similar differences exist for microbial growth efficiencies: 125 g kg^{-1} of digested carbohydrates in Scandinavian countries with 150 g kg^{-1} organic matter fermented in the rumen, and again variable values in the USA, the UK and Australia (Oldham, 1994).

In the majority of these systems, the factors are fixed and assumed to be additive. Requirements for and utilization of metabolizable protein are set at a point where energy requirements are exactly met and where the marginal response is approaching zero. This makes them rather rigid and poor in predicting responses, and more knowledge of the utilization of nutrients by organs and tissues preceding the mammary gland, such as the digestive tract and liver, is essential before improvements can be made in this area.

A recent development, partly driven by the development of rumen-protected amino acids, is that requirements and equations predicting supply have been formulated for individual amino acids (O'Connor *et al.*, 1993; Rulquin and Verité, 1993; Chalupa and Sniffen, 1994). More dynamic and mechanistic models describing the regulation of N metabolism in ruminants have also been developed, but these will be dealt with in other chapters. A good overview of existing models was presented recently by Hanegan *et al.* (1997)

CARBOHYDRATES

An important contribution to energy in diets for dairy cows is from carbo-hydrates, which can be separated into three groups, i.e. structural or fibrous carbohydrates, starches and soluble sugars. The role of structural carbo-hydrates in ruminant nutrition has long been limited to its description in crude fibre and non-structural carbohydrates as N-free extractives. During the last 25 years, crude fibre has been replaced by neutral detergent fibre (NDF), but only recently has adequate attention been paid to differences between fibre sources and differences between non-fibrous carbohydrates.

Physical Structure and Effective Fibre

Structural carbohydrates in diets for dairy cows serve two purposes. Their fermentation may contribute significantly to the energy supply of rumen microorganisms, and their fermentation end-products, in turn, yield useful nutrients for organs and tissues. In addition to its role as a provider of nutrients, fibre plays an important role in stabilizing rumen fermentation. It is an important adhering agent for rumen microorganisms, it stimulates rumination and, as a result, buffering through enhancing saliva production, and it stimulates rumen contractions. Because fibre takes up a lot of space in the rumen, it may limit feed intake; minimally required concentrations in dairy diets have, therefore, to be established. This has led to the introduction of the concept of 'effective fibre' (Mertens, 1985), which is arbitrarily defined as that NDF which does not pass through a 1 mm screen. According to Van Soest (1994), this is an oversimplification. When feeding adequate coarse fibre, the finer fibre will become effective because it is included in the rumen mat; if inadequate coarse fibre is fed, the finer fibre will remain ineffective. Minimum contributions recommended by Mertens (1985) are 35–38% of the dry matter, whereas the NRC (1989) suggests that 25–28% is adequate.

In Belgium, a system was recently developed and evaluated based on measuring the critical proportion of physical structure in the diet (DeBrabander *et al.*, 1996). This critical proportion is defined as the proportion of roughage at which no symptoms of deficiencies, characteristic of the start of a milk fat depression, develop. As for the energy and protein evaluation systems, a unit was needed to express dietary supply and the animal's demand. For practical reasons, the requirement of a multiparous cow, producing 25 kg of milk and to which concentrates are provided in two portions per day, was set at 1. Requirements vary with the level of milk production (increase or decrease of 0.8% per kg of milk above or below 25 kg day^{-1}), age of the animal (requirements are 5 and 13% lower in cows in the fourth and fifth lactation as compared with those in their first, second or third lactation) and, finally, with frequency of concentrate supply (10% less if concentrates are supplied in six or more portions). Values for the physical structure of feeds ranged between −0.34 (molasses) and +4.20 (straw). A summary of values is given in Table 6.4. The results show values for grass hay, grass silage and maize silage to vary with their NDF content. In the case of cassava and molasses, negative values were established due to the negative effect which these products have on rumen fermentation because of their rapid degradation in the rumen.

Starch

In addition to structural carbohydrates, non-structural carbohydrates, notably starch and soluble sugars, are given much consideration in diet formulation. Like fibre, starch and sugars are subject to microbial degradation in the rumen and are converted into VFAs. During this process starch is converted to propionic acid to a larger extent than are fibrous carbohydrates. Generally starch is degraded much more rapidly in the rumen than structural carbohydrates, but the susceptibility of different starches to microbial degradation in the rumen varies and, as a result, a variable proportion of dietary starch

Table 6.4. Values for physical structure.

Feed	Physical structure	Feed	Physical structure
Straw	4.20	Soya hulls	0.55
Hay	−0.214 + 0.0070 × NDF	Beet pulp	0.46
Grass silage	−0.200 + 0.0065 × NDF	Malt sprouts	0.38
Maize silage	−1.200 + 0.0075 × NDF	Milocorn	0.33
Fresh grass (autumn)	2.60	Corn	0.30
Fresh grass (spring)	1.80	Corn gluten feed	0.30
Fodder beets	1.00	Soybean meal	0.13
Pressed beet pulp	1.00	Wheat	0.00
Ensiled brewer's grains	1.00	Barley	0.00
Raw potatoes	0.75	Cassava	−0.14
Ensiled corn cob mix	0.75	Molasses	−0.34

escapes rumen fermentation and is subjected to enzymatic digestion in the small intestine (Nocek and Tamminga, 1991; Sauvant *et al.*, 1994). Feeds relatively high in rumen-resistant starch are maize, sorghum, millet, potato and rice. Feeds high in rumen-degradable starch are wheat, oats, barley and tapioca. Starch in legume seeds, such as peas and beans, has a rumen degradative behaviour in between the two groups mentioned, whereas that of maize silage is highly variable (Klop and De Visser, 1994). Most methods of feed processing, except toasting, result in an increased rate of starch degradation and tend to shift the site of starch degradation towards the rumen where it may interfere with the degradation of fibrous feed components.

The capacity of the small intestine to digest starch and absorb the glucose resulting from it is limited. In case of excess starch, significant quantities may be fermented in the hind gut. Moreover, an excess may limit the energy available for microbial protein synthesis in the rumen. The fate of glucose absorbed from the small intestine is not clear: part of it is oxidized by the intestinal wall and may spare amino acids from oxidation, while some of it appears in the portal blood, is subsequently transported to the liver and contributes to the pool of *glucogenic* nutrients.

In The Netherlands recommendations for the inclusion of total starch and sugars as well as rumen-resistant starch in concentrate feeds for different categories of dairy cows (early lactation, mid-lactation, late lactation) have been formulated for different forages (Table 6.5).

INTERACTIONS BETWEEN ENERGY AND PROTEIN

Using separate systems for energy and protein is a complicated and confusing situation. Not only are they expressed in completely different units, but protein is also a component of energy. Besides, in ruminants, the protein supply of the animal depends very much on the energy supply to the rumen

Table 6.5. Non-structural carbohydrates recommended in compound feeds for dairy cows.

Basal diet	Min/Max	SU + ST[a]	SU + RDST[b]	RRST[c]
Maize silage	Min	0	0	0
	Max	250	200	50
Grass silage (young)	Min	100	50	25
	Max	400	300	100
Grass silage (old)	Min	200	100	25
	Max	350	250	100
Mixed silage[d]	Min	0	0	0
	Max	325	250	75
Fresh grass	Min	100	0	25
	Max	250	150	100

[a]Sugars + starch; [b]Sugars + rumen-degradable starch; [c]Rumen-resistant starch; [d]50/50 mixture of corn silage and grass silage.

microorganisms. Moreover, energy is used for a number of completely different purposes; it not only provides the precursors needed for the synthesis of animal products differing widely in nature, but also yields the fuel (ATP) needed to link these precursors together in the desired composition, form, structure and conformation

Microbial Fermentation in the Fore-stomachs

Microbial growth in the rumen requires ATP as fuel and precursors for the synthesis of proteins, nucleic acids, lipids and carbohydrates. The form in which precursors are needed is simple: apart from carbohydrates as a source of carbon (C), simple forms of precursors containing nitrogen (N), phosphorus (P), sulphur (S) and a number of minerals and trace elements (notably cobalt), iso-acids and fat-soluble vitamins or their precursors are sufficient. Factors limiting microbial growth are either energy (carbohydrates) or nitrogen, and most modern protein evaluation systems take this into account. The ultimate supply of microbial protein to the small intestine may, however, not depend primarily on the availability of energy or other essential nutrients, or even on their degree of synchronization. The ratio of the different groups of microorganisms, each with their own specific residence time in the rumen, and the extent to which some predate on others to obtain nutrients for their own growth and development, are also important. Our knowledge of both these aspects is still limited (Theodorou and France, 1994).

Microbial activity in the fore-stomachs is responsible for redistribution between the different forms of energy present in the feed. In a typical mixed diet for high producing dairy cows, formulated to meet energy and protein requirements, energy (MJ) is distributed between carbohydrates (glucogenic), proteins (aminogenic) and lipids (lipogenic) in an approximate ratio of 70:22:8. After fermentation and digestion, the three groups of energy-yielding nutrients are absorbed in a ratio of approximately 24:20:56, whereas when excreted in milk they occur in the ratio 24:27:49. The shift between the ratio of absorbed nutrients and nutrients excreted in milk suggests that for maintenance, as well as for deposition in body reserves, a preference exists for lipogenic nutrients.

In recent years, much attention has been paid to ways to manipulate rumen fermentation. Important aspects are the ratio between microbial mass and VFAs and, within the total VFAs, the distribution between HAc, HPr and HBu. Synchronization of the supply of rumen-degradable protein and carbohydrates has also been recommended, but in practice this provides only limited possibilities (Chamberlain and Choung, 1995).

The ratio in which nutrients are absorbed and in which they are excreted in milk is relatively similar, which may explain why it is difficult to manipulate milk composition by feeding. Altering this situation would necessitate firstly the manipulation of absorbed nutrients either by going beyond the process of microbial fermentation in the rumen or by manipulating the microbial fermentation itself; as indicated earlier, both are possibilities, but only to a limited extent. Secondly, the capacity of the animal's organs and tissues to handle widely varying ratios of nutrients and ways in which to influence this are not adequately understood.

In order to accommodate the limitations of energy and protein evaluation systems presently used in the field, the Cornell net carbohydrate and protein system (CNCPS) was developed (Fox *et al.,* 1992; Russell *et al.,* 1992; Sniffen *et al.,* 1992). The system has a limited number of mechanistic components, like a sub-model that provides quantitative estimates of fermentation end-products (the ME from VFA production, microbial protein and ammonia) and materials that escape degradation in the rumen (carbohydrates, protein and undegraded peptides). Advantages of the CNCPS are that it integrates energy and protein and that it deals with rumen fermentation in a somewhat dynamic way. Conceptually, it has attractions, and some of its principles possibly can be used for the further development of nutrient-based feed evaluation systems for ruminants. A severe limitation is that the post-absorptive part of the model is based largely on empirical relationships and therefore needs further development (Hanigan *et al.,* 1997). It is likely, therefore, that updated user-friendly versions of really mechanistic models, like that developed in Davis, California (Baldwin *et al.,* 1987), will ultimately prove to be superior.

Metabolic Interactions

Before nutrients are made available for synthesis in tissues or in products considered valuable as a human food, i.e. milk and meat in dairy cows, they may have been used for other purposes by organs and tissues positioned earlier in the transport line. Prior to entering the portal vein, nutrients pass through the metabolically highly active tissues and organs associated with the digestive tract, such as the intestinal wall and pancreas. After entering the portal vein, nutrients pass through the liver, probably the most active organ of all.

Oxygen uptake by the portal drained viscera (digestive tract, spleen and pancreas) was estimated to account for 15% of the ME (Lindsay, 1993). It has also been estimated that the liver uses 15–17% of the total entry energy, and the mammary gland uses approximately 12% of its energy intake (Baldwin and Kim, 1993). The total maintenance requirement for energy of a 30 kg-producing dairy cow in energy balance is approximately 57.5 MJ of ME day^{-1}, which accounts for only 67.6% of the heat loss not accounted for by losses in fermentation heat or in the mammary gland. The liver is also the site where gluconeogenesis occurs, necessary to provide the mammary gland with sufficient glucose; precursors for gluconeogenesis are propionate and glucogenic amino acids. Recent estimates suggest that the contribution of amino acids to gluconeogenesis is between 15 and 25% in dairy cows (Kelly *et al.,* 1993; Lindsay, 1993). The significance of this figure is not clear; glucogenic amino acids may be supplied in excess of what is needed for protein synthesis and they are therefore available for gluconeogenesis but, on the other hand, some glucogenic amino acids (like methionine) are also required for other purposes which further aggravates their situation of being limiting.

In early lactation, the dairy cow's highest priority is milk production, regardless of the amount of nutrients supplied from feed intake. Utilization of energy extracted by the mammary gland from the blood is high and close to 90% (Baldwin and Kim, 1993), but differences have been recorded for the different classes of nutrients. Utilization of glucose, extracted from the blood for the synthesis of lactose and glycerol, and non-esterified (long chain) fatty acids (NEFAs), acetate and β-hydroxybutyrate used in triacylglycerides are close to 85%, while the utilization of extracted amino acids is close to 94%. Efficiency of conversion of NEFAs to triacylglycerides is 93–96%, whereas acetate is converted into milk fat with a lower efficiency of about 75–80% (Baldwin and Kim, 1993).

Milk composition depends to some extent on the ratio in which total nutrients are supplied from feed intake and the mobilization of body tissues; this is particularly true for the proportion of milk energy excreted as protein. In addition, the physiological stage (negative balance in early lactation versus positive balance in late lactation) is important, because it influences nutrient supply to the mammary gland, either by supplying extra nutrients from body reserves in early lactation, or by taking away nutrients to be deposited in the body in late lactation.

Results on the magnitude of protein mobilization in early lactation in dairy cows are conflicting. According to the AFRC (1993), body weight change comprises 19 MJ kg^{-1} and 150 g of protein. With an assumed energy content of 23.8 MJ kg^{-1} for protein, and 39.7 MJ kg^{-1} for lipids, this suggests that 19% of the energy is protein, about twice the amount reported by others (Waldo *et al.*, 1991).

Comparing the empty body weight of ten cows slaughtered 7 days pre-partum with that of seven cows slaughtered 63 days post-partum, Andrew *et al.* (1994) concluded that the effect of lactation on protein metabolism was restricted to a redistribution. Carcass protein remained the same while uterus protein was decreased by 4 kg, of which 2.5 kg went to organs and 1.5 kg to the digestive tract. The possibility exists, however, that, in the first weeks of lactation, protein became mobilized from the carcass but that after 63 days this had been replaced again. Gibb *et al.* (1992) showed, in an experiment in which 24 dairy cows were slaughtered between 0 and 8 weeks post-partum, that while protein was mobilized from the carcass, other tissues and organs remained the same or protein was even deposited, notably in liver and gut tissue. Contrary to this, fat was mobilized from all organs and tissues. Protein losses from the carcass were 171 g day^{-1} in the first 2 weeks, rapidly declining to 114 g day^{-1} in weeks 3–5 to 80 g day^{-1} in weeks 6–8.

Taminga *et al.* (1997) recently analysed a large data set from production trials with dairy cows, in which feed intake and milk output as net energy and live weight change had been measured weekly. Energy balance was calculated and it was further assumed that the water to protein ratio in mobilized tissue is 3.4. Live weight change was corrected to empty body weight change with the assumption that each kg change in DMI in early lactation causes a live weight change of 4 kg (INRA, 1989); the composition of empty body weight change for animals in a negative energy balance was calculated weekly. The results showed that in the first weeks after calving, protein was mobilized, but protein reserves were exhausted within 4 weeks. Fat mobilization continued until 8 weeks after calving, whereas between 4 and 8 weeks after calving protein was deposited again. Calculated total mobilized fat amounted to 18 kg whereas about 6 kg of protein was mobilized, a figure quite close to that found by Gibb *et al.* (1992).

After reaching energy equilibrium, dairy cows move into a positive energy balance and deposit fat and protein. The moment when this happens is difficult to predict; it may vary between 6 and 12 weeks post-partum (De Visser, 1993) and is influenced by the composition and quality of the diet and other factors. The efficiency with which lipogenic plus glucogenic nutrients are deposited in body fat is probably high and close to 80% (Baldwin and Kim, 1993). The efficiency of utilization of aminogenic energy for protein deposition in the body is influenced by protein turnover, which varies with both the site of deposition (carcass versus organs) and the ratio in which the different forms of energy are needed or supplied. Since the magnitude of body protein deposition compared with milk protein excretion is small, it is probably of little practical consequence.

PRACTICAL IMPLICATIONS AND CONCLUSIONS

Utilization of energy and protein in dairy cows is complicated by a limited predictability of feed intake. Regardless of this lack of predictability, the most severe limitation for milk production in early lactation is an inadequate feed (energy) intake. In early lactation, emphasis should therefore be on attempts to maximize feed intake, for which feed quality in terms of energy density, particularly that of the roughage part of the diet, is important. At the same time, optimal rumen function and microbial fermentation should be maintained. A further balancing of the ratio in which glucogenic, aminogenic and lipogenic nutrients are supplied seems possible, but this requires knowledge of the site of digestion (fore-stomachs versus intestines) as well as the fermentation pattern (microbial biomass versus VFAs) and the ratio of non-glucogenic and glucogenic nutrients. Our knowledge in these areas is still severely lacking; not only is more knowledge needed, but accurate measuring methods are also required to make attempts in this direction more predictable and successful. Further improvements may become possible, at least theoretically, by supplementing dairy cow diets with rumen-protected limiting amino acids such as lysine and methionine. To make such strategies successful, a better prediction of the amino acid composition of protein absorbed from the small intestine (both microbial protein and feed protein escaping rumen degradation) is necessary.

REFERENCES

AFRC, Agricultural and Food Research Council (1993) *Energy and Protein Requirements of Ruminants*. AFRC Technical Committee on Responses to Nutrients. CAB International, Wallingford, UK.

Andrew, S.M., Waldo, D.R. and Erdman, R.A. (1994) Direct analysis of body composition of dairy cows at three physiological stages. *Journal of Dairy Science* 77, 3022–3033.

Armstrong, D.G. and Hutton, K. (1975) Fate of nitrogenous compounds entering the small intestine. In: McDonald, I.W. and Warner, A.C.I. (eds), *Digestion and Metabolism in the Ruminant*. The University of New England Publishing Unit, Armidale, Australia, pp. 432–447.

Baldwin, R.L. and Kim, W.Y. (1993) Lactation. In: Forbes, J.M. and France, J. (eds), *Quantitative Aspects of Ruminant Digestion and Metabolism*. CAB International, Wallingford, UK, pp. 433–452.

Baldwin, R.L., France, J. and Gill, M. (1987) Metabolism of the lactating cow. I. Animal elements of a mechanistic model. *Journal of Dairy Research* 54, 77–105.

Bequette, B.J. and Backwell, C.F.R. (1997) Amino acid supply and metabolism by the ruminant mammary gland. *Proceedings of the Nutrition Society* 56, 593–605.

Blaxter, K.L. (1989) *Energy Metabolism in Animals and Man*. Cambridge University Press, Cambridge, UK.

Broster, W.H. (1972) Effect on milk yield of the cow of the level of feeding during lactation. *Dairy Science Abstracts* 34, 265–288.

Chalupa, W. and Sniffen, C. (1994) Carbohydrate, protein and amino acid nutrition of dairy cows. In: Garnsworthy, P.C. and Cole, D.J.A. (eds), *Recent Advances in Animal Nutrition 1994*. Nottingham University Press, UK, pp. 265–275.

Chamberlain D.G. and Choung, J.-J. (1995) The importance of rate of ruminal fermentation of energy sources in diets for dairy cows. In: Garnsworthy, P.C. and Cole, D.J.A. (eds), *Recent Advances in Animal Nutrition 1995*. Nottingham University Press, UK, pp. 3–28

Chen, X.B. and Gomez, M.J. (1992) Estimation of microbial protein supply in sheep and cattle based on urinary excretion of purine derivatives: an overview of the technical details. *Rowett Research Institute Occasional Publication no. 1*. International Feed Resources Centre.

Chilibroste, P., Tamminga, S. and Boer, H. (1997) Effect of length of grazing session, rumen fill and starvation time before grazing on dry matter intake, ingestive behaviour and dry matter rumen pool sizes of grazing lactating dairy cows. *Grass and Forage Science* 52, 249–257.

Chilibroste, P., Tamminga, S., Van Bruchem, J. and Van der Togt, P.L. (1998) Effect of allowed grazing time, inert rumen bulk and length of starvation before grazing, on the weight, composition and fermentative end-products of the rumen contents of lactating dairy cows. *Grass and Forage Science* 53, 146–156.

CVB (1990) *Voedernormen landbouwhuisdieren*. [Feed allowances for farm animals, in Dutch]. Centraal Veevoeder Bureau, Lelystad, The Netherlands.

Dado, R.G. and Allen, M.S. (1995) Intake limitations, feeding behaviour and rumen function of cows challenged with rumen fill from dietary fiber or inert bulk. *Journal of Dairy Science* 78, 118–133.

DeBrabander, D.L., DeBoever, J.L., DeSmet, A.M., Vanacker J.M. and Boucqué, Ch.V. (1996) *Structuurwaardering in de melkveevoeding*. [Structural value in dairy nutrition]. Report no. 967. Rijksstation voor de Veevoeding, 9090 Melle, Gontrode, The Netherlands.

De Visser, H. (1993) Influence of carbohydrates on feed intake, rumen fermentation and milk performance in high yielding dairy cows. PhD Thesis, Agricultural University, Wageningen, The Netherlands.

Dulphy, J.P. and Demarquilly, C. (1994) The regulation and prediction of feed intake in ruminants in relation to feed characteristics. *Livestock Production Science* 39, 1–12.

Forbes, J.M. (1995) *Voluntary Feed Intake and Diet Selection in Farm Animals.* CAB International, Wallingford, UK.

Fox, D.G., Sniffen, C.J., O'Connor, J.D., Russell, J.B. and Van Soest, P.J. (1992) A net carbohydrate and protein system for evaluating cattle diets. III. Cattle requirements and diet adequacy. *Journal of Animal Science* 70, 3578–3596.

Gibb, M.J., Ivings, W.E., Dhanoa, M.S. and Sutton, J.D. (1992) Changes in body composition of autumn-calving Holstein–Friesian cows over the first 29 weeks of lactation. *Animal Production* 55, 339–360.

Gill, M. and Romney, D. (1994) The relationship between the control of meal size and the control of daily intake in ruminants. *Livestock Production Science* 39, 13–18.

Grovum, W.L. (1995) Mechanisms explaining the effects of short chain fatty acids on feed intake in ruminants: osmotic pressure, insulin and glucagon. In: Von Engelhardt, W., Leonhardt-Marek, S., Breves, G. and Giesecke, D. (eds), *Ruminant Physiology: Digestion, Metabolism, Growth and Reproduction.* Enke Verlag, Stuttgart, Germany, pp. 173–197.

Hanigan, M.D., Dijkstra, J., Gerrits, W.J.J. and France, J. (1997) Modelling post-absorptive protein and amino acid metabolism in the ruminant. *Proceedings of the Nutrition Society* 56, 631–643.

Henneberg, W. and Stohmann, F. (1864) *Begründung einer rationellen Fütterung der Wiederkauer.* Schwetske and Söhne, Braunschweig, Germany.

IKC (1993) *Handboek voor de Rundveehouderij.* [Handbook of Dairy Husbandry]. Informatie en Kenniscentrum Veehouderij, Ede, The Netherlands.

Ingvartsen, K.L. (1994) Models of voluntary feed intake in cattle. *Livestock Production Science* 39, 19–38.

Institut National de la Recherche Agronomique (1989) In: Jarrige, R. (ed.), *Ruminant Nutrition: Recommended Allowances and Feed Tables.* John Libey, Eurotext, Montrouge, France.

Jarrige, R., Demarquilly, C., Dulphy, J.P., Hoden, A., Robelin, J., Béranger, C., Geay, Y., Journet, M., Malterre, C., Micol, D. and Petit, M. (1986) The INRA 'fill unit' system for predicting the voluntary intake of forage based diets in ruminants. *Journal of Animal Science* 63, 1737–1758.

Kellner, O. (1912) *Die Ernährung der Landwirtschaftlichen Nutztiere.* Paul Parey, Berlin.

Kelly, J.M., Park, H., Summers, M. and Milligan, L.P. (1993) Interactions between protein and energy metabolism. In: Forbes, J.M. and France, J. (eds), *Quantitative Aspects of Ruminant Digestion and Metabolism.* CAB International, Wallingford, UK, pp. 341–362.

Ketelaars, J.J.M.H. and Tolkamp, B.J. (1992a) Towards a new theory of feed intake regulation in ruminants. 1. Causes of differences in voluntary feed intake, critique of current views. *Livestock Production Science* 30, 269–296.

Ketelaars, J.J.M.H. and Tolkamp, B.J. (1992b) Towards a new theory of feed intake regulation in ruminants. 3. Optimum feed intake, in search of a physiological background. *Livestock Production Science* 31, 235–258.

Klop, A. and De Visser, H. (1994) Afbraak van snijmaissilage, maiskolvensilage en corn cob mix in de pens van melkkoeien. [Rumen degradation of fodder maize, husk corn meal and corn cob mix in dairy cows, in Dutch.] *IVVO-DLO report* 262, Lelystad.

Kristensen, V.F. and Ingvartson, K.L. (1986) Prediction of feed intake. In: Neimann-Sorensen, A. (ed) *Agriculture: New Developments and Future Perspectives in Research on Rumen Function.* Commission of European Communities, Luxembourg, pp. 157–182.

Lindsay, D.B. (1993) Metabolism of the portal drained viscera. In: Forbes, J.M. and France, J. (eds), *Quantitative Aspects of Ruminant Digestion and Metabolism.* CAB International, Wallingford, UK, pp. 267–290.

Mertens, D.R. (1985) Factors influencing feed intake in lactating dairy cows: from theory to application using neutral detergent fibre, In: *Proceedings of the Georgia Nutrition Conference.* University of Georgia, Athens, Georgia, pp. 1–18

Minson, D.J. (1990) *Forage in Ruminant Nutrition.* Academic Press, San Diego, California.

Morrison, F.B. (1936) *Feeds and Feeding*, 20th edn. Morrison Publishing Company, Ithaca, New York.

Nocek, J.E. and Tamminga, S. (1991) The prediction of nutrient supply to dairy cows from rate and extent of ruminal degradation of ration components. *Journal of Dairy Science* 74, 3598–3629.

National Research Council (1989) *Nutrient Requirements for Dairy Cattle, Update 1989.* National Academy Press, Washington, DC.

O'Connor, J.D., Sniffen, C.J., Fox, D.G. and Chalupa, W. (1993) A net carbohydrate and protein system for evaluating cattle diets. IV. Predicting amino acid adequacy. *Journal of Animal Science* 71, 1298–1311.

Oldham, J.D. (1996) Protein requirement systems for ruminants. In: Phillips, C.J.C. (ed.), *Progress in Dairy Science.* CAB International, Wallingford, UK, pp. 3–27.

Østergaard, V. (1979) Strategies for concentrate feeding to attain optimum feeding level in high yielding dairy cows. *Report 482, National Institute of Animal Science.* Copenhagen.

Rulquin, H. and Verité, R. (1993) Amino acid nutrition of dairy cows. In: Garnsworthy, P.C. and Cole, D.J.A. (eds), *Recent Advances in Animal Nutrition 1993.* Nottingham University Press, UK, pp. 55–77.

Russell, J.B., O'Connor, J.D., Fox, D.G., Van Soest, P.J. and Sniffen, C.J. (1992) A net carbohydrate and protein system for evaluating cattle diets. I. Ruminal fermentation. *Journal of Animal Science* 70, 3551–3561.

Sauvant, D., Chapoutot, P. and Archimède, H. (1994) La digestion des amidons par les ruminants et ses consequences. *INRA Production Animal*, 7, 115–124.

Sniffen, C.J., O'Connor, J.D., Van Soest, P.J., Fox, D.G. and Russell, J.B. (1992) A net carbohydrate and protein system for evaluating cattle diets. II. Carbohydrate and protein availability. *Journal of Animal Science* 70, 3562–3577.

Tamminga, S. (1993) Influence of feeding management on ruminant fiber digestibility. In: Jung, H.-J.G. Buxton, R.D., Hatfield, R.D. and Ralph, J. (eds), *Forage Cell Wall Structure and Digestibility.* American Society of Agronomy, Crop Science Society of America and Soil Science Society of America, Madison, Wisconsin, pp. 571–602.

Tamminga, S., Van Vuuren, A.M., Van der Koelen, C.J., Ketelaar R.S. and Van der Togt, P.J. (1990) Ruminal behaviour of structural carbohydrates, non-structural carbohydrates and crude protein from concentrate ingredients in dairy cows, *Netherlands Journal of Agricultural Science* 38, 513–526.

Tamminga, S., Luteijn, P.A. and Meijer, R.G.M. (1997) Changes in composition and energy content of liveweight loss in dairy cows with time after parturition. *Livestock Production Science* 52, 31–38.

Thaer. A. (1809) *Grundsatze der Rationellen Landwirtschaft,* Realschulbuchhandlung, Berlin.

Theodorou, M.K. and France, J. (1994) Rumen microorganisms and their interactions. In: Forbes, J.M. and France, J. (eds), *Quantitative Aspects of Ruminant Digestion and Metabolism.* CAB International, Wallingford, UK, pp. 145–164.

Tolkamp, B.J. and Ketelaars, J.J.M.H. (1992) Towards a new theory of feed intake regulation in ruminants. 2. Costs and benefits of feed consumption: an optimization approach. *Livestock Production Science* 30, 297–317.

Van der Honing, Y. and Alderman, G. (1988) Feed evaluation and nutritional requirements. III.2. Ruminants. *Livestock Production Science* 19, 217–278.

Van Soest, P.J. (1994) *Nutritional Ecology of the Ruminant*, 2nd edn. Cornell University Press, Ithaca, New York.

Waldo, D.R., Andrew, S.M. and Erdman, R.A. (1991) Protein and fat changes in empty body of growing and lactating Holstein cattle. In: Eggum, B.O., Boisen, S., Borsting, C., Danfaer, A. and Hvelplund, T. (eds), *Protein Metabolism and Nutrition.* National Institute of Animal Science, Foulum, Denmark, Vol. II, pp. 181–183.

7

Feeding Systems for Beef Cattle

J.G. Buchanan-Smith[1] and D.G. Fox[2]

[1]Department of Animal and Poultry Science, University of Guelph, Guelph, Ontario, Canada; [2]Department of Animal Science, Cornell University, Ithaca, New York, USA

INTRODUCTION

Among agricultural animals and fish, beef cattle are marked by diversity. There are over 100 breeds of cattle in the world and well over 20 of these make a significant contribution to beef production. Genetics accounts for much of the diversity in cattle type, but their constitution may be markedly affected in other ways, e.g. through periods of feed restriction, imposed as a consequence of financial constraints, feed shortages or optimization of local feed resources. Beef cattle are managed under extensive environments, wherever grass grows, as well as under intensive conditions, exemplified by the feeding of high grain diets to cattle in feedlots; there is therefore a wide variation in the quality of feedstuffs they consume. These diversities have made it challenging to prescribe nutrient requirements and rationalize feeding systems that are generally applicable beyond specified geographical regions. It is often difficult to predict performance of cattle grazing under very extensive conditions. However, the wealth of knowledge about feeding cattle should be sufficient to provide a sound conceptual framework to develop feeding systems for all types of cattle in all environments, including extensive grazing conditions. Many scientific committees have now capitalized on this new knowledge and prescribed considerable detail in feeding systems that take many of the diversities in animal type and environment into account. Van der Honing and Alderman (1988) reviewed feed evaluation and nutritional requirements presented in different systems prior to 1988. In this chapter, systems proposed since the late 1980s by the French (Institut National de la Recherche Agronomique; INRA, 1989), Australians (Commonwealth Scientific and Industrial Research Organization; CSIRO, 1990), in the UK (Agricultural and Food Research Council; AFRC, 1993) and in North America (National Research Council; NRC, 1996) will be covered.

©CAB *International 2000. Feeding Systems and Feed Evaluation Models*
(eds M.K. Theodorou and J. France)

Until recently, feeding systems for beef cattle have been described by only using empirical relationships between variables. Some of the information has been integrated using a factorial or incremental description, wherein nutrient requirements for different physiological functions are added together and the total amounts required for different animal classes at specified levels of performance are presented in tables. These methods are animal driven, and are not only limited by relative inflexibility but can be quite imprecise in predicting animal performance when the particular feedstuff, as distinct from the nutrients that the feedstuff contains, affects the outcome. This situation arises, for example, in ruminants where the feed and the amount eaten affects metabolic heat production and hence the performance of and response to energy eaten by cattle in cold environments. It may also arise as a result of interactions between nutrients, exemplified by the limitations of feed energy on rumen microbial protein synthesis. A feed-driven model to predict animal performance overcomes these limitations, and NRC (1996) is the first feeding system for beef cattle, prescribed by a national committee, to take this approach to analyse the problem. In order to formulate diets, however, it is useful to have nutrient requirements expressed without regard to the feed eaten, and so the factorial approach of presenting nutrient requirements was retained in the report of the NRC (1996) committee. Thus it is envisaged that a user of NRC (1996) could formulate a diet by conventional processes but go on to evaluate animal performance, and refine nutrient requirements, using the feed-driven model. Two feed-driven models were presented by NRC (1996) to contrast one method, where feed values are expressed on a conventional nutrient concentration basis, with another method where feed composition is expressed on the basis of chemical entities. In another comparison of the feeding systems, NRC (1996) has dealt in the greatest detail with cattle and environmental diversity. Thus, in this chapter, the conceptual framework, together with some of the detail for the NRC (1996) system, will be presented first. Modifications and additions to NRC (1996), considered in other systems, are presented subsequently. Emphasis is placed on energy and protein, as these nutrients are most affected by using a feed-driven model and they illustrate the role and potential of a feed-driven model quite clearly. The purpose of this chapter is not to compare practical recommendations from the different systems but rather to contrast methods for predicting how a wide variety of beef cattle should perform in different environments on different feedstuffs.

NRC (1996) NUTRIENT REQUIREMENTS OF BEEF CATTLE

The models use inputs from each farm that describes the animal, the animal's environment and the feed composition. These inputs are used to compute maintenance requirements that vary with breed type; previous level of nutrition and environmental conditions; growth requirements that vary with expected weight at slaughter or mature weight; energy reserves in cows that vary with body condition score; and finally lactation requirements that vary with level of milk production and month of lactation (Fig. 7.1). Each of the two models predict energy and protein available to meet requirements after digestion. Choice of model depends on the completeness of

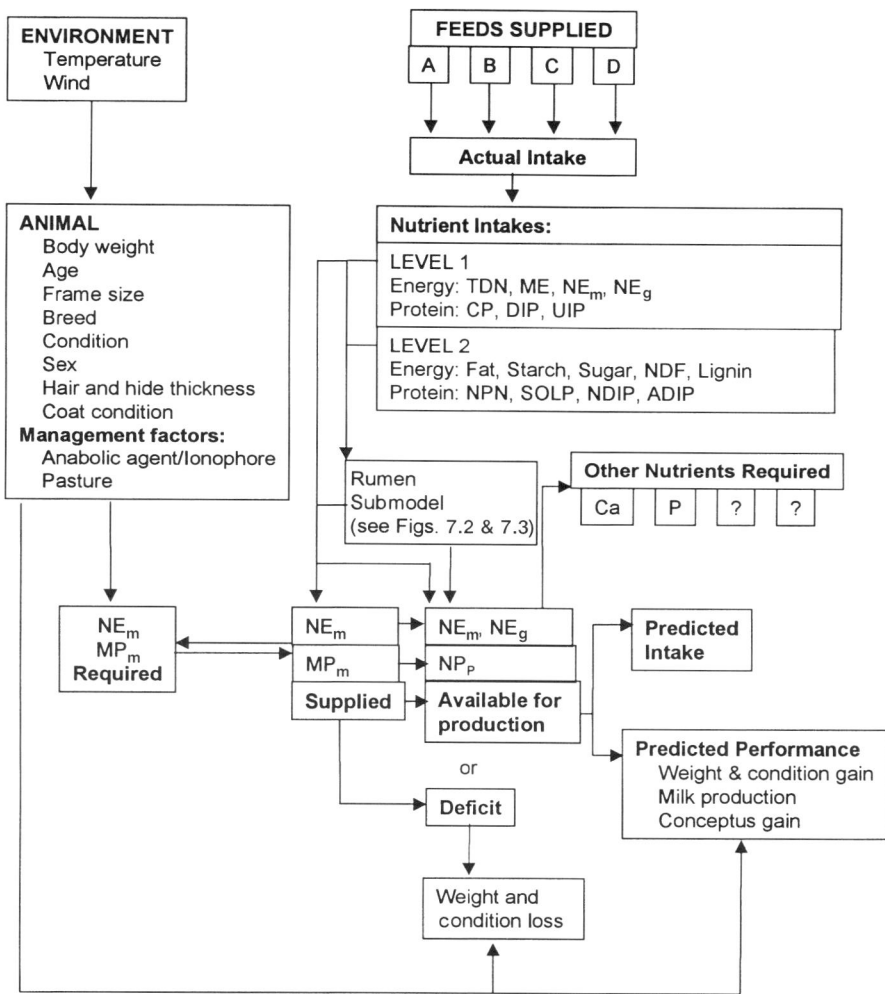

Fig. 7.1. The NRC (1996) model of nutrient requirements of beef cattle. TDN, total digestible nutrients; ME, metabolizable energy; NE$_m$, net energy for maintenance; NE$_g$, net energy for gain; CP, crude protein; DIP, degradable intake crude protein; UIP, undegradable intake crude protein; NDF, neutral detergent fibre; NPN, non-protein nitrogen; SOLP, soluble true protein; NDIP, protein insoluble in neutral detergent; ADIP, protein insoluble in acid detergent; MP$_m$, metabolizable protein for maintenance; NP$_p$, net protein for production.

information on composition of the feed and knowledge of the user. The first model uses traditional energy values, crude protein and ruminal protein degradability, to evaluate the adequacy of the diet in meeting requirements (level 1, Fig. 7.2). The second model computes feedstuff net energy and protein values for each situation, depending on the content of feed carbohydrate and protein fractions (level 2, Fig. 7.3). A mix of empirical and mechanistic representations of physiological functions was used to develop the models, depending on data available for developmental validation, inputs available to drive the model and risk of use.

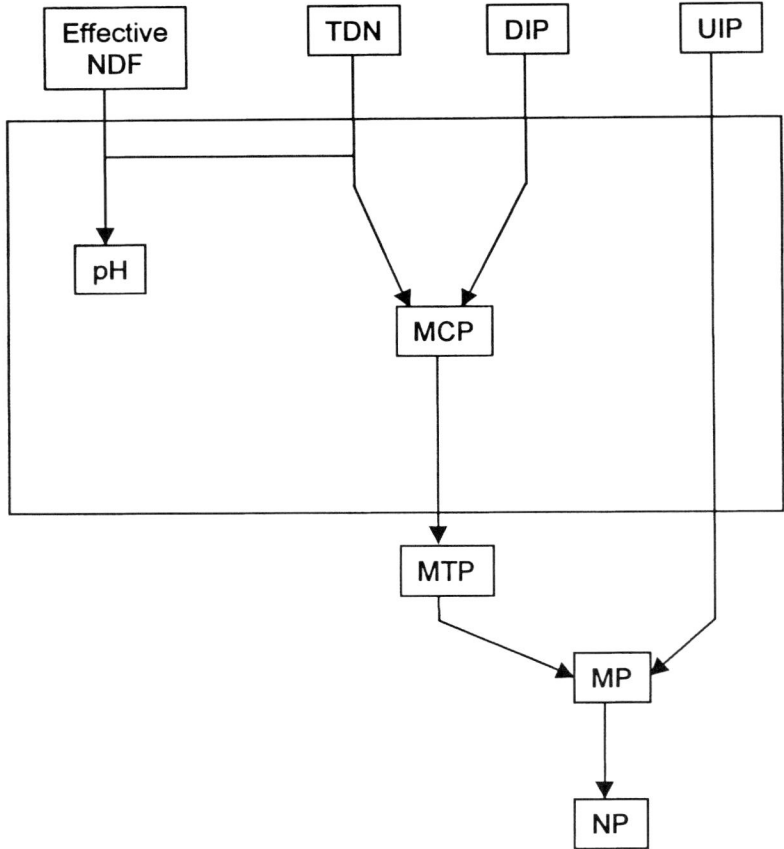

Fig. 7.2.   The NRC (1996) level 1 model of nitrogen digestion and metabolism in beef cattle. NDF, neutral detergent fibre; TDN, total digestible nutrients; DIP, degradable intake crude protein; UIP, undegradable intake crude protein; MCP, microbial crude protein; MTP, microbial true protein; MP, metabolizable protein; NP, net protein.

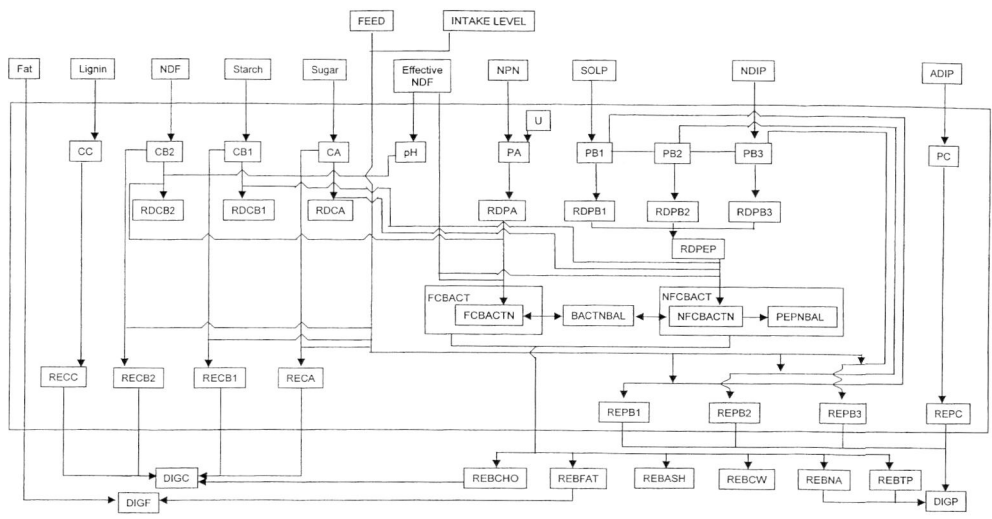

Fig. 7.3. The NRC (1996) level 2 model of nitrogen digestion in beef cattle.
NDF, neutral detergent fibre; NPN, non-protein nitrogen; SOLP, soluble true protein;
NDIP, protein insoluble in neutral detergent; ADIP, protein insoluble in acid detergent;
U, urea recycled from the body to the rumen; CC, unavailable fraction of carbohydrates;
CB2, a slowly degradable fibre fraction of carbohydrates; CB1, a slowly degradable fraction
of non-fibre carbohydrates; CA, a quickly degradable fraction of non-fibre carbohydrates;
PA, an instantaneously available fraction of crude protein; PB1, a rapidly available fraction
of true protein; PB2, a fraction of feed protein that is soluble in neutral detergent and of
intermediate availability; PB3, a fraction of protein that is insoluble in neutral detergent
but soluble in acid detergent and is slowly available; PC, protein insoluble in acid detergent
that is unavailable in the rumen; RDCB2, rumen-degradable CB2 carbohydrate;
RDCB1, rumen-degradable CB1 carbohydrate; RDCA, rumen-degradable CA carbohydrate;
RDPA, rumen-degradable PA protein; RDPB1, rumen-degradable PB1 protein;
RDPB2, rumen-degradable PB2 protein; RDPB3, rumen-degradable PB3 protein;
RDPEP, rumen-degraded peptides and amino acids; FCBACT, fibre-digesting microbes;
NFCBACT, non-fibre carbohydrate-digesting microbes; FCBACTN, nitrogen in FCBACT;
BACTNBAL, microbial N balance; NFCBACTN, nitrogen in NFCBACT; PEPNBAL, peptide
and amino acid N balance; RECC, rumen escape CC carbohydrate; RECB2, rumen escape
CB2 carbohydrate; RECB1, rumen escape CB1 carbohydrate; RECA, rumen escape CA
carbohydrate; REPB1, rumen escape PB1 protein; REPB2, rumen escape PB2 protein;
REPB3, rumen escape PB3 protein; REPB3, rumen escape PB3 protein; DIGC, digestible
carbohydrate; REBCHO, carbohydrate in microbes, REBFAT, fat in microbes;
REBASH, ash in microbes; REBCW, microbial cell wall protein; REBNA, microbial nucleic
acid N; REBTP, microbial true protein; DIGF, digestible fat, DIGP, digestible protein.
From the output of this model, TDN and faecal excretion of feed components can be
predicted. To provide net protein from metabolizable protein (MP), the same equations
are used in model 2 as in model 1. MP is the true protein component of the DIGP.

Characterization of Feedstuffs and Provision of NE and NP (NRC 1996 Model Level 1)

Fractionation of feedstuffs to quantify nutrient intake depends on the modelling level chosen. For level 1, the net energy (NE) system (net energy maintenance, NE_m and gain, NE_g) proposed by Lofgreen and Garrett (1968) is retained to describe feed energy. However, total digestible nutrient (TDN) data are required to predict microbial protein yield in the rumen, and metabolizable energy (ME) intake is required to take heat production into account in cold environments. In application of the system for feeding high-grain diets to cattle in North America, NE values reported in the literature are generally reliable, although users sometimes make minor adjustments, when feed quality is reduced, and these adjustments are based on experience or information in the literature. For all forages and some grains, NE_m and NE_g are usually predicted from digestible energy (DE) or TDN. Although many nutritionists in North America predict DE from one assessment of fibre, normally acid detergent fibre (ADF), these predictions can be quite unreliable (e.g Berthiaume *et al.*, 1996), and more complex equations, incorporating neutral detergent fibre (NDF) and lignin (Van Soest, 1994), should be more satisfactory. NRC (1996) provides equations to predict NE from DE or TDN, and these are based on ME being 0.82 of the DE and the heat increment being negatively related to digestibility of energy. Users of the system may consider using an alternative to 0.82 to predict ME from DE. However, the suggested equations to predict NE from ME should be retained since they are linked to the comparative slaughter method from which NE values on feedstuffs and requirements for animals were established (Lofgreen and Garrett, 1968). For level 1, feed crude protein is fractionated as to rumen degradability so that degradable (DIP) and, by difference, undegradable (UIP) crude protein intake can be estimated. In the level 1 model, DIP and UIP are static and based on values published from various sources and by various methods. These are limitations overcome in the level 2 model of NRC (1996) as well as in other recent systems.

Operation of the digestive sub-model in level 1 (Fig. 7.2) uses TDN and DIP to estimate potential microbial crude protein (MCP) synthesis. As with many other systems, nitrogen absorbed directly from the rumen is assumed equal to nitrogen recycled to the rumen, hence DIP is the only nitrogen available to rumen microorganisms. For many situations, MCP is estimated as 0.13 of the TDN intake; however, efficiency of MCP synthesis is decreased when high- or all-grain diets are fed. This is calculated using a correction factor for effective fibre, i.e. fibre that stimulates rumen motility and saliva secretion sufficiently to maintain rumen pH above 6.4 (Beauchemin, 1991). The concept of effective fibre became established largely as a result of research with dairy cattle. Effective fibre is related to particle size and may be determined through dry sieving where the NDF retained on a 1.18 mm screen is determined (Smith and Waldo, 1969; Mertens, 1985). Based upon *in vitro* studies, microbial yield is decreased by 2.5% for every 0.01 decrease in effective NDF, below 0.2 of the diet (Russell *et al.*, 1992). This relationship has been validated by at least one *in vivo* study (Spicer *et al.*, 1986). The efficiency of microbial protein synthesis is probably also reduced when feed quality is quite low, which may reflect low passage rates of digesta and a high proportion of energy used by micro-organisms for maintenance. No specific relationship was recommended by NRC (1996) to take this into account in level 1 operation of the model.

In level 1, it is assumed that 0.8 of the MCP is true protein and 0.8 of both the MCP true protein and the undegradable feed protein is digestible. Thus, metabolizable protein (MP) is estimated from 0.64 of the MCP plus 0.8 of the UIP. Protein requirements are expressed in terms of MP for maintenance and as net protein (NP) for productive functions. For conversion of MP to NP, coefficients of 0.65 were chosen for lactation and pregnancy. For growth, a variable coefficient based upon the following equation was adopted:

$$\text{Efficiency (NP/MP, \%)} = 83.4 - (0.114 \text{ EQSBW}) \tag{7.1}$$

where EQSBW (equivalent shrunk body weight) is the body weight of the animal after adjustment to the weight of a standard reference animal. EQSBW is determined in the process to estimate energy and protein content in the animal's gain (see below). From this equation, efficiency declines from 78.8% in a newborn 40 kg calf to 49.2% for an animal with an EQSBW of 300 kg. The minimum value of 49.2% is used for all growing and fattening cattle with an EQSBW greater than 300 kg. This equation was developed by Ainslie *et al.* (1993), using data reported by INRA (1989), and was validated from feeding trial results (Ainslie *et al.*, 1993; Wilkerson *et al.*, 1993). The form of declining efficiency in MP conversion to NP, as cattle grow, is in agreement with the principle that the efficiency of protein deposition in animals (the ratio of protein deposited to protein synthesized) declines as animals fatten (Lobley, 1988). The linear form of this equation may have to be adjusted in future as data become available, since a decreasing rate of decline in efficiency occurs as animals age.

Characterization of Feedstuffs and Components Available for Digestion in the Rumen (NRC 1996 Model level 2)

For level 2, total carbohydrate is estimated by subtracting the sum of fat, ash and crude protein from feed dry matter. Similarly, non-fibre carbohydrate (NFC) is estimated by subtracting NDF from feed total carbohydrate. NFC is fractionated into starch and, by difference from the total, a residue, referred to as 'sugar', that includes polymeric carbohydrates other than starch as well as monomeric sugars but also, in the case of fermented feeds, organic acids. To determine fibre available to rumen microorganisms, neutral detergent-insoluble protein (NDIP) must be subtracted from NDF to correct for protein, already accounted for in the CP fraction. This process of feed fraction-ation is summarized in Fig. 7.4.

With respect to protein for level 2, crude protein is fractionated into soluble protein, NDIP and acid detergent-insoluble protein (ADIP). The soluble protein is fractionated further into non-protein nitrogen (NPN) and

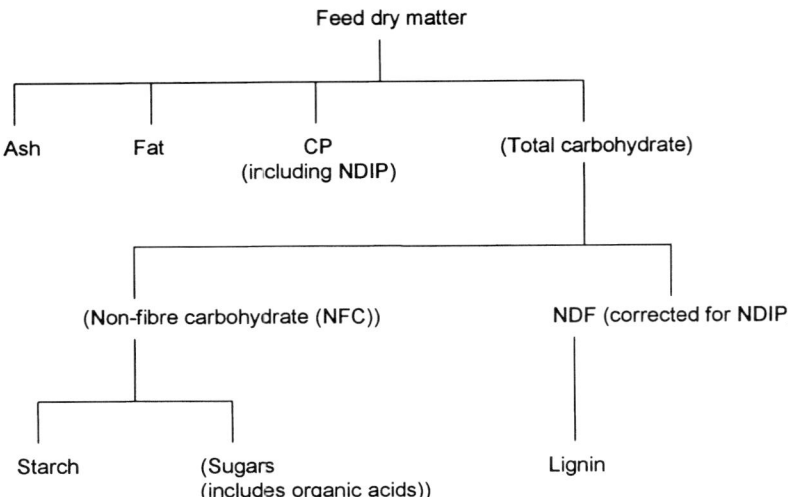

Fig. 7.4. Feed dry matter fractionation to determine carbohydrate components (NRC 1996 model level 2). CP. crude protein; NDIP, protein insoluble in neutral detergent; NDF, neutral detergent fibre. Lignin is the lignin insoluble in 72% sulphuric acid. Components in parentheses indicate that they are estimated by difference of the other components at the same level from the total.

soluble true protein (Fig. 7.5). Further details on the carbohydrate and protein fractionation used for level 2 of NRC (1996) may be found in Sniffen *et al.* (1992); recently the methods for protein fractionation have been standardized (Licitra *et al.*, 1996).

Operation of the rumen sub-model for level 2 is based on the concept that feedstuffs contain carbohydrate and protein fractions, each with unique first-order rate constants for digestion, as described by Sniffen *et al.* (1992). With respect to carbohydrates, there is an A component, two B components and one C component. The A component (CA) is derived from the sugar fraction and this component has a digestion rate that usually exceeds 1 h^{-1}. The first B component (CB1) is derived from starch with a digestion rate that varies between 0.1 and 0.5 h^{-1}, whereas the second B component (CB2) represents available fibre and has a digestion rate constant usually between 0.015 and 0.12 h^{-1}. The CB2 component is estimated by difference after the C component of carbohydrate (CC), or the unavailable fibre, has been estimated. The CC component is estimated by the amount of lignin, multiplied by 2.4. This may overestimate the unavailable fibre in feeds with low lignification but there are insufficient data available to define the amount of overall feedstuffs more precisely. Recently, Traxler *et al.* (1998) evaluated this method for predicting CC. A logarithmic equation predicting unavailable cell wall from the concentration of lignin in the NDF had lower prediction error and less bias than the

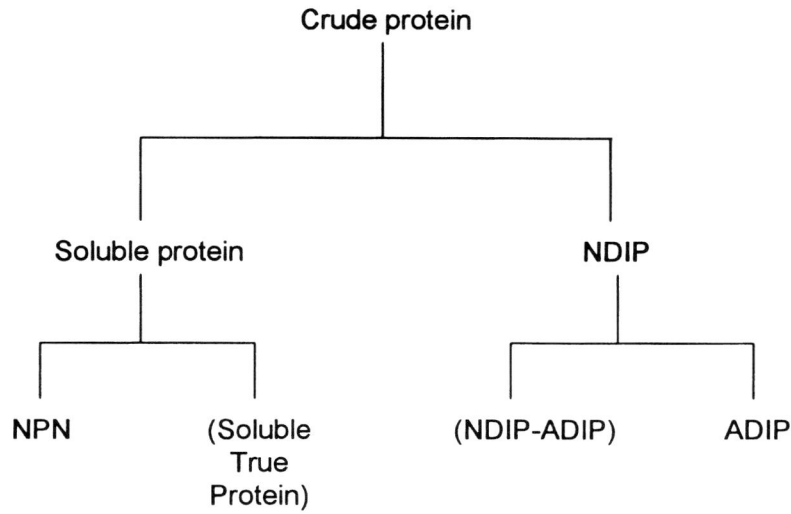

Fig. 7.5. Feed crude protein fractionation to determine protein components (NRC 1996 model level 2). NDIP, protein insoluble in neutral detergent; NPN, non-protein nitrogen; ADIP, protein insoluble in acid detergent. Components in parentheses indicate that they are estimated by difference of the other components at the same level from the total.

linear function based on lignin, multiplied by 2.4. However, the linear function, multiplying lignin by 2.4, gave the least bias in predicting animal performance.

The A component of protein (PA) is represented by NPN and is assumed to be entirely and instantaneously available for fermentation in the rumen. The PA component is augmented with an estimate of urea recycled from the body to the rumen. There are three B components. PB1 is represented by the soluble true protein and is degraded in the rumen with a rate constant varying between 1 and $4\,h^{-1}$, depending on the feedstuff. NDIP less ADIP represents the PB3 component and, in many grains, this corresponds to prolamines; the digestion rate for this fraction is very low, less than $0.1\,h^{-1}$. PB2 is NDIP, determined by difference once the other components are quantified, and corresponds to glutelins in grains; the PB2 fraction has a digestion rate between 0.1 and $0.4\,h^{-1}$. Unavailable protein is designated PC and is represented by ADIN. Animal proteins do not contain fibre and present a problem with this representation of protein availability. Thus, for the NRC (1996) model, animal proteins have been assigned ADIN, and hence values for the CC component, that correspond to average unavailable protein due to heat damage and the keratins.

Rumen Digestion and Output (NRC 1996 Model Level 2)

The underlying concept of digestion is that the rumen contains fibre-digesting (FCBACT) and non-fibre-digesting (NFCBACT) microorganisms, with differing responses to nitrogen in the form of ammonia or peptides, as described by Russell *et al.* (1992). The organization of the model is presented in Fig. 7.3.

he general form of the equations used to calculate microbial yield (Y) is derived using a Lineweaver–Burke derivation of the format used to describe enzyme kinetics, as follows:

$$1/Y = (K_M/K_d) + (1/YG) \qquad (7.2.)$$

where K_M is a maintenance coefficient for microorganisms, i.e. 0.05 g FCBACT g FCBACT^{-1} h^{-1} and 0.15 g NFCBACT g NFCBACT^{-1} h^{-1}. K_d is the degradation rate of the carbohydrate components utilized by the specified microorganism and YG is the theoretical maximum yield (0.4 g microorganisms g micro-organisms^{-1} for both FCBACT and NFCBACT).

The yield of FCBACT in the rumen is dependent on the size of the PA component and excess nitrogen not required by NFCBACT. Yield is modified according to the rate of digestion of CB2 and the proportion of effective NDF in the feed, that causes rumen pH to be adjusted, as described for operation of level 1. The degradation rate of CB2, being the fibre component of the feed, is dependent on rumen pH and is reduced when the pH is less than 6.46, to reach zero if the pH falls to 5.7.

The yield of NFCBACT is dependent on the sum of the rumen-degradable PB components and is adjusted according to the degradation rate of the CA and CB1 components and effective fibre, as described. PB components of protein, being true protein, degrade in the rumen to give peptides or amino acids (RDPEP) as intermediates on the path to ammonia. Research has shown that peptides or amino acids stimulate yield of NFCBACT over what is achieved through the provision of nitrogen in the form of ammonia (Russell *et al.*, 1983). Thus, in the NRC (1996) model, the yield of NFCBACT is stimulated by up to 18.7% when peptides and amino acids are available. The response to peptides and amino acids is curvilinear, when RDPEP is expressed as a proportion of total rumen-degradable non-protein organic matter (RDCA + RDCB1 + RDCB2) (Russell *et al.*, 1992). In the NRC (1996) model, it is recognized that NFCBACT do not obtain more than two-thirds of their nitrogen from peptides and amino acids (Russell *et al.*, 1983) and this constraint is included in the prediction of nitrogen utilization by NFCBACT. When RDPEP is deficient in providing the nitrogen requirements of NFCBACT, then additional nitrogen may come from RDPA, equivalent to the PA component of the diet. When RDPEP is in excess of the nitrogen requirements of the NFCBACT, then it can fulfil the nitrogen requirements of the FCBACT. A key output of the rumen sub-model of NRC (1996) is to estimate the balance of RDPEP (RDPEPBAL) as well as microbial nitrogen (BACTNBAL), so that a deficiency or excess of NPN or rumen-degradable feed true protein in the diet may be determined.

As indicated, yield of rumen microorganisms ultimately depends on the proportion of carbohydrate and protein digested in the rumen; therefore, a fraction of each carbohydrate and protein component that passes or escapes rumen digestion is estimated in the model. Passage rates through the rumen are estimated from the level of food intake expressed in proportion to body weight. The relationship of food intake level to passage rate is dependent on the feedstuff, being different between forages and concentrates and affected by particle size, which is reflected in the values for effective NDF. The general form to estimate the proportion of feedstuff components that escape digestion in the rumen (RECA, RECB1, etc.) is to calculate the rate of passage as a proportion of the sum of digestion and passage rates in the rumen.

For the calculation of escape, it is assumed that the C components of both carbohydrate and protein (CC and PC) escape rumen digestion in their entirety but, by contrast and as discussed above, none of the A fraction of protein (PA) escapes digestion.

From calculations of ruminal digestion and passage, as well as microbial yields, it is possible to define components leaving the rumen and being available for digestion in the lower portions of the gastrointestinal tract. Microbial matter (from both FCBACT and NFCBACT) is assumed to be 21% carbohydrate, 12% fat, 4.4% ash and 62.5% crude protein. Of the crude protein or nitrogen (N), 60% is taken to be true protein N, 25% cell wall N and 15% nucleic acid N. Digestible carbohydrates and proteins are estimated from the totals of digestible feed and microbial components. In the calculations, it is assumed that the entire microbial true protein and nucleic acid nitrogen portions are digested, but none of the cell wall. Microbial carbohydrate is assumed to be 95% digestible. For ruminal escape portions of feed carbohydrates, it is assumed that all of the RECA fraction is digested post-ruminally but only 0.2 of the fibre (REB2). For starch (REB1), true intestinal digestibilities depend on the grain and the method of processing and vary from 0.5 to 0.97 of the amount entering the intestines. With respect to feed protein that escapes rumen digestion, all of the neutral detergent-soluble true protein (PB1 and PB2) is taken as digestible, whereas a digestibility coefficient of 0.8 is ascribed to the portion originating from NDIP. All of the dietary fat is presumed to escape rumen fermentation, and this source of fat as well as the microbial fat is assigned a post-ruminal digestibility coefficient of 0.95. Thus, it is possible to predict TDN and MP from analysis of the feed and animal effects, e.g. feed intake level (Fig. 7.3), and by inference estimate the faecal output of fat, carbohydrate and protein of either feed or microbial origin. The derivation of TDN and MP from level 2 of the model implies that NE_m, NE_g and NP can be estimated so that remaining functions in model 2 are similar to those of model 1.

There is one other important feature, however, for level 2 operation of the model, and this is that the framework of rumen digestion given above is expanded to include amino acids. Although the detail of information here was too complex to include in Fig. 7.3, the general description on how these calculations are made is relatively straightforward. Ten amino acids considered necessary in the MP for ruminants are considered, namely, arginine, histidine, isoleucine, leucine, lysine, methionine, phenylalanine, tryptophan, tyrosine and valine. The amino acid composition of insoluble feed protein is applied to the amounts of ruminal escape feed protein components (REB1, REB2 and REB3), together with the corresponding digestion coefficient for these components, to estimate amino acids derived in the MP, from UIP, for the animal. For microorganisms, the amino acid composition of the cell wall and true protein components was characterized, based primarily on data from Clark *et al.* (1992), and the corresponding digestion coefficients for these components were applied so as to estimate the contribution of each amino acid to the microbial portion of the MP.

Description of Animal and Management Factors Affecting Maintenance (NRC 1996 Model Levels 1 and 2)

Beef cattle are described on the basis of several factors (Fig. 7.1), designed to apportion their requirements for NE_m and MP for maintenance (MP_m) and to rationalize performance expected to be gained or lost from additional or lesser amounts of feed provided (Fig. 7.1). The most significant factor affecting maintenance is live weight, used primarily to estimate the NE_m requirement (Mcal) as:

$$= 0.077(SBW)^{0.75} \tag{7.3}$$

where SBW is the shrunk body weight of the animals (kg). The convention in North America is to calculate shrink, usually 0.04 of full body weight, and subtract this amount from actual body weight to obtain a shrunk weight from which to estimate NE_m requirements. The relationship of body weight to NE_m requirements was first established by Lofgreen and Garrett (1968) and confirmed by additional studies (Garrett, 1980). Shrunk body weight is used in many other areas as well, notably to predict energy and protein composition of the weight gain and dry matter intake. Although animal age probably does affect the coefficient used to estimate NE_m requirements, being greater in the newborn and declining with age, assessment of energy intake in the suckling beef calf is rarely performed so an age adjustment on NE_m requirements is not considered significant. Animal age is required in the NRC (1996) model, however, to estimate core body insulation for the prediction of lower critical temperature and to adjust the lactation curves of cows to estimate the production of milk.

Cattle breed is necessary as an input variable to account for differences among breeds in maintenance energy expenditure and peak milk production. Generally, *Bos indicus* breeds have lower maintenance energy expenditures, expressed per unit of metabolic body weight, than *Bos taurus* and, within *B. taurus* breeds, breeds of cattle capable of high levels of milk production have greater expenditure than traditional beef breeds. This suggests a positive correlation between cattle productivity and maintenance energy expenditure (e.g. Ferrell and Jenkins, 1987). Adjustment factors from 1.0 for traditional beef breeds of *B. taurus* to 0.9 for *B. indicus* breeds and 1.2 for dairy breeds, with intermediate values for crossbreeds, are used in the NRC (1996) model.

Hair and hide thickness are other factors in cattle that affect heat transfer to the environment, and they are discussed with other environmental factors in the next section. Data on energy requirements for muscular work, such as walking, are limited for cattle in North America. An equation proposed by CSIRO (1990) that takes dry matter intake, feed quality, terrain, feed availability and body weight into account, and increases requirements for NE_m, was adopted by NRC (1996).

In order to express maintenance protein requirements, reflected by metabolic and endogenous losses of nitrogen in faeces and urine, NRC (1996) estimated a value of 3.8 g of MP kg^{-1} $BW^{0.75}$ day^{-1}. This form of expressing maintenance protein requirements has been adopted in other systems and is discussed further below.

Description of Environmental Factors (NRC 1996 Model Levels 1 and 2)

Much of the information to model the impact of environmental conditions on energy status in cattle was based on the model published by Fox *et al.* (1992), which used data from a previous NRC publication (NRC, 1981). Throughout the temperature range, regardless of the thermoneutral zone, the basic metabolic rate of cattle, adapted to the environment, is affected by ambient temperature. Hence, the NE_m required is increased proportionately 0.1 for every 10°C below 20°C and decreased by a corresponding amount for higher temperatures. In addition to this adjustment, effects of cold on cattle in ambient temperatures below the lower critical temperature (LCT) are considered. The equations used by NRC (1996) envisage the cattle body as a physical unit, where insulation is provided by the hair and hide as well as the external fat. External insulation is reduced through exposure to wind and when the animal's coat is wet. When cattle are exposed to ambient temperatures below the LCT and to account for energy required to maintain body temperature, then the caloric equivalent of the heat lost to the environment is in terms of metabolizable energy (ME). Thus, an adjustment taking efficiency of ME conversion to NE_m is necessary to express the effect of cold on energy requirements.

To account for heat stress, the NE_m requirement is adjusted for the work of panting. Additional effects of cold and heat on feed intake, presented below, account for effects of extreme temperatures on animal performance.

Calculation of Predicted Animal Performance (NRC 1996 Model Levels 1 and 2)

From the provision of NE_m and MP_m in the feed, it is straightforward to estimate energy and protein available for production or, in the event of a deficit in meeting maintenance requirements, an effect on weight and condition loss from the body (Fig. 7.1). The NE_g requirements for tissue gain in growing and fattening cattle are derived from the California data for medium frame steers (Garrett, 1980) and NP_g requirements are predicted from the energy concentration in the weight gain, according to NRC (1984). Limitations in the applicability of these relationships are elaborated upon in a later part of this chapter. Requirements for feed energy for milk production are expressed as milk energy secreted, in terms of NE_m, since the metabolizability of ME for NE_m or NE_l are similar. To assist predictions of milk production in suckling beef cows, an equation to enable calculation of milk production at any day of lactation, based on time of peak milk yield, level of milk production at peak and milk composition, is presented. Energy and protein requirements for fetal development in gestating beef cows are expressed as NE_m and NP. Both requirements are based on expected calf birth weight and stage of pregnancy, and were based on measurements taken on the gravid uterus of Hereford heifers bred to Hereford bulls (Ferrell *et al.*, 1976).

Variations in size and breed of cattle have presented considerable challenges in the past for the effective prediction of responses to nutrients by various types and sizes of cattle, particularly growing and fattening cattle. With most systems in the past, cattle have been described as small, medium and large frame, where adjustments to predict responses to nutrients were based on studies with representative animals of each frame size. In North America, such a categorical description of cattle is ineffective, because of so much crossbreeding among smaller and larger breeds and changes in cattle size to meet changes in market demands. For CSIRO (1990), the Australian committee had adopted the principle that the body composition of cattle, hence the composition of weight loss or gain, is constant at any proportion of mature size, with maturity being defined as the weight at 25% body fat. Energy requirements are given for each proportion of mature size, and mature weights are given for various breeds. This concept requires a data set on body fatness and composition for one type of animal from which to adjust others. The shrunk weight at which the standard animal achieves slaughter end-point is referred to as the standard reference weight (SRW). For NRC (1996), SRWs were established from the database on energy retention in cattle of medium frame size, which was used by NRC (1984). Three SRWs were selected to correspond to different degrees of fatness, currently being achieved in the North American beef industry and familiar to users. These SRWs were 435, 462 and 478 kg, corresponding to cattle grading standard (A), select (AA) and choice (AAA) in the USA (and Canada), respectively. These grades are defined by cattle with increasing body fatness and projected intramuscular fat or marbling in the meat. In preparation for using the NRC (1996) model on any given growing animal, the user must estimate the shrunk weight at which it will achieve a similar level of fatness (FSBW). From these two body weights, an equivalent shrunk body weight (EQSBW) is established:

$$EQSBW = SBW(SRW/FSBW) \qquad (7.4)$$

where SRW and FSBW are defined above and SBW is the current shrunk body weight of the animal being evaluated.

$$NE_g = 0.0635 \ EQEBW^{0.75} \ EBG^{1.097} \tag{7.5}$$

where NE_g is net energy requirement for gain or retained energy (Mcal), EQEBW is 0.91 EQSBW (kg) and EBG (empty body gain) is 0.956 SWG (shrunk weight gain, kg).

This method for predicting energy retention (NE_g) in a diverse set of cattle was validated with three distinctly different sets of data. With the proposed system, 0.94 of the variation in observed NE_g was explained and the bias was 0.02 of the mean. This was a major improvement over using NRC (1984) equations for discrete groups of cattle. There is a major onus on the user to provide appropriate FSBW values for test animals. As animal gender, anabolic agents, previous plane of nutrition and the energy level of the diet of feedlot cattle can all affect market end-points or FSBW, users are given guidelines to help them establish adjustments which they might want to make before running the programme.

One might question the suitability of using size scaling to a standard reference animal to standardize the composition of body growth across animals of all breeds. In fact, CSIRO (1990) did adopt two equations to predict body fat and protein, one for most breeds and one for large European breeds, where protein content of the gain is unusually high. From an analysis of the growth rate of bulls of different breeds, Webster *et al.* (1982a) inferred that there are slow and fast maturing large frame and small frame breeds of cattle, which affirms the CSIRO (1990) approach. The NRC (1996) approach should account for most composition of gain differences in cattle, as evidenced by the validations discussed above. The validation database included wide ranges in diet, gender, growth rate and mature body size, as well as *B. taurus* (beef and dairy breeds) and *B. indicus* cattle.

In many beef cattle systems, cows are underfed during winter or dry periods but make up for loss in weight and condition at other times during the year. Thus, prediction of nutritional effects on weight and condition gain or loss in cows is useful. Although cow condition can be predicted objectively using ultrasound to measure fat over the loin or rump, visual assessment is made most commonly. Different scoring systems are used, with a 1–5 or 1–9 scale most common. In NRC (1996), data collected over many years for cows of different sizes and conditions at the Meat Animal Research Center of the USDA were used to develop equations to predict changes in body condition score from diet energy balance. From these data, it was found that the energy content of 1 kg of weight reserves in a mature cow was constant at 5.82 Mcal. Each unit change in condition score (1–9 scale) for cattle in the NRC (1996) data corresponded to a change in proportional body fat of 0.0377. From this information, body weight and condition score changes for any cow size may be used to compute feed required to gain or lose condition.

Prediction of Dry Matter Intake (NRC 1996 Model Levels 1 and 2)

Dry matter intake (DMI) is predicted by NRC (1996) using empirical equations that take into account metabolic body size and energy concentration of the diet (NE_m). The relationships of NE_m to DMI were curvilinear for growing cattle, with highest intake for intermediate NE_m, but relatively linear and positive for cows. For growing cattle, a different intercept is proposed in the relationship for heifers and yearling steers compared with steer calves under 12 months of age. Also, for beef cows, a different intercept was established for pregnant compared with non-pregnant cows and, in addition, an allowance was made for milk production. Adjustments to the above predictions for intake, that account for additional observed effects of body fat content, breed, use of feed additives or anabolic agents, temperature and mud in the environment, were recommended. Only temperature and mud affect intake of cows, whereas all of the factors affect intake of growing cattle. These adjustments were based on information presented by Fox *et al.* (1988) and confirmed by NRC (1987). The adjustment for body fat content is required to express the fact that intake of growing cattle declines as they approach the desired levels of body fat for market conditions (1.0 of estimated intake at proportional 0.213 body fat declining to 0.73 of estimated intake at proportional 0.315 body fat). The adjustment for breed accounts for proportional 0.08 higher intakes with Holsteins versus growing cattle of other breeds.

With regard to ambient temperature, intake is predicted to decline to 0.65 of normal when temperatures rises from between 15 and 25°C to reach more than 35°C, and to increase by proportional 0.16 when temperatures fall to below −5°C. Muddy conditions cause a decline in feed intake.

Finally, intake is adjusted for grazing animals when the forage allowance is less than four times the predicted DMI or when the available herbage mass is less than 1150 kg of DM ha⁻¹. Prediction of intake by grazing cattle is a challenging area and has been addressed by one of the other committees in greater depth, this is discussed later in this chapter.

Use of Anabolic Agents and Ionophores (NRC 1996 Model Levels 1 and 2)

Anabolic agents are used quite widely in North America and many other parts of the world, and were once used in Europe, but they are now banned in that continent due to the fact that they are hormones. They repartition growth towards muscle and away from fat. NRC (1996) concluded that the impact of anabolic agents could be described through their effects on feed intake and frame size. It was predicted that DMI would decline proportionately by 0.06 when an anabolic agent was not being used. Anabolic agents increase the market weight of cattle at constant fatness, hence, by upward adjustment in FSBW to predict NE_g and NP_g, the effect of the agents in reducing fat and enhancing protein deposition at any given live weight can be accounted for.

Ionophores are widely used in the world in beef cattle diets to improve feed efficiency. However, they may reduce feed intake. Experimental observations have found that ionophores enhance the NE_m values of feed-stuffs. The recommendation of NRC (1996) is to increase feedstuff NE_m values proportionately by 0.12 when an ionophore is included in a diet for growing and fattening cattle. Simulations that made this adjustment and took effects of ionophore on feed intake into account generated predictions of ionophore effects on animal performance in agreement with observed results.

SIGNIFICANT DIFFERENCES IN OTHER FEEDING SYSTEMS FOR BEEF CATTLE

Feeding systems for beef cattle from outside North America have many similar features to NRC (1996). Understandably, different ideas exist among different committees, and idiosyncrasies of different regions of the world necessitate that recommendations in one area may not suit or may need to be modified for another area. This section will examine some of the important differences between NRC (1996) and other systems.

One important area where differences exist is in feed characterization. This is the subject of earlier chapters of this book and will not be dealt with extensively here. There is general agreement that feedstuffs for beef cattle should be analysed in fractions according to their availability for rumen digestion. Although, the approach taken by NRC (1996) for energy and carbo-hydrates is quite similar to others (e.g. AFRC, 1987, 1993), different approaches to the analyses of protein fractions have been taken. Notably, *in sacco* methods have been adopted (e.g. INRA, 1989). Results from these methods are compromised by the difficulty in standardization of procedures, although considerable effort has been made to overcome this problem. Enzymes to mimic rumen digestion have not been widely used, and newer technologies, such as near infrared reflectance spectroscopy (NIRS), although promising, have not gained widespread acceptance. Use of the *in vitro* gas production method to estimate digestion rates in the rumen (Pell and Schofield, 1993; Theodorou *et al.*, 1994) shows considerable promise, but this procedure has not yet been adopted by national committees on feeding standards. One of the most important considerations for feeding beef cattle is the analysis of dry matter. For fermented feeds, volatiles should be captured and accounted for; failure to do so can lead to significant errors in important economic parameters such as efficiency of feed utilization (Goodrich and Meiske, 1971).

Energy Assessment

All systems for beef cattle use a currency to evaluate energy that takes the heat increment into account. CSIRO (1990) and AFRC (1993) use ME as the currency, and coefficients of ME conversion to NE, usually denoted as k_m (maintenance) and k_f (growth and fattening), are applied to account for animal utilization of energy. All systems require accurate information on ME utilization for maintenance, milk production and tissue accretions. AFRC (1993) have based their system on studies using animal calorimetry, whereas NRC (1996) is based on the results of comparative slaughter trials. Both systems give good prediction of energy utilization in beef cattle, but a caveat is not to interchange components of one system with another as this can lead to more significant errors. One system with one methodology to describe energy metabolism in beef cattle across the world should be possible. This might come about when a mechanistic description of energy metabolism in cattle that is based primarily on chemical entities in the feed, but including its physical and biological characteristics, has been developed.

Two aspects of more immediate concern are a consideration of the effect of feed intake level and associative effects between feeds on feed DE values. Digestibility of feeds in ruminants is usually depressed as feed intake increases. This fact has been considered in UK feeding systems for ruminants for some time (e.g. AFRC, 1993) and is taken into account through the calculation of feed units (FU) for the French system (INRA, 1989). In North America, many beef cattle nutritionists have argued that a feed intake correction is not required because most cattle are fed *ad libitum* and DE data are at this feed intake level. However, cattle are not always fed *ad libitum*, for example pregnant non-lactating cows in winter. The aspect of associative effects between feeds is challenging because it is usually very difficult to predict. With the level 2 model, NRC (1996) has included components that account for a number of associative effects, including the effect of fibre and feed particle size on rumen pH and passage rate, the effect of rumen pH on fibre digestion and microbial yield, and the effect of nitrogen source on microbial growth.

Protein Digestion and Metabolism

All systems for feeding beef cattle recognize the need to consider rumen digestion of protein. The format of the NRC (1996) level 1 model is very similar to that of other systems. The French system designates protein values for feedstuffs (PDI values) that are equivalent to MP. However, two values are expressed for microbial protein, one that is based on degradable N (PDIMN) and the other on available energy (PDIME) (INRA, 1989). A PDI value is obtained by summing PDIA (undegraded digestible true protein) and the lesser of PDIMN and PDIME. This simplifies calculations but may not allow as much precision in balancing the rumen fermentation to minimize loss of excess degradable N. Degradability data for feedstuffs are obtained either from tabular values or through analysis of feed protein that is based on rumen incubations, usually *in sacco*. The relationship, developed by Webster *et al.* (1982b), that predicts degradability of forages from modified ADF (MADF) and crude protein analysis, was suggested by CSIRO (1990) as a simpler alternative. The form of this equation is:

$$\text{Degradability} = [CP - (0.1\ MADF)]/CP \tag{7.6}$$

where CP and MADF are expressed as g kg of DM^{-1}.

Protein degradability estimated from MADF and CP data should take other factors into account. For example, CSIRO (1990) recognized the need to reduce degradability when there is a high content of condensed tannins present.

In the AFRC (1993) system, degradability is estimated from *in sacco* incubations using the model proposed by Orskov and McDonald (1979). One quickly (a) and one slowly (b) degradable protein fraction are assumed. Only 0.8 of the 'a' fraction and all of the degradable 'b' fraction are assumed to be available to the rumen microorganisms. AFRC (1993) also adopted the Orskov and McDonald (1979) model to incorporate rate of passage to estimate 'effective' rumen degradability. Level of food intake is suggested as a means to predict rate of passage using the following formula:

$$\text{Rate of passage} = -0.024 + 0.179[1 - e^{(-0.278L)}] \tag{7.7}$$

where L is level of feeding in proportions of maintenance. This equation predicts passage rates ranging from 0.019 h^{-1} at maintenance to 0.104 h^{-1} at 4.5 times maintenance, and corresponding reductions in effective degradability are significant, 0.20 or more of the total protein.

All feeding systems adopt some measure of rumen available energy to estimate microbial protein synthesis. INRA (1989) recommends using fermentable organic matter, which is equivalent to digestible organic matter less ether extract, fermentation products, e.g. silage acids, and undegradable protein. CSIRO (1990) uses digestible organic matter or ME, and AFRC (1993) uses fermentable ME. Fermentable ME is found by subtracting ME in fat and fermentation products from the original ME value. Despite the different forms of available energy used, most systems generate similar estimates of microbial synthesis. Because fat can be a significant component of TDN, the NRC (1996) level 1 model, based on TDN, may not be as precise as other systems when dietary fat is elevated.

Digestibility coefficients applied to microbial protein are usually in the range of 0.80–0.85. However, there is greater variation in coefficients applied to undegradable protein, and these coefficients should take into account any protein that is rendered indigestible through heat treatment. AFRC (1993) estimate digestible undegradable protein as 0.9 times the undegradable protein minus the ADIN expressed as crude protein. INRA (1989) proposed a correction based on indigestibility values measured over the entire digestive tract.

Maintenance protein requirements of beef cattle have been estimated in several of the systems in terms of MP. The value used by NRC (1996) of 3.8 g kg^{-1} $BW^{0.75}$ day^{-1} compared with a corresponding value of 3.25 for INRA (1989). AFRC (1993) factorialized the maintenance requirement for protein into basal endogenous nitrogen and dermal losses in scurf and hair and quantified these components as equivalent to 2.1875 and 0.1125 g of MP kg $BW^{0.75}$ day^{-1}, respectively. CSIRO (1990) estimated endogenous urinary protein as a function of unadjusted live weight and endogenous faecal protein as a function of DM intake. Losses of nitrogen in skin and hair were taken as a function of metabolic body size.

Generally, lower efficiencies of utilization of MP for protein accretion in tissues than for milk and pregnancy (as in NRC, 1996) are assumed by the other systems. AFRC (1993) proposed an efficiency coefficient of 0.59 for all physiological functions, whereas the coefficient proposed by INRA (1989) varies between 0.4 and 0.68.

Composition of Growth (NE$_g$)

Energy and protein in body tissue accretions of cattle have been estimated in all current feeding systems from large sets of data, using wide ranges in cattle weights. Data used for the NRC (1996), AFRC (1993) and INRA (1989) are each different, reflecting beef cattle populations typical to North America, the UK and France, respectively. As mentioned previously, CSIRO (1990) proposed the concept of 'standard reference weight' (SRW) to adjust cattle of various breeds and/or frame sizes to provide more uniformity over cattle types, and this approach was taken by NRC (1996). CSIRO (1990) concluded that two sets of equations were necessary to cover all breeds, with Charolais, Blonde d'Aquitaine, Limousin, Chianina, Maine Anjou and Simmental being distinguished from all other *B. taurus* and *B. indicus* breeds. NRC (1996) concluded that classification into breeds was unnecessary. The main challenge with NRC (1996) is that the onus is placed on the user of this feeding system to classify cattle, at any stage of growth post-weaning, correctly with respect to mature size. The classification must take gender, use of anabolic agents, age and any dietary effect into account. The NRC (1996) and CSIRO (1990) approach has a decided advantage over other systems of unifying cattle across frame size and genetics. As presented above, NRC (1996) evaluated their approach on sets of data from the literature where body composition at various live weights, hence mature size, was known. In practice, where live weight at maturity must be estimated, this may be a challenge, although many users are sufficiently familiar with cattle that they can estimate mature live weight quite accurately. Scientists should pursue the approach taken by CSIRO (1990) and NRC (1996). Nutritionists in France and the UK, on the other hand, may find that the INRA (1989) and AFRC (1993) approach of providing descriptions of cattle in concert with data in tables or equations is quite sufficient. The classification of cattle into 16 groups by INRA (1989) may be particularly helpful to French users of the system but does not work well in North America because of extensive use of crossbreeding.

All of the current databases on composition of tissue accretions in cattle are based on empirical calculations. In Baldwin's book on modelling ruminant digestion and metabolism, Baldwin and Baldwin (1995) present a strong case for taking a more mechanistic approach to analysing growth in future models and one that should be based on DNA and protein synthesis, as in Oltjen *et al.*

(1986). In a validation, involving heifers and steers with or without an anabolic agent, NRC (1996) presented evidence that this system accounted for more of the variation and less bias in predicting cattle growth performance than either the previous edition of NRC (NRC, 1984) or Oltjen *et al.* (1986). Thus, for prediction purposes, empirical relationships may be better than mechanistic explanations of growth. In this regard, a limitation of the NRC (1996) equation is that it is based on a comparative slaughter approach, where there is a moderate range in initial and final weights and the average weight of cattle on trial must be used for definition. This implies that the NRC (1996) model may not predict composition of cattle at the extremes of body weight, generally approximating less than 300 and over 600 kg, respectively, as well as it does for cattle in between these weights. Further, the NRC (1996) model cannot be applied with confidence to cattle under 250 kg, which includes most pre-weaned beef calves.

Prediction of Feed Intake

The NRC (1996) approach to feed intake prediction takes body weight and energy concentration of the diet, productive level and several environmental factors into account. Other systems have adopted a similar approach with modifications. INRA (1989) have proposed a system that defines a forage fill value (FV), based on feed units (FU), to describe feedstuffs and a feed intake capacity (FIC) to describe the potential intake of livestock. Feed consumed is the ratio of FIC to FV. For growing and fattening cattle, FIC is defined by an exponential relationship to live weight. The exponent of the equation varies according to the diet, rising from 0.6 to 0.9 as diet quality declines. The coefficient of live weight is defined according to animal age, fatness, previous management and breed. For cows, FIC is defined according to metabolic body weight with an allowance for lactation and adjustment for body condition and parity. Most of the FU values for fresh forages and hays have been obtained through feeding trials using sheep, and this should be adequate since the FU is a relative term and defines the feedstuff rather than the animal. For silages, adjustments were made to account for feed intake depression, noted in cattle, relative to when the corresponding fresh forage was fed. Adjustments for feed intake depression with silage feeding were not considered so important by NRC (1996), because silages in North America are generally high dry matter and feed intake depression is small. For INRA (1989), additional adjustments to intake are made for feeding concentrates and take into account their substitution of forages in the diet.

AFRC (1993) have proposed an equation to predict the silage intake in mixed silage–concentrate diets for growing and fattening cattle that is based on silage dry matter and digestible organic matter in the dry matter, the ammonia concentration of the silage and the amount of concentrate fed, all in addition to the metabolic live weight of the cattle. For maize silage diets, specific recommendations including a constant substitution rate of 1 kg concentrate per 0.6 kg maize dry matter are given. For suckler cows, equations to predict intake are those established for cattle of dairy breeds.

CSIRO (1989) proposed that the voluntary food intake is the product of the potential and the relative intake, with the latter expressed as a fraction of the potential intake. Potential intake depends on the particular animal being defined and is based upon the mature weight as well as the current size of the animal and its energy demand, but disease or thermal stress can reduce it. The concept of relative intake was established to account for observed intakes of grazing cattle in particular. This is a situation largely ignored by other systems, although NRC (1996) have predicted reduced intakes by beef cattle when available herbage mass is less than 1150 kg ha^{-1} or when the forage allowance is less than four times the estimated dry matter intake. In the CSIRO (1989) system, selective grazing by cattle on pasture is simulated on the assumption that cattle eat the highest quality forage first and proceed through classes of descending quality in order. The relative intake of each class of forage is defined by its digestibility and whether the pasture consists of tame or native grasses and contains any legumes. Relative intake of a particular class of forage in the pasture is constrained by its availability.

Many other models to define selective grazing of cattle exist and different types of models for cattle, as well as other herbivores, under a variety of grazing systems have been described by Laca and Demment (1997). Future improvements in the prediction of intake by grazing beef cattle on a wide variety of pastures should be possible.

CONCLUSIONS

Considerable progress has been made towards a system of describing nutrient requirements of beef cattle and evaluating diets that should be applicable across diverse geographical regions. Scientists, teachers and advisers, students as well as some farmers, should use one or more of the current systems as one tool to integrate input information on feeds, cattle and their environment. This should enable them to predict animal performance more accurately than is otherwise possible and gain an understanding of why an animal may be performing below expectations. Greater use of these feeding systems in research, teaching and practice will improve progress towards success and profitability in beef cattle production.

NRC (1996) is the first of the systems recommended by national committees that has used a feed-driven model to predict animal performance. From a theoretical basis, as discussed in the Introduction, a feed-driven model should improve on traditional models, using factorial methods of developing nutrient requirements, to predict and evaluate performance of beef cattle. To date, there have been no independent published studies on the reliability of the NRC (1996) model. As cited in this chapter, the NRC (1996) committee did establish good agreement between model predictions and animal performance in tests that they undertook. It is hoped that this chapter will inspire others, both in North America and elsewhere, to evaluate the NRC (1996) model and improve feed-driven models as a primary component of formulating and evaluating beef cattle diets.

By aggregating steps and processes of digestion and metabolism in the animal, the NRC (1996) model is not nearly as complex as more mechanistic descriptions, when model building is part of the scientific method to improve our understanding of the processes themselves. The NRC (1996) model was developed to avoid undue complexity whilst retaining predictive accuracy, which can be a weakness with very mechanistic models. However, it would be very useful to know, and remains to be seen, if a more complex and mechanistic model could improve accuracy of predicting beef cattle performance. With the NRC (1996) model of the rumen, for example, would it be desirable to describe microbial fermentation in greater detail, by taking protozoa and nutrient recycling in the rumen into account? Would it be helpful if the uptake of peptides and ammonia by non-structural carbohydrate fermenters were represented so that peptides and ammonia could be taken up simultaneously, rather than sequentially? Dealing with another aspect of modelling technique, it would be useful to know that if time were specifically represented in the model, would it improve prediction of the performance of cattle over a specified feeding period, compared with what is possible by representing the animals at the mid-point of the period only?

Without even changing the modelling technique, some improvements in describing existing digestive and metabolic functions could improve the value of any of the feeding systems discussed in this chapter. For example, limitations of most systems exist with reference to evaluating compensatory growth and its effect on feed utilization and animal performance. In this case, the fact that observed responses seem to depend so much on the length and severity of feed restriction makes it difficult to resolve. For all systems, a more detailed description of animal metabolism would be useful. Dietary-induced thermogenesis in cattle has been an enigma for a considerable time (Orskov and MacLeod, 1990). Representation of thermogenesis, using metabolizability of the diet as a predictor, is an empirical process and is covering up considerable knowledge we now have on details about heat production in different tissues and why it is produced. With respect to NRC (1996), improvement here could incorporate representation of the volatile fatty acids produced during microbial fermentation as well as their utilization in metabolism. Integration of new information in this area with knowledge on protein and amino acid metabolism should also provide a more satisfactory representation of protein utilization. These are some of the gaps in existing systems; the recent systems have incorporated much that is now known about utilization of nutrients by beef cattle, and continuing effort on this scale will continue to benefit everyone.

REFERENCES

Agricultural and Food Research Council (1987) Technical Committee on Responses to Nutrients (1987) Report No. 1. Characterization of feedstuffs: Energy. *Nutrition Abstracts and Reviews Series B* 57, 507–523.

Agricultural and Food Research Council (1993) *Energy and Protein Requirements of Ruminants.* CAB International, Wallingford, UK.

Ainslie, S.J., Fox, D.G., Perry. T.C., Ketchen, D.J. and Barry, M.C. (1993) Predicting amino acid adequacy of diets fed to lightweight Holstein steers. *Journal of Animal Science* 71, 1312–1319.

Baldwin, R.L. and Baldwin, R.L., VI (1995) Biology and algebraic models of growth. In: Baldwin, R.L. (ed.), *Modeling Ruminant Digestion and Metabolism.* Chapman & Hall, London, pp. 247–266.

Beauchemin, K.A. (1991) Ingestion and mastication of feed by dairy cattle. *Veterinary Clinics of North America: Food Animal Practice* 7, 439–463.

Berthiaume, R., Buchanan-Smith, J.G., Allen, O.B. and Viera, D.M. (1996) Prediction of liveweight gain by growing cattle fed silages of contrasting digestibility, supplemented with or without barley. *Canadian Journal of Animal Science* 76, 113–119.

Clark, J.H., Klusmeyer, T.H. and Cameron, M.R. (1992) Microbial protein synthesis and flows of nitrogen fractions to the duodenum of dairy cows. *Journal of Dairy Science* 75, 2304–2323.

Commonwealth Scientific and Industrial Research Organization (1990) *Feeding Standards for Australian Livestock. Ruminants.* CSIRO Publications, East Melbourne, Australia.

Ferrell, C.E., Garrett, W.N. and Hinman, N. (1976) Growth, development and composition of the udder and gravid uterus of beef heifers during pregnancy. *Journal of Animal Science* 42, 1477–1489.

Ferrell, C.E. and Jenkins, T.G. (1987) Influence of biological type on energy requirements. In: *Proceedings of the Grazing Livestock Nutrition Conference.* Oklahoma Agricultural Experiment Station, Oklahoma State University, Stillwater, Oklahoma, pp. 1–7.

Fox, D.G., Sniffen, C.J. and O'Connor, J.D. (1988) Adjusting nutrient requirements of beef cattle for animal and environmental variations. *Journal of Animal Science* 66, 1475–1495.

Fox, D.G., Sniffen, C.J., O'Connor, J.D., VanSoest, P.J. and Russell, J.B. (1992) A net carbohydrate and protein system for evaluating cattle diets. III. Cattle requirements and diet adequacy. *Journal of Animal Science* 70, 3578–3596

Garrett, W.N. (1980) Energy utilization by growing cattle as determined in seventy-two comparative slaughter experiments. In: Mount, L.E. (ed.), *Energy Metabolism.* European Association of Animal Production Publication No. 26. Butterworths, London, pp.3–7.

Goodrich, R.D. and Meiske, J.C. (1971) Methods for improving the interpretation of experimental feedlot trials. *Journal of Animal Science* 33, 885–890.

Institut National de la Recherche Agronomique (1989) *Ruminant Nutrition.* Libbey Eurotext, Montrouge, France.

Laca, E.A. and Demment, M.W. (1997) Foraging strategies of grazing animals. In: Hodgson, J. and Illius, A.W. (eds), *The Ecology and Management of Grazing Systems.* CAB International, Wallingford, UK, pp 137–158.

Licitra, G., Hernandez, T.M., and Van Soest, P.J. (1996) Standardization of procedures for nitrogen fractionation of ruminant feeds. *Animal Feed Science and Technology* 57, 347–358.

Lobley, G.E. (1988) Protein turnover and energy metabolism in animals: interactions in leanness and obesity. In: LeClercq, B. and Whitehead, C.C. (eds), *Leanness in Domestic Birds. Genetic, Metabolic and Hormonal Aspects.* Butterworths, London, pp. 331–361.

Lofgreen, G.P. and Garrett, W.N. (1968) A system for expressing net energy requirements and feed values for growing and finishing cattle. *Journal of Animal Science* 27, 793–806.

Mertens, D.R. (1985) Effect of fiber on feed quality for dairy cows. In: *Proceedings of the 46th Minnesota Nutrition Conference.* University of Minnesota, St Paul, Minnesota, pp. 209–224.

National Research Council (1981) *Effect of Environment on Nutrient Requirements of Domestic Animals.* National Academy Press, Washington, DC.

National Research Council (1984) *Nutrient Requirements of Beef Cattle*, 6th edn. National Academy Press, Washington, DC.

National Research Council (1987) *Predicting Feed Intake of Food-Producing Animals.* National Academy Press, Washington, DC.

National Research Council (1996) *Nutrient Requirements of Beef Cattle*, 7th edn. National Academy Press, Washington, DC.

Oltjen, J.W., Bywater, A.C., Baldwin, R.L. and Garrett, W.N. (1986) Development of a dynamic model of beef cattle growth and composition. *Journal of Animal Science* 62, 86–97.

Orskov, E.R. and McDonald, I. (1979) The estimation of protein degradability in the rumen from incubation measurements weighted according to rate of passage. *Journal of Agricultural Science, Cambridge* 92, 499–503.

Orskov, E.R. and MacLeod, N.A. (1990) Dietary-induced thermogenesis and feed evaluation in ruminants. *Proceedings of the Nutrition Society* 49, 227–237.

Pell, A.N. and Schofield, P. (1993) Computerized monitoring of gas production to measure forage digestion *in vitro. Journal of Dairy Science* 76, 1063–1073.

Russell, J.B., Sniffen, C.J. and Van Soest, P.J. (1983) Effect of carbohydrate limitation on degradation and utilization of casein by mixed rumen bacteria. *Journal of Dairy Science* 66, 763–775.

Russell, J.B., O'Connor, J.D., Fox, D.G., Van Soest, P.J. and Sniffen, C.J. (1992) A net carbohydrate and protein system for evaluating cattle diets: I. Ruminal fermentation. *Journal of Animal Science* 70, 3551–3561.

Smith, L.W. and Waldo, D.R. (1969) Method for sizing forage cell wall particles. *Journal of Dairy Science* 52, 2051–2059.

Sniffen, C.J., O'Connor, J.D., Van Soest, P.J., Fox, D.G. and Russell, J.B. (1992) A net carbohydrate and protein system for evaluating cattle diets: II. Carbohydrate and protein availability. *Journal of Animal Science* 70, 3562–3577.

Spicer, L.A., Theurer, C.B., Sorne, J. and Noon, T.H. (1986) Ruminal and post-ruminal utilization of nitrogen and starch from sorghum grain-, corn- and barley-based diets by beef steers. *Journal of Animal Science* 62, 521–530.

Theodorou, M.K., Williams, B.A., Dhanoa, M.S., McAllan, A.B. and France, J. (1994) A simple gas production method using a pressure transducer to determine the fermentation kinetics of ruminant feeds. *Animal Feed Science and Technology* 48, 185–197.

Traxler, M.J., Fox, D.G., Van Soest, P.J., Pell, A.N., Lascano, C.E., Lanna, D.P.D., Moore, J.E., Lana, R.P., Velez, M. and Flores, A. (1998) Predicting forage indigestible neutral detergent fiber from lignin concentration. *Journal of Animal Science* 76, 1469–1480.

Van der Honing, Y. and Alderman, G. (1988) Feed evaluation and nutritional requirements. 2. Ruminants. *Livestock Production Science* 19, 217–278.

Van Soest, P.J. (1994) *Nutritional Ecology of the Ruminant*, 2nd edn. Cornell University Press, Ithaca, New York.

Webster, A.J.F., Ahmed, A.A.M. and Frappell, J.P. (1982a) A note on growth rates and maturation rates in beef bulls. *Animal Production* 365, 281–284.

Webster, A.J.F., Simmons, I.P. and Kitcherside, M.A. (1982b) Forage protein and the performance and health of the dairy cow. In: Thomson, D.J., Beever, D.E. and Gunn, R.G. (eds), *Forage Protein in Ruminant Animal Production*. British Society of Animal Production, Occasional Publication No. 6, pp. 89–95.

Wilkerson, V.A., Klopfenstein, T.J., Britton, R.A., Stock, R.A. and Miller, P.S. (1993) Metabolizable protein and amino acid requirements of growing cattle. *Journal of Animal Science* 71, 2777–2784.

8 Feeding Systems for Sheep

L.A. Sinclair and R.G. Wilkinson

Harper Adams University College, Edgmond, Newport, Shropshire, UK

INTRODUCTION

In the majority of ruminant feeding systems used throughout the world, the principles governing energy and protein supply have generally been determined from work conducted with sheep. As a result, most systems take a common approach to the prediction of energy and protein supply from feeds for both sheep and cattle. However, work concerned with the prediction of animal requirements has tended to focus on individual species, and it is here that major differences between sheep and cattle become apparent. Over the last 20 years, research on the energy and protein supply and requirement for sheep has developed to incorporate nutrient digestion and absorption, and there has been a growing awareness of the effects of tissue metabolism on modifying the effects of energy and protein supply. This chapter will review some of the major energy and protein systems in use for sheep and pay particular attention to methods of predicting the nutrient requirements for maintenance and production. In addition, some of the areas where current systems are lacking will be identified and the results of recent research that could be incorporated into sheep feeding systems will be discussed.

ENERGY SYSTEMS

In all major feeding systems for sheep, energy is considered to be the first limiting nutrient, and rations initially are formulated to satisfy energy requirements. However, there is no common way of calculating energy requirements or supply. The Australian and UK energy systems (CSIRO, 1990; AFRC, 1990, respectively) are based on metabolizable energy (ME), where the ME intake from the feed and the net energy (NE) derived or retained in animal products from that feed are connected by an efficiency factor (k). In the USA, energy

©CAB *International* 2000. *Feeding Systems and Feed Evaluation Models*
(eds M.K. Theodorou and J. France)

supply is expressed in terms of total digestible nutrients (TDN) which is related to digestible energy and then ME by the use of constant conversion factors (NRC, 1985a). In contrast, the French have adopted a net energy system with each feedstuff having two net energy values: one for milk production (maintenance and production) expressed in UFL (Unité Fourragère Lait) and one for meat production (maintenance and body weight gain) expressed in UFV (Unité Fourragère Viande) (INRA, 1989).

Net Energy Requirements for Maintenance

The net energy requirement for maintenance (NE_m) in sheep represents the unavoidable energy losses associated with vital body processes such as essential muscular activity, protein turnover, active transport and enzyme/hormone synthesis. Essentially all systems are similar in that NE_m requirements are based on fasting metabolism (FM) derived either directly by calorimetry (CSIRO, 1990; AFRC, 1993) or indirectly by regression analysis (NRC, 1985a):

AFRC (1993) NE_m (MJ day^{-1}) = C [0.25 (W/1.08)$^{0.75}$] + A <1.0 years (8.1)

NE_m (MJ day^{-1}) = C [0.23 (W/1.08)$^{0.75}$] + A >1.0 years (8.2)

CSIRO (1990) NE_m (MJ day^{-1}) = (C M) [0.28 W$^{0.75}$ e$^{(-0.03a)}$]
+ [(0.1 $\dot{k}_m$) ME$_p$] (8.3)

INRA (1989) NE_m = 0.033 UFL W$^{-0.75}$ (for dry ewes) (8.4)

NRC (1985a) NE_m (kJ day^{-1}) = 4.185 × 56 W$^{0.75}$ (8.5)

where W is live weight (kg), C is a correction for sex (1.15 for ram lambs and 1.0 for females and castrates), A is an activity allowance, M is a correction for milk-fed lambs {1 + [0.26 − (0.015 × week of life)]}, a is age in years (maximum value 6 years), ME$_p$ is ME requirement for production, k_m is the efficiency of ME utilization for maintenance and UFL is Unité Fourragère Lait.

The AFRC (1993) increments FM with an allowance of 0.0067 or 0.0106 MJ kg^{-1} live weight to allow for the activity of housed and outdoor lambs, respectively, whereas CSIRO (1990) include an activity allowance as part of the FM term. Both these systems recognize that the FM of intact males is higher (0.15) than that of females and castrates and that FM decreases with age. The AFRC (1993) simplify this by adopting different FM values for animals less than or greater than 1 year of age. In contrast to the AFRC (1993), CSIRO (1990) takes account of the evidence, although somewhat limited, to suggest that the FM of lambs fed liquid diets is approximately 0.23 greater than that of lambs of the same weight offered solid food (Walker and Faichney, 1964; Graham *et al.*, 1974) and the considerable body of evidence to suggest that FM varies directly with plane of nutrition (Foot and Tulloh, 1977; Ledger and Sayers, 1977; Webster, 1978; Gingins *et al.*, 1980). The NRC (1985a) make no allowances for activity, sex, diet or plane of nutrition in their calculation of NE_m.

Net Energy Requirements for Live Weight Gain

The net energy requirement for gain (NE_g) represents the daily energy retained by an animal and is calculated from the product of the energy value of each kilogram of gain (EV_g) and the daily live weight gain (LWG). The EV_g is calculated using the heats of combustion of fat (39.3 MJ kg^{-1}) and protein (23.6 MJ kg^{-1}) and is dependent on the relative proportions deposited, which in turn are known to vary with animal live weight (age), breed size, sex and rate of gain. All systems account for an effect of live weight on body composition and EV_g, but rely on the effects of sex to different extents.

AFRC (1993)
Non-merino males

$$EV_g \text{ (MJ } kg^{-1}) = (2.5 + 0.35 \text{ W}) \tag{8.6}$$

Castrates

$$EV_g \text{ (MJ } kg^{-1}) = (4.4 + 0.32 \text{ W}) \tag{8.7}$$

Females

$$EV_g \text{ (MJ } kg^{-1}) = (2.1 + 0.45 \text{ W}) \tag{8.8}$$

where W is liveweight (kg).

CSIRO (1990)

$$EV_g \text{ (MJ } kg^{-1} \text{ EBG)} = (6.7 + \text{R}) + [(20.3 - \text{R})/(1 + e^{[-6(\text{P} - 0.4)]})] \tag{8.9}$$

$$EV_g \text{ (MJ } kg^{-1}) = EV \text{ g (MJ } kg^{-1} \text{ EBG)} \times 0.92 \tag{8.10}$$

where EBG is the empty body gain, R is an adjustment for rate of gain, and P is the live weight divided by a standard reference weight (kg:SRW).

NRC (1985a)
Small ram breeds (95 kg)

$$EV_g \text{ (kJ } kg^{-1}) = 4.185\text{C} (318 \text{ W}^{0.75}) \tag{8.11}$$

Medium ram breeds (115 kg)

$$EV_g \text{ (kJ } kg^{-1}) = 4.185\text{C} (276 \text{ W}^{0.75}) \tag{8.12}$$

Large ram breeds (135 kg)

$$EV_g \text{ (kJ } kg^{-1}) = 4.185\text{C} (234 \text{ W}^{0.75}) \tag{8.13}$$

where C is an adjustment for sex (0.82 for ram lambs and 1.0 for females and castrates).

NB Data were not readily available to permit a comparison with INRA (1989).

The AFRC (1993) provide separate equations for males, females and castrates, and the NRC (1985a) suggest that the EV_g of male lambs is 0.82 that of females and castrates. Both these systems acknowledge that the EV_g of castrates is lower than that of females. However, the NRC (1985a) suggest that the difference is not well established and make no recommendation as to what it should be, whilst the AFRC (1993) state that the correction for female lambs is inadequate. In contrast to AFRC (1993) which makes no adjustment for the effect of breed, the NRC (1985a) account for the effects of breed by relating EV_g to ram mature size. For every 10 kg increase in ram mature size, then EV_g increases by 88 kJ $kg^{-0.75}$. The AFRC (1990) state that 'there is some evidence of a relationship between mature size and body composition at a fixed live weight' and 'that it would appear reasonable to include stage of maturity as a factor in order to try and improve the predictability of the system'. The most comprehensive system with regard to its ability to account for the effect of sex and breed is that of CSIRO (1990). In this system, sex and breed are accounted for by reference to a 'standard reference weight' (SRW).The SRW for an animal of a particular breed size and sex is defined as 'the live weight that would be achieved by an animal when skeletal development is complete and the empty body contains 250 g kg^{-1} of fat'. In addition, the CSIRO (1990) is the only system that makes an allowance for the effects of rate of gain on EV_g, a factor which is widely accepted as important with growing cattle.

Net Energy Requirements for Pregnancy and Lactation

Both AFRC (1993) and CSIRO (1990) describe the rate of accretion of energy by the gravid uterus over the final 12 weeks of gestation with Gompertz equations that are based on ARC (1980) recommendations:
Total energy content at time t (E_t, MJ) for a 4 kg lamb:

$$\log_{10} (E_t) = 3.322 - 4.979e^{-0.00643t} \tag{8.14}$$

and energy retention (E_c)

$$E_c \text{ (MJ day}^{-1}) = 0.25 \ W_0 \ (E_t \times 0.07372e^{-0.00643t}) \tag{8.15}$$

where: t is the number of days from conception and W_0 is the total weight of lambs at birth (kg).
NRC (1985a) requirements are based on the results of Rattray *et al.* (1974a) and do not consider energy requirements prior to day 100 of gestation, whilst INRA (1989) also consider that nutrient requirements are only important in the final 6 weeks of gestation. However, as lamb body composition alters with litter size (as discussed by ARC, 1980), energy concentration is adjusted for heavy (e.g. singles weighing 4.5 kg and above) and light (multiples weighing below 2.5 kg) lambs (INRA, 1989).
Most of the major feeding systems for sheep predict the net energy required for milk (NE_l) from the milk yield and the composition of the ewes milk (INRA, 1989; CSIRO, 1990; AFRC, 1993) or from a standard milk composition and the number of days of lactation (CSIRO, 1990; AFRC, 1993).

$$\text{AFRC (1993)} \quad NE_l \text{ (MJ day}^{-1}) = 0.04194 \times F + 0.01585 \times P + 0.02141 \ L \tag{8.16}$$

$$\text{CSIRO (1990)} \quad NE_l \text{ (MJ day}^{-1}) = 0.0381 \times F + 0.0245 \times P + 0.0165 \ L \tag{8.17}$$

$$\text{INRA (1989)} \quad \text{NE}_l \text{ (MJ day}^{-1}) = 0.00588 \times \text{F*} + 0.265 \quad\quad (8.18)$$

where, F = fat, P = protein, L = lactose (g kg^{-1} , except * = g l^{-1}). The net result of these equations is a relatively small difference between predictions, for example for a 70 kg ewe producing 2.5 litres of milk per day the daily net energy requirements vary from 16.7 MJ (INRA, 1989) to 17.9 MJ day^{-1} (CSIRO, 1990). The AFRC (1993) recommend that the energy content of the live weight change in an adult ewe should be taken as 23.85 MJ kg^{-1} resulting in the energy value of the loss, assuming an efficiency factor of 0.84, of 20.0 MJ kg^{-1} and equates to a dietary ME equivalent (assuming k_l = 0.62) of 32.2 MJ. The CSIRO (1990), based on the findings of MAFF (1984), suggest a value of 28 MJ kg^{-1} loss and 33 MJ kg^{-1} gain of ME in lactating ewes. INRA (1989) relate energy mobilization to a fraction of maintenance energy requirements, which varies from 0.68 maintenance in early lactation to 0.36 maintenance energy requirements in late gestation.

Net Energy Requirements for Wool Production

Based on the ARC (1980) recommendations, both the AFRC (1993) and CSIRO (1990) state that the energy value of wool is 0.023 MJ g^{-1} which, assuming a daily wool growth of 5.5 g day^{-1}, gives a net energy requirement for wool growth (NE$_w$) of 0.13 MJ day^{-1}. The ARC (1980) do not provide a value for the efficiency of ME utilization for wool growth (k_w), but CSIRO (1990) suggest that k_w is in the order of 0.18. This would suggest that the ME requirement for wool growth is approximately 0.72 MJ day^{-1}, a not inconsiderable amount in relation to the maintenance requirement of growing lambs. However, all systems (NRC, 1985a; INRA, 1989; CSIRO, 1990; AFRC, 1993) maintain that the energy requirement for wool growth is small and in practice can be ignored.

Having reviewed the information provided by the ARC (1980), CSIRO (1990) point out that many of the determinations of the efficiency of conversion of ME into NE for maintenance and growth (k_m and k_g) have involved sheep that have been gaining about 6.0 g of wool day^{-1}. Thus k_m and k_g inherently allow for the daily energy cost of up to 6.0 g of wool; an ME requirement for wool growth should only be considered if fleece growth exceeds this amount and can be calculated as:

$$\text{ME wool (MJ day}^{-1}) = 0.13 \text{ (F1} - 6) \quad\quad (8.19)$$

where F1 is the greasy fleece growth (g day^{-1}).

Efficiency of ME Utilization

As stated above, in all ration formulation systems the energy requirements of sheep are measured in terms of NE, whereas energy supply is often measured in terms of ME. The two are linked by efficiency constants which vary with the productive process (maintenance or growth). In all systems these efficiency constants are calculated from diet quality, expressed in terms of ME (NRC,

1985a) or q, the metabolizability (ME/GE) (INRA, 1989; CSIRO, 1990; AFRC, 1993) and processes such as maintenance and lactation are less affected by diet quality than growth.

AFRC (1993) and CSIRO (1990) $k_m = 0.35q + 0.503$ (8.20a)

$$k_g = 0.78q + 0.006 \qquad (8.20b)$$

$$k_l = 0.35q + 0.420 \qquad (8.20c)$$

INRA (1989) $k_m = 0.29q + 0.554$ (8.21a)

$$k_g = 0.78q + 0.006 \qquad (8.21b)$$

$$k_l = 0.24q + 0.463 \qquad (8.21c)$$

where k_m, k_g and k_l are the efficiencies of utilization of ME for maintenance, growth and lactation, respectively, and q is the metabolizability of the feed.

NRC (1985a) $NE_m = (1.37 \text{ ME}) - (0.138 \text{ ME}^2) + (0.0105 \text{ ME}^3) - 1.12$ (8.22a)

$$NE_g = (1.42 \text{ ME}) - (0.174 \text{ ME}^2) + (0.0122 \text{ ME}^3) - 1.65 \quad (8.22b)$$

Thus

$$k_m = NE_m/ME \qquad (8.22c)$$

With growing lambs, all systems recognize the curvilinear relationship between ME intake and energy retention, which is associated with a reduction in ME supply which is partly caused by an increased rate of passage at higher levels of feeding. However, they account for it using different approaches. The AFRC (1993) chose to regard this curvilinearity as being due to a decrease in the efficiency of ME utilization above a constant maintenance requirement by adopting the exponential function derived by Blaxter and Boyne (1970) as follows:

$$R = B (1 - e^{-kI}) - I \qquad (8.23)$$

where R is the retention of net energy, I is the intake of metabolizable energy, $B = k_m/(k_m - k_g)$ and $k = k_m \times \ln (k_m/k_g)$.

The alternative approach, adopted by CSIRO (1990), is to allow for a variable maintenance requirement as discussed previously and assume that the remaining energy available for growth is used with a constant efficiency.

The NRC (1985a) take a different approach to the calculation of efficiency constants. In contrast to the AFRC (1993) and CSIRO (1990), values for k_m and k_g calculated using this system show a curvilinear increase with increasing diet quality. They are also approximately 0.1 lower than equivalent values calculated using the AFRC (1993) and CSIRO (1990) systems. In pregnant sheep, there is little evidence to relate the efficiency of conversion of ME into NE with diet quality, and most systems assume an efficiency of 0.13–0.14, although this estimate ignores the deposition of nutrients in the mammary gland and the fact that the energy value of the feed is lower during late pregnancy due to an increased rumen outflow rate (Gonzalez *et al.*, 1985b; Ngongoni *et al.*, 1987). Both the AFRC (1993) and CSIRO (1990) assume k_l to be the same as for dairy cows and as published by ARC (1980), whilst INRA (1989) put less reliance on the metabolizability (q) of the feed, indicating that k_l is relatively unaffected by diet quality.

Metabolizable Energy Requirements

The total ME requirements of sheep are calculated using a factorial approach. For each body process, the NE requirement is divided by the appropriate efficiency constant. The total ME requirement is then calculated by summation of the ME requirement for each process:

$$ME_{mp} = NE_m/k_m + NE_g/k_g + NE_c/k_c + NE_l/k_l \qquad (8.24)$$

The total ME requirements of growing lambs calculated using various systems are presented in Fig. 8.1. In general, ME requirements predicted by the NRC (1985a) are slightly lower than those predicted by the AFRC (1993) and approximately 2.0 MJ day^{-1} lower than those predicted by CSIRO (1990) and INRA (1989). Differences between the AFRC (1993) and NRC (1985a) reflect a greater NE_m which is due to a slightly higher FM and the inclusion of a sex correction and activity allowance by AFRC (1993). In addition, AFRC (1993) predicts a slightly higher NE_g than NRC (1985a). The higher requirements predicted by CSIRO (1990) reflect a greater NE_m due to a higher FM and the inclusion of a correction for level of feeding. CSIRO (1990) also predicts a higher NE_g than AFRC (1993). With regard to the accuracy of each system, the AFRC (1990) evaluated two sets of data in order to validate the ARC (1980) model on which the AFRC (1993) system is based. They concluded that

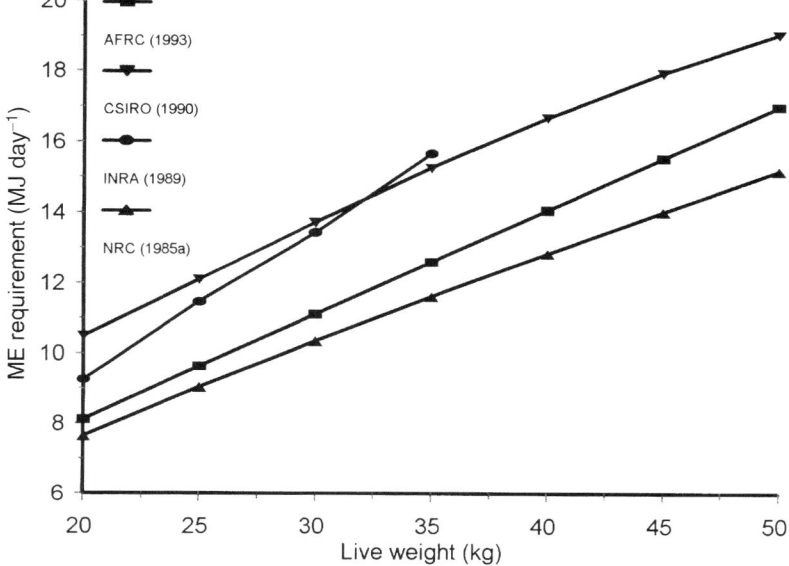

Fig. 8.1. Metabolizable energy (ME) requirements of male lambs of a medium sized breed gaining 200 g day^{-1} on a diet with a metabolizability of 0.6.

the two sets of data were in conflict with regard to predictive bias, and were unable to suggest whether ARC (1980) under- or over-predicts ME requirements.

The similarity in the ME requirements for pregnancy between AFRC (1993) and CSIRO (1990) reflects the use of the same prediction equations, with the differences between the two being attributable to a higher maintenance requirement of approximately 1 MJ day^{-1} predicted by CSIRO (1990) (Fig. 8.2). The use of a stepped system by INRA (1989), whilst reflecting practical feeding constraints, predicts a lower ME requirement in mid-gestation and consistently predicts a 2 MJ day^{-1} lower requirement in late pregnancy. These factors combined result in INRA (1989) predicting that during mid–late lactation the ME requirement of twin-bearing ewes is approximately 150 MJ less than that predicted by the other three systems, a value equivalent to approximately 5 kg body fat.

PROTEIN SYSTEMS

In all systems, the protein requirements of growing lambs are expressed in terms of net protein (NP) and protein supply is measured in terms of 'metabolizable protein' (MP) (AFRC, 1993), 'apparently digested protein leaving the stomach' (ADPLS) (CSIRO, 1990), 'absorbed protein' (AP) (NRC, 1985b) or 'protein truly digested in the small intestine' (PDI) all of which represent

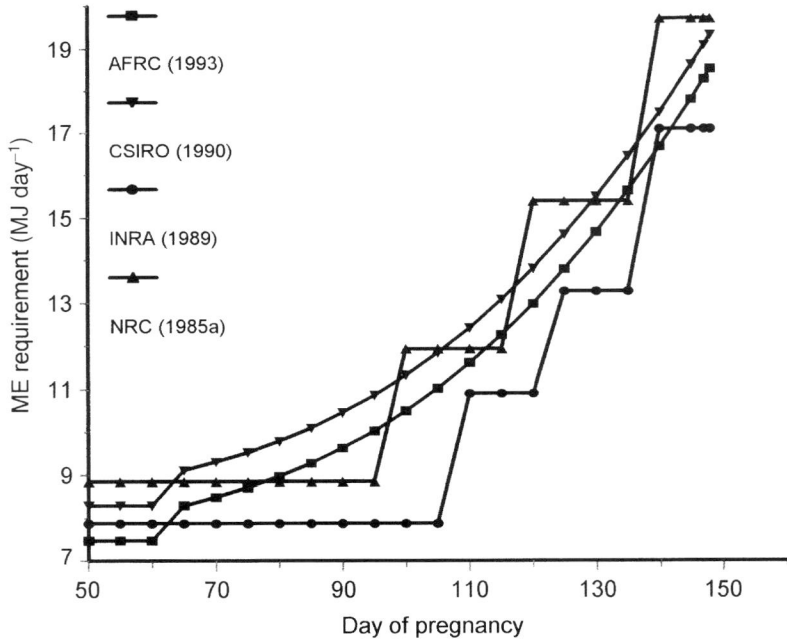

Fig. 8.2. Metabolizable energy (ME) requirements for a 70 kg, twin-bearing ewe in mid–late gestation.

'absorbed amino acids available to the animal'. Protein requirements and supply are linked by constants (kp) that vary with productive process.

Net Protein Requirements for Maintenance

The net protein requirement for maintenance (NP_m) represents the protein required to counterbalance inevitable N losses in urine and faeces. Essentially urinary N losses arise due to the inefficiency of protein turnover, and faecal N losses consist of indigestible microbial material, enzymes and cell debris arising from the digestive tract. Originally, NP_m requirements were based on the ARC (1980) estimates of endogenous urinary protein (EUP). No allowance was made for metabolic faecal protein (MFP) because it consists largely of microbial debris which was considered to be part of the inefficiency of microbial N utilization and therefore mainly exogenous in origin. However, based on estimates of total endogenous protein (TEP) (Ørskov and Grubb, 1979; Ørskov and Macleod, 1982) derived using animals nourished entirely by intragastric infusion, the AFRC (1993) include an allowance for both EUP and MFP in their calculation of NP_m.

$$\text{AFRC (1993)} \quad NP_m \text{ (g day}^{-1}) = 2.187 \ W^{0.75} \tag{8.25}$$

$$\text{CSIRO (1990)} \quad NP_m \text{ (g day}^{-1}) = \text{EUP} + \text{MFP} \tag{8.26a}$$

$$\text{EUP (g day}^{-1}) = 0.147 \ W + 3.375 \tag{8.26b}$$

$$\text{MFP (g day}^{-1}) = 15.2 \text{ g kg DMI}^{-1} \tag{8.26c}$$

$$\text{INRA (1989)} \quad \text{PDI (g day}^{-1}) = 2.50 \ W^{0.75} \tag{8.27}$$

$$\text{NRC (1985)} \quad NP_m \text{ (g day}^{-1}) = \text{EUP} + \text{MFP} \tag{8.28a}$$

$$\text{EUP (g day}^{-1}) = 1.125 \ W^{0.55} \tag{8.28b}$$

$$\text{MFP (g day}^{-1}) = 60.3 \text{ g kg}^{-1} \text{ IDM} \tag{8.28c}$$

$$\text{IDM} = \text{DMI} \ [1 - (0.92 \ \text{TDN})] \tag{8.28d}$$

where NP_m is the net protein required for maintenance, EUP is the endogenous urinary protein, MFP is the metabolic faecal protein, DMI is the dry matter intake, IDM is the indigestible dry matter output, TDN is total digestible nutrients and PDI is the protein truly digested in the small intestine.

The approach taken by CSIRO (1990) and NRC (1985b) is somewhat different. CSIRO (1990) argue that the microbial debris component of MFP originates from microbial growth in the large intestine and that the N source for microbial growth must be endogenous in origin since the amount of fermentable exogenous N reaching this part of the digestive tract will be small. In addition, estimates of MFP derived from animals nourished by intragastric infusion are substantially lower than those obtained from normally fed animals by regression analysis. Therefore, MFP must be considered when calculating NP_m requirements. Varying MFP with dry matter (DM) intake or indigestible DM output is conceptually sound as increased feed intake generally is associated with enhanced protein deposition and turnover, which in turn is associated with more endogenous protein entering the digestive tract.

Net Protein Requirements for Live Weight Gain

The net protein requirement for live weight gain (NP_g) represents the daily protein gain of an animal and is calculated from the product of the protein content of each kilogram of gain and the daily live weight gain. As with energy, the protein content of gain is known to vary with animal live weight, breed, sex and rate of gain. Different systems account for these factors to different extents. Both the AFRC (1993) and NRC (1985b) adopt equations provided by the ARC (1980). These account for the effects of live weight and provide separate estimates for males and castrates and for females. The most comprehensive system with regard to its ability to account for the effects of animal sex, breed and rate of gain is that of the CSIRO (1990) which calculates the protein content of gain using a similar approach to that used for energy (see previously), with a comparable system being used by INRA (1989). The AFRC (1992) report on which the 1993 system is based recognizes that the approach taken by CSIRO (1990) is more logical and biologically attractive. They state that 'we see no reason why their data for calculating NP_g should not be used in our proposed system, if desired'.

AFRC (1993) and NRC (1985)
Males and castrates

$$NP_g \text{ (g day}^{-1}) = \text{LWG } (160.4 - 1.22 \text{ W} + 0.0105 \text{ W}^2) \tag{8.29}$$

Females

$$NP_g \text{ (g day}^{-1}) = \text{LWG } (156.1 - 1.94 \text{ W} + 0.0173 \text{ W}^2) \tag{8.30}$$

where W is the live weight, LWG is the live weight gain and NP_g is the net protein for live weight gain

CSIRO (1990)

$$\text{Protein (g kg}^{-1} \text{ EBG}) = (212 - 4R) - [(140 - 4R)/(1 + e^{[-6(P - 0.4)]})] \tag{8.31}$$

$$\text{Protein (g kg}^{-1}) = \text{protein (g kg}^{-1} \text{ EBG}) \times 0.92 \tag{8.32}$$

$$NP_g = \text{LWG (kg day}^{-1}) \times \text{protein (g kg}^{-1}) \tag{8.33}$$

where EBG is the empty body gain, R is an adjustment for rate of gain [(EBW change/4 $SRW^{0.75}$) − 1], P is the relative size (W/SRW) and SRW is the standard reference weight.

Net Protein Requirements for Pregnancy and Lactation

As with energy, net protein requirements for pregnancy (NP_c) used by AFRC (1993) and CSIRO (1990) are based on the findings of the ARC (1980) who describe requirements by a Gompertz equation where the net protein retention to produce a single lamb is given as:

$$NP_c \text{ (g day}^{-1}) = Tp_t \times 0.06744e^{-0.00601t} \tag{8.34}$$

where t is the number of days from conception and Tp_t is in g and calculated as

$$\log_{10}(Tp_t) = 4.928 - 4.873e^{-0.00601t} \tag{8.35}$$

Whilst the NRC (1985b) adopted the same equations as described in Equation 8.35, the NRC (1985a) proposed two steps of requirements for the gravid uterus: i.e up to 4 weeks prior to lambing and until parturition. MP requirements for milk are generally based on the true protein content of the milk divided by the efficiency of conversion of MP to NP.

Even under conditions of high planes of nutrition, lactating animals generally lose weight in early lactation. The AFRC (1993) assumed that the NP loss in lactating ewes equated to 119 g protein kg^{-1} lost and that protein mobilized equated to NP (i.e. a conversion of 1.0), whilst CSIRO suggest a figure for dairy cows of 135 g kg^{-1} live weight change and an efficiency of conversion to NP of 0.8 resulting in a figure of approximately 108 g of NP kg^{-1} live weight loss. NRC (1985b) do not make any definitive recommendations for sheep but discuss the findings of Rattray *et al.* (1974b) that the protein content of empty body weight changes in adult ewes ranged from a modest 50–70 g protein kg^{-1} empty body weight.

Net Protein Requirements for Wool

The net protein requirement for wool production (NP$_w$) represents the protein retained in the growing fleece. Both the AFRC (1993) and NRC (1985b) base their recommendations on data reviewed by the ARC (1980) which relate NP$_w$ retained by British breeds of sheep (excluding Merinos) to NP$_g$. The approach adopted by CSIRO (1990) is based on observations of the gross energetic efficiency of wool production and accounts for the fact that Merinos produce more wool than other breeds by reference to a standard fleece weight.

AFRC (1993) and NRC (1985b)	$NP_w \text{ (g day}^{-1}) = 3 + (0.1\ NP_g)$	(8.36)
CSIRO (1990)	$NP_w \text{ (g MJ}^{-1}\text{ MEI)} = 8 \text{ (SFW/SRW)}$	(8.37)
INRA (1989)	$PDI \text{ (g day}^{-1}) = 0.12\ W^{0.75}$	(8.38)

where NP$_w$ is the net protein for wool production, NP$_g$ is the net protein for live weight gain, SRW is the standard reference weight and SFW is the standard fleece weight corrected for age.

Efficiency of Net Protein Utilization

As stated above, the net protein requirements of sheep are linked to the supply of protein by the use of efficiency factors (kp) (Table 8.1). In general, these conversion factors are assumed to be constant and do not take into account the variable amino acid composition of absorbed protein.

In the AFRC (1993) system, the kp values proposed are calculated from two separate efficiency factors: (i) the efficiency with which an ideal amino acid mixture for a particular purpose is utilized ($kaai$) and (ii) the extent to which the amino acid mixture available differs from the ideal (RV). The AFRC (1992) accept that evidence for the adoption of certain $kaai$ and RV values is questionable and that more research is required in this area.

The kp values adopted by CSIRO (1990) and NRC (1985a) are somewhat different from those proposed by the AFRC (1993). Both systems state that kp values are difficult to measure and that there are few estimates for ruminants that are in a productive state. CSIRO (1990) argue that there is considerable uncertainty with regard to the efficiency of utilization of absorbed amino acids and that at present there is no well-founded alternative to the use of a single value for all processes except wool production. With regard to wool, CSIRO (1990) accept that the partial efficiency of utilization of absorbed amino acids for wool production is approximately 0.20, but suggest that it would be inappropriate to use this value in the factorial calculation of protein requirements because it assumes that 0.8 of this fraction will be eliminated by excretion. As the amino acids used for wool growth are drawn from the same pool as those used for other body processes, this is clearly not true. The kp_w value adopted assumes that absorbed amino acids are used for wool growth with an efficiency of 0.2 and the balance are used for other processes with an efficiency of less than 0.7. This pragmatic approach has been noted, but not adopted, by AFRC (1993). Whilst the AFRC (1993) and CSIRO (1990) assume that the efficiency of conversion of MP into NP for gestation (kp_c) is fixed at 0.85 and 0.70 respectively, INRA (1989) suggest that the efficiency of conversion of PDI to NP varies from 0.36 in early gestation to 0.51 in mid and 0.63 in late gestation, with an overall efficiency of 0.42.

Table 8.1. Efficiency factors (kp) for the various protein systems.

	kp_m	kp_g	kp_w	kp_l	kp_c
AFRC (1993)	1.00	0.59	0.26	0.68	0.85
CSIRO (1990)	0.70	0.70	0.60	0.70	0.70
INRA (1989)	—	—	—	0.58	0.42
NRC (1985a)	0.67	0.50	0.50	0.65	0.50

kp_m, kp_g, kp_w, kp_l and kp_c are the efficiencies of utilization for maintenance, growth, wool, lactation and pregnancy, respectively.

Metabolizable Protein Requirements

For simplicity, the absorbed amino acid requirement calculated using various systems will be expressed as metabolizable protein (MP). As with energy, the total MP requirement of sheep is calculated using a factorial approach. For each body process, the NP requirement is divided by the appropriate efficiency constant, and the total MP requirement is then calculated by summation of the MP requirement for each process.

$$\text{MP total} = \text{NP}_m/kp_m + \text{NP}_p/kp_g + \text{NP}_w/kp_w + \text{NP}_l/kp_l + \text{NP}_c/kp_c \qquad (8.39)$$

The total MP requirements of growing lambs calculated using various systems are presented in Fig. 8.3 and a breakdown of requirements is presented in Table 8.2. The MP requirements predicted by CSIRO (1990) and INRA (1989) are lower than those predicted by the AFRC (1993), which are considerably lower than those predicted by the NRC (1985). As all systems predict similar NP requirements, differences between systems mainly reflect differences in the efficiency constants adopted. The total MP requirements of ewes in mid–late gestation using the four systems is presented in Fig. 8.4. The lower kp_c value for gestation used by INRA (1989) results in a greater requirement for MP in late gestation with this system, whilst the higher requirement predicted by the AFRC (1993) in mid-gestation is mainly a reflection of the greater maintenance requirement, which itself is composed of 20.4 g of MP day^{-1} for wool growth, a figure substantially greater than that for the other systems.

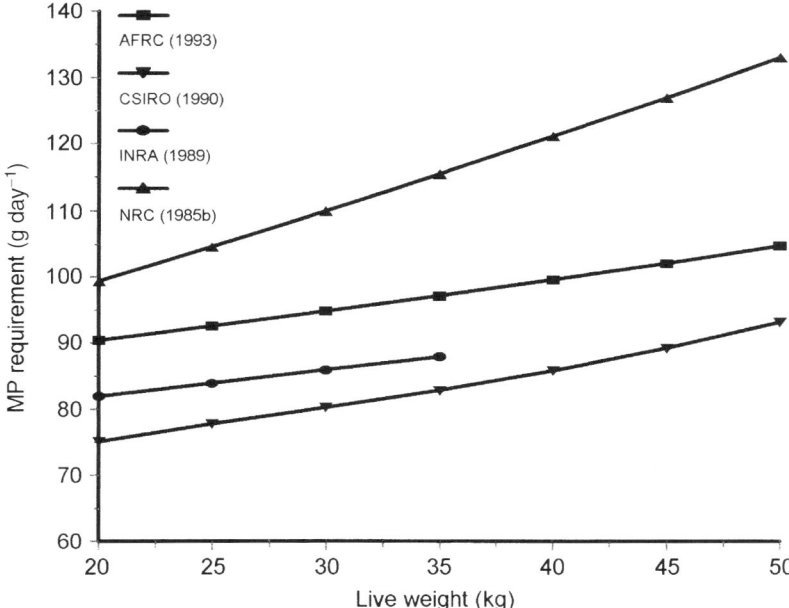

Fig. 8.3. Metabolizable protein (MP) requirements of male lambs of a medium sized breed gaining 200 g day^{-1} on a diet with a metabolizability of 0.6.

Table 8.2. Predicted protein requirements (g day^{-1}) of a 30 kg male lamb gaining 200 g day^{-1} assuming a medium sized breed and q = 0.6.

	AFRC (1993)	CSIRO (1990)	NRC (1985)
Maintenance			
EUP	—	7.8	7.3
MFP	—	18.2	23.1
Total	27.9	26.0	30.4
Production	26.6	24.7	26.6
Wool	5.7	4.7	5.7
Total	60.3	55.4	62.7
Metabolizable protein			
Maintenance	27.9	37.2	45.4
Production	45.2	35.3	53.3
Wool	21.8	7.8	11.3
Total	94.9	80.3	110.0

EUP, endogenous urinary protein; MFP, metabolic faecal protein.

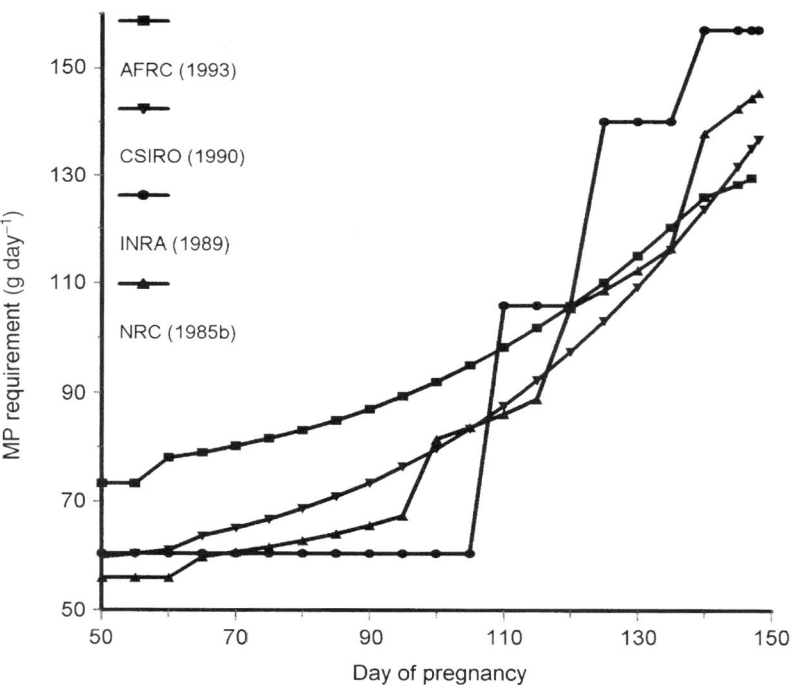

Fig. 8.4. Metabolizable protein (MP) requirements for a 70 kg, twin-bearing ewe in mid–late gestation.

Metabolizable Protein Supply

All four systems calculate the supply of digestible amino acids to the small intestine by different methods. A fuller discussion of factors affecting microbial protein and digestible undegradable protein supply is presented in previous chapters, but a summary of the main calculations involved with the four feeding systems discussed above is presented in Table 8.3. An example of the predicted MP requirements for a lactating ewe together with feed crude protein content and degradability is presented in Table 8.4. There is considerable variation in the predicted MP requirements for maintenance and milk production between the four systems, with the net effect being that NRC

Table 8.3. Summary of major components of four protein supply systems.

	AFRC (1993)	CSIRO (1990)	INRA (1989)	NRC (1985b)
Degradability estimates	*In situ*	*In situ*	*In situ*	*In situ*
Protein degraded in the rumen and available for microbial protein synthesis	0.8 soluble protein + slowly degradable protein	RDP	0.9 × RDP	0.9 × RDP + 0.15 protein intake
Microbial energy supply	FME = ME – energy in fats and volatiles	ME	FOM = digestible organic matter – ether extract – volatiles – bag UDP	TDN
Microbial growth (gN)	Variable: ERDP if limiting or FME × 11,10 or 9 depending on production level	RDP if limiting or 6.1, 8.4 or 11.0 g MJ^{-1} ME for different classes of feeds	0.9 × RDP if limiting or 145 g CP kg^{-1} FOM	Value of RDP if limiting or 6.25 × (23.04 TDN –1.29)
Amino acid content of microbes	0.75	0.8	0.8	0.8
Digestibility of microbial amino acids	0.85	0.7	0.8	0.8
UDP as a proportion of protein intake	1.0 – deg	1.0 – deg	(1 – deg)/1.11	1.0 – deg
Digestibility of undegraded protein	0.9 (UDP – ADIP)	0.7	Variable: 0.55–0.95	0.8

ERDP, effective rumen-degradable protein; RDP, rumen-degradable protein; ME, metabolizable energy; TDN, total digestible nutrients; FME, fermentable metabolizable energy; FOM, fermentable organic matter; deg, degradability (determined *in situ*); ADIP, acid detergent-insoluble protein.

Table 8.4. Estimated metabolizable protein requirement and supply for a housed, 70 kg ewe yielding 2.0 kg of milk day^{-1} containing 50 g kg^{-1} of protein, neither losing nor gaining weight and consuming 2.1 kg of DM and 23.1 MJ of ME day^{-1}.

	CSIRO (1990)	AFRC (1993)	INRA (1989)	NRC (1985b)
Requirements (g of MP day^{-1})				
Maintenance	54	53	60.5	77.7
Wool	12	20		13.6
Milk	143	147	172	153.8
Total	209	220	232.5	245.1
Dietary protein supply (g day^{-1})				
Microbial CP	194	203[a]	183[b]	207
Digestible microbial AA	109	129	117	132
Dietary RDP	194	225[c]	203	230
Digestible UDP required	100	91	115	113
UDP	143	130[d]	148[d]	141
Dietary CP				
Required (g kg^{-1} of DM)	160	169	167	154
Degradability	0.58	0.63	0.58	0.56

Assumptions [a]FME/ME = 0.8; [b]FOM = 600 g kg^{-1} of DM; [c]ERDP/RDP = 0.9; digestibility of UDP = 0.7.

(1985a,b) predict a requirement some 20% greater than CSIRO (1990), i.e. 245 g compared with 209 g of MP day^{-1} (Table 8.4). However, part of the discrepancy is negated by the high microbial protein synthesis predicted by NRC (1985a,b) and the incorporation of a recycling factor (0.15 of protein intake), resulting in the low crude protein content and degradability being predicted by this system. It is important to note that the protein supply systems of AFRC (1993) and INRA (1989) are affected by dietary alterations to fermentable energy, digestibility of the undegradable protein source and rumen outflow rate, all of which may result in predicted supplies being different from those presented in Table 8.4.

FUTURE SYSTEMS

Energy and Protein Requirements

Nutrient requirements predicted by current feeding systems are based on older empirical relationships and do not account for the dynamic interactions that take place between nutrient supply and tissue metabolism (Lindsay, 1993; Lobley, 1993). For example, current feeding systems predict that on a given diet the greatest rate and efficiency of gain will occur when animals are consuming the highest level of intake, an effect generally related to the dilution of maintenance requirements. However, recent work conducted by

Fluharty and McClure (1997) indicated a similar efficiency of gain in lambs when fed at a restricted level compared with those fed *ad libitum*. The effects were attributed to a reduction in visceral organ size in lambs fed the restricted level which therefore reduced maintenance energy requirements. The adaptation to a period of undernutrition through a reduction in visceral organ size and nutrient requirements in growing lambs has been reported in a number of other studies (Burrin *et al.*, 1989; Freetly *et al.*, 1995) and has been proposed as the major reason for the lower maintenance energy costs associated with restricted feeding in animals in environments that naturally undergo a period of nutrient deprivation. Differences between forage source, quality and level of concentrates on gut tissue mass and energy use have also been reported (Kouakou *et al.*, 1995; Goetsch *et al.*, 1997) and can result in peripheral tissue energy availability deviating from expectations based solely on intake and digestibility. Interestingly, in contrast to the growing lamb, the underfed adult non-pregnant, non-lactating ewe did not show any adaptation of energy metabolism during a period of undernutrition lasting 7 weeks (Ortigues and Vermorel, 1996), perhaps indicating that visceral metabolism is also age dependent. Additionally, whilst age and sex are included in calculating maintenance requirements, evidence suggests that the maintenance and deposition of protein is less energetically efficient than that of fat (Rattray and Joyce, 1976; Graham, 1980) and that differences in maintenance efficiency exist between genetically fat and lean lines (Afonso and Thompson, 1996). Use of single values for k_m and k_g may therefore overestimate gains when a high proportion of body tissue consists of protein. In a series of experiments, Wilkinson and Greenhalgh (1991, 1995) compared the growth of lambs offered fixed amounts of forage and concentrate by different feeding strategies with that predicted by the ARC (1980) model. The extent to which observed growth differed from that predicted varied with the feeding strategy adopted. However, the results demonstrated that predictions could be improved by the incorporation of a variable maintenance requirement, associative effects between feeds and separate efficiency constants for the deposition and maintenance of protein and fat.

In most current nutritional systems, estimates of maintenance energy requirements are based on sheep that are in a fasted state. However, it has been suggested that the energetic cost of proteolysis, lipolysis and ketogenesis in fasted animals may contribute considerably to their heat production. For example, Chowdhury *et al.* (1997b), when infusing volatile fatty acids (VFAs) into the abomasum of fasting lambs, reported that heat production did not increase despite the fact that 40% of VFA energy would be expected to be dissipated as heat. Whilst the extra N retention observed would be a compensating factor, the findings indicated that fasting is a particular metabolic adaptation which increases heat production. Therefore, the use of animals in a fasted state may be an inappropriate method of measuring the utilization of a nutritionally balanced diet.

Another example of interactions between nutrient supply and tissue metabolism relates to nitrogen retention. ARC (1980) proposed that N retention is a quadratic function of both energy and protein intake and, therefore, at very high levels of protein and low energy intakes, N retention could be negative. However, body fat has been demonstrated to be utilized effectively to support lean tissue growth during periods of undernutrition (Fattet *et al.*, 1984). Growing lambs would therefore appear to be able to meet their total energy demand by altering endogenous energy loss in response to dietary energy and protein supply (Chowdhury *et al.*, 1997a) although the anticipated effect of fatter animals being more able to maintain N deposition when dietary energy supply was restricted was not demonstrated (Chowdhury *et al.*, 1997b). The implication for feeding systems is that the classical nutritional concept of optimum energy:protein ratios in predicting responses to nutrient supply is only meaningful when both endogenous and exogenous supplies are considered (Chowdhury *et al.*, 1997a).

It has long been recognized that the molar proportions of the major VFAs in rumen fluid differ with the source of carbohydrate in the ration and that, in general, the efficiency of utilization of metabolizable energy for live weight gain (k_g) can be positively associated with the ratio of propionate in rumen fluid. In work conducted by intragastric infusion of lambs, Ørskov *et al.* (1979) measured k_g values for a range of VFA mixtures (Table 8.5). Only in rations that resulted in a considerable shift towards a propionate fermentation was the efficiency of utilization of ME for growth substantially altered, a dietary situation not encountered routinely in the majority of sheep feeding systems. In feeding experiments with lambs fed at twice maintenance, a propionate fermentation was also reported to be no more efficient for live weight gain than an acetate fermentation (MacRae *et al.*, 1993). Whilst a prediction of individual VFA absorption rates may therefore appear to have a limited effect on the interpretation of growth response in sheep, several authors have attempted to predict VFA production (e.g. Baldwin *et al.*, 1977; Dijkstra *et al.*, 1992) although a robust model that can be used in sheep nutrition has not yet evolved.

With respect to pregnant and lactating sheep, current feeding systems contain a protein requirement for the gravid uterus and for milk production. However, they do not consider the importance of maternal tissue protein as a reservoir of amino acids available to meet these requirements. Mobilized carcass tissue has been shown to be the major source of N for deposition in the gravid uterus when ewes are fed both above and below requirements (McNeill *et al.* 1997) and theoretical calculations indicate that up to 14% of the

Table 8.5. Molar proportions of volatile fatty acids (%) in lambs sustained by intragastric infusion and the efficiency of energy utilization for growth.

	Molar proportions					
Acetic	35	45	55	65	75	85
Propionic	55	45	35	25	15	5
Butyric	10	10	10	10	10	10
k_g	0.78	0.64	0.57	0.61	0.61	0.59

Source: Ørskov *et al.* (1979).

requirement for milk production can be met by carcass N. Additionally, current systems do not consider the consequences of nutrition during one phase of the productive cycle on performance in subsequent phases. For example, Zhang *et al.* (1995) fed differing levels of dietary protein to growing ewe lambs and, whilst they obtained no effect on mammary growth up to puberty, there was an increase in mammary weight and parenchymal tissue during lactation and a trend towards a greater milk yield. Supplementation of pregnant ewes in late gestation has also been shown to alter milk composition. In work with mature ewes, O'Doherty and Crosby (1997) reported a greater response in colostrum production to supplies of dietary energy than protein pre-lambing, even though protein supply was expected to be deficient. However, supplementation with protein pre-lambing increased the lambs' ability to absorb colostral IgG during the first 24 h and therefore potentially improved lamb viability.

As with growing lambs, body fat mobilization in ewes is dependent on the supply of absorbable amino acids. An estimate of the factors influencing the mobilization of body energy (and protein) reserves at different levels of nutrient supply is necessary to predict animal performance adequately. For example, at low levels of ME intake, additional increments of fish meal resulted in an apparent efficiency of conversion of dietary protein into milk protein of around 50% (Gonzalez *et al.*, 1985a). However, lower marginal efficiencies were evident when higher levels of ME were fed, an effect not accounted for in current feeding systems. In this respect, an accurate prediction of protein degradability (or, rather, undegradability) is required. Dove *et al.* (1984) reported significant increases in milk yield to additional undegradable protein supply with ewes consuming perennial ryegrass. However, in a trial designed to evaluate the UK metabolizable protein system for lactating ewes at grass, milk yield was not affected by additional DUP supply (Wilkinson *et al.*, 1996). Additionally, there was a greater marginal efficiency of milk protein response to additional ERDP (0.23) than MP (0.12), a value considerably lower than the 0.20 marginal efficiency of MP suggested for dairy cows (Webster, 1992). One of the major factors contributing to the larger effect of ERDP on milk protein yield reported in this work was the very low ERDP content of the grass at an ERDP:FME ratio of 6.6 g MJ^{-1}, a figure substantially lower than that commonly accepted for spring grass. These results highlight the importance of using degradability coefficients that are appropriate to the batch of feed being used and the need to develop a rapid, accurate method of routinely predicting the degradability coefficients of protein- and energy-yielding components.

Rumen Metabolism

A considerable amount of research over the last 20 years has been directed towards predicting the proportion of dietary protein that will be degraded in the rumen and available for microbial metabolism and the proportion that will pass through the rumen and can potentially be digested in the small intestine.

Whilst a number of feeding systems calculate the effective degradability of dietary N based on *in situ* measurements and estimated rumen outflow rate (as demonstrated in Table 8.4), energy supply to the rumen is considered as either a constant (ARC, 1984; NRC, 1985b; CSIRO, 1990; AFRC, 1992) or marginally affected by outflow rate (INRA, 1989). The apparent discrepancy in the way in which energy and protein supply to the rumen microorganisms is calculated could be rectified by the use of *in situ* degradability coefficients for organic matter or classes of carbohydrate (as presented in the Cornell net carbohydrate and protein system for cattle, Fox *et al.*, 1992). This would be a simple but conceptually more descriptive index of energy supply to the rumen and, as with protein, would be affected by changes in rumen outflow rate. Most feeding systems also make some reference to ensuring that energy and protein supply to the rumen within the day are balanced (or synchronous) although all are based upon the daily ratio of degradable protein and energy substrates. The inclusion of an index that describes the synchrony of energy and protein release has been argued to be unnecessary due to the effects of frequency of meal consumption and N recycling, both of which could have an ameliorating effect on the pattern of nutrients available to the rumen microorganisms within the day. The quantitative assessment of the rate of energy and protein supply to the rumen has also presented an obstacle to the evaluation of rumen synchrony in studies apart from those that have involved the infusion of soluble substrates into the rumen. Indeed, trials that have been conducted to assess the effects of rumen synchrony on microbial and whole body metabolism have produced conflicting results (Herrera-Saldana *et al.*, 1990; Hussein *et al.*, 1991; Henning *et al.*, 1993). Sinclair *et al.* (1993) defined a simple index, calculated from the sum of *in situ* degradability data of protein, organic matter or carbohydrate, which described the degree of synchrony between hourly energy and nitrogen supply in the rumen:

$$\frac{25 - \Sigma_{1-24} \dfrac{\sqrt{[(25 - \text{hourly N:OM})^2]}}{24}}{25} \tag{8.40}$$

where 25 = 25 g of N kg^{-1} of organic matter truly degraded in the rumen and is ssumed to be the ratio of nitrogen to energy supply that is optimal for efficient microbial growth (Sinclair *et al.*, 1993). An index of 1.0 indicates perfect synchrony between energy and nitrogen supply for each hour of the day, whilst values of less than 1.0 indicate the degree of asynchrony.

Sheep fed diets that had a higher synchrony index had an increased efficiency of microbial protein synthesis (g of N kg^{-1} of OM degraded) of between 10 and 20% compared with those fed the same level of nutrients but in an asynchronous pattern (Sinclair *et al.*, 1993, 1995). This quantitative approach to diet formulation was adopted by Witt *et al.* (1999a,b, 1997) and applied to the formulation of diets for growing lambs (Table 8.6). At a restricted level of feeding, the increased efficiency of nutrient utilization was exhibited in a greater live weight gain, whilst offering the diets *ad libitum* resulted in a reduction in intake (particularly with the fast rate of energy

Table 8.6. Effects of synchronizing dietary energy and protein supply to the rumen in diets with either a fast or slow rate of energy release on the performance of growing lambs.

	SS	SA	FS	FA	SED
Predicted synchrony index	0.84	0.51	0.87	0.51	
Witt *et al.* (1999a)					
Ewe lambs fed *ad libitum*					
Intake (kg day^{-1})	1.58	1.56	1.47	1.67	0.061
Daily live weight gain (g day^{-1})	238	219	259	272	17.7
Food conversion ratio (kg feed kg^{-1} gain)	0.151	0.141	0.176	0.163	0.0077
Witt *et al.* (1999b)					
Ram lambs: restricted level of feeding					
Daily live weight gain (g day^{-1})	126	108	137	107	8.1
Food conversion ratio (kg gain kg^{-1} feed)	0.132	0.114	0.145	0.112	0.0085

Where SS = slow rate of energy release, synchronous N; SA = slow rate of energy release, asynchronous N; FS = fast rate of energy release, synchronous N; FA = fast rate of energy release, asynchronous N.

release) with similar rates of gain being achieved. Interestingly, plasma β-hydroxybutyrate levels were significantly higher at all growth stages in lambs fed the asynchronous diets, perhaps indicating that body energy metabolism may have been partly responsible for the differences observed. Nitrogen balance studies with lambs fed the same diets revealed that there was little difference in the nitrogen retained in animals fed synchronous or asynchronous diets (Witt *et al.*, 1999b) although studies with sheep (Holder *et al.*, 1995) and cattle (Scollan *et al.*, 1997) have indicated that nitrogen recycling is greater in animals fed asynchronous diets. The net effects of increased N recycling on animal metabolism are not totally clear, although Lobley *et al.* (1995) and Mustvangwa *et al.* (1997) reported an increased amino acid deamination associated with increased urea recycling in lambs and isolated lamb hepatocytes, respectively, which could have an effect on net tissue supply of amino acids. In addition to the effects of synchrony on microbial metabolism, it has been demonstrated that growth of rumen microbes or, more specifically, individual microbial species, can be enhanced by a supply of amino acids (McAllan and Smith, 1983). The Cornell net carbohydrate and protein system for cattle (Fox *et al.*, 1992) incorporates the findings of Russell and Sniffen (1984) in that the yield of non-fibrilolytic bacteria is improved by approximately 19% as the ratio of peptides to non-structural carbohydrates increases from 0 to 14%. However, this is not included in systems currently used for the formulation of diets for sheep.

CONCLUSIONS

During the last 20 years, various feeding systems for sheep have evolved around the world, each adopting a slightly different approach to the way in which nutrient supply and requirements are calculated. Recent research has tended to concentrate on predicting nutrient supply, with animal requirements being derived using older empirical relationships. Additionally, there have been few experiments conducted which have been designed to compare the relative efficacy of the different systems for sheep. There is a need for more research into factors affecting animal requirements and a reassessment of the way in which requirements are calculated. Substantial advances in the accuracy of prediction of animal responses in different situations will only be made when the dynamic interactions between nutrient supply and tissue metabolism, both in terms of individual metabolites and the pattern by which they are absorbed and utilized, are taken into account.

REFERENCES

Afonso, J. and Thompson, J.M. (1996) Changes in body composition as sheep selected for high and low backfat thickness, during periods of *ad libitum* and maintenance feeding. *Animal Science* 63, 395–406.

Agricultural and Food Research Council (1990) Technical Committee on Responses to Nutrients, Report No. 5, Nutritive Requirements of Ruminant Animals: Energy. *Nutrition Abstracts and Reviews Series B* 60, 729–804.

Agricultural and Food Research Council (1992) Technical Committee on Responses to Nutrients, Report No. 9, Nutritive Requirements of Ruminant Animals: Protein. *Nutrition Abstracts and Reviews, Series B* 62, 787–835.

Agricultural and Food Research Council (1993) *Energy and Protein Requirements of Ruminants.* An advisory manual prepared by the AFRC Technical Committee on Responses to Nutrients. CAB International, Wallingford, UK.

Agricultural Research Council (1980) *Nutrient Requirements of Ruminant Livestock.* Commonwealth Agricultural Bureaux, Farnham Royal, Slough, UK.

Baldwin, R.L., Koong, L.J.and Ulyatt, M.J. (1977) A dynamic model of ruminant digestion for evaluation of factors affecting nutritive value. *Agricultural Systems* 2, 255–288.

Blaxter, K.L. and Boyne, A.W (1970) A new method of expressing the nutritive value of feeds as sources of energy. In: Schurch, A. and Wenk, C. (eds), *Energy Metabolism of Farm Animals. 5th Symposium of the European Association of Animal Production.* Vol. 13, pp. 9–13.

Burrin, D.G., Ferrell, C.L., Eisemann, J.H., Britton, R.A. and Nienaber, J.A. (1989) Effect of level of nutrition on splanchnic blood flow and oxygen consumption in sheep. *British Journal of Nutrition* 62, 23–34.

Chowdhury, S.A., Ørskov, E.R., DeB. Hovell, F.D., Scaife, J.R. and Mollison, G. (1997a) Protein utilization during energy under nutrition in sheep sustained by intra gastric infusion: effects of protein infusion level, with or without sub-maintenance amounts of energy from volatile fatty acids, on energy and protein metabolism. *British Journal of Nutrition* 77, 565–576.

Chowdhury, S.A., Ørskov, E.R., DeB. Hovell, F.D., Scaife, J.R. and Mollison, G. (1997b) Protein utilization during energy under nutrition in sheep sustained by intra gastric infusion: effect of body fatness on the protein metabolism of energy restricted sheep. *British Journal of Nutrition* 78, 273–282.

Commonwealth Scientific and Industrial Research Organization (1990) *Feeding Standards for Australian Livestock. Ruminants.* CSIRO Publications, East Melbourne, Australia.

Dijkstra, J., St. C. Neal, H.D., Beever, D.E. and France, J. (1992) Simulation of nutrient digestion, absorption and outflow in the rumen: model description. *Journal of Nutrition* 122, 2239–2256.

Dove, H., Milne, J.A., Lamb, C.S., McCormack, H.A. and Spence, A.M. (1984) The effect of supplementation on non-ammonia nitrogen flows at the abomasum of lactating grazing ewes. *Proceedings of the Nutrition Society* 44, 63A.

Fattet, I., De B. Hovell, F.D., Orskov, E.R., Kyle, D.J., Pennie, K. and Smart, R.I. (1984) Under nutrition in sheep. The effect of supplementation with protein on protein accretion. *British Journal of Nutrition* 52, 561–574.

Fluharty, F.L. and McClure, K.E. (1997) Effects of dietary energy intake and protein concentration on performance and visceral organ mass in lambs. *Journal of Animal Science* 75, 604–610.

Foot, J.Z. and Tulloh, N.M. (1977) Effects of two paths of liveweight change on the efficiency of feed use and on body composition of Angus steers. *Journal of Agricultural Science, Cambridge* 88, 135–142.

Fox, D.G., Sniffen, C.J., O'Connor, J.D., Russell, J.B. and Van Soest, P.J. (1992) A net carbohydrate and protein system for evaluating cattle diets: III. Cattle requirements and diet adequacy. *Journal of Animal Science* 70, 3578–3596.

Freetly, H.C., Ferrell, C.L. Jenkins, T.G. and Goetsch, A.L. (1995) Visceral oxygen consumption during chronic feed restriction and realimentation in sheep. *Journal of Animal Science* 73, 843–852.

Gingins, M., Bickel, H. and Schurch, A. (1980) Efficiency of energy utilisation in under-nourished and realimentated sheep. *Livestock Production Science* 7, 465–471.

Goetsch, A.L., Patil, A.R., Galloway, D.L., Kouakou, B, Wang, Z.S., Park, K.K. and Rossi, J.E. (1997) Net flux of nutrients across splanchnic tissues in wethers consuming grasses of different sources and physical forms *ad libitum. British Journal of Nutrition* 77, 769–781.

Gonzalez, J.S., Robinson, J.J. and McHattie, I. (1985a) The effect of level of feeding on the response of lactating ewes to dietary supplements of fishmeal. *Animal Production* 40, 39–45.

Gonzalez, J.S., Robinson, J.J. and Fraser, C. (1985b) The effect of physiological state on digestion in the ewe and its influence on the quantity of protein reaching the abomasum. *Livestock Production Science* 12, 59–68.

Graham, N.McC. (1980) Variations in energy and nitrogen utilisation in sheep between weaning and maturity. *Australian Journal of Agricultural Research* 31, 335–345.

Graham, N.McC., Searle, T.W. and Griffiths, D.A. (1974) Basal metabolic rate in lambs and young sheep. *Australian Journal of Agricultural Research* 25, 957–971.

Henning, P.H., Steyn, D.G. and Meissner, H.H. (1993) Effect of synchronisation of energy and nitrogen supply on ruminal characteristics and microbial growth. *Journal of Animal Science* 71, 2516–2528.

Herrera-Saldana, R., Gomez-Alarcon, R., Torabi, M. and Huber, J.T. (1990) Influence of synchronising protein and starch degradation in the rumen on nutrient utilization and microbial protein synthesis. *Journal of Dairy Science* 73, 142–148.

Holder, P., Buttery, P.J. and Garnsworthy, P.C. (1995) The effect of dietary asynchrony on rumen nitrogen recycling in sheep. *Animal Science* 60, 528.

Hussein, H.S., Jordan, R.M. and Stern, M.D. (1991) Ruminal protein metabolism and intestinal amino acid utilization as affected by dietary protein and carbohydrate sources in sheep. *Journal of Animal Science* 69, 2134–2146.

Institut National de la Recherche Agronomique (1989) *Ruminant Nutrition: Recommended Allowances and Feed Tables.* INRA, Paris.

Kouakou, B., Goetsch, A.L., Patil, A.R., Galloway, D.L., Sr, Park, K.K. and West, C.P. (1995). Effects of grass source and maturity on performance and visceral organ mass of growing wethers. *Journal of Animal Science* 73, (Suppl. 1), 261.

Ledger, H.P. and Sayers, A.R. (1977) The utilization of dietary energy by steers during periods of restricted food intake and subsequent realimentation. 1. The effect of time on the maintenance requirements of steers held at constant liveweight. *Journal of Agricultural Science, Cambridge* 88, 11–26.

Lindsay, D.B. (1993) Metabolism of the portal drained viscera. In: Forbes, J.M. and France, J. (eds), *Quantitative Aspects of Ruminant Digestion and Metabolism.* CAB International, Wallingford, UK, pp. 267–289.

Lobley, G.E. (1993) Protein metabolism and turnover. In: Forbes, J.M. and France, J. (eds), *Quantitative Aspects of Ruminant Digestion and Metabolism.* CAB International, Wallingford, UK, pp. 313–339.

Lobley, G.E., Connell, A., Lomax, M.A., Brown, D.S., Milne, E., Calder, A.G. and Farningham, D.A.H. (1995) Hepatic detoxification of ammonia in the ovine liver: possible consequences for amino acid catabolism. *British Journal of Nutrition* 73, 667–685.

MacRae, J.C., Walker, A., Brown, D. and Lobley, G.E. (1993) Accretion of total protein and individual amino acids by organs and tissues of growing lambs and the ability of nitrogen balance techniques to quantitate protein retention. *Animal Production* 57, 237–245.

McAllan, A.B. and Smith, R.H. (1983) Factors influencing the digestion of dietary carbohydrates between the mouth and abomasum of steers. *British Journal of Nutrition* 50, 445–454.

McNeill, D.M., Slepetis, R., Ehrhardt, R.A., Smith, D.M. and Bell, A.W. (1997) Protein requirements of sheep in late pregnancy: partitioning of nitrogen between gravid uterus and maternal tissues. *Journal of Animal Science* 75, 809–816.

Ministry of Agriculture, Fisheries and Food (1984) *Energy Allowances and Feeding Systems for Ruminants.* ADAS reference booklet 433, HMSO, London.

Mustvangwa, T., Buchanan-Smith, J.G. and McBride, B.W. (1997) Effects of ruminally degradable nitrogen intake and *in-vitro* addition of ammonia and propionate on the metabolic fate of L-[1-^{14}C]alanine and L-[^{15}N]alanine in isolated sheep hepatocytes. *Journal of Animal Science* 75, 1149–1159.

Ngongoni, N.T., Robinson, J.J., Kay, R.N.B., Stephenson, R.G.A., Atkinson, T., Grant, I. and Henderson, G. (1987) The effect of altering the hormone status of ewes on the outflow rate of protein supplements from the rumen and so on protein degradability. *Animal Production* 44, 395–404.

National Research Council (1985a) *Nutrient Requirements of Sheep,* 6th Edn. National Academy Press, Washington, DC.

National Research Council (1985b) *Ruminant Nitrogen Usage.* National Academy Press, Washington, DC.

O'Doherty, J.V. and Crosby, T.F. (1997) The effect of diet in late pregnancy on colostrum production and immunoglobulin absorption in sheep. *Animal Science* 64, 87–96.

Ortigues, I. and Vermorel, M. (1996) Adaptation of whole animal energy metabolism to under nutrition in ewes: influence of time and posture. *Animal Science* 63, 413–422.

Ørskov, E.R. and Grubb, D.A. (1979) The minimal nitrogen metabolism of lambs. *Proceedings of the Nutrition Society* 38, 24A

Ørskov, E.R. and Macleod, N.A. (1982) The flow of N from the rumen of cows and steers maintained by intra-ruminal infusion of volatile fatty acids. *Proceedings of the Nutrition Society* 41, 76A.

Ørskov. E.R., Grubb, D.A., Smith, J.S., Webster, A.J.F. and Corrigall, W. (1979) Efficiency of utilization of volatile fatty acids for maintenance and energy retention by sheep. *British Journal of Nutrition* 41, 541–551.

Rattray, P.V. and Joyce, J.P. (1976) Utilisation of metabolisable energy for fat and protein deposition in sheep. *New Zealand Journal of Agricultural Research* 19, 299–305.

Rattray, P.V., Garrett, W.N., East, N.E. and Hinman, N. (1974a) Efficiency of utilisation of metabolisable energy during pregnancy and energy requirements for pregnancy in sheep. *Journal of Animal Science* 38, 383–393.

Rattray, P.V., Garrett, W.N., East, N.E. and Hinman, N. (1974b) Growth, development, and composition of the ovine conceptus and mammary gland during pregnancy. *Journal of Animal Science* 38, 613–626.

Russell, J.B. and Sniffen, C.J. (1984) Effect of carbon-4 and carbon-5 volatile fatty acids on growth of mixed rumen bacteria *in vitro. Journal of Dairy Science* 67, 987–994.

Scollan, N.D., Kim, E.J., Dhanoa, M.S., Gooden, J.M., Neville, M.A., Evans, R.T., Dawson, J.M. and Buttery, P.J. (1997) The effect of diet asynchrony on portal drained viscera metabolism. *Proceedings of the British Society of Animal Science*. 1.

Sinclair, L.A., Garnsworthy, P.C., Newbold, J. and Buttery, P.J. (1993) Effect of synchronizing the rate of dietary energy and nitrogen release on rumen fermentation and microbial protein synthesis in sheep. *Journal of Agricultural Science, Cambridge* 120, 251–263.

Sinclair, L.A., Garnsworthy, P.C., Newbold, J. and Buttery, P.J. (1995) Effects of synchronizing the rate of dietary energy and nitrogen release in diets with a similar carbohydrate composition on rumen fermentation and microbial protein synthesis in sheep. *Journal of Agricultural Science, Cambridge* 124, 463–472.

Walker, D.M. and Faichney, G.J. (1964) Nitrogen balance studies with milk fed lamb. 3) Effects of different nitrogen intakes on growth and nitrogen balance. *British Journal of Nutrition* 18, 295–306.

Webster, A.J.F. (1978) Prediction of the energy requirements for growth in beef cattle. *World Review of Nutrition and Dietetics* 30, 189–226.

Webster, A.J.F. (1992) The metabolizable protein system for ruminants. In: Garnsworthy, P.C., Haresign, W. and Cole, D.J.A. (eds), *Recent Advances in Animal Nutrition*. Butterworths, London, pp. 3–110.

Wilkinson, R.G. and Greenhalgh, J.F.D. (1991) Growth of lambs offered fixed amounts of roughage and concentrate either simultaneously or sequentially. *Journal of Agricultural Science, Cambridge* 116, 125–134.

Wilkinson, R.G. and Greenhalgh, J.F.D. (1995) Growth of lambs offered fixed amounts of roughage and concentrate either simultaneously, progressively or separately. *Journal of Agricultural Science, Cambridge* 124, 301–311.

Wilkinson, R.G., Sinclair, L.A., Powles, J. and Minter, C.M. (1996) The effect of ERDP:FME ratio and DUP supply from concentrates on the performance of lactating ewes at grass. *Animal Science* 62, 664–665.

Witt, M.W., Sinclair, L.A., Wilkinson, R.G. and Buttery, P.J. (1997) Effects of rate of energy and nitrogen supply to the rumen in diets with two rates of carbohydrate release on growth and metabolism of male lambs. *Proceedings of the British Society of Animal Science*, 2.

Witt, M.W., Sinclair, L.A., Wilkinson, R.G. and Buttery, P.J. (1999a) The effects of synchronising the rate of dietary energy and nitrogen supply to the rumen on the production and metabolism of sheep. Feed characterization and growth and metabolism of ewe lambs given food *ad libitum. Animal Science* 69, 223–235.

Witt, M.W., Sinclair, L.A., Wilkinson, R.G. and Buttery, P.J. (1999b) The effects of synchronising the rate of dietary energy and nitrogen supply to the rumen on the metabolism and growth of ram lambs fed at a restricted level. *Animal Science* (in press).

Zhang, J., Grieve, D.G., Hacker, R.R. and Burton, J.H. (1995) Effects of dietary protein percentage and β-agonist administered to pre-pubertal ewes on mammary gland growth and hormone secretions. *Journal of Animal Science* 73, 2655–2661.

9 Feeding Systems for Pigs

L.I. Chiba

Department of Animal and Dairy Sciences, Auburn University, Alabama, USA

INTRODUCTION

In commercial pig production, the main objective of diet formulation and feeding strategy is to maximize profits, which does not necessarily imply maximal animal performance. To maximize the economic efficiency, therefore, supplying indispensable nutrients as close as possible to meeting, but not exceeding, the requirements of the pig is advantageous. Such optimum feeding strategies involve consideration of a multitude of factors such as genetic variations in the pig (sex and genotypes), alternative feed ingredients, variability, availability and stability of nutrients in feed ingredients, interactions among the nutrients and non-nutritive factors, voluntary feed intake, physical and social environment, and others. In addition, there must be an effective means to incorporate all the necessary information to formulate efficient diets in a convenient and economical manner. Because of declining profit margins in recent years, offering diets containing just enough nutrients to satisfy the needs of the pig would have a significant impact on the profitability and success of the pig enterprise. Furthermore, it will have a positive impact on today's environmentally conscious society by reducing the excretion of unutilized nutrients.

It is reasonable to assume that pigs with different sex and genotypes can express their genetic potential for growth, production and/or reproduction if they are provided with the opportunity to consume the optimum amounts of all nutrients in just the proportions required to satisfy their needs. Vitamins and minerals are obviously important nutrients for pigs to perform optimally, but their requirements can be met with relatively little cost. On the other hand, amino acids and energy together account for more than 90% of total feed costs (SCA, 1987). From a practical standpoint, the efficient utilization of these nutrients is essential for economical production of animal products. In this chapter, therefore, emphasis will be placed on energy and protein or amino

©CAB *International* 2000. *Feeding Systems and Feed Evaluation Models*
(eds M.K. Theodorou and J. France)

acid nutrition, although brief discussions on minerals and vitamins are included. As for arguably the most critical nutrient, water, and its closely related area of electrolytes and/or electrolyte balance, the readers are referred to excellent reviews on the subjects including those by Brooks and Carpenter (1993), Frazer *et al.* (1993) and Patience (1993).

Although the nutrient requirements of pigs have been published by several organizations in various countries including Australia, France, Germany and The Netherlands, the most well established are those of the UK's Agricultural Research Council and the USA's National Research Council (Lewis, 1991). The nutrient requirements established by those two organizations are, therefore, included in this chapter. Extensive reviews on establishing the nutrient requirements, factors influencing the requirement and satisfying the needs and other pertinent topics in pig nutrition have been presented by ARC (1981), SCA (1987), NRC (1988), Whittemore (1993) and others, including many excellent articles in books edited by Cole and Haresign (1985), Miller *et al.* (1991) and Cole *et al.* (1993).

In this chapter, the discussion will be limited to some areas that the author feels pertinent to feeding the pig satisfactorily. The nutrient requirements established by the NRC (1988) and ARC (1981) are discussed first, with the emphasis on energy and amino acids. In the subsequent sections, the estimation of the requirements and satisfying the nutrient needs are discussed, followed by discussion on some factors that can have a great impact on the nutrient requirements and satisfying the needs of the pig. A brief discussion on pig wastes and environment is also included.

ENERGY REQUIREMENTS OF GROWING PIGS AND SOWS

Energy Systems

Defining the energy requirement and energy content of feedstuffs in terms of digestible energy (DE), rather than gross energy (GE), enhances the accuracy of diet formulation because the variation in digestibility is the major factor affecting the efficiency of energy utilization from various feed sources. Metabolizable energy (ME) takes into account the energy lost in the urine and combustible gases produced in the digestive tract. Net energy (NE) is the difference between ME and heat increment. Because it is the energy ultimately required by the animal for maintenance or production, the NE would be the best measure of the energy that is available to the animal. The estimation of the NE is, however, difficult and imprecise and affected by many factors (NRC, 1988). Therefore, the use of NE might be too sensitive to be of practical use (Wiseman and Cole, 1985) and unlikely to provide any greater precision in formulating diets or predicting responses compared with the ME or DE system (Whittemore, 1993).

The amount of energy lost in the urine is not a constant, and the ME content of the diet decreases with poor quality and excess protein relative to the pig's needs because of the increased excretion of nitrogen (N) as urea

(Filmer and Curran, 1985). The following equation (NRC, 1979) can be used to take into account the quantity of crude protein (CP):

$$ME = DE [96 - (0.202\% \ CP)]/100 \qquad (9.1)$$

Other equations describing the relationship between the ME and DE have been reported by ARC (1981). The published ME to DE ratios range from 0.92 to 0.98 (ARC, 1981; Filmer and Curran, 1985; NRC, 1988). For practical purposes, a commonly accepted value of 0.96 is regarded as an appropriate factor for a wide range of ingredients and diets. The loss of energy as combustible gases in pigs is generally ignored because the losses are negligible and difficult to measure (NRC, 1988). Furthermore, the variation in the relationship between DE and ME is more of a function of the animal than of the feed ingredient itself (Whittemore, 1993). In addition, DE values for the pig are available for most of the commonly used feed ingredients. For these reasons, it is preferable to use DE values, which can be determined much more easily and precisely, rather than the ME, to express the requirements of pigs and describe the energy value of feed ingredients and diets.

Requirements of Growing Pigs and Sows

Growing pigs

The DE requirements of the growing pig can be estimated by a factorial approach, and the following equation can be used for that purpose:

$$DE = DE_m + DE_{pr} + DE_{fr} + DEH_c \qquad (9.2)$$

where DE_m, DE_{pr}, DE_{fr} and DEH_c are the requirements for maintenance, protein retention, fat retention and cold thermogenesis, respectively (NRC, 1988). Similarly, ARC (1981) placed an emphasis on the factorial approach, but they also used an empirical method in their estimation of the requirements. In addition, the use of the factorial approach was extended to allow prediction of basic performance responses to be assessed against existing empirical observations (Close and Fowler, 1985).

The energy requirement for maintenance is usually expressed as a function of body weight to the power of 0.75, though other exponents may be more appropriate (Whittemore, 1993). Many factors can influence the energy needs for maintenance, but most of the estimated DE_m requirements fall between 0.418 and 0.523 MJ kg body weight (BW)$^{-0.75}$ day^{-1}, with the mean of 0.460 (NRC, 1988). Estimates for the energy costs of protein accretion range from 29.7 to 61.1 MJ of DE kg^{-1} with an average of 52.7, whereas the estimates for fat accretion range from 39.7 to 68.2 MJ of DE kg^{-1} with an average of 52.3 (Tess *et al.*, 1984). Lean muscle tissues, however, contain only 20–22% protein, thus the energy cost per unit of lean muscle is considerably less than that for fat accretion (NRC, 1988). On the other hand, the energy lost as heat in protein turnover may reduce the efficiency of protein accretion as compared with fat accretion. The energetic efficiencies of protein and fat accretion have been reported to be 0.54 and 0.74, respectively, with corresponding estimated DE requirements of 45.7 MJ kg^{-1} for protein and 55.7 MJ kg^{-1} for fat accretion (ARC, 1981).

The estimates of DE requirements for maintenance and growth apply to pigs kept in thermoneutral conditions, and less energy would be available for growth when ambient temperature (T) falls below the critical temperature (T_c). The lower and upper critical temperatures can be defined as the temperatures below and above ambient temperatures where energy must be used to maintain optimum body temperature. The energy cost of cold thermogenesis can be derived from the following equation:

$$DEH_c \text{ (kJ of DE day}^{-1}) = (1.365 \text{ BW} + 98.96) \ (T_c - T) \qquad (9.3)$$

where BW is in kg and T_c and T are expressed in degrees Celsius (ARC, 1981). Assuming a partial efficiency of energy utilization below T_c to be 0.8, the increase in energy requirements can be estimated (ARC, 1981). The estimates of DE requirements for maintenance and daily DE intake of growing pigs are presented in Tables 9.1 and 9.2, respectively.

Pregnant and lactating sows

The factorial approach to estimate the energy requirements of the sow during pregnancy has been summarized by NRC (1988). An energy intake during gestation is generally restricted to control weight gain and maintain appropriate condition of the sow, because the greater the weight gain during pregnancy, the greater the weight loss during lactation (ARC, 1981). The DE_m requirements have been estimated to be 0.402–0.699 MJ with a mean of 0.460 MJ kg BW$^{-0.75}$ day^{-1}, whereas the DE requirement for maternal protein and fat gain is

Table 9.1. Estimates of digestible energy for maintenance (DE_m) based on the two equations using different exponents of live weight (BW$^{0.63}$ or BW$^{0.75}$)[a,b].

Live weight (kg)	DE_m (MJ day^{-1})	
	BW$^{0.63}$	BW$^{0.75}$
5	2.06	1.59
10	3.20	2.69
20	4.95	4.51
30	6.39	6.11
40	7.66	7.58
50	8.80	8.97
60	9.88	10.28
70	10.89	11.54
80	11.84	12.76
90	12.75	13.94

[a]Based on ARC (1981) and Close and Fowler (1985).
[b]Estimates derived from the equation $DE_m = 0.749$ BW$^{0.63}$, and the estimates from the same data if the equation is constrained to the exponent of 0.75 ($DE_m = 0.477$ BW$^{0.75}$, where BW = live weight in kg).

Table 9.2. Daily digestible energy (DE) intake of the growing pig (MJ day^{-1}).

Body weight (kg)	NRC[b]	ARC[a]	
		4 × M[c]	Asymptotic[d]
1–5	3.6		
5–10	6.5		
10–20	13.5		
20–50	27.0		
50–110	44.2		
20		19.8	18.4
30		25.5	25.1
40		30.6	30.6
50		35.2	35.2
60		39.5	38.8
70		43.6	41.8
80		47.4	44.2
90		51.0	46.2

[a]ARC (1981).
[b]NRC (1988).
[c]Daily DE intake based on 4 × maintenance (M) requirement, where maintenance (MJ) = 0.749 BW$^{0.63}$ and BW = live weight in kg.
[d]Daily DE intake based on an asymptotic equation, DE (MJ) = 55(1 − e$^{-0.0204BW}$).

assumed to be 52.3 MJ kg^{-1} gain. If the composition of maternal gains consists of about 25% fat and 15% protein, the energy cost of maternal gain can be calculated as 20.9 MJ of DE kg^{-1}. With a desirable net weight gain of 25 kg during pregnancy, the daily DE requirement is 4.60 MJ. The daily energy requirement for uterine gain has been estimated to be 0.79 MJ of DE, thus the daily DE requirement during gestation is 5.40 MJ. Assuming an energetic efficiency of 0.80, the DE$_m$ requirement would be increased by 19.8 kJ kg BW$^{-0.75}$ for each 1°C decrease in temperature below the T_c (Close and Fowler, 1985).

The estimates of the energy requirements during lactation have been based on the requirement for maintenance, milk production and the contribution from the mobilization of tissue resulting from the inevitable loss of body weight (Close and Fowler, 1985; Noblet *et al.*, 1990). The DE$_m$ has been reported to be 0.460 MJ kg BW$^{-0.75}$ day^{-1}, but it may be 5–10% greater than that reported for the pregnant sow (NRC, 1988). The energy requirement for milk production is 8.4 MJ of DE kg^{-1} milk based on the GE content of 5.4 MJ of DE kg^{-1} milk and an efficiency of utilization of 65% (Close and Fowler, 1985). The contribution derived from body weight is assumed to represent the loss of lipids (Close and Fowler, 1985). Based on the conversion of body fat to milk fat, which has an energy value of 41.0 MJ of DE kg^{-1} fat with the conversion efficiency of 0.85, body weight loss can contribute 48.7 MJ of DE kg^{-1} (ARC, 1981).

Daily DE requirements of both pregnant gilts and sows and lactating sows are presented in Table 9.3. It has been demonstrated that additional feed during late gestation (Cromwell *et al.*, 1989) and supplementation of the diets with lipids during late gestation or lactation (Moser and Lewis, 1980) can improve the reproductive performance of the sow. Depending on their cost-effectiveness, therefore, these management practices may be beneficial in increasing the sow productivity.

Table 9.3.   Daily digestible energy (DE) requirements of pregnant and lactating sows.

Class and body weight (kg)	NRC[b]	ARC[a] Low/medium[c]	High[d]
Pregnant gilts and sows[e]			
142.5	24.4		
162.5[f]	26.3		
182.5	28.3		
120		25.5	29.8
140		27.6	31.9
160		29.7	33.9
Lactating sows[g]			
145	60.7		
165	73.2		
185	85.8		
140		62.8	71.2
160		64.8	73.2
180		66.8	75.1
200		68.6	77.0

[a]ARC (1981).
[b]NRC (1988).
[c]Based on the requirements for pregnant gilts and sows on a net gain of 20 kg during pregnancy (low), whereas the requirements for lactating sows are based on medium milk yield (medium; 6.25 kg day^{-1}).
[d]Based on the requirements for pregnant gilts and sows on a net gain of 40 kg during pregnancy (high), whereas the requirements for lactating sows are based on high milk yield (high; 7.25 kg day^{-1}).
[e]Based on the NRC requirements for a 25 kg maternal weight gain and 20 kg conceptus gain, and the weights indicated are mean gestation weights; for the ARC, feed intake should be increased by between 3 and 4% per 1°C decrease in temperature.
[f]The same DE requirement for adult boars (162.5 kg) as the requirement for a 162.5 kg sow.
[g]For the NRC, based on the requirements for the assumed milk yield of 5.0, 6.25 and 7.5 kg day^{-1} for 145, 165 and 185 kg sows, respectively; for the ARC, based on the requirements indicated for a 21 day weaning.

PROTEIN AND AMINO ACID REQUIREMENTS

Protein and Non-protein Nitrogen

The difference between crude and true protein, and the role of non-protein nitrogen (NPN) in pig nutrition have been reviewed by Lewis (1991). The use of a single factor (6.25) to convert N to protein works well in expressing the protein content of feed ingredients or diets because typical diets contain mixtures of protein and mean N contents that are usually close to 16%. Similarly, most of the NPN in many feed ingredients is in the form of amino acids, which can be utilized efficiently by pigs. In general, therefore, the use of crude protein (CP) values is sufficiently accurate for most instances in pig nutrition.

To be of nutritional value, NPN must be in a form that can be converted to protein within the pig. Pigs are capable of absorbing amino acids synthesized from NPN by intestinal bacteria (Niiyama *et al.*, 1979), and can incorporate them into tissue proteins (Grimson et al., 1971). These results indicate that pigs have the ability to utilize inexpensive NPN sources, and overall utilization of diets may be improved by the inclusion of NPN (Chiba *et al.*, 1995). However, it is generally assumed that the amount of NPN utilized by pigs would be too small to elicit substantial beneficial effects on the performance because of their anatomical and metabolic limitations (NRC, 1976).

Protein Quality and Amino Acid Balance

The body uses mixtures of amino acids collectively for protein synthesis, thus the balance of amino acids is very important for an optimum utilization of protein (Whittemore, 1993). Any departure from a desirable pattern of amino acids may lead to a reduction in pig performance, at least in terms of the efficiency of protein utilization (Lewis, 1991). In addition, it may result in acute neurological aberrations and even death (D'Mello, 1994). The topic of amino acid disproportions, imbalance, antagonism and toxicity, has been reviewed in detail by Harper *et al.* (1970). Adverse effects of the imbalance and antagonism can be alleviated by supplying the limiting amino acid(s) and structurally related amino acid(s), respectively. Amino acid toxicity is not likely to be a problem in practical pig nutrition, and it is usually associated with errors in formulation and/or mixing of diets using crystalline amino acids.

Ideal protein
Proteins serve to supply each individual indispensable amino acid and an adequate amount of dispensable amino N as a whole to meet the needs of the animal. Ideal protein can be described operationally as protein that cannot be improved by any substitution of a quantity of one amino acid for the same quantity of another (Fuller and Chamberlain, 1985); it has been derived from individual estimates of amino acid requirements, the composition of sow's milk and body tissues (Lewis, 1991), and the results of experiments designed specifically for that purpose (Wang and Fuller, 1989). The ARC (1981) adopted

the ideal protein concept in their estimation of the amino acid requirements for growing pigs. On a practical basis, the proportions of threonine, tryptophan and methionine + cysteine relative to lysine are perhaps the most important because these amino acids are likely to be limiting after lysine (Knabe, 1996).

In general, pigs seem to be able to tolerate quite wide variations in the pattern of amino acids, provided all the indispensable amino acid requirements are satisfied (Lewis, 1991). Furthermore, excess or moderate oversupply of some individual amino acids is unlikely to have clear adverse effects on the performance of pigs, as mentioned by Chiba *et al.* (1991a). Nevertheless, consideration of the ideal protein concept in diet formulation would be beneficial in terms of efficient utilization of dietary amino acids. The patterns of amino acids in ideal protein for growing pigs proposed by various research groups are summarized in Table 9.4.

Requirements of Growing Pigs and Sows

The subject of dispensable, indispensable and conditionally dispensable amino acids has been reviewed by Lewis (1991) and Fuller (1994). Pigs do not have a requirement for protein as such, but they need adequate amounts of indispensable amino acids and a sufficient amount of non-specific N in the diet to synthesize dispensable amino acids. It is generally assumed that if the indispensable amino acid supply is adequate for pigs fed a conventional diet, then the amount of dispensable amino acids would be sufficient. With increased use of crystalline amino acids, however, this assumption may not be valid. Sufficient amounts of both fractions must be provided in the diet to support an optimum performance of pigs.

Table 9.4. Proportions of each amino acid in ideal protein relative to lysine.

Amino acid	Cole (1978)	ARC (1981)	NRC (1988)[a]	Wang and Fuller (1989)	Chung and Baker (1992)[b]
Lysine	100	100	100	100	100
Arginine	—	—	43	—	42
Histidine	40	33	26	—	32
Isoleucine	50	55	54	60	60
Leucine	100	100	71	111	100
Methionine + cysteine	50	50	49	63	60
Phenylalanine + tyrosine	100	96	79	120	95
Threonine	60	60	57	72	65
Tryptophan	18	15	14	18	18
Valine	70	70	57	75	68

[a]Based on lysine:tryptophan and arginine:tryptophan ratios of 7.0 and 3.0, respectively.
[b]The proportions of amino acids for pigs weighing 5–20 kg.

The NRC (1988) amino acid requirements have been derived from the results of various experiments designed for that purpose, and certain theoretical principles of amino acid nutrition have been considered in the process. A summary of extensive reviews on research designed to establish each amino acid requirement of growing pigs and gestating and lactating sows has been presented by the NRC (1988).

Growing pigs

Fuller and Chamberlain (1985) summarized the factorial and empirical approaches used by the ARC (1981) in establishing the protein requirements of growing pigs. The ARC (1981) indispensable amino acid recommendations are expressed in terms of their relationships to the lysine content of ideal protein, and they assumed that such an ideal balance is equally applicable for pigs from birth to 90 kg. Similarly, the NRC (1988) assumed that the proportion of amino acids required is relatively consistent throughout the growth phases.

The amino acid requirements of growing pigs established by the ARC (1981) and NRC (1988) are presented in the Table 9.5. An adequate amino acid intake, along with other indispensable nutrients, is necessary to maximize lean accretion in growing pigs. Therefore, making appropriate adjustments according to various factors, some of which will be discussed later, is very important in establishing the allowances for a population of pigs.

Table 9.5. Amino acid requirements of very young and growing pigs (g kg^{-1}).

	Pig weight (kg) or age (weeks)								
	NRC (1988)					ARC (1981)[a]			
Amino acid	1–5	5–10	10–20	20–50	50–110	0–3[b]	3–8[b]	15–50	50–90
Lysine	14.0	11.5	9.5	7.5	6.0	15.8	13.8	11.0	7.8
Arginine	6.0	5.0	4.0	2.5	1.0	—	—	—	—
Histidine	3.6	3.1	2.5	2.2	1.8	5.2	4.5	3.6	2.6
Isoleucine	7.6	6.5	5.3	4.6	3.8	8.6	7.5	6.0	4.3
Leucine	10.0	8.5	7.0	6.0	5.0	15.8	13.8	11.0	7.8
Methionine + cysteine	6.8	5.8	4.8	4.1	3.4	7.9	6.9	5.5	3.9
Phenylalanine + tyrosine	11.0	9.4	7.7	6.6	5.5	15.1	13.3	10.4	7.5
Threonine	8.0	6.8	5.6	4.8	4.0	9.4	8.3	6.5	4.7
Tryptophan	2.0	1.7	1.4	1.2	1.0	2.3	2.0	1.6	1.2
Valine	8.0	6.8	5.6	4.8	4.0	11.0	9.7	7.7	5.5

[a]Based on the requirements for pigs weighing 15–90 kg on the assumption that the diet contains 13.0 MJ kg^{-1}.
[b]Age in weeks; based on the requirements on the assumption that the diet contains 14.1 MJ kg^{-1}, which is similar to the energy density recommended by the NRC (1988) for young pigs.

Pregnant and lactating sows

The NRC (1988) requirements for pregnant gilts and sows have been estimated experimentally, or based on the amount needed for satisfactory N retention during the late phase of pregnancy. Similarly, the requirements for lactating sows have been estimated experimentally or extrapolated from the published requirements for maintenance and the amount needed to support milk production. Although the ARC (1981) indicated that the evidence is insufficient to test the validity of using ideal protein concept for adults, they expressed the amino acid requirements in terms of a balance relative to lysine along with daily intake of amino acids and dietary percentages. The patterns of amino acids for pregnant and lactating sows proposed by several research groups have been summarized by Fuller (1994).

The amino acid requirements of sows established by the ARC (1981) and NRC (1988) are presented in the Table 9.6. Several researchers (Stahly *et al.*, 1990; Johnston *et al.*, 1993) reported that sows producing large amounts of milk and nursing large litters responded to higher amino acid levels than those recommended by the NRC (1988). Sow productivity has increased in recent years for various reasons (Knabe *et al.*, 1996). It is important, therefore, to consider the genetic potential of lactating sows, as well as other factors, in establishing the allowances.

Table 9.6. Amino acid requirements of pregnant gilts and sows and lactating sows (g kg^{-1}).

Amino acid	Pregnancy[a]		Lactation[b]	
	NRC[c]	ARC[d]	NRC[c]	ARC[d]
Lysine	4.3	4.3	6.0	6.3
Arginine	0.0	—	4.0	4.2
Histidine	1.5	1.3	2.5	2.5
Isoleucine	3.0	3.7	3.9	4.4
Leucine	3.0	3.2	4.8	7.2
Methionine + cysteine	2.3	2.9	3.6	3.4
Phenylalanine + tyrosine	4.5	3.3	7.0	7.2
Threonine	3.0	3.6	4.3	4.4
Tryptophan	0.9	0.7	1.2	1.2
Valine	3.2	4.6	6.0	4.4

[a]Based on the requirements of feed intakes of 1.9 kg (or 26 MJ of DE) and 2.0 kg day^{-1} (or 25 MJ of DE day^{-1}) for the NRC and ARC, respectively.
[b]Based on the requirements of feed intakes of 5.3 kg (or 74 MJ of DE) and 5.25 kg day^{-1} (or 66 MJ of DE day^{-1}) for the NRC and ARC, respectively.
[c]NRC (1988).
[d]ARC (1981).

MINERAL AND VITAMIN REQUIREMENTS

Mineral Requirements

The importance of mineral elements for normal life processes was recognized even in ancient times. Minerals have extremely diverse functions, including their roles in skeletal structure, homeostasis, cell membranes, enzymes and hormones. Pigs require at least 13 known mineral elements including calcium, chlorine, copper, iodine, iron, magnesium, manganese, phosphorus, potassium, selenium, sodium, sulphur and zinc (NRC, 1988). Cobalt is also required for the synthesis of vitamin B_{12}, but there is no convincing evidence that non-ruminant species need this element with adequate vitamin B_{12} in the diet. Pigs may also need other elements such as arsenic, boron, bromine, cadmium, chromium, fluorine, lead, lithium, molybdenum, nickel, silicon, tin and vanadium, but these elements are required at such low levels that their dietary essentiality has not been proven (NRC, 1988).

Based on the corn–soybean meal diet formulated to satisfy the lysine requirement of a 25 kg pig (NRC, 1988), corn and soybean meal can supply only few minerals in adequate amounts to satisfy the needs, and others are clearly deficient or marginal at best. In recent years, a supplementation of diets with pharmacological levels of zinc (2500–3000 p.p.m.) has been shown to enhance growth performance and reduce scouring of weanling pigs (Hahn and Baker, 1993; Poulsen, 1995), even though, there are some conflicting results (e.g. Schell and Kornegay, 1994). Similarly, the beneficial effects of chromium supplementation on the carcass quality of grower–finisher pigs (Page *et al.*, 1993a; Lindemann *et al.*, 1995) and the reproductive performance of sows (Lindemann *et al.*, 1995) have been reported. Several minerals are toxic at very low concentrations, and the toxicity and tolerance of those mineral elements have been reviewed by the NRC (1980).

Vitamin Requirements

Vitamins constitute only a small fraction of the diet but they are important in the health, well-being and productivity of the animal. The fat-soluble vitamins include vitamins A, D, E and K, whereas water-soluble vitamins include thiamin, riboflavin, niacin, vitamin B_6, pantothenic acid, biotin, folacin, vitamin B_{12}, choline and ascorbic acid. The water-soluble vitamins are relatively non-toxic, but excesses of dietary vitamin A and D have been shown to cause some toxic effects (NRC, 1987). Pigs fed chemically defined diets have a dietary requirement for all of the fat- and water-soluble vitamins except ascorbic acid.

As with the minerals, the corn–soybean meal diet formulated to satisfy the lysine requirement of a 25 kg pig (NRC, 1988) can supply only few vitamins in adequate amounts. It is generally assumed that feed ingredients and microbial synthesis in the gastrointestinal tract are adequate to satisfy the requirements for vitamin B_6, thiamin, folic acid and biotin, but the beneficial effects of biotin supplementation on reproductive performance of sows have been reported over the years (Kornegay, 1986). In addition, folic acid supplementation is likely to improve litter size, especially among sows with a high ovulation rate, as Lindemann (1993) concluded in his review. Similarly, although further research is needed, there might be an opportunity to increase the reproductive performance of sows by short-term administration of β-carotene or vitamin A at critical stages of the reproductive process (Brief and Chew, 1985; Coffey and Britt, 1992).

The mineral and vitamin requirements are presented in Tables 9.7 and 9.8, respectively. Recent findings on beneficial effects of minerals may indicate that some mineral elements play a critical role(s) in not only achieving acceptable pig performance, but realizing the full genetic potential for growth, production and/or reproduction. A similar conclusion can be drawn for some vitamins. Further research is needed to elucidate fully the effect of those nutrients on the performance of pigs and the economical feasibility of using those nutrients.

ESTABLISHING AND SATISFYING THE NUTRIENT REQUIREMENTS

Comparison of the NRC and ARC Requirements

The similarities and differences between the NRC (1979) and ARC (1981) in estimating the energy and protein requirements of growing pigs have been reviewed by Lewis (1993). Although there were some differences, the approaches used by the NRC (1988) in their latest revision and the ARC (1981) in arriving at the recommendations seem to be very similar. The revised nutrient requirements of the NRC (1988) and those recommended by the ARC (1981), however, indicate the differences in the final estimation for some nutrients. Although the estimates of amino acid requirements for sows during gestation and lactation are quite similar, there are clearly large differences in the estimates for growing pigs, with the ARC values being greater for all stages (Lewis, 1991). The NRC (1988) requirements are based on gilts and castrates fed *ad libitum*, whereas the ARC (1981) requirements are based on both *ad libitum* and restricted feeding regimens, and include boars, which may partly account for the differences as mentioned by Chiba (1989). It is not reasonable to assume that there are such large differences in the requirements of pigs, indicating a need for further research in this area of establishing nutrient requirements (Lewis, 1991).

Establishing the Nutrient Requirements

Empirical and factorial approaches
In this section, the discussion on the estimation of the requirements is generally limited to protein or amino acids, but, to a large extent, the general premises presented here would be equally applicable to other nutrients. Nutrient requirements can be determined by the empirical and/or factorial

Table 9.7. Mineral requirements of growing pigs, and pregnant and lactating sows.

Mineral (U kg⁻¹)	NRC[a]							ARC[b]				
	1–5	5–10	10–20	20–50	50–110	Pregnancy	Lactation	Up to 20	20–55[c,d]	55–90	Up to 90	Sows
Calcium (g)	9.0	8.0	7.0	6.0	5.0	7.5	7.5	9.9	8.1	7.2	—	8.1
Phosphorus (g)	7.0	6.5	6.0	5.0	4.0	6.0	6.0	8.1	6.3	5.4	—	6.3
Available phosphorus (g)	5.5	4.0	3.2	2.3	1.5	3.5	3.5	—	—	—	—	—
Sodium (g)	1.0	1.0	1.0	1.0	1.0	1.5	2.0	—	1.2	—	—	—
Chlorine (g)	0.8	0.8	0.8	0.8	0.8	1.2	1.6	—	1.4	—	—	—
Magnesium (g)	0.4	0.4	0.4	0.4	0.4	0.4	0.4	0.4	0.4	—	—	—
Potassium (g)	3.0	2.8	2.6	2.3	1.7	2.0	2.0	2.3	2.3	—	—	—
Copper (mg)	6.0	6.0	5.0	4.0	3.0	5.0	5.0	—	—	—	3.6	—
Iodine (mg)	0.14	0.14	0.14	0.14	0.14	0.14	0.14	—	—	—	0.14	0.45
Iron (mg)	100	100	80	60	40	80	80	54	—	—	—	—
Manganese (mg)	4.0	4.0	3.0	2.0	2.0	10.0	10.0	—	—	—	4–14	9.0
Selenium (mg)	0.3	0.3	0.25	0.15	0.10	0.15	0.15	—	—	—	0.14	—
Zinc (mg)	100	100	80	60	50	50	50	—	—	—	45	—

[a]NRC (1988); based on the requirements of feed intakes of 1.9 kg (or 26 MJ of DE) and 5.3 kg day⁻¹ (or 74 MJ of DE day⁻¹) for pregnant and lactating sows, respectively.

[b]ARC (1981); converted to 'air dry' basis (900 g kg⁻¹ dry matter).

[c]For chlorine and sodium, the requirements for pigs weighing 20–35 kg.

[d]For potassium and magnesium, the requirements for pigs weighing up to 45 and 55 kg, respectively.

Table 9.8. Vitamin requirements of growing pigs, and pregnant and lactating sows.

| | Pig weight (kg) or class | | | | | | | | | | | |
| | NRC[a] | | | | | | | ARC[b] | | | | |
Vitamin (U kg⁻¹)[c]	1–5	5–10	10–20	20–50	50–110	Pregnancy	Lactation	Up to 20[d]	Up to 40[e]	20–90[f]	Up to 90	Sows
Vitamin A (IU)	2200	2200	1750	1300	1300	4000	2000	—	1800	1200	—	2100
Vitamin D (IU)	220	220	200	150	150	200	200	126	—	108	—	—
Vitamin E (IU)	16	16	11	11	11	22	22	10.7	—	6.7	—	8.1
Vitamin K (mg)	0.5	0.5	0.5	0.5	0.5	0.5	0.5	—	0.27	—	—	—
Biotin (mg)	0.08	0.05	0.05	0.05	0.05	0.20	0.20	—	—	—	—	—
Choline (g)	0.6	0.5	0.4	0.3	0.3	1.25	1.00	0.71	—	<0.9	—	0.9–1.7
Folacin (mg)	0.3	0.3	0.3	0.3	0.3	0.30	0.30	—	—	—	—	—
Niacin (mg)	20.0	15.0	12.5	10.0	7.0	10.0	10.0	18.0	—	12.6	—	—
Pantothenic acid (mg)	12.0	10.0	9.0	8.0	7.0	12.0	12.0	—	—	—	9.0	9.0
Riboflavin (mg)	4.0	3.5	3.0	2.5	2.0	3.75	3.75	—	—	—	2.3	2.7
Thiamin (mg)	1.5	1.0	1.0	1.0	1.0	1.0	1.0	—	—	—	1.4	1.4
Vitamin B$_6$ (mg)	2.0	1.5	1.5	1.0	1.0	1.0	1.0	—	—	—	2.3	1.4
Vitamin B$_{12}$ (μg)	20.0	17.5	15.0	10.0	5.0	15.0	15.0	16.2	—	9.0	—	13.5

[a]NRC (1988); based on the requirements of feed intakes of 1.9 kg (or 26 MJ of DE) and 5.3 kg day⁻¹ (or 74 MJ of DE day⁻¹) for pregnant and lactating sows, respectively.

[b]ARC (1981); Converted to 'air dry' basis (900 g kg⁻¹ dry matter).

[c]For the ARC estimates, assumed 1 IU of vitamin A, D and E = 0.30 μg of retinol, 0.025 μg of vitamin D$_3$ and 0.67 mg of D-α-tocopherol, respectively.

[d]For niacin, the requirement for pigs weighing up to 10 kg.

[e]For vitamin K, the requirement for pigs weighing up to 30 kg.

[f]For niacin and vitamin A, the requirements are for pigs weighing between 10 and 70 kg and between 40 and 90 kg, respectively.

method, both of which are considered by the ARC (1981) and NRC (1988). The empirical method involves the evaluation of the responses of pigs to different concentrations or intakes, and the requirement is the concentration or intake up to which there is a favourable response and beyond which there is no further response. There are many difficulties associated with this method including: (i) observation of diminishing marginal responses without defining clearly a point of maximum response; (ii) the difference in the estimates depending on the criterion of response used; and (iii) the effect of many factors such as some aspects of the diet, animal or environment on the response pattern (ARC, 1981; Whittemore, 1993). The requirements estimated by the empirical method, therefore, may have limited usefulness because the results may be applicable to the specific conditions under which the experiment was conducted (Knabe, 1996).

The factorial method, in its simplest form, assumes that the requirement is the sum of the requirements for maintenance, tissue protein accretion and/or milk protein secretion, with some considerations for the efficiency of protein or amino acid utilization (Fuller and Chamberlain, 1985). Unlike the empirical method, this approach is more flexible, and takes into account various factors that can affect the requirements (Knabe, 1996) and allows requirements to be estimated for animals differing in their productive state (Fuller, 1994). However, present knowledge of the factors contributing to protein requirements is little more than rudimentary (Fuller and Chamberlain, 1985). For instance, one of the limitations in the application of this approach is a paucity of information on the efficiency of amino acid utilization for various components constituting the requirements (Fuller, 1994). There is, therefore, nearly as much empiricism in the factorial method as in the empirical method (Fuller and Chamberlain, 1985). The accuracy of the projected requirements depends on the validity of the data used in the estimation, and many assumptions themselves may need verification.

Modelling approach as an alternative

Estimating the nutrient requirements with accuracy and precision is rather difficult because a multitude of factors can affect the requirements, implying that no single set of estimates would satisfy the nutrient requirements of all animals. Although not complete, the published information necessary to formulate effective diets for pigs and to establish their likely responses to changes in dietary or environmental conditions seems to be available. However, the interactions among many factors that determine growth and productivity make it difficult to apply all the information in practice (SCA, 1987), and a modelling approach might be the only defensible means of incorporating all the necessary information (Ferguson *et al.*, 1994). Several pig simulation models have been developed and/or described in detail, including those by Black *et al.* (1986), SCA (1987), Pettigrew *et al.* (1992), Whittemore (1993) and Ferguson *et al.* (1994); see Chapter 16 of this book.

As indicated by Whittemore (1993), the main purpose of a model is to show the direction and magnitude of the response and the sensitivity of the production system to a tactical or strategic change so that more effective financial and management decisions can be made. This implies that it is not necessary for a model to come to a single optimum solution. Instead, it would be better to provide the options and guidance as to the likely outcome of taking those options to a decision-maker. Because of its flexibility, modelling would be useful in addressing the nutrient requirements in terms of economics rather than simply achieving the biological production targets (Whittemore, 1993). The modelling approach would allow nutritionists and producers to be involved in 'catering' for the situations, thus having a significant impact on profitability of the pig enterprise.

Satisfying the Requirements and Nutrient Variability and Availability

Under commercial conditions, many factors such as biological variations in both the pig and nutrient sources, bioavailability and stability of nutrients in feed ingredients, interactions among the nutrients and non-nutritive factors, stress, physical and social environment, infectious diseases, parasite infestations and others can influence nutrition of the pig. Therefore, satisfying the nutrient needs of a population(s) of pigs, rather than individual animals, is a challenging task. Furthermore, it is conceivable that not only energy and amino acids, but also some vitamins and minerals, may play critical roles in pigs to express fully their genetic potential for growth, production or reproduction, as mentioned before. The ARC (1981) and NRC (1988) recommendations are generally designed to prevent nutrient deficiency signs and/or satisfy the requirements of average pigs. For the optimum performance of pigs, therefore, it is necessary to make appropriate adjustments to those recommendations according to various factors, including economical factors.

The formulation of pig diets that will satisfy the nutrient needs economically depends on the knowledge of: (i) the nutrient requirements; (ii) the nutrient contents of feed ingredients; and (iii) the availability of the nutrients in feed ingredients. The latter two are associated predominantly with the evaluation of feed ingredients, but they cannot be treated in isolation from that of the requirements. For instance, to evaluate protein quality, it is first necessary to establish the requirement for indispensable amino acids because the protein quality of any feed ingredient is simply a reflection of the limitations imposed by the amino acid composition and/or availability.

Nutrient variability

Considerable variation in the nutritional value of cereal grains, which form the basis of typical pig diets throughout the world, and other feed ingredients exists because of various factors (Wiseman and Cole, 1985). For instance, compared with the value of 85 g kg^{-1} reported by the NRC (1988), the CP content of more than one-half of corn samples contained less than 80 g kg^{-1} on an air dry basis (Reese, 1986). Similar variations in the CP content of cereal grains (and protein supplements) have been reported (Patience, 1996). Cereal grains are not only main sources of energy, but are also main sources of protein or amino acids, and generally supply 40–50% of the protein in the diet of growing pigs. Thus, their amino acid contents are very important (Lewis, 1991). It is reasonable to assume that the variability is associated with other nutrients in cereal grains, as well as many nutrients in other feed ingredients (Wiseman and Cole, 1985; NCR-42, 1993). In addition, the variability associated with various laboratories and analytical techniques (NCR-42, 1993; Patience, 1996; NCR-42 and S-145, 1997) may have to be considered.

Besides dealing with the variation in the content of nutrients in feed ingredients, nutritionists must face the issue of nutrient availability because not all of the energy, amino acids and others are available to the pig (Patience, 1996). For instance, amino acids may not be available because of an incomplete protein hydrolysis by proteolytic enzymes, suppression of enzymatic activity by inhibitors and/or inhibition of absorption. Similarly, mineral elements may be bound to phytate or fiber, or form complexes with others, and vitamins can exist as either precursor compounds or as coenzymes that may be bound or complexed in some manner, which render them unavailable to the pig.

The term bioavailability or availability can be defined as the degree to which an ingested nutrient in a particular source is absorbed in a form that can be utilized in the metabolic process by the animal (Ammerman *et al.*, 1995); for amino acids, this involves the digestion, absorption, and utilization by the tissue after absorption. It is important to note that the availability influences not only dietary requirements but also tolerance of a nutrient as well. A number of comprehensive reviews have been written on the subject of nutrient availability including various articles on amino acids, minerals and vitamins that comprise the book edited by Ammerman *et al.* (1995).

Formulation of diets based on available nutrients

Simply because pigs can utilize only those nutrients available to them, it is reasonable to assume that expressing the requirements and formulating diets based on the available nutrients, rather than the total, would be more effective in precisely meeting the pig's needs. Most of the data on ME values have been derived mathematically from DE. Therefore, these two systems can be used interchangeably to a large extent, and the use of more easily determinable DE is preferred in pig diet formulation, as mentioned before. As for amino acids, in practice, most investigators have used apparent ileal digestibilities as an index of amino acid availability in feed ingredients. Although the values determined in such a manner are not perfect (Knabe, 1996), they are similar to values determined by other methods such as growth assays (NRC, 1988).

The NRC (1988) recommends the formulation of pig diets based on the available amino acids rather than the total when the availability of limiting amino acids in the ingredients is less than 70% or more than 90%. They also expressed the phosphorus requirements in terms of available phosphorus. Consideration of the availability of phytate phosphorus is important in pig nutrition because phosphorus is the third most expensive nutrient in the pig diet after energy and amino acids. Microbial phytase has been shown to be effective in improving the utilization of phytate phosphorus, as well as other mineral elements (Kornegay, 1996).

The competition between humans and animals for quality sources of nutrients is likely to increase continuously because of the ever-increasing world population, indicating the importance of exploring the potential of all sources as feed ingredients. Alternative feed ingredients have different feeding values because of variations in the nutrient contents and other factors such as palatability and handling property. Obtaining accurate information on the feed ingredients is, therefore, necessary to make appropriate adjustments for the formulation of diets that meet the nutrient standard in a cost-effective way. Apparent ileal digestibility of selected amino acids and availability of phosphorus in common feed ingredients are presented in Table 9.9.

Although other nutrients are equally important, consideration of the availability of energy, amino acids and phosphorus in diet formulation would contribute greatly to the efficiency and economics of pig production and a reduction of the release of unutilized nutrients into the environment. It is, however, questionable whether there is sufficient information on the nutritive value of individual feed ingredients (Close and Fowler, 1985). Therefore, there is very little agreement on how to address the availability issue in a day-to-day diet formulation, and there is also a question regarding whether this practice will improve the precision of diet formulation sufficiently to meet the needs of the industry (Patience, 1996). Further progress must be made in developing procedures to describe the true nutritional value of feed ingredients so that practical, convenient, cost-effective and environmentally friendly pig diets can be formulated.

FACTORS AFFECTING NUTRIENT REQUIREMENTS AND/OR EFFICIENCY OF NUTRIENT UTILIZATION

Voluntary Feed Intake

In North America, growing pigs and lactating sows generally are allowed to consume feed *ad libitum*, whereas feed intakes of developing gilts, boars and pregnant females are restricted to accommodate their nutritional needs. Voluntary feed intake is expressed in terms of DE because this is the energy introduced into the biological system, and the DE intake (DE_i; kJ day^{-1}) can be predicted by the following equations (NRC, 1986):

Suckling pig	$DE_i = 46.9\ D - 634.7$	(9.4)
Weanling pig	$DE_i = 1933\ BW - 40.7\ BW^2 - 6397$	(9.5)
Growing pig	$DE_i = 55071\ (1 - e^{-0.0176\ BW})$	(9.6)
Lactating sow	$DE_i = (56067 + 2494\ D) - 72.0\ D^2$	(9.7)

where D is age of the pig and the day of lactation for the suckling pig and lactating sows, respectively, and BW is in kg for the weanling and growing pig.

Feed intake of the pig is determined largely by the energy density of the diets, and it is generally assumed that pigs can adjust a voluntary feed intake

Table 9.9. Apparent ileal digestibilities of selected amino acids and estimates of biological availability of phosphorus in pig feed ingredients[a].

Feed ingredient	Amino acids (%)[b]				Phosphorus (%)[c]	
	Lysine	Tryptophan	Threonine	Methionine	Average	Range
Alfalfa meal	—	—	—	—	100	—
Barley, grain	73	73	70	82	31	—
Beans, broad						
(*Vicia faba*)	82	68	75	73	—	—
Blood meal	81	—	82	—	—	—
Bone meal	—	—	—	—	82	—
Canola meal	75	—	67	84	—	—
Corn, grain	80	70	73	89	15	9–29
Corn, high						
moisture	—	—	—	—	49	42–58
Cottonseed meal	65	73	63	70	21	0–42
Defluorinated						
rock phosphate	—	—	—	—	87	83–90
Dicalcium						
phosphate	—	—	—	—	100	—
Fish meal	80	—	76	84	100	—
Meat and bone						
meal	64	53	56	73	93	—
Oat groats	82	81	78	89	—	—
Oats	58	59	53	75	30	23–36
Peanut meal	79	—	—	—	12	—
Rice bran	—	—	—	—	25	—
Rye, grain	68	62	62	80	—	—
Sorghum, grain	80	75	73	85	22	19–25
Sorghum, high						
moisture	—	—	—	—	43	42–43
Soybean meal	87	81	77	86	38	36–39
Soybean meal,						
dehulled	85	78	74	88	25	18–35
Triticale	82	—	74	85	—	—
Sunflower meal	72	—	71	84	—	—
Wheat, grain	80	78	74	85	50	40–56
Wheat bran	—	—	—	—	35	—
Wheat middlings	—	—	—	—	45	34–55

[a]Based on NRC (1988).
[b]Values represent the percentage of the total amino acid contained in the feed ingredient that has disappeared from the digestive tract of growing swine when digesta arrive at the terminal ileum.
[c]Relative to the availability of phosphorus in monosodium phosphate (100%).

to achieve a constant energy intake (NRC, 1986). Such control mechanisms, however, may not work under extreme conditions because of a physical limitation and a lack of gut fill with diets of low and high nutrient densities, respectively. Besides the dietary energy density and ambient temperatures, the adequacy of dietary protein or amino acids (Henry, 1985) and amino acid disproportions (Pond *et al.*, 1969) have also been shown to influence feed intake, indicating the importance of amino acids in this regard.

By definition, appetite of the animal is a function of the nutrient requirement (Whittemore, 1993). Voluntary feed intake dictates the amounts of indispensable nutrients consumed by the animal; consequently, it is a major factor determining the growth, health, production and/or reproduction of the pig. Manipulating a feed intake and/or understanding various factors that are known to influence the feed intake and making appropriate adjustments would, therefore, contribute greatly to successful pig production.

Amino Acid and Energy Interrelationships in Growing Pigs

The subject of energy and amino acid relationships has been reviewed thoroughly by the SCA (1987), Chiba (1989) and Edwards and Campbell (1993). Nutritionally, energy and amino acids are very closely related, and it is almost impossible to discuss one without considering the other. Considering the effect of dietary energy on the intake of pigs given *ad libitum* access to feed, it seems logical that the amino acid levels in the diet should be related to its energy density. However, investigations on the need to adjust amino acids or protein according to changes in the energy density yielded conflicting results (Allee and Hines, 1972; Tribble *et al.*, 1979; Chiba *et al.*, 1985, 1987). Factors that may have been responsible for the inconsistent responses have been summarized by Chiba (1989). Chiba *et al.* (1991a,b) conducted research to verify the need to adjust dietary amino acids according to the dietary energy content by using three amino acid levels that were either adjusted for five DE levels or unadjusted for three DE levels. Their results demonstrate the need to adjust dietary amino acid levels in concert with changes in energy densities so that pigs can consume adequate amounts of amino acids.

Simply because an energy intake can influence the rate of protein accretion in the pig, expressing the amino acid requirements in terms of dietary energy density would be beneficial. The concept of using a constant amino acid to energy value is, however, only valid if the relationship between energy intake and rate of protein accretion is linear (SCA, 1987). This approach has been adopted by ARC (1981), and their recommendations imply that this concept holds under conditions of protein adequacy. The relationship seems to be linear for young pigs (Campbell *et al.*, 1983), but not for pigs weighing 45–90 kg (Campbell *et al.*, 1985). As indicated by the SCA (1987), the growing pig's potential for protein growth from birth to 50 kg seems to lie beyond the upper limit of appetite, whereas it lies within the limits of appetite for pigs weighing 50–100 kg. This implies that diets of high energy density can be offered *ad libitum* to pigs weighing up to 50 kg without excessive fat accretion or reducing feed efficiency, but *ad libitum* feeding of such diets to pigs weighing 50–100 kg would cause undesirable results. Although it depends on sex and genotypes, the amino acid to energy value would decline continuously once pigs reach the phase where an energy intake exceeds maximum protein accretion.

The relationship of energy and amino acids in a pig's diet seems to be less critical because pigs can grow well with various energy and amino acid levels. However, in terms of protein or lean accretion, which is the primary objective of today's pig production, it is not an appropriate generalization. Conceivably, there is a combination of energy, amino acids and other conditions that can lead to maximum utilization of nutrients to produce desirable animal products.

Effects of Environmental Factors

Temperature is perhaps the most important component of the environment. Assuming the feed intake at 15°C as 100 and the temperature increased from 5 to 30°C, a reduction in the feed intake can be predicted by the following equation (NRC, 1986):

$$\% \text{ change} = 126.3 - 1.65 \ T \tag{9.8}$$

where percent change reflects the deviation from a feed intake at 15°C, and T is the reported ambient temperature in degrees Celsius. The effect of BW on the optimum temperature (T_o) can be described by the following equation (NRC, 1986):

$$T_o = 26 - 0.0614 \ \text{BW} \tag{9.9}$$

where T_o is in degrees Celsius and BW in kg. The temperature can be modified by several factors, and the term effective ambient temperature (EAT) has been used to describe the temperature that the animal actually experiences; the EAT can be affected by many factors (NRC, 1986). With an assumption that the deviation from T_o influences the feed intake, the correction in DE_i between 5 and 30°C can be described by the following equation (NRC, 1986):

$$\% \text{ change in } DE_i = (T_o - \text{EAT})0.0165. \tag{9.10}$$

Temperatures below the critical temperature will stimulate a feed intake, but the additional feed consumed would be used for the increased energy demands, thus reducing the efficiency of feed utilization (Holmes and Close, 1985). On the other hand, warm or hot temperatures decrease voluntary feed intake and would be likely to reduce the rate and efficiency of growth. To compensate for a reduced feed intake at high temperatures, it would be necessary, therefore, to increase the concentrations of certain components of the diet, such as amino acids and energy. In addition, it is well known that floor space allowance and group size can influence a feed intake and growth rate, but their impact on the nutrient content of the diet is not clear at this time (e.g. Brumm and Miller, 1996). Although the description or consideration of the animal's environment has focused mainly on the aspect of physical environment, other dimensions such as the social and infectious environment may have to be considered when determining the pig's nutritional needs in the future (Kyriazakis, 1996).

Effect of Sex and Genotypes

It is well known that boars have a greater potential for lean accretion than gilts (Campbell *et al.*, 1983, 1985) and castrates (Campbell and King, 1982), while gilts have a higher protein accretion rate than castrates (Just, 1984). Similarly, carcass quality of boars is generally considered superior to that of gilts and barrows (Taverner *et al.*, 1977), thus it is likely that their nutrient requirements would be different (Campbell and King, 1982; Campbell *et al.*, 1983, 1985). Because of these differences, it might be advantageous to consider split-sex feeding and/or management of pigs (Fuller, 1985) to optimize profitability of the pig enterprise. However, neither the ARC (1981) nor the NRC (1988) differentiates possible variations in their recommendations, possibly due to an inconsistent sex effect on the pig performance, especially with *ad libitum* feeding (Taverner *et al.*, 1977), indicating that further research is required in this area.

Similarly, differences in growth rate and body composition have been recognized among different strains and breeds of pigs. Pigs selected for or against subcutaneous fat or growth rate show physiological and metabolic alterations (e.g. Steele *et al.*, 1974; Steele and Frobish, 1976; Pond *et al.*, 1980). It is, therefore, reasonable to assume that pigs with distinct genotypes may differ in their responses to the level of nutrients, efficiency of nutrient utilization, interrelationships among the nutrients, etc. In some herds, optimum production may be achieved by feeding diets containing marginal levels of nutrients, thus reducing feed costs. Conversely, in some instances, pigs may respond to higher levels of nutrients than those normally recommended, thus improving the productivity and possibly the efficiency. Therefore, nutritional management and/or recommendations may have to reflect the genetic potential of pigs for growth and protein accretion in order to achieve overall productivity and efficiency of pigs, which can lead to greater profitability of the pig enterprise.

Effects of Nutritional History

Most of the feed for the whole pig enterprise is consumed by finisher pigs. Therefore, any improvement in the efficiency of amino acid and energy utilization during the finisher phase contributes greatly to the overall efficiency of pig production. It has been demonstrated over the years that the growth performance of weanling and grower pigs can be improved by offering diets containing various special ingredients and high levels of energy and amino acids, respectively. Compensatory growth responses after a period of feed (e.g. Prince *et al.*, 1983) or protein restrictions (e.g. Chiba, 1994, 1995) in young pigs have been reported. Possible reasons for the compensatory response (e.g. Chiba, 1994) and the importance of the nutritional status of pigs in the subsequent phase on the ability of pigs to exhibit compensatory growth (Campbell and Dunkin, 1983; Kyriazakis *et al.*, 1991) have been suggested.

These findings indicate that the early nutritional history may have little importance in terms of overall rate and efficiency of growth. The available evidence is, however, far from conclusive, as exemplified by recent reports (Henry, 1995; Chiba *et al.*, 1997). It is likely that the extent of compensatory response is dependent on the age of pigs and the degree and duration of dietary energy and/or amino acid restrictions. Further research needs to be conducted to evaluate fully the potential implications of nutritional history on the optimum production of pigs.

PIG WASTES AND THE ENVIRONMENT

Pollution of the environment with N and phosphorus is a major problem in many parts of the world (Lenis, 1989), and the management of wastes and odours has become a major issue facing the pig industry. Large amounts of N excreted in animal wastes can lead to the leaching of nitrate from the wastes to the surface, ground and drinking water, and to odorous emissions because many odorous compounds originate from undigested dietary protein and other nitrogenous compounds. Likewise, excess phosphorus can lead to undesirable eutrophication of surface water.

Nutritional means to manipulate microorganisms in the intestinal tract by using various feed additives, particular carbohydrates or decreasing dietary CP, may be effective in reducing the odour emissions from pig wastes (Cromwell, 1996). There are several potential ways to reduce N excretion in the pig: (i) avoid too generous safety margins in dietary protein; (ii) reduce excesses of unneeded individual amino acids; (iii) supplement limiting amino acid(s) to satisfy the needs (Henry *et al.*, 1992; Page *et al.*, 1993b); and (iv) phase and split-sex feeding (Fuller, 1985; Lenis, 1989); together with other management practices that improve the overall efficiency of feed utilization.

Phosphorus excretion can be reduced by using similar approaches, e.g. avoiding generous safety margins, phase- and split-sex feeding, and by enhancing the utilization of phytate phosphorus. Phytate phosphorus can be utilized only after a hydrolysis by phytase, and microbial phytase has been shown to be very effective in improving phosphorus availability of plant feed ingredients in pigs and poultry (Kornegay, 1996). Because phytates are known to limit the availability of multivalent cations and possibly protein, the use of phytase is beneficial not only in enhancing phosphorus utilization and reducing its excretion, but also in improving the availability of other mineral elements and amino acids. However, a practical application of microbial phytase in pig nutrition depends on many factors, including the need to reduce the excretion of phosphorus and other mineral elements and the availability and cost of phytase.

Because of its impact on the air, land and water, the pig industry seems to be especially vulnerable to environmental criticism. Public concerns about environmental issues are, in part, a result of the concentration of the animal agriculture on a relatively small land area (i.e. fewer, but larger, intensified operations), and will continue to increase in the future. Environmental issues cannot be ignored, and exploring all possible avenues and implementing effective methods to alleviate public concerns are, therefore, essential for a sustainable pig industry.

CONCLUSIONS

The main goal of diet formulation and feeding strategy in commercial pig production is to maximize profits, which does not necessarily imply maximal animal performance. In recent years, management of wastes and odours associated with intensive animal production has become a major issue facing animal agriculture. The pig industry is especially vulnerable to the environmental criticism because of its impact on the air, land and water. Therefore, exploring all possible avenues and implementing effective methods to reduce potential pollutants, as well as maximizing the economic efficiency, can have a significant impact on ensuring successful pig production in the future.

Offering diets containing just enough nutrients to meet, but not exceed, the needs of the pig would be beneficial in: (i) maximizing the economic efficiency; (ii) optimizing the utilization of quality sources of nutrients, for which pigs compete with humans; and (iii) alleviating public environmental concerns by reducing the excretion of unutilized nutrients. Such optimum feeding strategies involve consideration of a multitude of factors such as genetic potential, nutritional history and voluntary feed intake of the pig, variability, bioavailability and stability of nutrients in feed ingredients, various types of stress, the physical and social environment and possible interactions among those factors. Considering the effects of those factors, estimating the nutrient requirements with accuracy and precision, or satisfying the needs of a population(s) of pigs, rather than individual animals, is a challenging task. This contention implies that no single set of estimates is likely to satisfy the requirements of all animals.

Nutritional management and/or recommendations should take into account various factors known to influence nutrition of the pig, and appropriate adjustments must be made to achieve overall productivity and efficiency of pig production. A modelling approach, therefore, might be the only defensible means of incorporating all the information necessary to establish the nutrient requirements and/or to formulate efficient, practical, economical and environmentally friendly diets. Such an approach would provide the direction and magnitude of the response and sensitivity of the production system to a tactical or strategic change so that more effective financial and management decisions can be made. Because of its flexibility, modelling would be useful in defining nutrient requirements and satisfying the needs of pigs in terms of the profitability and environmental accountability; this is essential for a sustainable pig industry, as opposed to simply achieving biological production targets.

REFERENCES

Allee, G.L. and Hines, R.H. (1972) Influence of fat level and calorie:protein ratio on performance of young pigs. *Journal of Animal Science* 35, 210.

Agricultural Research Council (1981) *The Nutrient Requirements of Pigs*. Commonwealth Agriculture Bureaux, Slough, UK.

Ammerman, C.B., Baker, D.H. and Lewis, A.J. (eds) (1995) *Bioavailability of Nutrients for Animals: Amino Acids, Minerals, and Vitamins*. Academic Press, San Diego, California.

Black, J.L., Campbell, R.G., Williams, I.H., James, K.J. and Davies, G.T. (1986) Simulation of energy and amino acid utilisation in the pig. *Research and Development in Agriculture* 3, 121–145.

Brief, S. and Chew, B.P. (1985) Effects of vitamin A and β-carotene on reproductive performance in gilts. *Journal of Animal Science* 60, 998–1004.

Brooks, P.H. and Carpenter, J.L. (1993) The water requirement of growing–finishing pigs – theoretical and practical considerations. In: Cole, D.J.A, Haresign, W. and Garnsworthy, P.C. (eds), *Recent Development in Pig Nutrition 2*. Nottingham University Press, Loughborough, UK, pp. 179–200.

Brumm, M.C. and Miller, P.S. (1996) Response of pigs to space allocation and diet varying in nutrient density. *Journal of Animal Science* 74, 2730–2737.

Campbell, R.G. and Dunkin, A.C. (1983) The influence of protein nutrition in early life on growth and development of the pig. 1. Effects on growth performance and body composition. *British Journal of Nutrition* 50, 605–617.

Campbell, R.G. and King, R.H. (1982) The influence of dietary protein and level of feeding on the growth performance and carcass characteristics of entire and castrated male pigs. *Animal Production* 35, 177–184.

Campbell, R.G., Taverner, M.R. and Curic, D.M. (1983) The influence of feeding level from 20 to 45 kg live weight on the performance and body composition of female and entire male pigs. *Animal Production* 36, 193–199.

Campbell, R.G., Taverner, M.R. and Curic, D.M. (1985) Effect of sex and energy intake between 48 and 90 kg live weight on protein deposition in growing pigs. *Animal Production* 40, 497–503.

Chiba, L.I. (1989) Amino acid and energy interrelationships in pigs weighing 20 to 50 kilograms. PhD dissertation. University of Nebraska, Lincoln, Nebraska.

Chiba, L.I. (1994) Effects of dietary amino acid content between 20 and 50 kg and 50 and 100 kg live weight on the subsequent and overall performance of pigs. *Livestock Production Science* 39, 213–221.

Chiba, L.I. (1995) Effects of nutritional history on the subsequent and overall growth performance and carcass traits of pigs. *Livestock Production Science* 41, 151–161.

Chiba, L.I., Peo, E.R., Jr, Lewis, A.J., Brumm, M.C., Fritschen, R.D. and Crenshaw, J.D. (1985) Effect of dietary fat on pig performance and dust levels in modified-open-front and environmentally regulated confinement buildings. *Journal of Animal Science* 61, 763–781.

Chiba, L.I., Peo, E.R., Jr and Lewis, A.J. (1987) Use of dietary fat to reduce dust, aerial ammonia and bacterial colony forming particle concentrations in swine confinement buildings. *Transactions of the American Society of Agricultural Engineers* 30, 464–468.

Chiba, L.I., Lewis, A.J. and Peo, E.R., Jr (1991a) Amino acids and energy interrelationships in pigs weighing 20 to 50 kilograms. I. Rate and efficiency of weight gain. *Journal of Animal Science* 69, 694–707.

Chiba, L.I., Lewis, A.J. and Peo, E.R., Jr (1991b) Amino acids and energy inter-relationships in pigs weighing 20 to 50 kilograms. II. Rate and efficiency of protein and fat deposition. *Journal of Animal Science* 69, 708–718.

Chiba, L.I., Ivey, H.W., Cummins, K.A. and Gamble, B.E. (1995) Effects of urea as a source of extra dietary nitrogen on growth performance and carcass traits of finisher pigs. *Nutrition Research* 15, 1029–1036.

Chiba, L.I., Ivey, H.W., Cummins, K.A., Gamble, B.E., Owsley, W.F., Carroll, M.W. and Tyler, P.J. (1997) Effect of amino acid content of finisher diets on growth performance of pigs subjected to dietary restriction during the grower phase. *Journal of Animal Science* 75 (Suppl. 1), 187.

Chung, T.K. and Baker, D.H. (1992) Ideal amino acid pattern for 10-kilogram pigs. *Journal of Animal Science* 70, 3102–3111.

Close, W.H. and Fowler, V.R. (1985) Energy requirements of pigs. In: Cole, D.J.A. and Haresign, W. (eds), *Recent Development in Pig Nutrition*. Butterworths, London, pp. 1–16.

Coffey, M.T. and Britt, J.H. (1992) Enhancement of sow reproductive performance by β-carotene or vitamin A. *Journal of Animal Science* 71, 1198–1202.

Cole, D.J.A. (1978) Amino acid nutrition of the pig. In: Haresign, W. and Lewis, D. (eds), *Recent Advances in Animal Production*. Butterworths, London, pp. 59–72.

Cole, D.J.A. and Haresign, W. (eds) (1985) *Recent Development in Pig Nutrition*. Butterworths, London.

Cole, D.J.A., Haresign, W. and Garnsworthy, P.C. (eds) (1993) *Recent Development in Pig Nutrition 2*. Nottingham University Press, Loughborough, UK.

Cromwell, G.L. (1996) Nutritional practices that influence odor in swine manure. In: *Proceedings of 1996 Pork Academy*. Pfizer Animal Health in cooperation with the National Pork Producers Council, Des Moines, Iowa, pp. 96–110.

Cromwell, G.L., Hal, D.D., Calwson, A.J., Combs, G.E., Knabe, D.A., Maxwell, C.V., Noland, P.R., Orr, D.E., Jr and Prince, T.J. (1989) Effects of additional feed during late gestation on reproductive performance of sows: a cooperative study. *Journal of Animal Science* 67, 3–14.

D'Mello, J.P.F. (1994) Amino acid imbalances, antagonisms and toxicities. In: D'Mello, J.P.F. (ed.), *Amino Acids in Farm Animal Nutrition*. CAB International, Wallingford, UK, pp. 63–97.

Edwards, A.C. and Campbell, R.G. (1993) Energy-protein interactions in pigs. In: Cole, D.J.A, Haresign, W. and Garnsworthy, P.C. (eds), *Recent Development in Pig Nutrition 2*. Nottingham University Press, Loughborough, UK, pp. 30–46.

Ferguson, N.S., Gous, R.M. and Emmans, G.C. (1994) Preferred components for the construction of a new simulation model of growth, feed intake and nutrient requirements of growing pigs. *South African Journal of Animal Science* 24, 10–17.

Filmer, D.G and Curran, M.K (1985) Climatic environment and practical nutrition of the growing pig. In: Cole, D.J.A. and Haresign, W. (eds), *Recent Development in Pig Nutrition*. Butterworths, London, pp. 41–58.

Frazer, D., Patience, J.F., Phillips, P.A. and McLeese, J.M. (1993) Water for piglets and lactating sows: quantity, quality and quandaries. In: Cole, D.J.A, Haresign, W. and Garnsworthy, P.C. (eds), *Recent Development in Pig Nutrition 2*. Nottingham University Press, Loughborough, UK, pp. 201–224.

Fuller, M.F. (1985) Sex differences in the nutrition and growth of pigs. In: Cole, D.J.A. and Haresign, W. (eds), *Recent Development in Pig Nutrition*. Butterworths, London, pp. 1177–1189.

Fuller, M.F. (1994) Amino acid requirements for maintenance, body protein accretion and reproduction in pigs. In: D'Mello, J.P.F. (ed.), *Amino Acids in Farm Animal Nutrition*. CAB International, Wallingford, UK, pp. 155–184.

Fuller, M.F. and Chamberlain, A.G. (1985) Protein requirements of pigs. In: Cole, D.J.A. and Haresign, W. (eds), *Recent Development in Pig Nutrition*. Butterworths, London, pp. 85–96.

Grimson, R.E., Bowland, J.P. and Milligan, L.P. (1971) Use of nitrogen-15 labelled urea to study urea utilization by pigs. *Canadian Journal of Animal Science* 51, 103–110.

Hahn, J.D. and Baker, D.H. (1993) Growth and plasma zinc responses of young pigs fed pharmacologic levels of zinc. *Journal of Animal Science* 71, 3020–3024.

Harper, A.E., Benevenga, N.J. and Wouhlhueter, R.M. (1970) Effect of ingestion of disproportionate amounts of amino acids. *Physiological Reviews* 50, 428–558.

Henry, Y. (1985) Dietary factors involved in feed intake regulation in growing pigs: a review. *Livestock Production Science* 12, 339–354.

Henry, Y. (1995) Influence d'un déficit ou d'un déséquilibre alimentaire en acides aminés pendant une phase initiale de la croissance sur les performances du porc en finition. *Annales de Zootechnie* 44, 3–28.

Henry, Y., Colléaux, Y. and Séve, B. (1992) Effects of dietary level of lysine and of level and source of protein on feed intake, growth performance, and plasma amino acid pattern in the finishing pig. *Journal of Animal Science* 70, 188–195.

Holmes, C.W. and Close, W.H. (1985) The influence of climatic variables on energy metabolism and associated aspects of productivity in the pig. In: Cole, D.J.A. and Haresign, W. (eds), *Recent Development in Pig Nutrition*. Butterworths, London, pp. 18–40.

Johnston, L.J., Pettigrew, J.E. and Rust, J.W. (1993) Responses of maternal-line sows to dietary protein concentration during lactation. *Journal of Animal Science* 71, 2151–2156.

Just, A. (1984) Nutritional manipulation and interpretation of body compositional differences in growing swine. *Journal of Animal Science* 58, 740–752.

Knabe, D.A. (1996) Optimizing the protein nutrition of growing–finishing pigs. *Animal Feed Science and Technology* 60, 331–341.

Knabe, D.A., Brendemuhl, J.H., Chiba, L.I. and Dove, C.R. (1996) Supplemental lysine for sows nursing large litters. *Journal of Animal Science* 74, 1635–1640.

Kornegay, E.T. (1986) Biotin in swine production: a review. *Livestock Production Science* 14, 65–89.

Kornegay, E.T. (1996) Natuphos™ phytase in swine diets: digestibility, bone and carcass characteristics. In: *Proceedings of the BASF Symposium, 1996 Carolina Swine Nutrition Conference*. November 12, 1996, Raleigh, North Carolina, pp. 28–67.

Kyriazakis, I. (1996) A solution to the problem of predicting the response of an animal to its diet. *Proceedings of the Nutrition Society* 55, 155–166.

Kyriazakis, I., Stamataris, C., Emmans, G.C. and Whittemore, C.T. (1991) The effects of food protein content on the performance of pigs previously given foods with low or moderate protein content. *Animal Production* 52, 165–173.

Lenis, N.P. (1989) Lower nitrogen excretion in pig husbandry by feeding: current and future possibilities. *Netherlands Journal of Agricultural Science* 37, 61–70.

Lewis, A.J. (1991) Amino acids in swine nutrition. In: Miller, E.R., Ullrey, D.E. and Lewis, A.J. (eds), *Swine Nutrition*. Butterworths-Heinemann, Boston, Massachusetts, pp. 147–164.

Lewis, A.J. (1993) Comparison of ARC and NRC recommended requirements for energy and protein in growing pigs. In: Cole, D.J.A, Haresign, W. and Garnsworthy, P.C.

(eds), *Recent Development in Pig Nutrition 2*. Nottingham University Press, Loughborough, UK, pp. 47–459.

Lindemann, M.D. (1993) Supplemental folic acid: a requirement for optimizing swine production. *Journal of Animal Science* 71, 239–246.

Lindemann, M.D., Wood, C.M., Harper, A.F., Kornegay, E.T. and Anderson, R.A. (1995) Dietary chromium picolinate additions improve gain:feed and carcass characteristics in growing–finishing pigs and increase litter size in reproducing sows. *Journal of Animal Science* 73, 457–465.

Miller, E.R., Ullrey, D.E. and Lewis, A.J. (eds) (1991) *Swine Nutrition*. Butterworth-Heinemann, Boston, MA

Moser, B.D. and Lewis, A.J. (1980) Adding fat to sow diets – an update. *Feedstuffs* 52, 36–62.

NCR-42 Committee on Swine Nutrition (1993) Variability among sources and laboratories in selenium analysis of corn and soybean meal. *Journal of Animal Science* 71 (Suppl. 1), 67.

NCR-42 and S-145 Regional Committees (1997) Variability in mixing efficiency and in laboratory analysis of diets at 25 experiment stations. *Journal of Animal Science* 75 (Suppl. 1), 196.

Niiyama, M., Deguchi, E., Kagota, K. and Namioka, S. (1979) Appearance of ^{15}N-labeled intestinal microbial amino acids in the venous blood of the pig colon. *American Journal of Veterinary Research* 40, 716–718.

Noblet, J., Dourmad, J.Y. and Etienne, M. (1990) Energy utilization in pregnant and lactating sows: modelling of energy requirements. *Journal of Animal Science* 68, 562–572.

National Research Council (1976) *Urea and Other Nonprotein Nitrogen Compounds in Animal Nutrition*. National Academy of Sciences, Washington, DC, pp. 88–94.

National Research Council (1979) *Nutrient Requirements of Domestic Animals, No. 2: Nutrient Requirements of Swine*, 8th edn. National Academy of Sciences–National Research Council, Washington, DC.

National Research Council (1980) *Mineral Tolerance of Domestic Animals*. National Academy of Sciences, Washington, DC.

National Research Council (1986) *Predicting Feed Intake of Food-Producing Animals*. National Academy Press, Washington, DC.

National Research Council (1987) *Vitamin Tolerance of Animals*. National Academy Press, Washington, DC.

National Research Council (1988) *Nutrient Requirements of Swine*, 9th edn. National Academy Press, Washington, DC.

Page, T.G., Southern, L.L., Ward, T.L. and Thompson, D.L., Jr (1993a) Effects of chromium picolinate on growth and serum and carcass traits of growing–finishing pigs. *Journal of Animal Science* 71, 656–662.

Page, T.G., Southern, L.L. and Watkins, K.L. (1993b) Threonine supplementation of low-protein, lysine-supplemented diet, sorghum–soybean meal diets for growing–finishing pigs. *Livestock Production Science* 34, 153–162.

Patience, J.F. (1993) The physiological basis of electrolytes in animal nutrition. In: Cole, D.J.A, Haresign, W. and Garnsworthy, P.C. (eds), *Recent Development in Pig Nutrition 2*. Nottingham University Press, Loughborough, UK, pp. 225–242.

Patience, J.F. (1996) Precision in swine feeding programs: an integrated approach. *Animal Feed Science Technology* 59, 137–145.

Pettigrew, J.E., Gill, M., France, J. and Close, W.H. (1992) Evaluation of mathematical model of lactating sow metabolism. *Journal of Animal Science* 70, 3762–3773.

Pond, W.G., Yen, J.T., Lindvall, R.N. and Hill, D. (1980) Dietary alfalfa meal for genetically obese and lean growing pigs: effect on body weight gain and on carcass and gastrointestinal tract measurements and blood metabolites. *Journal of Animal Science* 51, 367–373.

Pond, W.G., Devilat, J., Miller, P.D. and Walker, E.F., Jr (1969) Amino acid balance for growing–finishing swine: effect of diet composition on diet preference, performance and blood components. In: *Proceeding of the Cornell Nutrition Conference for Feed Manufacturers*. Cornell University, Ithaca, New York, pp. 80–86.

Poulsen, H.D. (1995) Zinc oxide for weanling piglets. *Acta Agriculture Scandinavica, Section A, Animal Science* 45, 159–167.

Prince, T.J., Jungst, S.B. and Kuhlers, D.L. (1983) Compensatory responses to short-term feed restriction during the growing period in swine. *Journal of Animal Science* 56, 846–852.

Reese, D.E. (1986) Improving on-farm manufactured feed quality. In: *Proceedings of Nebraska Whole Hog Days*. Institute of Agriculture and Natural Resources, University of Nebraska, Lincoln, Nebraska, pp. 88–97.

SCA (1987) *Feeding Standards for Australian Livestock. Pigs*. Commonwealth Scientific and Industrial Research Organization, East Melbourne, Australia.

Schell, T.C. and Kornegay, E.T. (1994) Effectiveness of zinc acetate injection in alleviating postweaning performance lag in pigs. *Journal of Animal Science* 72, 3037–3042.

Stahly, T.S., Cromwell, G.L. and Monegue, H.J. (1990) Lactational responses of sows nursing large litters to dietary lysine levels. *Journal of Animal Science* 68 (Suppl. 1), 369.

Steele, N.C., Frobish, L.T. and Keeney, M. (1974) Lipogenesis and cellularity of adipose tissue from genetically lean and obese swine. *Journal of Animal Science* 39, 712–719.

Steele, N.C. and Frobish, L.T. (1976) Selected lipogenic enzyme activities of swine adipose tissue as influenced by genetic phenotype, age, feeding frequency and dietary energy source. *Growth* 40, 369–378.

Taverner, M.R., Campbell, R.G. and King, R.H. (1977) The relative protein and energy requirements of boars, gilts and barrows. *Australian Journal of Experimental Agriculture and Animal Husbandry* 17, 574–580.

Tess, M.W., Dickerson, G.E., Nienaber, J.A., Yen, J.T. and Ferrell, C.L. (1984) Energy cost of protein and fat deposition in pigs fed *ad libitum*. *Journal of Animal Science* 58, 111–122.

Tribble, L.F., Ingram, S.H., Gaskins, C.T. and Ramsey, C.B. (1979) Evaluation of added fat and lysine to sorghum–soybean meal diets for swine. *Journal of Animal Science* 48, 541–546.

Wang, T.C. and Fuller, M.F. (1989) The optimum dietary amino acid pattern for growing pigs. *British Journal of Nutrition* 62, 77–89.

Whittemore, C.T. (1993) *The Science and Practice of Pig Production*. Longman Scientific and Technical, Essex, UK.

Wiseman, J. and Cole, D.J.A. (1985) Energy evaluation of cereals for pig diets. In: Cole, D.J.A. and Haresign, W. (eds), *Recent Development in Pig Nutrition*. Butterworths, London, pp. 246–262.

10   Feeding Systems for Poultry

S. Leeson and J.D. Summers

Department of Animal and Poultry Science, University of Guelph, Guelph, Ontario, Canada

INTRODUCTION

Poultry diets are usually formulated to meet the needs of a given class of bird under specific environmental situations. This is not to imply that one can formulate diets to meet exacting nutrient requirements. Indeed, we are a long way from achieving this goal. However, there has been development to the point where one given set of requirement values is no longer acceptable for all strains within a given class of poultry housed under a range of environmental conditions. A single requirement value under such conditions is not only inefficient as far as nutrient utilization is concerned, but in many cases does not allow the true genetic potential of the bird to be expressed.

This chapter outlines diet nutrient recommendations and examples of diet formulations for poultry. It is not suggested that these are the only dietary situations that should be considered, since it is realized that specific environmental and management practices dictate the need for flexibility. It is hoped that the reader will look through the discussion on factors influencing feeding programmes prior to making final decisions on diet specifications. Nutrient specifications are not intended to be all encompassing or to suit conditions in all environments and geographical locations. Rather, they are intended to show the relative balance and contribution of selected nutrients and ingredients, and how these can be manipulated readily.

IMMATURE EGG STRAIN PULLETS

It is generally agreed that most Leghorn and brown egg strains have changed over the last 5–10 years and, because of this, nutritional management is becoming more critical. In essence, these changes relate to age at maturity, although it is questionable that this has changed suddenly in just a few years.

In fact, what has been happening is that age of maturity has been decreasing slowly, by about 1 day per year. Unfortunately, many producers are just now becoming aware of earlier maturity because their conventional programmes are no longer working, and this is especially true for many strains of brown egg pullets. The key to successful nutritional management today is through maximizing the body weight of the pullet. Pullets that are on target or slightly above target weight at maturity will inevitably be the best producing birds for the shell-egg market. Diet specifications for Leghorn birds are shown in Table 10.1. The traditional concern with early maturity has been that it results in small egg size (Leeson and Summers, 1981). Results from our early studies indicate the somewhat classical effect of early maturity in Leghorns without regard to body weight (Table 10.2).

There seems little doubt that body weight and/or body composition are the major factors influencing egg size both at maturity and throughout the remainder of the laying period. Although it is fairly well established that body

Table 10.1. Diet specifications for growing pullets.

	Chick starter		Chick grower		Pre-lay
Approximate CP level (%)	18.0	20.0	15.0	17.0	17.0
Amino acids (% of diet)					
Arginine	0.94	1.03	0.78	0.92	0.80
Lysine	0.90	1.00	0.72	0.85	0.70
Methionine	0.41	0.45	0.34	0.39	0.35
Methionine + cysteine	0.66	0.72	0.55	0.65	0.60
Tryptophan	0.18	0.19	0.16	0.18	0.17
Histidine	0.33	0.36	0.28	0.32	0.30
Leucine	1.16	1.28	0.95	1.10	1.00
Isoleucine	0.62	0.68	0.51	0.61	0.55
Phenylalanine	0.58	0.64	0.48	0.57	0.51
Phenylalanine + tyrosine	1.13	1.24	0.93	1.11	1.00
Threonine	0.56	0.62	0.47	0.55	0.50
Valine	0.69	0.76	0.67	0.68	0.67
ME (MJ kg^{-1})	11.92	12.13	11.92	12.34	11.92
Calcium (%)	1.0	1.0	0.85	0.90	2.0
Available phosphorus (%)	0.40	0.42	0.37	0.39	0.43
Sodium (%)	0.18	0.18	0.18	0.18	0.18

Table 10.2. Pullet maturity and egg characteristics.

	Egg production (%)		Egg size (% large, >57 g)	
Age at housing	18–20 weeks	Mean to 35 weeks	30 weeks	63 weeks
15 weeks	32	92	17	44
18 weeks	12	92	21	65
21 weeks	0	91	37	69

weight is an important criterion for early production, there is still insufficient evidence regarding optimum body structure and composition. Relating frame size to tibia length is now frequently included in breeder management guides as a means of monitoring skeleton development. It is known that most (90%) of the frame size develops early and so, by 12–16 weeks of age, the so-called size of the pullet is fixed. While this parameter is useful as a monitoring tool, and its measurement should be encouraged, there has been little success in affecting frame size without also affecting body weight. It therefore seems very difficult to produce, by nutritional modification, pullets that are below target weight yet above average frame size, and vice versa (Leeson and Caston, 1993).

The key to solving many of the present industry problems would therefore seem to be the attainment of heavy pullets at the desired age of maturity. In this instance, heavy refers to the weight and condition that will allow the bird to progress through maturity with optimum energy balance. It is likely that to obtain such conditions in a flock one must consider stocking density, environmental temperature, feather cover, etc. Unfortunately, attainment of desired weight-for-age has not always proven easy, especially where earlier maturity is desired or when adverse environmental conditions prevail. Leeson and Summers (1981) suggested that the energy intake of the pullet is the limiting factor influencing growth rate since, regardless of diet specifications, pullets seem to consume similar quantities of energy.

Studies indicate that growth rate is more highly correlated with energy intake than with protein intake (Leeson and Summers, 1989). This does not mean to say that protein (amino acid) intake is not important to the growing pullet. Protein intake is very important, but there does not seem to be any measurable return from feeding more than 800 g of protein to the pullet beyond 18 weeks of age. On the other hand, it seems as though the more energy consumed by the pullet, the larger the body weight at maturity. Obviously, there must be a fine line between maximizing energy intake and creating an obese pullet.

Maximizing Nutrient Intake

If one calculates expected energy output in terms of egg mass and increase in body weight, and relates this to feed intake, it becomes readily apparent that the Leghorn must consume at least 90 g bird^{-1} day^{-1} and the brown egg bird close to 100 g bird^{-1} day^{-1} at peak production for diets of around 11.9 MJ of metabolizable energy (ME) kg^{-1}. With egg-type stock, feeding is to appetite and so management programmes must be geared to stimulating appetite. The practical long-term solution is to rear birds with optimum body weight and body reserves as they begin production. This situation has been aggravated in recent years, with the industry trend of attempting to rear pullets on minimal quantities of feed. Unfortunately, this move has coincided with genetically smaller body weights and hence smaller appetites, together with earlier sexual maturity.

In order to maximize nutrient intake, one must consider relatively high nutrient-dense diets, although these alone do not always ensure optimum growth. Relatively high protein [16–8% crude protein (CP)] with adequate methionine (2% of CP) and lysine (5% of CP) levels together with high energy levels (11.7–12.6 MJ kg^{-1}) are usually given to Leghorn hens, especially in hot weather situations (McNaughton *et al.*, 1977). However, there is some evidence to suggest that high-energy diets are not always helpful under such heat stress conditions and intake of other nutrients such as protein and amino acids must be given priority (Martin *et al.*, 1994) during formulation. The Leghorn pullet eats for energy requirement, albeit with some imprecision, and so the energy:protein balance is critical. All too often there is inadequate amino acid intake when high-energy corn-based diets are used, the result of which is pullets that are both small and fat at maturity.

One of the most important concepts in pullet feeding today is to offer diets according to body weight and condition of the flock, rather than according to age. For example, traditional systems involve feeding starter diets for about 6 weeks, followed by grower, and then perhaps developer, diets. This approach does not take into account individual flock variation, and today this can be most damaging to underweight flocks. It is becoming more difficult to attain early weight-for-age, and this means that flocks are often underweight at 4–6 weeks of age. This can be for a variety of reasons such as suboptimal nutrition, heat stress, disease, etc. The worse thing that can happen to these flocks is an arbitrary introduction of a grower diet merely because the flock has reached some set age. Higher nutrient-dense starter diets must be fed until the target weight is reached. In some instances, this can mean feeding higher protein starter diets for up to 10–12 weeks of age. Some producers and especially contract pullet growers are sometimes reluctant to accept this type of programme, since they correctly argue that feeding a high protein diet for 10–12 weeks will be more expensive. Depending upon local economic conditions, feeding an 18% protein starter diet for 12 compared with 6 weeks of age, will cost the equivalent of two eggs. A bird in ideal condition at maturity will produce far in excess of these two eggs relative to a small underweight bird at maturity (Leeson and Summers, 1997).

Suggested Feeding Programme

The following schedule is recommended for growing pullets to maturity:

Starter 18–19% CP; 11.5–12.1 MJ of ME kg^{-1}

(from day old to target body weight)

Grower 15–16% CP; 11.5–12.1 MJ of ME kg^{-1}

(from target weight to mature body size)

Pre-lay or layer 16–18% CP; 11.5–12.1 MJ of ME kg⁻¹

(from mature body size to first egg)

As previously indicated, there are no recommendations regarding age or even the body weight at which diet changes should occur. Rather the recommendations dictate the need for flexibility and the treatment of each flock as an individual case. For example, the starter diet is to be used until target weight-for-age is achieved; hopefully this will be at around 450 g when the Leghorn bird is 6–8 weeks of age. However, each flock will be subjected to varying environmental conditions, and so this may vary. The time to change to a lower nutrient-dense diet is when a desired weight-for-age is achieved, which is a weight that will be towards the top of the breeder's growth curve. Changing at a specific weight or a specific age in isolation can lead to disastrously underweight flocks.

The lower nutrient-dense grower diets are then to be fed from this target weight-for-age up until the desired mature body size is achieved. Again, a specific mature body weight is not being dictated since this may be varied at the desire of the pullet grower. Pre-lay diets should only be used in an attempt at conditioning the calcium metabolism of the bird (see following section) and not as a means of initiating catch-up growth. Such growth spurts rarely occur at this age and, as such, pre-lay diets are being used as a crutch for poor rearing management. The actual body weights to be achieved during rearing will obviously vary with breed and strain. Most Leghorn strains should weigh around 400, 900 and 1300 g at 6, 12 and 18 weeks, respectively. Similarly, brown egg birds should weigh around 500, 1000 and 1500 g at these ages. As a rule of thumb, these weights for age can be used as guidelines for anticipated diet change.

Pre-lay Nutrition

Pre-lay diets are often used to try and manipulate body size or to bring about a transitional change in the bird's calcium metabolism prior to maturity. There is still considerable confusion and variation practised in the levels of calcium given to birds prior to egg production. During the laying cycle, the bird utilizes its medullary bone reserves, in the long bones of the leg, to augment its diet supply when a shell is being formed. Because egg production is an all-or-none event, the production of the first egg obviously places a major strain on the bird's metabolism when it has to contend with a sudden 2 g loss of calcium from the body. Some of this calcium will come from the medullary bone (Clunies *et al.*, 1992), and so the concept has arisen of building up this bone reserve prior to the first egg. This obviously means higher levels of calcium are needed in pre-lay diets.

In terms of calcium metabolism, the most effective pre-lay programme is early introduction of the high-calcium layer diet. Such high-calcium diets allow sustained production of even the earliest maturing birds. As previously mentioned, higher calcium diets fed to immature birds lead to reduced percentage retention, although absolute retention is slightly increased (Leeson *et al.*, 1986).

Feeding layer diets containing 3.5% calcium, prior to the first egg, therefore results in a slight increase in calcium retention of about 0.16 g day^{-1} relative to birds fed 0.9% calcium grower diets at this time. Over a 10 day period, however, this increased accumulation is equivalent to the output in one egg.

Early introduction of layer diets is therefore beneficial in terms of optimizing the calcium balance of the bird. However, there has been some criticism levelled at this practice. There is the argument that feeding excess calcium prior to lay imposes undue stress on the bird's kidneys, since this calcium is in excess of the immediate requirement and must be excreted. Recent evidence suggests that pullets must be fed a layer diet from as early as 6–8 weeks of age before any adverse effect on kidney structure is seen. It seems likely that the high levels of excreta calcium shown in Table 10.3 reflect faecal calcium, suggesting that all excess calcium may not even be absorbed into the body, merely passing through the bird with the undigested feed. This is perhaps too simplistic a view, since there is other evidence to suggest that the immature bird may absorb excess calcium at this time. Such evidence is seen in the increased water intake and excreta water content of birds fed layer diets prior to maturity. In summary, the calcium metabolism of the earliest maturing birds in a flock should be the criterion for selection of calcium levels during the pre-lay period. Prolonged feeding of low-calcium diets is not recommended. Early introduction of layer diets is ideal, although, where wet manure may be a problem, a 2% calcium pre-lay diet is recommended. There seems to be no problem with the use of 2% calcium pre-lay diets as long as birds are consuming a high-calcium layer diet not later than 1% egg production.

In recent years, there has been interest in some countries in so-called 'pre-pause' feeding programmes. The idea behind these programmes is to withdraw feed, or feed a very low nutrient-dense diet, at the time of sexual maturity. This somewhat unorthodox programme is designed to pause the normal maturation procedure and at the same time to stimulate greater egg size when production resumes after about 10–14 days. This type of pre-lay programme is therefore most beneficial where early small egg size is undesirable. Pre-pause can be induced by simply withdrawing feed, usually at

Table 10.3. Effect of percentage diet calcium fed to birds, immediately prior to lay, on calcium retention.

Diet Ca (%)	Daily Ca retention (g)	Excreta Ca (% dry matter)
0.9	0.35	1.4
1.5	0.41	3.0
2.0	0.32	5.7
2.5	0.43	5.9
3.0	0.41	7.5
3.5	0.51	7.7

around 1% egg production; under these conditions, pullets immediately lose weight and fail to realize their normal weight-for-age when re-fed. Egg production and feed intake quickly normalize, although there is a 1–1.5 g increase in egg size. The most noticeable effects of using a pre-pause diet, such as wheat bran, are very rapid attainment of peak egg production and an increase in egg size once re-feeding commences. This management system could therefore be used to better synchronize onset of production (due to variance in body weight), to improve early egg size or to delay production for various management-related decisions. The use of such pre-pause management will undoubtedly be affected by local economic considerations.

Lighting Programmes

Light can have a dramatic influence on the growth and body composition of the growing pullet, and so light programmes must be taken into account in the selection of feeding programmes. In terms of pullet management, day length has two major effects, namely the development of reproductive organs and a change in feed intake. It is well known that birds reared on a step-up or naturally increasing day length will mature earlier than those reared on a constant day length. Similarly if birds are subjected to a step-down day length much after 12 weeks of age, they will probably exhibit delayed sexual maturity. The longer the photoperiod, the longer the time that birds have to eat feed, and this usually results in heavier birds (Leeson and Summers, 1985).

Longer photoperiods may be beneficial in hot weather situations where a depressed feed intake in pullets is often a problem. As the day length for the growing pullet is increased, we can expect a reduction in age at maturity. Research data suggest earlier maturity with constant rearing day lengths up to 16–18 h per day, although longer day lengths such as 20–22 h per day seem to delay maturity. Another potential problem with longer day length during rearing is that it allows less potential for light stimulation when birds are moved to laying cages. However, in parts of the world around equatorial regions where maximum day length fluctuates between 11 and 13 h, many birds are managed without any light stimulation. In fact, in such hot weather, high light intensity conditions, excessive stimulation often results in prolapse. The step-down programme has the advantage of allowing the young pullets to eat feed for considerably longer times each day during their early development. In hot weather conditions, this long day length means that birds are able to eat more feed during the cooler parts of the day. The system should not be confused with older step-down lighting programmes that continued step-down until 18–20 weeks; these older programmes were designed to delay maturity. With this newer programme, maturity will not be affected as long as the step-down regime is stopped by 10–12 weeks of age, i.e. before the pullet becomes very sensitive to changes in day length. The step-down lighting programme is one of the simplest ways of stimulating appetite and increasing growth rate in pullets, and is practical with both controlled environment and open-sided buildings.

FEEDING PROGRAMMES FOR ADULT LAYING HENS

There is a considerable range of daily feed intake patterns shown by laying hens and, therefore, it is important to select and formulate diets based on expected feed intake, so that the daily intake of specified nutrients is achieved. The large range of daily feed intakes encountered with laying hens is caused by variation in age at sexual maturity, inherent body weight and environmental effects such as temperature and bird density. Most Leghorn strains of bird will now start to mature on intakes of 80–85 g day^{-1}, and it is quite difficult to formulate diets for these birds that will ensure adequate intakes of all nutrients. Meeting the bird's energy needs is perhaps most critical at this time. Through the period of peak egg numbers it is important that the bird is not deficient in energy and so high- rather than low-energy diets are usually preferred. However, energy level can be altered, within reasonable limits, and the bird will adjust its feed intake accordingly. Maintaining the balance of other nutrients to energy is therefore an important concept in layer nutrition (Zhang and Coon, 1994). Diet specifications are shown in Table 10.4. (Leeson and Summers, 1997).

It is well known that, under normal environmental and management conditions, feed intake will vary with the egg production and/or age of bird, and this must be taken into account when formulating diets. While Leghorns may adjust intake according to diet energy levels, there is no evidence to suggest that such precision occurs with other nutrients. Table 10.5 shows the

Table 10.4.   Diet specifications for layers.

	Feed intake day^{-1} (g) (Approximate % CP level)			
	110 (15.5)	100 (17.0)	90 (19.0)	80 (20.5)
Amino acids (% of diet)				
Arginine	0.68	0.75	0.82	0.90
Lysine	0.63	0.70	0.77	0.84
Methionine	0.34	0.37	0.41	0.47
Methionine + cysteine	0.58	0.64	0.71	0.80
Tryptophan	0.14	0.15	0.17	0.18
Histidine	0.15	0.17	0.19	0.21
Leucine	0.82	0.91	1.00	1.09
Isoleucine	0.57	0.63	0.69	0.73
Phenylalanine	0.42	0.47	0.52	0.57
Phenylalanine + tryosine	0.75	0.83	0.91	0.99
Threonine	0.57	0.63	0.69	0.73
Valine	0.63	0.70	0.77	0.82
ME (MJ kg^{-1})	11.3	11.7	11.9	11.9
Calcium (%)	3.25	3.50	3.60	3.00
Available phosphorus (%)	0.40	0.40	0.42	0.45
Sodium (%)	0.18	0.18	0.19	0.20

Table 10.5. Daily nutrient recommendations.

Crude protein	17 g
ME	1.17 MJ
Methionine	360 mg
Methionine + cysteine	640 mg
Lysine	720 mg
Calcium	3.5 g
Available phosphorus	0.4 g
Sodium	0.18 g

daily intakes of nutrients suggested under ideal management and environmental conditions.

However, as feed intake changes, specifications must be modified in order to maintain this intake of nutrients. Knowledge of feed intake, and the factors that influence it, are therefore essential for any feed management programme. To a degree, the energy level of the diet will influence feed intake, although one should not assume the precision of this mechanism to be perfect. In general, birds overconsume energy with higher energy diets, and they will have difficulty maintaining normal energy intake when diets of less than 10.5 MJ of ME kg^{-1} are offered. In most instances, underconsumption rather than overconsumption is the problem, and so use of higher energy diets during situations such as heat stress will help to minimize energy insufficiency.

There is little doubt that body weight at maturity is a major factor influencing feed intake and, therefore, the economic performance of laying hens. Body weight differences seen at maturity are maintained throughout lay almost regardless of the nutrient profile of layer diets. It is therefore difficult to attain satisfactory nutrient intakes with small birds. Conversely, larger birds will tend to eat more and this may become problematic in terms of the potential for obesity and/or too large an egg towards the end of lay. Phase feeding of nutrients can overcome some of these problems, although a more simplistic long-term solution is control over body weight at maturity.

Phase Feeding

Phase feeding refers essentially to reductions in the protein and amino acid level of the diet as the bird progresses through a laying cycle. The concept of phase feeding is based on the fact that as birds get older their feed intake increases while their egg production decreases. For this reason, it should be economical to reduce the nutrient concentration of the diet. If nutrient density is to be reduced, this should not occur immediately after peak egg numbers but rather after peak egg mass has been achieved. There are two reasons for reducing the level of dietary protein and amino acids during the latter stages of egg production, namely to reduce feed costs and to reduce egg size. The advantages of the first point are readily apparent if protein costs are high, but the advantages of the second point are not so easily defined and will vary depending upon the price of eggs.

It is difficult to give specific recommendations as to the decrease in dietary protein level that can be made to reduce egg size without decreasing the level of production. The appropriate reduction in protein level will depend on the season of the year (effect of temperature on feed consumption), age and production of the bird, and energy level of the diet. Hence, it is necessary to consider every flock on an individual basis before a decision is made to reduce the level of dietary protein. As a guide, it is recommended that protein intake be reduced from 17 g day^{-1} to 16 g day^{-1} after the birds have dropped to 80% production, and to 15 g day^{-1} after they have dropped to 70% production. With an average feed intake of 100 g day^{-1}, this would be equivalent to diets containing 17, 16 and 15% protein. It must be stressed that these values should be used only as a guide after all other factors have been properly considered. If a reduction in the level of protein is made and egg production declines, then the decrease in intake has been too severe and it should be increased immediately. If, on the other hand, production is held constant and egg size is not reduced, then the decrease in protein intake has not been severe enough and it can be reduced still further. The amino acid to be considered in this exercise is methionine since this is the amino acid that has the greatest effect on egg size. Phase feeding of phosphorus has also been recommended as a method of halting the decline in shell quality often seen with older birds. Using this technique, available phosphorus levels may be reduced from approximately 0.45% at peak production to slightly less than 0.3% at the end of lay.

A major criticism of phase feeding is that birds do not actually lay percentages of an egg. For example, if a flock of birds is producing at 75% production, does this mean that 100% of the flock is laying at 75% or is 75% of the flock laying at 100% production? If the latter is true, then the concept of phase feeding may be harmful. If a bird lays an egg on a specific day, it can be argued that its production is 100% for that day, and so its nutrient requirements are the same regardless of the age of the bird. Alternatively, it can be argued that many of the nutrients in an egg, and especially the yolk, accumulate over a number of days and so this concept of 100% production, regardless of age, is misleading.

Nutrition and Shell Quality

Layer diets usually contain all the calcium needed by the bird under most conditions. However, if egg shell quality is a problem during hot weather, or if the pullets have come into production at a fairly young age and have peaked very quickly, it may be advisable to increase the levels of calcium by at least 0.4% (Roland, 1995). Research has indicated that a marked improvement in shell quality can be obtained by feeding part of the dietary calcium as oyster shell or limestone chips. This is especially true if limestone flour rather than a granular source of limestone is used. The hen's requirement for calcium is relatively low except at the time of the day when egg shell formation is taking place. The greatest rate of shell deposition occurs in the dark phase when birds are not actively eating feed. The source of calcium during this period then becomes residual feed in the digestive tract and the labile medullary bone reserve.

When feeding limestone chips or oyster shell it is recommended that the diet contain 1–1.5% calcium and that the remainder be supplied by the supplemental source. The ideal time to feed this calcium supplement would be in the afternoon since this is when the hen normally has a high calcium requirement. Since separate feeding of calcium is not very practical, the only apparent solution is to have the calcium supplement mixed in the feed. The hen has the opportunity of leaving the oyster shell or limestone chips until the latter part of the day when it is required. The feeding of limestone or oyster shell on a continuous free-choice basis, or on top of a diet containing the full calcium requirement, is not recommended. It has been shown that egg shells with chalky deposits and rough ends are probably a direct result of feeding too much calcium to laying hens (Roland, 1986). Feeding birds oyster shell *ad libitum* can also result in the production of soft-shelled eggs; this unusual circumstance is due to a deficiency of phosphorus. If too much calcium is ingested, it must be excreted, usually as soluble calcium phosphate. This can lead to a deficiency of phosphorus, which results in no medullary bone being re-deposited between successive periods of calcification.

Calcium is the nutrient most often considered when shell quality problems occur, although it is realized that deficiencies of vitamin D_3 and phosphorus can also result in weaker shells. Vitamin D_3 is required for normal calcium absorption, and so if inadequate levels are fed, induced calcium deficiency quickly occurs. Diets devoid of synthetic vitamin D_3 are diagnosed quickly because there is a dramatic loss in shell weight. However, a more serious problem occurs with suboptimal levels of vitamin D_3, where changes in shell quality are subtler but nevertheless of economic significance. A major problem with deficiency of vitamin D_3 is that this nutrient is very difficult to assay in complete feeds. It is only at concentrations normally found in vitamin pre-mixes that meaningful assays can be carried out, and so, if D_3 problems are suspected, access to the vitamin pre-mix is usually essential. In addition to uncomplicated deficiencies of D_3, problems can arise due to the effect of certain mycotoxins. Compounds such as zearalenone, that are produced by *Fusarium* moulds, have been shown effectively to tie up vitamin D_3, resulting in poor egg shell quality. Under these circumstances, dosing birds with 300 IU of D_3 per day, for three consecutive days, with water-soluble D_3 may be advantageous (Leeson *et al.*, 1995).

Minimizing phosphorus levels is also advantageous in maintaining shell quality, especially under heat stress conditions. Because phosphorus is a very expensive nutrient, high inclusion levels are not usually encountered, yet limiting these within the range of 0.3–0.4%, depending upon flock conditions, seems ideal in terms of shell quality. Periodically, unaccountable reductions in shell quality occur and it is possible that some of these situations may be related to nutrition. As an example, vanadium contamination of phosphates causes an unusual shell structure, and certain weed seeds, such as those of the *Lathyrus* species, cause major disruptions of the shell gland.

Up to 10% reduction in egg shell thickness has been reported for layers fed saline drinking water, and a doubling in incidence of total shell defects has been seen with water containing 250 mg of salt l^{-1}. If a laying hen consumes 100 g of feed and 200 ml water per day, then water at 250 mg of salt l^{-1} provides only 50 mg of salt compared with intake from the feed of around 400 mg of salt. The salt intake from saline water therefore seems minimal in relation to total intake, but nevertheless shell quality problems often occur under these conditions. It appears that saline water results in limiting the supply of bicarbonate ions to the shell gland and that this is mediated via reduced activity of the enzyme carbonic anhydrase in the mucosa of the shell gland. However, it is still unclear why saline water has this effect, since much more salt is provided by the feed. There seems to be no effective method of correcting this loss of shell quality in established flocks, although for new flocks the adverse effect can be greatly reduced by adding 1 g of vitamin C l^{-1} of drinking water.

Diet and Egg Size

Increasing the hen's intake of balanced protein will result in an increase in egg size, while feeding higher levels of protein at the onset of production may help to increase egg size more rapidly. For strains of birds that produce many extra large eggs during the latter part of their egg production cycle, lowering the level of dietary protein during this period will result in slightly smaller and more uniform eggs. In these situations, when considering changes to the level of dietary protein, the energy content of the diet must also be taken into account. If diets are suboptimal in energy, little increase in egg size will be noted by increasing the level of protein because the hen will utilize protein to meet requirements for energy. Indeed, one of the main factors limiting early egg size is that energy intake is suboptimal.

Over the last few years, there has been considerable research involving the source of methionine. When comparing DL-methionine with Alimet® (a methionine hydroxy analogue), Harms and Russell (1994) showed the classical response of egg weight to both methionine sources (Table 10.6). There has

Table 10.6.   Effect of methionine source on layer performance[a].

Diet methionine (%)	Exp. 1, Egg weight (g)		Exp. 2, Egg weight (g)	
	DL	Alimet®	DL	Alimet®
0.228 (basal)	54.5	54.5	51.5	51.5
0.256	55.2	55.3	53.2	52.7
0.284	56.8	56.8	55.1	56.2
0.311	57.6	57.2	55.9	55.7
0.366–0.378	58.0	57.5	57.0	56.8

[a]Mean 80% egg production. Taken from Harms and Russell (1994).

been a suggestion that L-methionine may in fact be superior to any other source; this compound is not usually produced commercially, because routine manufacture of methionine produces a mixture of D- and L-methionine. This is the only amino acid where there is apparently 100% efficacy of the D-isomer; most research data indicate no difference in the potency of L- versus DL-methionine sources.

Attempts at reducing or tempering egg size later in the production cycle by phase feeding of protein or methionine have met with only limited success, probably because producers are reluctant to use very low protein diets. Our studies indicate that protein levels around 13% and less are necessary to bring about a meaningful reduction in egg size (Table 10.7). However, with protein levels much less than this, loss in egg numbers often occurs. Waldroup and Hellwig (1995) recently outlined estimates of methionine and methionine + cysteine requirements both for egg production and for egg weight/mass (Table 10.8). During peak egg mass output (38–45 weeks), the methionine requirement for egg size is greater than for egg numbers, while the latter requirement peaks at 51–58 weeks of age. If these data are verified in subsequent studies, they suggest that care should be taken in reducing methionine levels much before 60 weeks of age.

Table 10.7. Effect of reducing dietary protein level on egg size of 60-week-old layers (average for two, 28-day periods).

Dietary protein level (%)	Egg production (%)	Average feed intake day^{-1} (g)	Egg weight (g)	Daily egg mass (g)	Average protein intake day^{-1} (g)
17	78.8	114	64.8	51.0	19.4
15	77.5	109	64.3	49.7	16.4
13	78.3	107	62.2	49.1	13.9
11	72.7	108	61.7	45.1	11.9
9	54.3	99	58.2	36.1	8.9

All diets 11.7 MJ of ME kg^{-1}.

Table 10.8. Estimated methionine and methionine + cysteine requirements (mg day^{-1}).

	Bird age (weeks)	Egg no.	Egg weight	Egg mass
Methionine	25–32	364[b]	356[b]	369[b]
	38–45	362[b]	380[a]	373[b]
	51–58	384[a]	364[a]	402[a]
	64–71	374[ab]	357[b]	378[b]
Methionine + cysteine	25–32	608[b]	610[ab]	617[b]
	38–45	619[b]	636[a]	627[b]
	51–58	680[a]	621[ab]	691[a]
	64–71	690[a]	601[b]	676[a]

Adapted from Waldroup and Hellwig (1995).
Means followed by different superscript letters are significantly different ($P < 0.05$).

BROILER CHICKENS

The broiler chicken continues to show significant improvement in growth and overall feed efficiency as a standard market weight is achieved some 0.5 days earlier each year. This is due, in part, to improved understanding of nutrient requirements, and the realization of continual change in the proportion of nutrients directed towards growth versus maintenance. Perhaps one of the most striking features of the broiler to materialize over the last few years is its ability to respond adequately to a range of diet situations. For example, varying the protein:energy ratio of a diet seems to have less of an effect today than was recorded some 15–20 years ago. In large part, the adaptability of the broiler chicken is due to its voracious appetite and the fact that feed intake seems to be governed both by physical satiety and by cues related to specific nutrients. However, as will be discussed later, attempting to reduce the cost of broiler diets through the use of lower protein/amino acid levels, while not having major effects on gross performance, leads to subtle changes in carcass composition. Feed programmes may therefore vary depending upon the goals of the producer compared with the processor. Another major change in broiler nutrition that has occurred over the last 5 years is the realization that maximizing nutrient intake is not always the most economical situation, at least for certain times in the grow-out period. A time of so-called undernutrition, which slows down early growth rate, appears to result in dramatic reduction in the incidence of metabolic disorders such as sudden death syndrome and the various skeletal abnormalities. A period of slower initial growth, followed by compensatory growth, is almost always associated with improved feed efficiency because less feed is directed towards maintenance. Table 10.9 shows nutrient specifications for broilers, while Table 10.10 indicates examples of feed allocation based on bird age.

Feeding Programmes

Various types of feeding programme currently are being considered by broiler producers and feed manufacturers, and these may be thought of as speciality feeds. These programmes may involve low nutrient-dense diets as a means of simply reducing feed cost, or diets of higher protein/amino acid content used in an attempt to reduce carcass fat content. Alternatively, there is now interest in feed programmes involving feed restriction or diet dilution.

Low-nutrient density programmes
By offering low-protein, low-energy diets, it is hoped to reduce feed costs and so make feeds more attractive to customers. However, it is obvious that the birds will necessarily consume more of these diets and that birds may also take longer to reach market weight; these two factors result in reduced feed efficiency. Surprisingly, broiler chickens seem to perform quite reasonably with low nutrient-dense diets, and in certain situations these may prove to be

Table 10.9. Broiler diet specifications.

| | Approximate protein level (%) | | | | | |
| | Starter | | Grower | | Finisher/withdrawal | |
	22	20	20	18	18	16
Amino acids (% of diet)						
Arginine	1.20	1.10	1.05	0.95	0.90	0.85
Lysine	1.20	1.05	1.10	0.90	0.90	0.80
Methionine	0.48	0.42	0.44	0.38	0.37	0.36
Methionine + cysteine	0.82	0.75	0.73	0.65	0.64	0.61
Tryptophan	0.20	0.18	0.17	0.15	0.14	0.13
Histidine	0.40	0.35	0.32	0.30	0.28	0.27
Leucine	1.40	1.20	1.10	1.00	1.00	0.90
Isoleucine	0.75	0.60	0.55	0.50	0.47	0.45
Phenylalanine	0.75	0.65	0.60	0.55	0.53	0.50
Phenylalanine + tyrosine	1.40	1.20	1.10	1.00	1.00	0.90
Threonine	0.70	0.62	0.60	0.55	0.55	0.50
Valine	0.80	0.70	0.65	0.60	0.58	0.55
ME (MJ kg^{-1})	12.8	12.1	13.2	12.6	13.4	12.8
Calcium (%)	0.95	0.95	0.92	0.90	0.90	0.90
Available phosphorus (%)	0.42	0.42	0.40	0.40	0.38	0.38
Sodium (%)	0.18	0.18	0.18	0.18	0.18	0.18

Table 10.10. Feed allocation for regular type broiler diets (kg bird^{-1}).

| | | Starter | Grower | Finisher 1 | Finisher 2 | Total |
		CP 22%; ME 12.8 MJ kg^{-1}	CP 20%; ME 13.2 MJ kg^{-1}	CP 18%; ME 13.4 MJ kg^{-1}	CP 16%; ME 13.4 MJ kg^{-1}	
Bird age/type						
28 day Cornish	(F)	0.5	0.8	0.6		1.9
35 day Cut-up	(M)	0.7	1.1	0.7		2.5
	(F)	0.6	0.9	0.8		2.3
Mixed sex		0.7	1.0	0.7		2.4
42 day Whole bird	(M)	0.7	1.9	1.3		3.9
	(F)	0.5	1.7	1.5		3.7
Mixed sex		0.6	1.8	1.4		3.8
49 day Whole bird	(M)	0.6	2.0	2.3		4.9
	(F)	0.5	2.0	2.0		4.5
Mixed sex		0.6	2.0	2.1		4.7
56 day Whole bird	(M)	0.5	2.0	2.3	1.1	5.9
60 day Whole bird	(M)	0.4	2.1	2.4	1.9	6.8
70 day Roaster	(M)	0.4	2.3	2.8	2.8	8.3

the most economical programme. If diets of low energy level are fed, the broiler will eat more feed (Table 10.11); in this study, only the energy level was changed and the broiler adjusted reasonably well in an attempt to maintain constant energy intake. Diet energy levels from 13.8 to 11.3 MJ of ME kg^{-1} had no significant effect on body weight, and this suggests the bird is still eating for its energy need.

Obviously these data on growth rate are confounded with the intake of all nutrients other than energy. For example, birds offered the diet with 11.3 MJ of ME kg^{-1} increased their protein intake in an attempt to meet energy needs. Using these same diets, but controlling feed intake at a constant level for all birds shows that energy intake *per se* is a critical factor in affecting growth rate (Leeson *et al.*, 1996) (Table 10.12).

With low-energy diets, therefore, we can expect slightly reduced growth rate because normal energy intake is rarely achieved, and this fact is the basis for programmes aimed at reducing early growth rate. However, live body weight is often not the end-point of consideration for broiler production since carcass weight and carcass composition are becoming of prime consideration. From the point of view of the processor or integrator, these cheaper diets may be less attractive. Carcass weight and meat yields are often reduced, and this is associated with increased deposition of carcass fat, especially in the abdominal region. Low-protein diets are therefore less attractive when one considers feed cost per kg of edible carcass or feed cost per kg of edible meat. This consideration of carcass composition leads to development of diets that maximize lean meat yield.

Table 10.11. Performance of broilers fed diets of variable energy content.

Diet ME (MJ kg^{-1})	Body weight (g)		Feed intake (g bird^{-1})		
	25 days	49 days	0–25 days	25–49 days	0–49 days
13.8	1025	2812	1468	3003	4471
13.0	1039	2780	1481	3620	5101
12.1	977	2740	1497	3709	5206
11.3	989	2752	1658	3927	5586

Table 10.12. Performance of broilers given fixed quantities of feed.

Diet ME (MJ kg^{-1})	Body weight (g)		Feed intake:body weight gain
	25 days	49 days	0–49 days
13.8	825[a]	2558[ab]	1.84[c]
13.0	818[a]	2599[a]	1.82[c]
12.1	790[b]	2439[b]	1.94[b]
11.3	764[b]	2303[c]	2.05[a]

Means followed by different superscript letters are significantly different ($P < 0.05$).

Feed restriction and compensatory growth

Most broiler chickens are given unlimited access to feed or, at most, have limited access during brief periods of darkness. It is generally assumed that the faster that birds reach market weight, the better the feed conversion since maintenance requirement should be reduced. While this is usually true, there may be some potential for modifying the growth pattern of the bird in favour of an even greater reduction in maintenance requirement (Plavnik and Hurwitz, 1985, 1989). If broiler growth rate could be reduced during early life, and this is followed by compensatory growth so as to achieve the same market weight-for-age, then maintenance requirements must be reduced, implying improved feed efficiency. This concept raises the question of restricted feeding and/or reduced nutrient intake during early life. If it is accepted that feed conversion in its classical sense (digestibility, metabolizability, etc.) has improved little over the years, then improvements that we continue to see in feed utilization must be associated with the reduction in maintenance requirement. In addition to improving feed utilization, there is also interest in manipulating growth because of mortality associated with metabolic conditions. The broiler chicken shows exceptionally fast early growth rate when fed high nutrient-dense diets without any form of restriction.

If growth rate is to be reduced, then, based on needs to optimize feed usage, such restriction must occur early in the grow-out period. As the bird gets older, a greater proportion of nutrients are used for maintenance and less are used for growth. Therefore, reducing nutrient intake in the first 7 days will have little affect on feed efficiency because only 8–12% of feed is directed towards maintenance. At 8 weeks of age, a feed restriction programme would be more costly because, with a 20% restriction, there would be likely to be no growth since 80% of nutrients must go towards maintenance. Early feed restriction programmes therefore make sense from an energetic efficiency point of view, and also are the most advantageous in programmes aimed at reducing the incidence of metabolic disorders.

In order to allow potential for compensatory growth, while maintaining carcass quality, some means of maintaining the correct balance of amino acids to energy must be achieved (Cabel and Waldroup, 1990). Physical feed restriction, or diet dilution, using conventional type diets best accommodates this. There is current interest in diet dilution of young broilers as a means of controlling fat deposition, because it is assumed that fat cell numbers increase most rapidly in the very young bird (Cherry *et al.*, 1984). Controlling fat cell growth at this age may therefore place an upper limit on the subsequent fatness of the bird. Improvements in feed efficiency with such systems are claimed to be related to production of leaner birds, although such early qualitative feed restriction does imply compensatory growth. In some early studies, we fed broiler chickens conventional starter diets to 6 days of age and then the same diet diluted with up to 55% rice hulls from 6 to 11 days. After this time, the conventional starter was re-introduced, followed by regular grower and finisher diets. Table 10.13 indicates the amazing ability of the modern broiler chicken to compensate for this drastic reduction in nutrient intake from 6 to 11 days of age (Zubair and Leeson, 1994a).

Table 10.13. Effect of diet dilution with rice hulls from 6 to 11 days of age, on compensatory growth of male broiler chickens.

Treatment	Body weight (g)				Feed:gain		ME kg^{-1} gain
	21 days	35 days	42 days	49 days	21–35 days	0–49 days	0–49 days
1. Control	733	1790	2390	2890	1.84	2.01	6.21
2. 50% dilution 6–11 days	677	1790	2380	2950	1.70	1.93	5.90

Adapted from Zubair and Leeson (1994a).

Growth compensation was complete by 35 days of age, and this was associated with improvement in feed efficiency and a 5% improvement in energy efficiency. Such improvement in feed efficiency probably relates to a more favourable growth rate being induced and/or that birds utilize nutrients more efficiently during the period of compensation. Although not significantly different, there was also an indication that these birds deposit less fat, which is another factor that will improve feed efficiency. A practical problem with the type of diet dilution described in this study is potential for wet litter conditions related to high fibre intake. Diet dilution is not always a practical approach to reducing nutrient intake because birds will compensate with increased feed intake and diluents are very expensive per unit of energy provided. An alternative approach is physically to limit feed intake, and most experiments involve restriction down to the level of maintenance energy need which is around 6.3 kJ g^{-1} BW$^{0.67}$.

The reasons for improvement in feed efficiency with compensatory growth feeding programmes are not entirely clear at this time. Based on classical studies with other animals, it has been suggested that birds have reduced maintenance energy needs, even during *ad libitum* re-feeding up to market weight. However, our studies with the broiler chicken do not confirm this hypothesis (Zubair and Leeson, 1994b). A more likely reason for improved feed utilization is that there is simply a reduction in overall maintenance energy needs associated with the bird being smaller for a significant part of the grow-out period. This hypothesis suggests that complete growth compensation should be delayed for as long as possible and that ideally birds would not achieve normal weight-for-age until the day of marketing.

Most research to date has not been able to duplicate the dramatic reductions in carcass fatness attributed to a compensatory growth programme as originally described by Plavnik and Hurwitz, (1985). This may be due to other researchers not imposing a severe enough degree of undernutrition, since Plavnik and Hurwitz (1985) used a feed allowance that accounted only for maintenance. In practice, this means an exceptionally low feed intake (10–12 g bird^{-1} day^{-1} for 1–4 days) which is difficult to calibrate with commercial equipment, and difficult to simulate with diet dilution. However, the potential benefits of a compensatory growth programme seem exciting, and presumably any degree of modification of the growth curve will be beneficial.

The broiler chicken therefore appears able to benefit from a period of early undernutrition in that subsequent compensatory growth results in no overall loss of market weight and should be associated with improved feed utilization. Depending on the method used to impose such undernutrition, there is potential for these birds to be as lean or leaner than conventionally fed birds. The only potential problem with this technique is that it may shift the mortality/morbidity peak to later in the grow-out period. For example, leg problems and sudden death syndrome are both related to fast growth rate, rather than body weight *per se*. With early undernutrition, we are shifting the time of most rapid growth (compensation period) to later in the cycle (3–5 weeks) and so growth-related problems may be more prominent in this period. Such effects have not been recorded with experimental flocks, although research involving larger numbers of birds needs to be conducted. A period of compensatory growth may therefore be beneficial to grow-out of commercial broilers and, as long as good quality finisher diets are employed, there are interesting potential economic advantages to this technique. Even greater benefits may apply to roaster birds. Another very practical problem with diet dilution or feed restriction is deciding upon levels of anticoccidials and other pharmacological compounds. With diet dilution, birds will eat much more feed. If, for example, feed intake is doubled due to a 50% dilution, should the level of anticoccidial be reduced by 50%? With 50% feed restriction on the other hand, does there need to be an increase in the concentration of these additives? This general area needs careful consideration and results may well vary with the chemical compounds under consideration due to potential toxicity at critical levels.

Lighting programmes and feed intake

Many broiler chickens are grown under 23 or 24 h light each day, because it is thought that unlimited access to feed is required for maximum growth rate. However, there may be some potential for modifying the bird's feeding activity through lighting so as to improve the efficiency of feed utilization. Such programmes involve periods of darkness for varying lengths of time throughout the day. The idea behind an intermittent programme of, for example, 1 h light:3 h darkness, repeated six times each day, is that birds will actively seek feed during the light period, and subsequently rest during the dark period. If enough feed has been consumed in the 1 h of light, then feed efficiency should be improved because birds will be quite docile during the dark period and so expend less energy on maintenance. After this period of darkness, the upper digestive tract should be empty and the bird is again ready to eat. If this type of programme is used, it is important to ensure that adequate feeder space is available such that all birds can eat during the light periods. These types of programmes have been shown to improve feed efficiency to 49 days of age, by about 0.05 units. However such intermittent programmes do not slow down growth rate and so do not help in reducing late cycle

Table 10.14.   Step-down lighting programme for broilers.

	Bird age (days)	Hours of light
Example 1	0–4	23
	4–10	8
	10–14	10
	14–18	14
	18–23	18
	23 to market	23
Example 2 (open-sided building)	0–4	23
	4–14	Natural day length
	14–18	18
	18 to market	23

mortality associated with metabolic conditions. More recently there has been interest in so-called 'step-down/step-up' programmes for broiler chickens (Classen, 1988). Broilers are subjected to very short days for a 1–2 week period, ostensibly in an attempt to reduce the incidence of leg problems and occurrence of sudden death syndriome. In effect, the birds are likely to show reduced feed intake, and so this type of programme fits in well with the concept of restricted feeding detailed in the previous section. An example of this type of lighting programme being used commercially is shown in Table 10.14.

With very short periods of darkness, birds rarely eat in the dark, yet with the step-down programmes it has been shown that birds will consume up to 30–40% of their feed in the dark period. This has caused problems for those producers using a high stocking density, since birds apparently clamber over one another during the dark period in an attempt to get to the feeder. This can result in increased scratching of the skin and increased incidence of scabby-hip syndrome. At more liberal stocking densities (14.4 birds m^{-2}) such problems are less severe, and improved leg condition and lower incidence of early sudden death syndrome is noticed. If normal market weight-for-age is desired, then such step-down/step-up programmes must be ended and birds returned to full lighting by at least 20 days before expected market age. If reduced day length is continued for too long a period, then reduced growth rate will have to be accepted. If mortality due to metabolic disorders is excessive (> 8% in males to 45 days of age) then it may be economical to prolong the period of reduced day length, and accept the delayed market age.

Carcass Composition

The composition of the broiler carcass is now receiving considerable attention with the poultry industry's major thrust in further processing (Hickling *et al.*, 1990). While carcass portion yield is largely a factor of age and genetics, carcass composition can, to a large extent, be modified through diet choice. In general, diets high in energy produce fatter carcasses, and vice versa. On the other hand, high protein diets produce leaner carcasses. The situation is a little more complex than this since it is actually the balance of protein to energy that is important. If the bird consumes excess energy in relation to protein, a fatter bird develops, whereas feeding larger quantities of protein in relation to energy can produce a leaner bird. This manipulation of nutrients is sometimes referred to as changing the energy:protein ratio. Unfortunately, simple changes such as these are not economical, since the required degree of leanness in the carcass often only results from uneconomically high levels of protein.

When discussing the effect of diet protein or energy level on carcass composition, it is very important to appreciate the units of measurement. Often there is discussion about the effects of diet on percentage changes in composition but in some situations the percentage of a component in the carcass changes simply because there has been a corresponding change in the level of another component. This situation is clearly shown in Tables 10.15 and 10.16 (Jackson *et al.*, 1982); in this study, over the very wide range of energy levels used, there is the classic response for increase in percentage carcass fat and decrease in percentage carcass protein. However, only the actual quantity

Table 10.15. Proportional and absolute changes in carcass components in response to diet energy level.

Diet energy (MJ kg^{-1})[1]	Carcass fat		Carcass protein	
	g	%	g	%
10.9	161[a]	37.5[a]	221	51.9[e]
11.7	178[b]	39.3[b]	225	50.0[d]
12.6	208[c]	42.4[c]	229	47.1[c]
13.4	211[c]	42.6[c]	230	46.9[c]
14.2	239[d]	45.6[d]	229	44.7[b]
15.1	258[e]	47.9[e]	229	42.9[a]

[1]At constant protein.
Means followed different superscript letters are significantly different.

Table 10.16. Proportional and absolute changes in carcass components in response to diet protein level.

Diet protein level[1] (%)	Carcass fat		Carcass protein	
	g	%	g	%
16	252[d]	50.0[d]	202[a]	40.7[a]
20	237[c]	46.2[c]	227[b]	44.9[b]
24	210[b]	42.4[b]	233[b]	47.7[c]
28	189[a]	39.4[a]	233[b]	49.2[cd]
32	185[a]	39.2[a]	233[b]	50.3[d]
36	179[a]	38.3[a]	234[b]	50.7[d]

[1]At constant energy level.
Means followed by different superscript letters are significantly different ($P < 0.05$).

of fat changes (from 161 to 258 g) as energy level increases, with no change in grams of carcass protein. Producing a leaner carcass with low-energy diets is therefore achieved by reducing fatness rather than by increasing lean meat deposition. Changing the protein level of the diet has the same basic effect on tissue deposition (Table 10.16).

Except when very low protein diets are used, diet protein has no effect on the quantity of protein deposited in the carcass. Proportional changes in carcass protein are a consequence of less grams of fat being deposited as protein level increases. Consequently, producing leaner carcasses with higher protein diets is a consequence of there being less fat, rather than there being more protein.

Nutrition and Metabolic Disorders

Metabolic disorders today account for most of the mortality occurring in broiler chickens. The only common feature is a very fast growth rate, and, in most instances, limiting growth rate can control these conditions to some degree.

Sudden death syndrome

Also referred to as acute death syndrome or flip-overs, sudden death syndrome is most common in males, and especially when growth rate is maximized. Mortality may start as early as 3–4 days, but most often peaks at around 3–4 weeks of age, with affected birds invariably being found dead on their back. Mortality may reach as high as 1.5–2% in mixed-sex flocks; in male flocks, the condition is often the major single cause of mortality, with death rates as high as 4% being quite common. Any nutritional factors that influence growth rate will have a corresponding effect on sudden death syndrome. The syndrome can virtually be eliminated with diets of low nutrient density (18% CP, 10.0 MJ of ME kg^{-1}), although these may not be economical in terms of general bird performance. Research data suggest that diets based on pure glucose as an energy source result in a much higher incidence of sudden death syndrome compared with birds fed starch- or fat-based diets. It seems likely that some anomaly in electrolyte balance is involved in sudden death syndrome, although this has not been clearly defined. In part, this is due to the fact that metabolic changes occur rapidly after death, and hence blood profiles taken from syndrome birds are likely to vary depending upon sampling time following death.

Skeletal disorders

Abnormal skeletal development continues to be a major reason for mortality and/or downgrading in commercial meat birds. In North America, this loss probably equates to some $30 million per year. Since there have been numerous studies on skeletal conditions, it is obvious that the aetiology is complex and not outwardly related to a single nutritional or environmental factor. The most common skeletal abnormalities seen in meat-type birds are tibial dyschondroplasia and field rickets (Edwards, 1988). The fact that leg problems are more prevalent in broilers and turkeys than in egg-type birds has led to the speculation of growth rate and/or body weight as causative factors. On this basis, one is faced with numerous reports of general nutritional factors influencing leg problems. For example, it has been suggested that energy restriction in the first few weeks of growth halves the number of leg problems in broilers, while reduced protein intake results in fewer leg abnormalities. Similarly, restricting access to feeder space also seems to result in fewer leg defects. However, most recent evidence suggests that body weight *per se* is not a major predisposing factor for leg problems. From experiments involving harnessing weights to the backs of broiler chickens and poults, it is concluded that the severity of leg abnormalities is independent of body weight and that regular skeletal development is adequate to support loads far greater than normal body weight. There seems to be some disparity between the effects on skeletal development of (i) limiting the incidence by reducing the plane of nutrition and (ii) failing to aggravate the problem by artificially increasing body weight. This apparent dichotomy suggests that it is the rate of growth, related to higher levels of nutrient intake, that is important in precipitating the condition. Skeletal abnormalities invariably will occur if the diet is deficient in nutrients affecting bone and/or cartilage development. However, skeletal problems continue to occur in diets well fortified with these nutrients, and so obviously the aetiology is not simple and is likely to be related to deficiencies of nutrients at active bone growth plates.

Ascites

Ascites is rapidly becoming one of the major causes of mortality/morbidity in broiler chickens. Once only seen at high elevations, ascites now causes problems in fast growing birds in most areas. Ascites is characterized by the accumulation of fluid in the abdomen, and hence the basis for the common phrase of water-belly. Fluid in the abdomen is in fact plasma that has seeped from the liver, and this occurs as the end result of a cascade of events ultimately triggered by oxygen inadequacy within the bird. For whatever reason, the need to provide more oxygen to the tissues leads to increased heart stroke volume and ultimately to hypertrophy of the right ventricle. Such heart hypertrophy, coupled with malfunction of the heart valve, leads to increased pressure in the venous supply, and so pressure builds up in the liver, and often leads to the characteristic fluid leakage.

Because of the relationship to oxygen demand, ascites is affected and/or precipitated by factors such as growth rate, altitude (hypoxia) and environmental temperature. Of these factors, hypoxia was the initial trigger some years ago because the condition was first seen as a major problem in birds held at high altitude where mortality in male broilers of 20–30% was not uncommon. Today, ascites is seen commonly in fast growing lines of male broilers fed high nutrient-dense diets at most altitudes and where the environment is cool/cold for at least part of each day. Mortality seen with ascites is dictated by the number of stresses involved and hence the efficacy of the cardiopulmonary system in oxygenating tissues.

While growth rate *per se* is the major factor contributing to oxygen demand, the composition of growth is also influential, because oxygen need varies for metabolism of fats versus proteins. Oxygen need for nitrogen and protein metabolism is high in relation to that for fat, although it must be remembered that the chicken carcass actually contains little protein or nitrogen. The carcass does contain a great deal of muscle, but 80% of this is water. On the other hand, adipose tissue contains about 90% fat and so its contribution to oxygen demand is proportionally quite high. Excess fatness in birds will therefore lead to significantly increased oxygen needs for metabolism.

Environmental temperature, and associated oxygen/energy demand, is usually a factor in most cases of ascites. Keeping birds warm is perhaps the single most practical way of reducing the incidence of ascites. As environmental temperature changes, there is a change in the bird's oxygen requirement. If one considers the thermo-neutral zone following the brooding period to be 20–26°C, then temperatures outside this range cause an increase in metabolic rate, and so an increased need for oxygen. Low environmental temperatures are most problematic, since they are accompanied by an increase in feed intake with little reduction in growth rate. While there is an increased oxygen demand at high temperatures due to panting, etc., this is usually accompanied by a reduced growth rate, and so overall reduced oxygen demand. Under commercial farm conditions, cold environmental conditions are probably the major factor contributing to ascites. For example, at 10 versus 26°C, the oxygen demand of the bird is almost doubled. This dramatic increase in oxygen need, coupled with the need to metabolize increased quantities of feed, often leads to ascites It is interesting to note that birds maintained at high altitude under commercial conditions are often subjected to cool or cold night time temperatures.

Manipulation of the diet composition and/or feed allocation system can have a major effect on the incidence of ascites. In most instances, such changes to the feeding programme influence ascites via their effect on growth rate. However, there is also a concern about the levels of nutrients that influence electrolyte and water balance, the most notable being sodium. Feeding high levels of salt to broilers (> 0.5%) does lead to increased fluid retention, although ascites invariably occurs with diets containing a vast range of salt, sodium and chloride concentrations.

Apart from obvious nutrient deficiencies, or excesses as in the situation with sodium, the major involvement of the feeding programme with regard to its effect on ascites revolves around nutrient density and feed restriction. Ascites is more common when high-energy diets are used, especially when these are pelleted. Dale and Villacres (1988) grew birds on high-energy diets designed to promote rapid growth and likely to induce ascites. There was no correlation between 14 day body weight and propensity of ascites although birds fed 12.6–13.0 MJ of ME kg^{-1}, rather than 11.9–12.3 MJ of ME kg^{-1}, had twice the incidence of ascites. Unfortunately, Dale and Villacres (1988) did not show body weight data. In this study, the higher energy diets were produced essentially by increasing the level of supplemental fat. In a previous report,

Dale and Villacres (1986) support the concept that feed change *per se* is often the trigger to ascites but that the condition is also seen in single diet feeding programmes. In formulating diets of varying nutrient density and fat content, these workers show clear evidence of correlation between ascites and growth rate (Table 10.17). The highest incidence of ascites occurred when the highest energy level was fed, regardless of either the energy:protein or fat content of the diet. These data (Table 10.17) show little effect of added fat, although with both series of the energy:protein used, the greatest incidence of ascites occurred in the fastest growing birds.

Because the feeding programme, nutrient density and growth rate are all intimately involved in affecting the severity of ascites, there is invariably discussion on the possible advantages of feed restriction. Arce *et al.* (1992) carried out a series of interesting studies to record bird response to varying nutrient restriction programmes. As pointed out by these authors, the goal of such programmes is to reduce the incidence of ascites without adversely affecting economics of production. It is expected that nutrient restriction programmes will reduce final weight-for-age to some degree, and obviously there is a balance between the degree of feed restriction and commercially acceptable growth characteristics. Arce *et al.* (1992) conducted studies in Mexico at 1940 m or at 2500 m elevation. Birds were either fed on a skip-a-day schedule for varying periods from 7 to 28 days or allowed access to the feeders for just 8 h per day (7 a.m. to 3 p.m.). In all studies, control full-fed birds were the heaviest and exhibited the greatest ascites mortality (40 compared with 8–15%). As expected, the later in the grow-out cycle that skip-a-day feeding is introduced, the greater the reduction in ascites mortality, although this is accompanied by greater reduction in body weight. By restricting the feed access time to 8 h per day, ascites mortality was greatly reduced and this seems a very practical system, especially where hand-feeding is practised, and feeders can be raised easily during the restriction period. Restricting feed access time with mechanical feeders is more complicated and may not be a practical solution. It appears as though ascites mortality can be reduced through limiting feed intake and, depending upon the timing of this restriction programme, there will be about 100–200 g loss in weight-for-age. In most commercial operations, this will mean a 2–3 day delay in achieving 50–56 day market weight.

Table 10.17. Effect of nutrient density and diet composition on incidence of ascites.

Diet ME (MJ kg^{-1})	CP (%)	ME/CP	Diet fat (%)	49 day body weight (g)	Ascites mortality (%)
12.3	23	128	0	1800	8.8
12.3	23	128	4	1820	8.7
13.0	24	128	4	1830	15.8
12.3	21	140	0	1810	9.0
12.3	21	140	4	1810	8.5
13.0	22	140	4	1860	12.0

Adapted from Dale and Villacres (1986).

REFERENCES

Arce, J., Berger, M. and Coellc, C. (1992) Control of ascites syndrome by feed restriction techniques. *Journal of Applied Poultry Research* 1, 1–5.

Cabel, M.C. and Waldroup, P.W. (1990) Effect of different nutrient restriction programs early in life on broiler performance and abdominal fat content. *Poultry Science* 69, 652–660.

Cherry, J.A., Swartworth, W.J. and Siegel, P.B. (1984) Adipose cellularity studies in commercial broiler chicks. *Poultry Science* 63, 97–108.

Classen, H. (1988) The role of photoperiod manipulation in broiler chicken management. *Canada Poultryman* 75, 8–10.

Clunies, M., Emslie, J. and Leeson, S. (1992) Effect of dietary calcium level on medullary bone reserve and shell weight of Leghorn hens. *Poultry Science* 71, 1348–1356.

Dale, N. and Villacres, A. (1986) Nutrition influences ascites in broilers. *World Poultry.* Misset International, The Netherlands, p. 40.

Dale, N. and Villacres, A. (1988) Relationship of two-week body weight to the incidence of ascites in broilers. *Avian Diseases* 32, 556–560.

Edwards, H.M., Jr (1988) Effect of dietary calcium, phosphorus, chloride and zeolite on the development of tibial dyschondroplasia. *Poultry Science* 67, 1436–1446.

Harms, R.H. and Russell, G.B. (1994) A comparison of the bioavailability of DL-methionine and MHA for the commercial laying hen. *Journal of Applied Poultry Research* 3, 1–6.

Hickling, D., Guenter, W. and Jackson, M.E. (1990) The effects of dietary methionine and lysine on broiler chicken performance and breast meat yield. *Canadian Journal of Animal Science* 70, 673–678.

Jackson, S., Summers, J.D. and Leeson, S. (1982) The response of male broilers to varying levels of dietary protein and energy. *Nutrition Reports International* 25, 601–612.

Leeson, S. and Caston, L.J. (1993) Does environmental temperature influence body weight; shank length in Leghorn pullets? *Journal of Applied Poultry Research* 2, 253–258.

Leeson, S. and Summers, J.D. (1981) Effect of rearing diet on performance of early maturing pullets. *Canadian Journal of Animal Science* 61, 743–749.

Leeson, S. and Summers, J.D. (1985) Response of growing Leghorn pullets to long or increasing photoperiods. *Poultry Science* 64, 1617–1622.

Leeson, S. and Summers, J.D. (1989) Response of Leghorn pullets to protein and energy in the diet when reared in regular or hot-cyclic environments. *Poultry Science* 72, 1349–1358.

Leeson, S. and Summers, J.D. (1997) *Commercial Poultry Nutrition*, 2nd edn. University Books, Guelph, Canada.

Leeson, S., Julian, R.J. and and Summers, J.D. (1986) Influence of prelay and early-lay dietary calcium concentration on performance and bone integrity of Leghorn pullets. *Canadian Journal of Animal Science* 66, 1087–1096.

Leeson, S., Diaz G. and Summers, J.D. (1995) *Poultry Metabolic Disorders and Mycotoxins.* University Books, Guelph, Canada.

Leeson, S., Caston L.J. and Summers, J.D. (1996) Broiler response to diet energy. *Poultry Science* 75, 529–535.

Martin, P.A., Bradford, G.D. and Gous, R.M. (1994) A formula method of determining the dietary amino acid requirements of laying type pullets during their growing period. *British Poultry Science* 35, 709–724.

McNaughton, J.L., Kubena, L.F., Deaton, J.W. and Reece, F.N. (1977) Influence of dietary protein and energy on the performance of commercial egg-type pullets reared under summer conditions. *Poultry Science* 56, 1391–1398.

Plavnik, I. and Hurwitz, S. (1985) The performance of broiler chicks during and following a severe feed restriction at an early age. *Poultry Science* 64, 348–355.

Plavnik, I. and Hurwitz, S. (1989) Effect of dietary protein, energy, and feed pelleting on the response of chicks to early feed restriction. *Poultry Science* 68, 1118–1125.

Roland, D.A. (1986) Eggshell quality IV. *World Poultry Science Journal* 42, 166–171.

Roland, D.A. (1995) *The Egg Producers Guide to Optimum Calcium and Phosphorus Nutrition*. Mallinckordt International, Iowa, USA.

Waldroup, P.W. and Hellwig, H.M. (1995) Methionine and total sulfur amino acid requirements influenced by stage of production. *Journal of Applied Poultry Research* 3, 1–6.

Zhang, B. and Coon, C.N. (1994) Nutrient modelling for laying hens. *Journal of Applied Poultry Research* 3, 416–431.

Zubair, A.K. and Leeson, S. (1994a) Effect of varying period of early nutrient restriction on growth compensation and carcass characateristics of male broilers. *Poultry Science* 73, 129–136.

Zubair, A.K. and Leeson, S. (1994b) Effect of early feed restriction and realimentation on heat production and changes in sizes of digestive organs of male broilers. *Poultry Science* 73, 529–538.

11 Feeding Systems for Horses

D. Cuddeford

Department of Veterinary Clinical Studies, Royal (Dick) School of Veterinary Studies, The University of Edinburgh, Midlothian, UK

INTRODUCTION

In a seminal review of the nutrition of the horse in 1955, Olsson and Ruudvere summarized the energy requirements of the horse for maintenance computed according to different authors and the systems available. At that time, systems were based on starch equivalent, Scandinavian feed units, feed units for ruminants, Russian oat feed units and total digestible nutrients (TDN). Since that time, the TDN system developed in the USA (Morrison, 1937) has given way to one based on digestible energy (NRC, 1989) which is used throughout North and South America, the UK, Australia, New Zealand, South East Asia and parts of Southern Europe. Over the last 15 years, a new French net energy (NE) system has been developed and refined from that originally proposed (INRA, 1984). This system expresses the energy value of feeds in terms of horse feed units (unite fouragire cheval; UFC); one feed unit is the net energy content (9.42 MJ) of 1 kg barley for maintenance. Another NE system currently in use in The Netherlands has feed values based on NE for milk production in ruminants (Smolders, 1990). The Scandinavian feed unit (ScFU) continues to be used in Denmark, whilst in Finland, Iceland and Norway, the fattening feed unit (FFU) is in use. Both systems are based on digestibility trial values obtained with cattle, and ruminant digestible crude protein (DCP) values are used as well. Protein-rich feeds have a higher energy value calculated in ScFU than in FFU (Staun, 1990). However, a feed or diet containing about 110 g of digestible protein per ScFU has the same calculated energy value in both units. The absence of data derived from experiments with horses in the Nordic countries precludes the use of a special feed unit for horses. Sweden is exceptional in that it has relied on a digestible energy (DE) system and is now using metabolizable energy (ME). In the former USSR, the energy value of feedstuffs is based on the system derived by Kellner (1926) and expressed as Russian oat feed units; one unit (1 kg) corresponds to 0.6 kg starch equivalent.

The former USSR also uses energetic feed units (EFU), equivalent to 10.46 MJ of ME (Memedeikin, 1990), which have been derived experimentally with horses. Spain, Portugal and Greece do not use any particular system, whereas Italy has moved from using the Leroy fodder unit to adopting the French UFC system in its entirety (Miraglia and Olivieri, 1990). More recently, it has been proposed (Austbø, 1996) that the Nordic countries adopt the French system as well. It is thus quite conceivable that, in the near future, there will be only two systems of horse feeding practised and these will be based on DE according to the National Research Council (NRC, 1989) devised in the USA or NE (Vermorel and Martin-Rosset, 1997) as developed by INRA, France. In the meantime, the ScFU is still in use together with the Russian oat feed unit.

FEED ENERGY VALUES

Digestible Energy (DE)- and Net Energy (NE)-based Systems

The NRC originally used TDN to describe the energy content of feed for horses. These units can be converted to DE since 1 kg of TDN is equivalent to 18.4 MJ of DE (NRC, 1989). This conversion factor was based on work with ruminants, and the validity of its use has always been questionable and particularly in the light of results obtained in a small study with ponies. Barth *et al.* (1977) estimated 19.246 MJ kg^{-1} for a hay or hay–concentrate diet, which is 1.05 times greater than the proposed factor. Now, however, the current NRC tables use DE expressed as Mcal. The *raison d'être* for the NRC (1989) using DE is that few data exist for ME or NE values of horse feeds. Although it is preferable that DE values are determined by *in vivo* experimentation, equations are becoming available for the prediction of DE (see, for example, Pagan, 1994). More recently, new technologies, such as *in vitro* gas production techniques have been developed and the data used to predict feed values ($r^2 = 0.72$; Lowman *et al.*, 1997). Whatever predictive methods are used, they require validation; a relatively simple procedure for DE, but much more complicated for ME or NE.

The NE system was introduced in France during 1984 by INRA because the DE system was considered to overvalue high-fibre feeds. The classic experiments of Wolff *et al.* (1877a) showed that more digestible energy (15%) was required by the horse when fed a 75% hay diet compared with when fed a 75% cereal diet. Furthermore, Vermorel and Martin-Rosset (1997) showed that the DE requirements for maintenance and work were 25% higher when hay was fed compared with grain. Kronfeld (1996) developed a calorimetric model to derive heat production for the utilization of different diets by horses and showed clearly that fibre is thermogenic. Replacement of fibre with cereal reduced the yield of heat, and thus it follows that fibrous foods will be used less efficiently by the horse. The NE system relies on maintenance being the major component of energy expenditure in horses (Martin-Rosset *et al.*, 1994) and that the NE value of nutrients depends on free energy being produced by oxidative processes (Vermorel and Martin-Rosset, 1997).

Derivation of Feed Values

The NE values of feeds were calculated by determining their following characteristics.

Organic matter digestibility (OMD)

The OMD of 33 forages was determined simultaneously in light horses and wethers. Prediction equations were developed so that the OMD of feeds in horses could be derived from OMD values obtained with wethers. For example:

$$\%OMD \text{ (horses)} = -14.91 + 1.1544\%OMD \text{ (wethers)}; r^2 = 0.96 \qquad (11.1)$$

can be used to predict values for grass (Martin-Rosset *et al.*, 1984). OMD values for horses were predicted for 114 forages evaluated by INRA (1978) with wethers. Martin-Rosset *et al.*, (1994) took OMD for concentrates from pig data when crude fibre was less than 150 g kg^{-1} or extrapolated from pig and ruminant data.

Digestible energy (DE)

The DE of 75 feeds was measured in horses and included roughages, cereal by-products, cereals, oilseed meals and compounded feeds. Martin-Rosset *et al.* (1994) used these data together with OMD values (obtained with wethers) to derive an equation to predict (%) DE:

$$DE = 0.034 + \Delta + 0.9477 \text{ OMD}; r^2 = 0.994 \qquad (11.2)$$

($\Delta = -1.1$ for forages, 1.1 for concentrates).

Metabolizable energy (ME)

Vermorel and Martin-Rosset (1997) fed 12 different diets, at one or two feeding levels, and measured methane and urine energy losses. The former losses varied from 1.04 to 3.66% of gross energy (GE) and depended on dietary fibre content and on the individual horse. In contrast, urinary losses depended on both the fibre and protein content of the diet, and averaged 4.26% of GE. Insufficient data were available to predict gas and urine energy loss reliably, so instead ME:DE ratios were predicted from feed components in the following way:

$$\frac{ME}{DE}\% = 84.07 + 0.0165 \text{ CF} - 0.0276 \text{ CP} + 0.0184 \text{ CC}; r^2 = 0.45 \qquad (11.3)$$

where CF = crude fibre; CP = crude protein; CC = cytoplasmic carbohydrates (all values in g kg^{-1} of DM). This equation underestimated ME:DE ratios of protein-rich feeds (> 300 g of CP kg^{-1}) so the following equation was used:

$$\frac{ME}{DE}\% = 94.36 - 0.011 \text{ CF} - 0.0275 \text{ CP}; r^2 = 0.17. \qquad (11.4)$$

ME:DE ratios are variable and range from 78–80% (oil meals), to 84–88% (forages, legume seeds and cereal by-products) and to 90–95% (cereals). The important point that arises is that methane and urine energy losses account for a further 8% difference between cereals and forages and a 16% difference between cereals and oil meals; differences not apparent within the DE system.

Net energy (NE)

The nutrients available to the horse will reflect the chemical composition of the feed that it consumes and the site of digestion; feeds with the same or similar DE may contribute different absorbable nutrients. The quantitative nutrient supply from different feeds has been tabulated by Vermorel and Martin-Rosset (1997); some values were measured whilst others were assumed from studies with pigs and ruminants. Enzymatic digestion in the small intestine accounted for 0.85 of starch consumed, 0.05 of cell wall of dried grass and hays, 0.08 of fresh grass, 0.10 of lucerne and 0.15 of concentrates. Between 0.90 and 0.95 of apparently digested ether extract was long chain fatty acids (LCFAs) absorbed from the small intestine. Jarrige and Tisserand (1984) estimated true digestibility of true protein pre-caecally to be about 0.80 for cereals and oil meals, 0.70 for fresh grass, 0.60 for dehydrated lucerne and 0.30–0.45 for hays, the latter value depending on stage of growth at harvest. These authors suggested that undigested feed true proteins entering the large intestine had true digestibilities in this part of the tract of 0.90, 0.80 and 0.75 for concentrates, fresh grass and hays/silages, respectively.

Undigested feed residues that leave the ileum are subjected to microbial degradation. Based on ruminant data (Webster *et al.*, 1975), Vermorel and Martin-Rosset (1997) deduced that methane and heat of fermentation amounted to some 8% of the energy of fermented substrates. Furthermore, they estimated volatile fatty acid (VFA) production as follows:

$$\text{VFA (g kg}^{-1}\text{ DM)} = \text{OMD} - \text{OM digested in small intestine} \times 0.92 \qquad (11.5)$$

The molar proportions of VFA produced reflects the nature of the feed degraded; high-fibre feeds result in high molar proportions of acetate. Vermorel and Martin-Rosset (1997) were able to predict acetate (C_2) production as follows:

$$\text{Acetate (\%)} = 0.54\text{ CF (\% DM)} + 57;\ r^2 = 0.8 \qquad (11.6)$$

At 50 g of CF kg^{-1} DM, the energy content (kJ g^{-1} of VFA) and the efficiency of utilization (k_m) values were 18.29 and 0.68, respectively, compared with 17 and 0.64 at 350 g of CF kg^{-1} of DM. Vermorel and Martin-Rosset (1997) assert that k_m values for nutrients absorbed from the small intestine should be similar to those in monogastrics and for VFAs, the same as in ruminants. Glucose and lactate (GL) were ascribed a k_m value of 0.85, LCFAs 0.80, and amino acid (AA) ME 0.70. VFA k_m values varied, as shown above, between 0.64 and 0.68.

Absorbed energy will reflect the sum of the gross energies of the individual nutrients that are absorbed. Individual values (kJ g^{-1}) used for glucose/lactate, acetate, propionate, butyrate, amino acids and LCFAs were respectively 15.65, 14.60, 20.76, 24.94, 23.44 and 39.76. Absorbed end-products of digestion and their relative contribution in energy terms are shown in Table 11.1. It is clear that feeds degraded in the small intestine

Table 11.1. Estimated end-products of digestion (g kg^{-1} of DM), the proportion of absorbed energy (AE) they contribute and derived k_m values.

| Substrate | End-products of digestion | | | | |
	GL	AA	LCFA	VFA	k_m values
Maize	628	53	32	174	0.80
AE	0.63	0.08	0.08	0.21	
Oats	390	63	47	188	0.78
AE	0.48	0.11	0.15	0.26	
Good hay	74	49	12	387	0.65
AE	0.12	0.12	0.05	0.71	
Poor hay	40	21	5	348	0.61
AE	0.09	0.06	0.03	0.82	

GL, glucose; AA, amino acids; LCFA, long chain fatty acids; VFA, volatile fatty acids (adapted from Vermorel and Martin-Rosset, 1997).

contribute most energy via glucose and lactate, with a resultant high efficiency of utilization. In contrast, those feeds degraded in the large intestine provide energy via VFAs with a much lower overall efficiency of utilization; it should be noted that k_m is reduced further by the energy costs of eating roughages.

Based on the above values for absorbed energy, k_m (%) was calculated by Vermorel and Martin-Rosset (1997) as follows:

$$k_m = 0.85\ E_{GL} + 0.80\ E_{LCFA} + 0.70\ E_{AA} + (0.63 \text{ to } 0.68)\ E_{VFA} \qquad (11.7)$$

where E_{GL}, etc., represents the proportion of nutrient absorbed in energy terms. However, there remains the energy cost of eating to consider. Ørskov and MacLeod (1990) have shown in ruminants that this is a significant factor affecting the utilization of feed energy. Experiments with ponies (Vermorel and Mormède, 1991) and horses (Vernet *et al.*, 1995) showed that the energy cost of eating per kg of feed DM was much greater than when measured in sheep (Osuji *et al.*, 1975).

Because horses cannot ruminate, they spend at least twice as long as ruminants comminuting roughages. However, it is not necessarily true to say that horses expend more energy on comminution *per se* since rumination has an energy cost. KuVera *et al.* (1989) estimated this to be 9.3 J kg^{-1} of weight (W) min^{-1}. The physical work associated with the two activities is likely to be similar although the higher cost of eating in ruminants, at 32 J kg^{-1} of W min^{-1} (Adam *et al.*, 1984), may be due to changes in blood flow and activity of the viscera (Ørskov and MacLeod, 1990). The physical form and nature of the diet has a large effect on energy expenditure during eating; 300 kg steers fed long roughage expended 6 MJ day^{-1} compared with only 0.8 MJ day^{-1} when fed pelleted diets. In ponies, there is a reciprocal relationship between OMD and energy cost of eating; it decreased linearly from 5.8 to 0.4 kJ g^{-1} of DMI kg^{-1} of W whilst OMD increased from 0.32 to 0.88 (Vermorel and Mormède, 1991). Vernet *et al.* (1995) measured the cost of eating hays and pelleted maize and showed that the former averaged 0.10 of ME intake compared with only 0.01 for the maize. Not surprisingly, k_m values were lower than those calculated from the supply of nutrients. Thus, correction factors (Δk_m) were developed which depended on (i) digestibility of energy:

$$\Delta k_m = -0.14 \ (76.4 - DE\%) \tag{11.8}$$

or (ii) fibre content of the DM:

$$\Delta k_m = -0.20 \ CF\% + 2.50 \tag{11.9}$$

These correction factors are applied to Equation 11.6 for roughages in order to calculate k_m values; those values in Table 11.1 have been corrected in this manner.

Ninety five feeds of known chemical composition and digestibility, together with estimates of their digestion end-products, were used to predict k_m. Vermorel and Martin-Rosset (1997) used CF, CC, OM and OMD in equations for roughages. cereal–legume seeds, cereal by-products and compounds. Correlations (r^2) varied between 0.700 and 0.992 and, as expected, the highest and most consistent r^2 values were obtained with the seeds. Subsequently, UFC values (NE/9.42) were calculated for these feeds and then the relationships between these values and feed composition were elucidated; OMD and DE were used as independent variables. Correlations (r^2) for roughages varied between 0.832 and 0.996 and for concentrates, between 0.931 and 0.995. These prediction equations were used to determine the UFC values of most of the 149 horse feeds listed by Martin-Rosset *et al.* (1984). UFC values for compound feeds were predicted from the chemical composition of the raw materials commonly used according to their average usage rates; r^2 varied between 0.956 and 0.988.

ENERGY REQUIREMENTS

For convenience, the American DE-based system will be referred to as NRC and the French NE-based system as INRA.

Maintenance (NRC)

The system devised in the USA was based on digestible energy (DE), and full details of it were published most recently in 1989 (NRC, 1989). Maintenance requirements were based on the work of Pagan and Hintz (1986a) who used four mature male animals weighing 125, 206, 500 and 856 kg to derive values. The previous publication by the National Research Council (NRC, 1978) used metabolic body size ($W^{0.75}$) in the calculation of maintenance energy requirements. This power function of live weight was derived for interspecific comparisons (Brody, 1945; Schmidt-Nielson, 1984), and Kleiber (1961) proposed that the metabolic requirement of mammalian herbivores, MR (kJ day^{-1}) is related to live weight as follows:

$$MR = 4.185 \times 70 \ W^{0.75} \tag{11.10}$$

where W is the weight of the animal in kg. However, intraspecific comparisons produce exponents for this relationship which are quite different from 0.75 (Thonney *et al.*, 1976). For example, Blaxter (1962) stated that the exponent was greater than 0.75 and as high as 0.90 when measuring metabolism of mature animals within a species. Pagan and Hintz (1986a) demonstrated that $W^{0.87}$ produced the best relationship between energy intake and energy balance. However, they also showed that the regression relationship between resting maintenance DE intake and body weight had an r^2 value of 0.999. Much earlier, Benedict (1938) had shown a linear relationship between fasting heat metabolism and live weight. NRC (1989) uses the equation:

$$DE \ (MJ \ day^{-1}) = 4.185(0.975 + 0.021 \ W) \tag{11.11}$$

(where W is weight in kg) derived by Pagan and Hintz (1986a) for stalled animals in a thermoneutral environment.

Activity allowance

Pagan and Hintz (1986a) cited the work of others (Breuer, 1968; Stillions and Nelson, 1972; Anderson *et al.*, 1983) and practical experience to justify the values for a 500 kg horse proposed by NRC (1978). This requirement was 29.1% higher than that predicted by Equation 11.10 above, so Pagan and Hintz (1986a) arbitrarily increased the values predicted by this equation by multiplying them by a factor of 1.291 and so obtained a new equation:

$$DE \ (MJ \ day^{-1}) = 4.185(1.375 + 0.03 \ W) \tag{11.12}$$

which was adjusted in NRC (1989) to:

$$DE \ (MJ \ day^{-1}) = 4.185(1.4 + 0.03 \ W) \tag{11.13}$$

This equation allows for the energy required for 'normal' activity although the use of a single value across a range of live weights is questionable. Equally, although no mass exponent is used and a linear relationship exists between energy requirements and live weight, there is a large *y*-intercept (5.9 MJ) which will have an effect similar to that of a mass exponent.

Heavy horses

It has been proposed (Potter *et al.*, 1987) that Equation 11.13 overestimates the energy need of horses exceeding a mature weight of 600 kg. It is considered that they have a lower voluntary activity than lighter horses and are more docile. The equation for estimating maintenance requirements of horses weighing in excess of 600 kg is:

$$DE \ (MJ \ day^{-1}) = 4.185(1.82 + 0.0383W - 0.000015 \ W^2) \tag{11.14}$$

and the net result is that the energy requirements are reduced by 5, 10 and 15% at 700, 800 and 900 kg, respectively compared with values obtained with Equation 11.13. The somewhat arbitrary nature of these reductions is justified on the basis that the values obtained are consistent with other findings (NRC, 1989).

Environmental effects

An unprotected, equine mechanical model has been used to assess climatic energy demand under Scottish conditions (MacCormack and Bruce, 1991). Demand increased by only 9% when the model was sheltered and rugged, 18% when exposed and rugged and 26% when sheltered but unrugged; rugging appears to be the most effective way of reducing heat losses. McBride *et al.* (1985) have demonstrated that daily maintenance energy needs in outwintered, mature, 500 kg horses increased by 1.534 MJ of ME $°C^{-1}$ fall in effective ambient temperature (AT) below the lower critical temperature (LCT). If we assume that ME is 0.9 DE, then DE requirements increase by 1.704 MJ $°C^{-1}$ fall, or, for adult horses, maintenance DE intake must be increased by 2.48% $°C^{-1}$ fall below LCT. Accordingly, Equation 11.13 can be amended (Cymbaluk and Christison, 1990) to take account of the LCT and exposure to cold as follows:

$$\text{DE (MJ day}^{-1}) = 4.185\{(1.4 + 0.03 \text{ W}) + [(\text{LCT} - \text{AT}) \ 0.00082 \text{ W}]\} \qquad (11.15)$$

However, definition of LCT is fraught with difficulty since it is affected by insulation, food intake and acclimitization. It can vary from −15°C to 10°C (Cymbaluk and Christison, 1990) and is thus particular to the situation in which the animal is kept; cold-acclimatized horses have a much lower LCT than those that are not acclimatized.

It is of interest to note that transportation of equids by road can also increase their energy expenditure. Doherty *et al.* (1997) showed that Shetland ponies expended 3.04 J kg^{-1} s^{-1} when transported compared with 1.34 J kg^{-1} s^{-1} when standing; the difference was significant ($P < 0.001$).

Maintenance (INRA)

Vermorel *et al.* (1984) reviewed feeding trials conducted with horses over the last century (see, for example, Wolff *et al.*, 1888; Breuer, 1968; Wooden *et al.*, 1970; Stillions and Nelson, 1972) and concluded that maintenance energy requirements amounted to 586 kJ of DE, 502 kJ of ME or 351 kJ of NE kg^{-1} $W^{0.75}$. For a 500 kg horse, NRC (1989), Equation 11.10 gives a value of 48 MJ of DE compared with a value of 62 derived using the French data. Table 11.2 illustrates the difference between the systems over a range of live weights.

Activity allowance

Vermorel *et al.* (1984) suggest values whereby maintenance requirements should be increased depending on the type of horse and whether or not it has been working (see Table 11.3). It is generally accepted that the maintenance requirements of working animals are enhanced through an up-regulation of metabolic processes. Furthermore, temperament can affect requirements, and

Table 11.2. Energy requirements (MJ day^{-1}) for maintenance of horses of varying live weight using different energy systems and compared on a DE basis.

		Live weight (kg)		
Reference		400	600	800
NRC (1989)	Equation 11.10	39.2	56.8	74.4
	Equation 11.12	56.1	81.2	106.3
	Equation 11.13	—	—	95.8
Vermorel *et al.* (1984)		52.4	71.0	88.2

Table 11.3. Factors for adjusting maintenance (m) requirements (factor × m): dependent on horse type and activity.

	Horse type		
Horse activity	Draft	Riding	Bloodstock
At rest	1.00	1.05	1.10
In work	1.05	1.10	1.15
Stallion – at rest	1.10	1.15	1.20
– working	1.20–1.30	1.25–1.35	1.30–1.40

After Vermorel *et al.* (1984).

this can be seen from Table 11.3 where Vermorel *et al.* (1984) indicate that bloodstock/thoroughbreds have higher maintenance requirements.

Application of the factor 1.10 to the maintenance needs of a 600 kg riding horse as shown in Table 11.3 raises the value from 71 (Table 11.2) to 78.1 MJ of DE day^{-1}, which is close to the NRC (1989) value of 81.2 MJ of DE.

Heavy horses

No allowance is made for the reduced voluntary activity of these animals, although the factors proposed in Table 11.3 mean that values calculated for heavy horses are slightly below those obtained using the NRC (1989) equation, Equation 11.13. Jespersen (1949) showed that for draft horses weighing about 500 kg, differences between estimates of maintenance requirements were relatively small (< 10%). However, with heavier horses, the differences were greater, varying between 628 and 703 kJ kg^{-1} W$^{0.75}$ at 500 kg to between 582 and 715 kJ kg^{-1} W$^{0.75}$ at 700 kg.

Environmental effects

The French NE system assumes that animals are housed during the winter period (Martin-Rosset *et al.*, 1994) and no account is taken of environmental effects on maintenance energy needs.

Work (NRC)

A mobile, open-circuit, indirect calorimetry system was used to determine the energy expended by four geldings weighing between 433 and 520 kg (Pagan and Hintz, 1986b). Energy expended was exponentially related to speed and proportional to the mass that was moved. DE requirements above maintenance are given by the following equation:

$$\text{DE (kJ kg}^{-1}\text{h}^{-1}) = \frac{4.185[e^{(3.02+0.0065y)} - 13.92] \times 0.06}{0.57} \qquad (11.16)$$

where y is the speed in m min^{-1}. However, requirements could only be reliably predicted for horses up to medium canter (350 m min^{-1}) and the equation would therefore only be appropriate for horses working aerobically for extended periods of time, such as endurance horses. Earlier, Anderson *et al.* (1983) developed a quadratic equation to calculate the quantity of DE required for maintenance and work:

$$\text{DE (MJ day}^{-1}) = 4.185(5.97 + 0.021\ W + 5.03X - 0.48X^2) \qquad (11.17)$$

where $X = W$ (kg) $\times$ km $\times 10^{-3}$. The application of this equation is limited to situations where the workload (kg $\times$ km) is less than 3560 which was the upper limit defined experimentally. Thus, the use of this equation is limited to those animals performing short periods of intense work, such as sprint horses.

Analysis of feeding practices in racing yards, etc., indicates that horses consume up to 30 g of dry feed kg^{-1} W (Glade, 1983) and, based on the feeds used, it has been estimated (NRC, 1989) that horses in intense work consume up to twice their maintenance DE requirement. Unlike ruminant production systems, output, in terms of work, is poorly defined, and words such as 'light', 'medium' and 'hard' abound when describing 'work'. There is little prospect of this unsatisfactory state of affairs being resolved since most horses in training are fed to maintain their weight and condition; the balance between forage and concentrate in the daily ration is adjusted according to 'work' load which is judged in a fairly subjective fashion.

Based on the work of Brody (1945), NRC (1989) suggest increasing maintenance DE requirements by 10% for each hour of draught work performed.

Work (INRA)

The energy costs of locomotion (kJ kg^{-1} h^{-1}) at different speeds have been calculated from those data published by a number of different authors, including Brody (1945), Thomas and Fregin (1981) and Hörnicke *et al.* (1983). Their impact on DE requirements is shown in Table 11.4. There are no great differences between the NRC and French values, although the latter serve to illustrate the variation between individuals in relation to their work energy requirement.

Table 11.4. Effect of speed of movement (m min^{-1}) on DE requirement (kJ kg^{-1} h^{-1}) above maintenance.

Work	Speed	NRC (1978)	NRC (1989)	Vermorel *et al.* (1984)
Walking	110	2.1	12.3	8.6–14.2
Slow trotting	200	21.3	27.0	20–40
Fast trotting	300	52.3	57.3	52–80
Cantering	350	100.0	87.8	59–92

Martin-Rosset *et al.* (1994) considered that data obtained from feeding trials were more reliable than assessments based on oxygen consumption when estimating the energy costs of draught power. These authors relied on the data published by Olssen and Ruudvere (1955) who recalculated those obtained by Jespersen (1949) into Scandinavian feed units (ScFU); 0.5 and 1.0 units were required for 1 h of medium and very heavy work, respectively. One ScFU is equivalent to the net energy of 1 kg of barley and thus is equivalent to 1 UFC. For a 800 kg horse, NRC (1989) would increase the energy allowance by 9.58 MJ of DE, equating to about 0.74 UFC, for 1 h of draught work.

Growth (NRC)

Energy requirements for growth are directly related to the energy content of the gain, which increases as the animal gets older, reflecting the changing proportions of tissues. Furthermore, rapid live weight gain is usually associated with more energy being deposited per unit of gain. NRC (1989) estimates the DE requirement per kg of gain as:

$$\text{MJ DE kg}^{-1} \text{ LWG} = 4.185(4.81 + 1.17X - 0.023X^2) \text{ ADG} \qquad (11.18)$$

where X is the age in months, and ADG is the average daily gain in kg. This equation was developed from data reported by a number of different authors, including Ott and Asquith (1986) and Schryver *et al.* (1987). There is no optimum, defined rate of growth for horses or ponies, although excessive weight gain has been associated with the development of bone abnormalities (Thompson *et al.*, 1988).

Growth (INRA)

Two studies (Agabriel *et al.*, 1984; Bigot *et al.*, 1987) provided data on energy intake, live weight and resulting live weight gain in growing horses. This information was incorporated into a ruminant model:

$$\text{UFC kg}^{-1} \text{ W}^{0.75} \text{ day}^{-1} = a + bG^{1.4} \qquad (11.19)$$

and used to predict energy requirements for growth (Geay *et al.*, 1978; Robelin, 1979). The exponent value 1.4 was validated by the work of Agabriel *et al.* (1984); 'a' represents the coefficient of maintenance requirement, 'b' the coefficient of gain and 'G' is the average daily gain (kg day^{-1}). Table 11.5 gives the coefficients used for calculating UFC requirements.

Table 11.6 shows a comparison between NRC (1989) and Martin-Rosset *et al.* (1994) estimates of the DE requirements of growing horses with a mature weight of 500 kg.

NRC (1989) appears to produce higher estimates of requirement, although it should be noted that Equation 11.18 is age-dependent rather than weight-dependent; the latter is likely to be a more reliable basis for calculating requirements.

Pregnancy (NRC)

Most fetal development occurs during the last 3 months of gestation, and NRC (1989) relies on the data accumulated by Meyer and Ahlswede (1976) to recommend that requirements for months 9, 10 and 11 be estimated by multiplying maintenance values by 1.11, 1.13 and 1.20, respectively.

Table 11.5. INRA coefficients for calculating UFC requirements for growth.

Age (months)	Coefficients	
	a	b
6–12	0.0602	0.0183
18–24	0.0594	0.0252
30–36	0.0594	0.0252

Adapted from Martin-Rosset *et al.* (1994).

Table 11.6. A comparison between INRA[a] and NRC estimates of the DE requirements of growing horses (mature weight 500 kg).

Age (months)	Weight (kg)	Gain (kg day^{-1})	System	MJ of DE required day^{-1}		Total MJ of DE day^{-1}
				Maintenance	Growth	
6	215	0.85	NRC	32.80	39.13	71.96
			INRA	32.89	21.14	54.03
12	325	0.65	NRC	46.65	42.47	89.12
			INRA	44.84	26.10	70.94
18	400	0.35	NRC	56.07	26.77	82.84
			INRA	52.39	22.66	75.05

[a]Assumes 1 UFC ≡ 12.87 MJ of DE.
Adapted from Martin-Rosset *et al.* (1994).

Pregnancy (INRA)

Martin-Rosset and Doreau (1984) used the same source materials as NRC (1989), together with their own work, to devise energy requirements for pregnancy. They estimated the efficiency of energy utilization to be somewhere between that for cows and sows, at 25%, recommending that 1.1–1.3 times maintenance should be fed from the eighth month to term, respectively. This is very similar to the conclusions arrived at by the NRC.

Lactation (NRC)

The NE required for milk production is the product of the yield and gross energy (GE) content of each kg of milk. Doreau *et al.* (1988) showed that the latter is affected by stage of lactation, falling from 2.406 MJ kg^{-1} in the first month to 1.987 in the fourth month. However, these changes are relatively small compared with the change in milk yield. During the first 3 months of lactation, broodmares up to 500 kg secrete about 0.03 W as milk, and 0.02 W over the next 3 months (Doreau *et al.*, 1988). Ponies produce more milk *pro rata*; Neuhaus (1959) estimated they produce 0.04 W during early lactation and 0.03 W in later lactation. NRC (1989) assumed the same values as NRC (1978) for determining the DE requirements for milk production on the assumption that mares milk contains 1.987 MJ of GE kg^{-1} (Ullrey *et al.*, 1966) and that DE is converted into milk energy with an efficiency of 60%. Ultimately, a value of 3.314 MJ of DE kg^{-1} of milk was adopted.

Lactation (INRA)

It is assumed that energy is used with an efficiency of 65% for milk production, a value extrapolated from cow and sow data. DE requirements based on different feeding systems are shown in Table 11.7.

Some authors, for example Pagan *et al.* (1984), have suggested that the DE requirements for pony mares (NRC, 1978) were in fact too high. Whilst the INRA value appears generous, there is remarkable accord across the different systems.

FEED PROTEIN VALUES

Protein-based

Mature horses and ponies appear to tolerate low levels of protein as well as low-quality protein in their diets. Often, mature animals may be overwintered on hay containing only 60 g of CP kg^{-1} and yet do not manifest signs of protein malnutrition. It has been suggested (Martin *et al.*, 1996) that, when fed low-nitrogen diets, renal conservation of nitrogen occurs in horses.

Table 11.7. DE requirement of a 500 kg mare producing 15 kg of milk day^{-1} in the first month of lactation: a comparison of feeding systems.

	Maintenance		Lactation			Total
System	Basis day^{-1}	MJ of DE day^{-1}	MJ of GE kg^{-1}	NE/DE	MJ of DE day^{-1}	MJ of DE day^{-1}
Norway[a]	578 kJ kg^{-1} W$^{0.75}$	61.1	1.99	0.60	49.75	110.85
NRC (1978)	648 kJ kg^{-1} W$^{0.75}$	68.5	1.99	0.60	49.75	118.25
Germany[b]	585 kJ kg^{-1} W$^{0.75}$	61.9	2.30	0.66	52.27	114.17
INRA (1984)	1.05 (585 kJ kg^{-1} W$^{0.75}$)	64.9	2.30	0.55	62.73	127.63
NRC (1989)	(1.4 + 0.03 W) 4.184 MJ day^{-1}	68.5	1.99	0.60	49.75	118.25

[a]Nedkvitne (1976); [b]Löwe and Meyer (1982).
Adapted from Doreau *et al.* (1988).

Furthermore, more urea is lost from the blood and degraded in the digestive tract, suggesting a compensatory source of nitrogen when horses are fed low-nitrogen diets. In contrast, there is evidence that growing horses can be affected by protein of suboptimal quality. There are numerous reports (see, for example, Hintz *et al.*, 1971; Barton *et al.*, 1973; Ott *et al.*, 1979) that indicate that poor quality proteins, such as linseed meal, do not support such a high rate of growth as better quality proteins. Addition of lysine to low-quality protein resulted in live weight gains comparable with those achieved by feeding milk or soya protein. Apart from this type of work and the demonstrable need for lysine, there are very few data on the effect of protein quality and, thus, NRC (1989) consider it is impractical to suggest a requirement in these terms. Whilst it would appear that mature horses are unaffected by protein quality, there is a difference between protein sources in their ability to meet protein needs of the animal (NRC, 1989). However, this is probably a reflection of differences in digestibility between different protein sources (Reitnour and Salsbury, 1976) and of the protein content of the diet rather than the quality. NRC (1989) concludes that there are inadequate data on protein digestibility and, since the site of protein degradation greatly affects the efficiency of utilization, crude protein should be used.

MADC-based

MADC is the matières azotées digestibles cheval which is the DCP corrected for that part of the CP which does not supply amino acid to the animal. In most cases, DCP is predicted from CP (see Table 11.8).

Apparent digestibility data give no indication of the site of digestion nor of the nutrients absorbed. It is more beneficial to the horse if the crude protein contains little non-protein nitrogen (NPN) and that it is digested pre-caecally

Table 11.8. Prediction of DCP (g kg^{-1} DM).

Feed type	Equation	r^2
Fresh forage	= −27.33 + 0.8614 CP	0.967
	= −74.52 + 0.9568 CP + 0.1167 CF	0.980
Grass hay	= −25.96 + 0.8357 CP	0.968
Legume hay	= −29.95 + 0.8673 CP	0.933
All forages	= −27.57 + 0.8441 CP	0.964
Cereals	= −4.94 + 0.8533 CP	0.931

Adapted from Martin-Rosset *et al.* (1994).

so that amino acids may be absorbed from the small intestine. Martin-Rosset *et al.* (1994) assume that endogenous nitrogen entering the caecum amounts to 5 g kg^{-1} of DMI and that lost in the faeces equates to 3 g kg^{-1} of DMI. These data, together with that obtained from fistulated animals, enables the calculation of the true pre-caecal digestibility of nitrogen. Jarrige and Tisserand (1984) have estimated values of 0.45, 0.70 and 0.85 for grass hay, fresh grass and concentrates, respectively. Clearly, the stage of growth at which roughages are harvested will affect how much protein N is degraded pre-caecally. More recent work, using illeally fistulated ponies (Potter *et al.*, 1992) or the mobile-bag technique in horses (Macheboeuf *et al.*, 1995), has confirmed the estimates of Jarrige and Tisserand (1984). New work (Moore-Colyer *et al.*, 1998) has shown pre-caecal CP disappearances of 0.77, 0.52, 0.60 and 0.30 for an oat hull–naked oat mixture (2:1), hay, soya hulls and sugar beet pulp, the proportional disappearance depending on the botanical nature of the material.

Some feed protein and endogenous nitrogen will pass into the large intestine where some will be degraded and the end-products incorporated into microbial protein. Faecal nitrogen will be composed of indigestible feed nitrogen mostly bound to fibre, the so-called acid detergent-insoluble nitrogen (ADIN), microbial protein nitrogen and endogenous nitrogen. The different origins of faecal nitrogen make the estimation of the large intestinal contribution to absorbed nitrogen complex. However, Jarrige and Tisserand (1984) calculated large intestinal true protein digestibility to be 0.90 for concentrates, 0.80 for fresh grass and 0.75 for hays and silages. It was assumed that 0.15 of the digested protein was absorbed as amino acids. However, it appears that this assumption is flawed because *in vitro* studies (Bochröder *et al.*, 1994) have failed to demonstrate a transmucosal flux of lysine, histidine and arginine across colonic tissue. If these amino acids are released in the lumen of the large intestine by microbial activity, they will not be able to contribute to the plasma amino acid pool. These findings confirm those of others (Wysocki and Baker, 1975; Freeman *et al.*, 1989; Freeman and Donawick, 1991), but are at variance with the results of Slade *et al.* (1971). The latter group of workers suggested that lysine, in particular, was absorbed rapidly from the large intestine. However, the nature of their experiment would have allowed ^{15}N enrichment of ammonia which could have been absorbed from the caecum and colon and could then subsequently have labelled circulating lysine, giving the impression that lysine had been absorbed from the lumen of the gut. It is noteworthy that ammonia is transported across the colonic mucosa at rates equal to, or greater than those measured with sheep ruminal mucosa (Bochröder *et al.*, 1994) and will be available for incorporation into non-essential amino acids in the liver of the horse.

The MADC is the sum of the feed and microbial amino acids absorbed from the small and large intestine, respectively; the former has been measured, the latter assumed. Comparing MADC with DCP values shows that the latter overvalues forages by 0.10–0.30 (Martin-Rosset *et al.*, 1994). Thus, the appropriate MADC value is calculated from the DCP of forages by multiplying by a factor, K, which is 0.90, 0.85, 0.80 and 0.70 for green forages, hay/dehydrated forages, straws and good quality grass silages, respectively. The DCP value of concentrates, cereals, etc. is equivalent to MADC.

PROTEIN REQUIREMENTS

Maintenance (NRC)

Estimates of DCP requirements vary from 0.49 to 0.68 g kg^{-1} W day^{-1}, and NRC (1989) considered that a value of 0.60 g (2.8 g of DCP kg^{-1} $W^{0.75}$) was appropriate for most horses. To express requirements in terms of CP, NRC (1989) assumed that a horse at maintenance would consume a forage diet with a protein digestibility of 0.46. Thus, the CP requirement would become 1.3 g kg^{-1} W day^{-1} and, when divided by the energy requirement, it equates to 9.6 g of CP MJ^{-1} of DE. Amino acid requirements are not specified, although it is assumed that 'average' diets contain 3.5 g of lysine kg^{-1}.

Maintenance (INRA)

Using the same reference sources as NRC (1989), namely Slade *et al.* (1970) and Prior *et al.* (1974), Martin-Rosset *et al.* (1994) estimated maintenance requirements to be 2.4 g of DCP kg^{-1} $W^{0.75}$ or 2.8 g of MADC kg^{-1} $W^{0.75}$. Forage DCP values are downgraded 10–20% to express them in terms of MADC. Martin-Rosset *et al.* (1994) use K values that vary from 1.0 for concentrate to 0.7 for grass silages to convert DCP to MADC because not all DCP contributes amino acids to the horse; the proportion depends on the source of DCP. Examination of Table 11.9, which illustrates the maintenance protein needs of 200, 400 and 600 kg horses, clearly shows that the MADC system as proposed, increases food requirements by 1.78 times when silage is fed to a 200 kg horse. In contrast, provision of the same material to a 600 kg horse only increases the silage dry matter requirement by 1.35 times when rationed using the MADC system. Martin-Rosset *et al.* (1994) concur with NRC (1989) in that amino acid balance is of no significance in mature horses at maintenance.

Table 11.9. A comparison between NRC (1989) requirements (0.6g of DCP kg^{-1} W) and those of Martin-Rosset et al. (1994)[a] using MADC (2.8 g kg^{-1} W$^{0.75}$) in terms of meeting the protein requirements of 200, 400 and 600 kg horses at maintenance (kg food day^{-1}).

				Animal weights:	200		400		600	
				Source:	INRA	NRC	INRA	NRC	INRA	NRC
Requirement g day^{-1} of MADC or DCP:					149	120	250	240	339	360
	DCP (g kg^{-1}DM)	K	MADC (g kg^{-1} DM)		kg DM day^{-1}					
Green forage	130	0.90	117		1.27	0.92	2.14	1.85	2.90	2.77
Grass hay	50	0.85	42.5		3.51	2.40	5.88	4.80	7.98	7.20
Barley straw	30	0.80	24		6.21	4.00	10.42	8.00	14.13	12.00
Grass silage	150	0.70	105		1.42	0.80	2.38	1.60	3.23	2.40

[a]INRA (1984).

Work (NRC and INRA)

Neither NRC (1989) nor Martin-Rosset et al. (1994) stipulate protein require-ments for work, assuming that if the protein:energy ratio at maintenance is maintained, then the higher energy intake of the working horse will provide any extra nitrogen required. It is reasonable to expect that horses being trained for whatever activity will have an elevated nitrogen retention due to muscle hypertrophy. Freeman et al. (1988) confirmed an apparent increase in nitrogen retention in mature horses performing different levels of work. However, there are no data to justify feeding more than 9.6 g of CP MJ^{-1} of DE. Martin-Rosset et al. (1994) suggest that a supply of 60–65 g of MADC UFC^{-1} (4.66–5.05 g of MADC MJ^{-1} of DE) is satisfactory beyond maintenance for a mature horse in work. They assessed the maintenance energy requirement at 0.038 UFC kg^{-1} W$^{0.75}$ and the protein requirement at 2.8 g of MADC kg^{-1} W$^{0.75}$ (the same as for maintenance, see p. 254) which gives a protein:energy ratio of 74 g of MADC UFC^{-1} which is rather different from their proposal of 60–65 g MADC. Feeding high-protein diets to horses that perform hard work is not recommended since the heat of assimilation (Ha) of protein is 15–30% of the gross energy intake (IE) of protein (Kronfeld, 1996); protein is both thermogenic and acidogenic. It has been suggested (Glade, 1983) that excess protein depressed racehorse performance, although the conclusion was invalid because protein intake was confounded with energy intake. Generally, racehorses are fed protein in excess of published need; compounded feedstuffs for racehorses commonly contain 140 g of CP kg^{-1}, equivalent to 10.8 g of CP MJ^{-1} of DE. Field surveys of thoroughbreds (Gallagher et al., 1992a) and standardbreds (Gallagher et al., 1992b) have shown that, in practice, they may consume an excess of between 21 and 56% protein relative to NRC (1989) standards. Apart from the thermogenic disadvantage of protein, which is irrelevant in a sprint horse, high-protein diets may in fact confer a metabolic advantage, reducing plasma lactate, heart rate and glycogen metabolism (Pagan *et al.*, 1987).

Growth (NRC)

Being deprived of protein is probably no more damaging than being deprived of energy for horses as in other animals (Allden, 1970). Restricting protein intake will limit growth rate (Ellis and Lawrence, 1979), although foals born at the height of the grass-growing season in the UK will have available to them grass containing between 160 and 200 g of CP kg^{-1} of DM. Mare's milk, which can provide up to 0.4 kg of protein daily, together with good grass will support up to 1.5 kg gain day^{-1} in thoroughbred foals; it is unlikely that they will be protein-limited. Based on an analysis of published work, NRC (1989) recommends 12 and 10.8 g of CP MJ^{-1} of DE for weanlings and yearlings, respectively. It is generally considered that lysine is the most limiting amino acid and that 502 and 454 mg should be supplied per MJ of DE to weanlings and yearlings, respectively.

Growth (INRA)

Probably the most reliable way to estimate protein need in growing animals is to conduct serial slaughter studies and subsequent tissue analysis. Martin-Rosset *et al.* (1983) examined heavy breeds of horses during their growth phase and estimated protein requirements on the basis of tissue composition. They assumed a faster turnover of body protein and increased the maintenance protein requirement for growing animals by 125% of maintenance, i.e. 3.5 g of MADC kg^{-1} W$^{0.75}$. Incidentally, Meyer (1983) allows 4.5 g of DCP kg^{-1} W$^{0.75}$. For light horse breeds, Martin-Rosset *et al.* (1994) used a model similar to that used for determining energy requirements (Equation 11.19) to calculate total MADC requirements:

$$\text{g MADC day}^{-1} = a \ W^{0.75} + bG \tag{11.20}$$

where 'a' is the coefficient of maintenance, 'b' the coefficient of gain and G is the average daily gain (kg day^{-1}); these coefficients are shown in Table 11.10. Comparisons of estimated requirements for growth using the different systems (Table 11.11) shows that it is difficult to compare the values other than by

Table 11.10. INRA coefficients for calculating MADC requirements for growth.

Age (months)	Coefficients	
	a	b
6–12	3.5	450
18–24	2.8	270
30–36	2.8	270

Table 11.11. A comparison between INRA and NRC protein requirements of growing horses (mature weight 500 kg) together with estimated DCP requirements (g day^{-1}).

Age (months)	Weight (kg)	Gain (kg day^{-1})	System	CP (g)	DCP[a] (g)	MADC (g)	DCP[b] (g)
6	215	0.85	NRC	860	602[c]	—	—
			INRA	—	—	579	579[e]
12	325	0.65	NRC	958	575[d]	—	—
			INRA	—	—	560	609[f]
18	400	0.35	NRC	891	535[d]	—	—
			INRA	—	—	345	375[f]

[a]Estimated from NRC data; [b]estimated from INRA data; [c]assumes 0.70 CP is digestible; [d]assumes 0.60 CP is digestible; [e]assumes concentrate only; [f]assumes 50:50 concentrate–hay.

making assumptions about the diets fed and the digestibility of the protein contained therein. The protein content of gain declines as live weight increases; it is about 200 g kg^{-1} gain in young animals (3–6 months) and around 170 g kg^{-1} of gain in a 2-year-old animal with a mature weight of 500 kg.

Pregnancy (NRC)

It has been estimated (Meyer and Ahlswede, 1976) that 60–65% of fetal development takes place during months 9, 10 and 11 of pregnancy. Furthermore, these authors used the data they obtained from the composition of the fetal tissues to estimate a protein utilization efficiency of 60%. A 500 kg pregnant mare was calculated to need 127, 130 and 178 g of DCP day^{-1} above maintenance (0.6 g of DCP kg^{-1} W = 300 g DCP day^{-1}) for months 9, 10 and 11, respectively. NRC (1989) assumed that typical diets for mares in the last trimester of pregnancy would have an overall protein digestibility of 0.55. Thus, the crude protein requirements would amount to 776, 782 and 869 g of CP day^{-1} and equate to a protein:energy ratio of 10.5 g of CP MJ^{-1} of DE overall; this latter figure was the basis for calculating the protein requirements of pregnant mares.

Pregnancy (INRA)

Martin-Rosset and Doreau (1984) used the same source of data as NRC (1989) in order to calculate MADC requirements. Daily protein retention was calculated from the protein content of the fetus (130, 153 and 171 g kg^{-1} in months 9, 10 and 11, respectively) and the efficiency with which DCP was used for this purpose (0.50–0.55). The daily amount of protein deposited ranged from 5 g 100 kg^{-1} W at 8 months to 21 g 100 kg^{-1} at term. Thus, a

500 kg mare would be depositing 25 g day^{-1} at the beginning of the last trimester to 105 g at the end; this would equate to 40–45 g and 190–210 g of DCP, respectively. Table 11.12 contrasts various estimates of protein requirement and it can be seen that both NRC and INRA figures are well below those of Meyer (1983).

Lactation (NRC)

The protein content of mare's milk is lower than that found in cow's milk (33–36 g kg^{-1}) and is highest in early lactation, and decreases gradually thereafter (Table 11.13). It is perhaps surprising that NRC (1989) chooses to ignore the summary figures in its own appendix, shown in Table 11.13, and instead uses 21 g of protein kg^{-1} for early lactation and 18 g kg^{-1} for late lactation as the basis for calculating protein need.

The efficiency of utilization of digestible protein for milk protein production is 0.65, and NRC (1989) assumes that mixed diets fed to lactating mares will contain protein with an apparent digestibility coefficient of 0.55 and calculates dietary crude protein requirements on this basis. The values thus obtained exceed those of NRC (1978) by about 0.05 for early lactation and, in contrast, are 0.05 below the 1978 values for late lactation.

Lactation (INRA)

This system uses an efficiency of 0.55 for the conversion of DCP to milk protein, and Martin-Rosset *et al.* (1994) consider that the NRC (1978, 1989)

Table 11.12. Protein requirements (g day^{-1}) of a 500 kg mare during months 9, 10 and 11 of pregnancy.

Months	g of DCP[a]	g of MADC[b]	g of DCP[c]
9	427	340	500
10	430	460	528
11	478	485	561

[a]NRC (1989); [b]Martin-Rosset *et al.* (1994); [c]Meyer (1983).

Table 11.13. The composition and yield of milk from light horses.

Month	Yield (kg 100 kg^{-1} W)		Energy (MJ kg^{-1})		Protein (g kg^{-1})	
	Ref. 1	Ref. 2	Ref. 2	Ref. 3	Ref. 2	Ref. 3
1	3.0	3.0	2.406	2.427	24	27
2	2.5	3.0	2.092	2.218	24	22
3	2.5	3.0	2.092	2.092	24	18
4	2.0	2.0	1.987	—	21	—

Ref. 1: Doreau and Boulot (1989). Ref. 2: Doreau *et al.* (1988). Ref. 3: NRC (1989).

value of 0.65 is too high. They deduce that the g of DCP required per kg of milk is 44 for month 1, 38 for months 2 and 3, and 36 for month 4. However, these data do not accord with the tabulated values for milk protein shown in Table 11.13; application of a factor of 0.55 would give 44, 44, 44 and 38, respectively. Comparable figures using NRC (1989) milk proteins and the factor 0.65 would be 42, 34 and 28 g of DCP kg^{-1} of milk (Table 11.13 data) or 32 and 28 g of DCP kg^{-1} of milk using the recommended NRC (1989) values for early and late lactation. However, it is unlikely that lactating mares in the UK would ever receive less than 40–45 g of DCP kg^{-1} of milk produced, equivalent to 73–82 g of CP kg^{-1}. For example, a 500 kg mare offered a 50:50 hay (80 g of CP kg^{-1}) and compound (180 g of CP kg^{-1}) diet would consume at least 14.0 kg of dry food and, with an overall dietary CP of 130 g kg^{-1}, would obtain 1820 g of CP day^{-1}. The requirement for maintenance (650 g of CP) together with that for 15 kg of milk ($15 \times 80 = 1200$) totals 1880 g and this would be met by the above daily ration. The values obtained with the two systems agree fairly well.

AN OVERVIEW OF FEEDING SYSTEMS

For a feeding system to be effective, it must be possible to evaluate reliably the feeds available and then clearly define animal requirements. All that remains thereafter is to combine feeds in such a way that they provide nutrients in the quantity and proportion required by the animal.

Predicting Feed Value *In Vivo*

Metabolism studies are the classical means of determining apparently digestible nutrients; knowledge of endogenous losses enables the calculation of true digestibilities. Digestibility studies are easy to conduct and require minimal facilities. Collection of gas losses and urine from the animal enable the estimation of ME, but this type of procedure requires calorimeters and, therefore, few estimations have been made with horses. One such study was used to provide supportive data for the French horse NE system (Vermorel and Martin-Rosset, 1997) but relied on substantive assumptions in terms of the proportions of absorbed energy supplied by glucose and VFAs. Furthermore, it was assumed that DE and ME values of forages were similar when fed alone or in mixed diets. This assumption was based on the findings of Martin-Rosset and Dulphy (1987). Cuddeford *et al.* (1992) showed a non-linear increase in ration energy digestibility as alfalfa was substituted progressively with naked oats. However, this could have been a 'level of feeding effect' which might have affected the results because the rations used were adjusted on an energy basis rather than by equalizing DMI. Differences in the ME requirements for maintenance obtained in this study were explained on the basis of differences in the energy cost of eating, digestive tract metabolism and utilization of digestion end-products.

In vivo studies are the 'gold standard' against which other methods of evaluating feeds are judged. Simultaneous *in vivo* digestibility trials in horses and wethers were used by Martin-Rosset *et al.* (1984) to develop prediction equations so that *in vivo* ruminant data could be used to generate values for horses. The r^2 value for legumes was 0.71 with a residual standard deviation (RSD) of 2.6; comparable values for grasses were 0.96 and 2.3, respectively. Vander noot and Trout (1971) used *in vivo* studies in steers to predict values for horses, and found that, not surprisingly, crude fibre was the best predictor of DMD (r^2 = 0.81). However, predictions based on *in vivo* studies in different species are prone to error, and the INRA system is heavily dependent on such work.

Chemical Composition of Feeds

Vermorel and Martin-Rosset (1997) were unable to predict urine and methane losses reliably and so predicted ME from DE using equations (Equations 11.3 and 11.4) which relied on feed composition. However, very low r^2 values, 0.45 and 0.17, were associated with these equations, which must give cause for concern over their reliability. Martin-Rosset *et al.* (1996a) used multiple regression to predict OMD of forages from their chemical composition and found that the following relationship could be used:

$$OMD(\%) = 67.78 + 0.07088 \ CP - 0.000045 \ NDF^2 - 0.12180 \ ADL;$$
$$r^2 = 0.878 \tag{11.21}$$

These authors considered that this was a more reliable equation (RSD ± 2.5) than that used earlier by Martin-Rosset *et al.* (1984), which serves to illustrate the point that the INRA system can be improved.

In Vitro Systems

Applegate and Hershberger (1969) were probably the first to use *in vitro* digestion techniques to investigate forage digestibility in equids. Since that time, a number of different approaches have been developed.

Pepsin–cellulase
This method was used with 52 forages whose OMD had been determined *in vivo* by Martin-Rosset *et al.* (1996b). Using multiple regression, the following equation was derived:

$$OMD(\%) = 29.38 + X + 2.30315Y - 0.01384Y^2; \ r^2 = 0.927 \tag{11.22}$$

where X = 4.12 for green forages, 0 for grass hays and −2.61 for legume hays; Y = cellulase DMD%. This relationship had an RSD of 1.9 and was more reliable than that which depended on chemical composition above. Thus, for any feeding system, this enzymatic method deserves further study for the purpose of evaluating horse feeds.

Gas production

Macheboeuf *et al.* (1998) used the Menke and Steingass (1988) method with caecal fluid inocula to ferment the same feeds as used previously (see p. 260, Pepsin–cellullase). Two relationships were obtained, one for alfalfa hays and the other for green forages and grass hays. In the former, gas production after 24 h was the best predictor of OMD (r^2 = 0.76; RSD of 2.2), whereas in the latter case, gas production at 24 h together with CP was the best (r^2 = 0.87; RSD of 2.7). Using faeces as the source of inoculum improved the prediction of OMD for alfalfa hays (r^2 = 0.96; RSD of 0.9) from 'c', the rate of gas production (Macheboeuf and Jestin, 1998). The prediction for green forages and grass hays was as good with faecal inocula as with caecal inocula using the same parameters (r^2 = 0.86; RSD of 2.8). Lowman *et al.* (1997) used the method of Theodorou *et al.* (1994) with faecal inocula to predict dry matter digestibility values based on the *in vivo* values of 16 diverse feeds. The best predictive equation that was obtained used gas production parameters:

$$\text{DMD (g kg}^{-1}) = 155 + 6209 \text{ FRGP} + 1.505 \text{ GP}; \ r^2 = 0.72 \tag{11.23}$$

where FRGP is the fractional rate of gas production estimated when 50% of the gas has been produced, and GP is the total gas production.

It is clear that gas production methods can provide useful data although, so far, the enzymatic methods seem more reliable.

Near infrared spectrophotometry (NIRS)

NIRS is a routine laboratory procedure which is used extensively to evaluate forages for ruminants (see, for example, Baker and Barnes, 1990). Andrieu *et al.* (1996) applied the NIRS method to 52 forages that had been evaluated *in vivo* and obtained a prediction of OMD with an r^2 = 0.96 and an RSD of 1.8. They concluded that the NIRS method was as reliable as the enzymatic method. Thus, two techniques are available for determining OMD which is a crucial component in the calculation of UFC values.

Limitations

NRC (1989) maintenance energy requirements were calculated on the basis of experiments with four animals (Pagan and Hintz, 1986a) and from feeding trials which were the basis for NRC (1978). These feeding trials also provided the foundation for the INRA system (INRA, 1990) which subsequently has been validated using indirect calorimetry.

One UFC has a NE value of 9.414 MJ and is equivalent to 1 kg of 'standard barley' (sic) of 870 g of DM kg^{-1} given to horses at maintenance. The justification for using maintenance as the base is that 0.50–0.90 of total energy expended by horses is for maintenance purposes in lactating and pregnant mares, respectively (Doreau *et al.*, 1988), 0.60–0.90 in growing horses (Agabriel *et al.*, 1984) and 0.70–0.80 in working horses (Martin-Rosset *et al.*, 1994). In relation to growing animals, Vermorel and Martin-Rosset (1997) assert that only 0.20 of the total energy requirement is used for growth in light breeds, the balance being used for maintenance. The greatest differences in nutritive value between horse feeds are due to differences in OMD and are thus related to cell wall content or neutral detergent fibre (NDF). This is highly relevant since requirements are expressed in terms of DE or UFC; Table 11.14 illustrates the significances of this. It is clear that DE overvalues roughages; however, large effects are only seen at the extremes, cf. maize with straw. Vermorel and Martin-Rosset (1997) suggest that the NRC DE system overvalues cereal by-products, oil meals and hays by about 0.15, 0.25–0.30 and 0.30–0.35, respectively; starch-rich feeds are undervalued. In practice, the bias introduced by the DE system is not great when mixed diets are fed.

It is assumed that the horse's need for different functions, such as work and maintenance, are additive. Thus, the daily energy or protein need is arrived at using a factorial approach, summing the needs for maintenance, growth, etc. The INRA system (INRA, 1990) assumes that the efficiency with which ME is used for different functions is the same as that for maintenance (k_m).

This means, for example, that the utilization of ME for maintenance is the same as that for work ($k_m = k_w$); however, the increased heat production associated with work may well invalidate this assumption. In calculating requirements for pregnancy and lactation, k_c and k_l are assumed to be about 0.25 and 0.65, respectively. However, they have not been determined directly but rather have been extrapolated from pig and ruminant data. Using k_m to predict feed energy values for growing, pregnant or lactating animals will introduce errors. Vermorel and Martin-Rosset (1997) argue that in ruminants k_m/k_l is relatively constant whatever the feed. The relationship between the efficiency of utilization of ME for these processes may be similar, as in ruminants, but this remains to be proven in horses and, certainly, it is unlikely that nutrient utilization for growth will be particularly efficient. The coefficients adopted for use with a ruminant model (Agabriel *et al.*, 1984) do not different-iate between 18 and 36 months of age and do not account for the 0–6 month stage of growth.

The INRA system (INRA, 1990) assumes that NE requirements of horses in different situations have been described accurately, but this is not the case; validation experiments have only been conducted at or near maintenance

Table 11.14. DE (MJ kg^{-1} DM) and UFC (kg kg^{-1} DM) values of some common feeds and as a proportion of maize.

Feed	DE	DE feed/ DE maize	UFC	UFC feed/ UFC maize
Maize	16.1	1.00	1.33	1.00
Barley	14.5	0.90	1.16	0.87
Oats	12.5	0.78	0.99	0.74
Beet pulp	12.5	0.78	0.86	0.65
Lucerne	10.2	0.63	0.60	0.45
Grass hay	7.4	0.46	0.44	0.33
Barley straw	6.5	0.40	0.36	0.27

(Vermorel and Vernet, 1991). The INRA system assumes that UFCs are additive (Martin-Rosset and Dulphy, 1987) based on digestibility data. However, the system relies on estimating absorbed energy by predicting which nutrients are absorbed. No account is taken of feed interactions, in other words how different feeds and the way in which they are fed affect the site of digestion, extent of digestion and the resultant nutrient uptake. It has been demonstrated (Potter *et al.*, 1992) that both the origin of starch and the manner in which it has been processed affect the site of digestion and the magnitude of pre-caecal digestion; the latter will have a major impact on whether glucose or VFA is the substrate for subsequent ATP production. Recent work (McLean *et al.*, 1998) has shown that processing barley by micronizing or extrusion affects pre-caecal starch degradation, the molar proportions of VFAs in the caecum and caecal pH, all factors that will affect k_m values. Moore-Colyer *et al.* (1997a,b) have shown that considerable quantities of non-starch poly-saccharide (up to 149 g kg^{-1}) can disappear pre-caecally and, as a consequence, will affect the proportion of nutrients absorbed from the lumen of the hind-gut. Thus, until such time as those factors which affect the magnitude and site of nutrient uptake have been defined, the putative assumptions should be interpreted with care. In contrast, Vermorel and Martin-Rosset (1997) assert that large errors in the estimation of nutrient uptake have small effects on k_m (0.004), although how efficiency is affected in terms of productive function is not disclosed.

The NE of barley is deduced from the assumption that it contains 16.13 MJ of GE kg^{-1} and has a DM of 870 g kg^{-1}. The NE is determined by a step-wise procedure assuming the following relationships: DE/GE = 0.80, ME/DE = 0.931 and NE/ME = 0.785. The validity of using these conversion factors has been questioned (Harris, 1997) particularly as the NE value is assumed to be constant for whatever purpose the food energy is being used. The function of the animal, environment, nutrient content of the ration and other factors will affect the NE value of a food. For practical rations composed of 0.75 forage and 0.25 concentrate, Tisserand (1988) suggests the following relationships between energy values: DE/GE = 0.67–0.83, ME/DE = 0.83–0.91 and NE/ME = 0.63–0.80. With this magnitude of variation, it is difficult to accept that UFC values are additive (Martin-Rosset and Dulphy, 1987). Indeed, it seems to be a retrograde step to use a feed unit with no dimension in place of absolute energy values; the justification given is that feeds are compared more frequently in terms of substitution value.

In summary, the DE system relies on digestibility as being the most important factor for discriminating between feeds. The NE system separates concentrates and roughages and is based on end-product usage. As well as considering the efficiency of utilization of these end-products (glucose, lactate, VFAs, etc.), the NE system quantifies the impact of the energy costs of mastication, propulsion of food through the gut and heat of fermentation on k_m. For example, the k_m for barley might be 0.79 compared with only 0.60 for grass hay. Ultimately, however, the practical value of a feeding system will depend upon the reliability of feed evaluation methodologies and, in this context, roughages, succulents and fresh forages are poorly defined.

It is perhaps salutary to record the comments of Hintz and Cymbaluk (1994) who noted that direct comparisons between the NRC and INRA systems cannot be made easily or without several assumptions. Furthermore, the calculations, based on either system, when expressed in feed rather than in terms of DE, UFCs, etc. produced similar values. They took the example of a lactating 500 kg mare and, in spite of using different energy units, the final rations were similar. Frape (1998) compared the two systems for ration formulation and concluded that: (i) expected selection against poor hay by the NE system did not occur; (ii) the DE system generally assumed higher requirements for different functions which offset the higher values given to hay; and (iii) a change in assumed feed intake of working horses had a large effect on ration composition with either system.

The INRA system is flexible and can be modified and updated as new information comes to hand, although, at present, its complexity compared with NRC does not appear to confer any major advantage.

THE FUTURE

Whatever feeding system is used for horses, it cannot be effective if we do not understand the dynamic processes of digestion within the horse's gut. The recent application of ruminant techniques to investigate the *in situ* degradation of feeds in the caecum (Hyslop *et al.*, 1996a,b, 1997a,b; Stefansdottir, 1996; Stefansdottir *et al.*, 1996) has provided data which have been fitted to the exponential equation:

$$P = a + b\,(1 - e^{-ct}) \tag{11.24}$$

where 'a' is the rapidly soluble fraction, 'b' the slowly degradable fraction, 'c' is the fractional rate constant at which 'b' is degraded and P is the amount degraded at time t (Ørskov and McDonald, 1979). Effective degradability (ED) of feeds has been calculated at different fractional outflow rates (k) using the equation:

$$ED = a + bc\,(c + k) \tag{11.25}$$

Where a lag phase existed, degradation coefficients were fitted using the following equations:

$$y = a \qquad\qquad \text{for } t \leq \text{lag} \tag{11.26}$$

$$y = a + b\,(1 - e^{-c/t-\text{lag})}) \qquad \text{for } t > \text{lag} \tag{11.27}$$

where a, b and c are as defined before and lag is the period before degradation of b begins (Dhanoa, 1988). ED values were also calculated using the equation:

$$ED = ak + (a + b)c/(c + k) \tag{11.28}$$

This type of information on feed degradation together with knowledge of retention time of feed particles in different parts of the gut allows more precise predictions of the site and extent of digestion. The equine caecum is small, representing about 0.16 of the total tract, in contrast to the rumen, which can account for up to 0.70 (Argenzio, 1993). Information on digesta passage rate through the caecum is sparse and has been estimated at 20% h^{-1} ($k = 0.20$) (Howell and Cupps, 1950); typical rates through the rumen are 2–8% h^{-1} ($k = 0.02$–0.08) (AFRC, 1992). Two- to fourfold variations in daily caecal volume have been reported when animals were fed alfalfa (Goodson *et al.*, 1988) or a pelleted diet (Argenzio *et al.*, 1974). Clearly these changes will have a profound effect on degradation rates and nutrient uptake by the horse. Measurement of degradation rates *in situ* together with the use of markers to measure the rate of digesta passage will contribute to the production of a model for digestion in the horse.

The recent use of *in sacco* techniques (Hyslop and Cuddeford, 1996; Macheboeuf *et al.*, 1996; Longland *et al.*, 1997; Tomlinson, 1997) has complemented *in situ* studies and has contributed information on nutrient disappearances in different parts of the digestive tract. With these new methodologies, it is now possible to quantify the digestive process within the horse under different feeding regimes. The use of *in vitro* techniques that rely on faecal inocula (Lowman *et al.*, 1996) provide new, non-invasive methods for further characterizing feeds.

Against this background of development, it is feasible to produce 'designer' diets for which it will be possible to predict the site of digestion and subsequent nutrient availability. For example, horses used in flat-racing require glucose as a substrate for energy storage and release and would thus benefit from feedstuffs that are degraded pre-caecally. In contrast, horses used for endurance competitions require 'slow-release' energy which could be made available from the large intestine in the form of VFAs; appropriate mixes of raw materials could meet these diverse needs.

REFERENCES

Adam, I., Young, B.A., Nicol, A.M. and Degen, A.A. (1984) Energy cost of eating in cattle given diets of different form. *Animal Production* 38, 53–56.

Agabriel, J., Martin-Rosset, W. and Robelin, J. (1984) Croissance et besoins du poulain. In: Jarrige, R. and Martin-Rosset, W. (eds), *Le Cheval, Reproduction, Sélection, Alimentation, Exploitation*. INRA Publications, Versailles, France, pp. 371–384.

Agricultural and Food Research Council (1992) Technical committee on response to nutrients: nutritive requirements of ruminant animals. Report no. 9: protein. *Nutrition Abstracts and Reviews Series B*, 62, 787–835.

Allden, W.G. (1970) The effects of nutritional deprivation on the subsequent productivity of sheep and cattle. *Nutrition Abstracts and Reviews* 40, 1167–1184.

Anderson, C.E., Potter, G.D., Kreider, J.L. and Courtney, C.C. (1983) Digestible energy requirements for exercising horses. *Journal of Animal Science* 56, 91–95.

Andrieu, J., Jestin, M. and Martin-Rosset, W. (1996) Prediction of the organic matter digestibility (OMD) of forages in horses by near infra-red spectrophotometry (NIRS). In: *Proceedings of the 47th European Association of Animal Production Meeting*. Lillehammer, Norway, p. 299 (abstract).

Applegate, C.S. and Hershberger, T.V. (1969) Evaluation of *in vitro* and *in vivo* caecal fermentation techniques for estimating the nutritive value of forages for equines. *Journal of Animal Science* 28, 18–22.

Argenzio, R.A. (1993) Digestion, absorption and metabolism. In: Swanson, M.J. and Reece, W.O. (eds), *Duke's Physiology of Domestic Animals,* 11th edn. Comstock Publishing Associates, Ithaca, New York, pp. 325–335.

Argenzio, R.A., Lowe, J.E., Pickard, D.W. and Stevens, C.E. (1974) Digesta passage and water exchange in the equine large intestine. *American Journal of Physiology* 226, 1035–1042.

Austbø, D. (1996) Energy and protein evaluation systems and nutrient recommendations for horses in the Nordic countries. In: *Proceedings of the 47th European Association of Animal Production Meeting.* Lillehammer, Norway, p. 293 (abstract).

Baker, C.W. and Barnes, R. (1990) The application of near infra-red spectrometry to forage evaluation in the agricultural development and advisory service. In: Wiseman, J. and Cole, D.J.A. (eds), *Feedstuff Evaluation.* Butterworths, London, pp. 337–351.

Barth, K.M., Williams, J.W. and Brown, D.G. (1977) Digestible energy requirements of working and non-working ponies. *Journal of Animal Science* 44, 585–589.

Barton, A., Anderson, D.L. and Lyford, S. (1973) Studies of protein quality and quantity in the early weaned foal. In: *Proceedings of the 3rd Equine Nutrition and Physiology Symposium.* Gainesville, Florida, p. 19.

Benedict, F.G. (1938) *Vital Energetics.* Publication number 503, Carnegie Institute, Washington, DC.

Bigot, G., Trilland-Geyl, C., Jussiaux, M. and Martin-Rosset, W. (1987) Elevage du cheval de selle du sevrage au débourrage. *Bulletin Technical. Theix* 69, 45–53.

Blaxter, K.L. (1962) *The Energy Metabolism of Ruminants.* Hutchison and Co. Ltd, London.

Bochröder, B., Schubert, R. and Bödeker, D. (1994) Studies on the transport *in vitro* of lysine, histidine, arginine and ammonia across the mucosa of the equine colon. *Equine Veterinary Journal* 26, 131–133.

Breuer, L.H. (1968) Energy nutrition of the light horse. In: *Proceedings of the 1st Equine Nutrition and Physiology Symposium.* University of Kentucky, Lexington, Kentucky, p. 8.

Brody, S. (1945) *Bioenergetics and Growth.* Hafner Publishing Co., New York.

Cuddeford, D., Khan, N. and Muirhead, R. (1992) Naked oats – an alternative energy source for performance horses. In: *Proceedings of the 4th International Oat Conference.* Adelaide, South Australia, pp. 42–50.

Cymbaluk, N.F. and Christison, G.L. (1990) Environmental effects on thermoregulation and nutrition of horses. *Veterinary Clinics of North America: Equine Practice* 6, 355–372.

Dhanoa, M.S. (1988) On the analysis of dacron bag data for low degradability feeds. *Grass and Forage Science* 43, 441–444.

Doherty, O., Booth, M., Waran, N., Salthouse, C. and Cuddeford, D. (1997) Study of the heart rate and energy expenditure of ponies during transport. *Veterinary Record* 141, 589–592.

Doreau, M. and Boulot, S. (1989) Recent knowledge on mare milk production. *Livestock Production Science* 22, 213–235.

Doreau, M., Martin-Rosset, W. and Boulot, S. (1988) Energy requirements and the feeding of mares during lactation: a review. *Livestock Production Science* 20, 53–68.

Ellis, R.N.W. and Lawrence, T.L.J. (1979) Energy and protein under-nutrition in the weanling filly foal. *British Veterinary Journal* 135, 331–337.

Frape, D. (1998) *Equine Nutrition and Feeding*, 2nd edn. Blackwell Science Ltd, London.

Freeman, D.E. and Donawick, W.J. (1991) *In vitro* transport of cycloleucine by equine cecal mucosa. *American Journal of Veterinary Research* 52, 539–542.

Freeman, D.E., Kleinzeller, A., Donawick, W.J. and Topkis, V.A. (1989) *In vitro* transport of L-alanine by equine cecal mucosa. *American Journal of Veterinary Research* 50, 2138–2144.

Freeman, D.W., Potter, G.D., Schelling, G.T. and Kreider, J.L. (1988) Nitrogen metabolism in mature horses at varying levels of work. *Journal of Animal Science* 66, 407–412.

Gallagher, K., Leech, J. and Stowe, H. (1992a) Protein, energy and dry matter consumption by racing thoroughbreds: a field survey. *Journal of Equine Veterinary Science* 12, 43–48.

Gallagher, K., Leech, J. and Stowe, H. (1992b) Protein, energy and dry matter consumption by racing standardbreds: a field survey. *Journal of Equine Veterinary Science* 12, 382–388.

Geay, Y., Robelin, J., Beranger, C., Micol, D., Gueguen, L. and Malterre, C. (1978) Bovins en croissance et à l'engrais. In: Jarrige, R. (ed.), *Alimentation des Ruminants*. INRA Publications, Versailles, France, pp. 297–344.

Glade, M.J. (1983) Nutrition and performance of racing thoroughbreds. *Equine Veterinary Journal* 15, 31–36.

Goodson, J., Tyznik, W.J., Cline, J.H. and Dehority, B.A. (1988) Effects of an abrupt diet change from hay to concentrate on microbial numbers and physical environment in the caecum of the pony. *Applied and Environmental Microbiology* 58, 1946–1950.

Harris, P. (1997) Energy sources and requirements of the exercising horse. *Annual Review of Nutrition* 17, 185–210.

Hintz, H.F. and Cymbaluk, N.F. (1994) Nutrition of the horse. *Annual Review of Nutrition* 14, 243–267.

Hintz, H.F., Schryver, H.F. and Lowe, J.E. (1971) Comparison of a blend of milk products and linseed meal as protein supplements for young growing horses. *Journal of Animal Science* 33, 1274–1277.

Hörnicke, H., Meixner, R. and Pollmann, R. (1983) Respiration in exercising horses. In: Snow, D.H., Persson, S.G.B. and Rose, R.J. (eds), *Equine Exercise Physiology*, Granta Editions, Cambridge, UK, pp. 7–16.

Howell, C.E. and Cupps, P.T. (1950) Motility patterns of the caecum of the horse. *Journal of Animal Science* 9, 261–268.

Hyslop, J.J. and Cuddeford, D. (1996) Investigations on the use of the mobile bag technique in ponies. *Animal Science* 62, 647 (abstract).

Hyslop, J.J., Stefansdottir, G.J. and Cuddeford, D. (1996a) Degradation of protein and fibre in the caecum of ponies measured by the *in situ* technique. In: *Proceedings of the 47th European Association of Animal Production Meeting*. Lillehammer, Norway, p. 297 (abstract).

Hyslop, J.J., Stefansdottir, G.J. and Cuddeford, D. (1996b) The effect of incubation sequence on *in situ* degradation of concentrate feed components in the caecum of ponies. *Animal Science* 62 646 (abstract).

Hyslop, J.J., Jessop, N.S., Stefansdottir, G.J. and Cuddeford, D. (1997a) Comparative protein and fibre degradation measured *in situ* in the caecum of ponies and in the rumen of steers. In: *Proceedings of the British Society of Animal Science.* Scarborough, p. 121.

Hyslop, J.J., Jessop, N.S., Stefansdottir, G.J. and Cuddeford, D. (1997b) Comparative degradation *in situ* of four concentrate feeds in the caecum of ponies and the rumen of steers. In: *Proceedings of the 15th Equine Nutrition and Physiology Symposium.* Fort Worth, Texas. pp. 116–117.

Institut Nationale de la Recherche Agronomique (1978) *Alimentation des Ruminants.* Jarrige, R. (ed.), INRA Publications, Versailles, France.

Institut Nationale de la Recherche Agronomique (1984) Tables des apports alimentaires recommandés pour le cheval. In: Jarrige, R. and Martin-Rosset, W. (eds), *Le Cheval, Reproduction, Sélection, Alimentation, Exploitation.* INRA Publications, Versailles, France, pp. 645–660.

Institut Nationale de la Recherche Agronomique (1990) *L'Alimentation des Chevaux.* Martin-Rosset, W. (ed.), INRA Publications, Versailles, France.

Jarrige, R. and Tisserand, J.L. (1984) Métabolisme, besoins et alimentation azotée du cheval. In: Jarrige, R. and Martin–Rosset, W. (eds), *Le Cheval, Reproduction, Sélection, Alimentation, Exploitation.* INRA Publications, Versailles, France, pp. 277–302.

Jespersen, J. (1949) Normes pour les besoins des animaux: chevaux, porcs et poules. In: *Vème Congrès International de Zootechnie.* Paris, Vol. 2, Rapports Particuliers, pp. 33–43.

Kellner, O. (1926) *The Scientific Feeding of Farm Animals*, 2nd edn (translated by W. Goodwin). Duckworth, London.

Kleiber, M. (1961) *The Fire of Life.* John Wiley and Sons, Inc., New York.

Kronfeld, D.S. (1996) Dietary fat affects heat production and other variables of equine performance, under hot and humid conditions. *Equine Veterinary Journal* (Suppl. 22), 24–34.

KuVera, J.C., MacLeod, N.A. and Ørskov, E.R. (1989) Energy exchanges of cattle nourished by intragastric infusion of nutrients. In: van der Honing, Y. and Close, W.H. (eds), *Energy Metabolism of Farm Animals; Proceedings of the 11th Symposium.* European Association of Animal Production, Publication no. 43, Pudoc, Wageningen, The Netherlands, pp. 271–274.

Longland, A.C., Moore-Colyer, M.J.S., Hyslop, J.J., Dhanoa, M.S. and Cuddeford, D. (1997) Comparison of the *in sacco* degradation of the non-starch polysaccharide and neutral detergent fibre fractions of four souces of dietary fibre by ponies. In: *Proceedings of the 15th Equine Nutrition and Physiology Symposium.* Fort Worth, Texas, pp. 120–121.

Löwe, H. and Meyer, H. (1982) *Pferdesucht und Pferdefütterung.* Eugen Ulmer, Stuttgart, Germany.

Lowman, R.S., Theodorou, M.K., Longland, A.C. and Cuddeford, D. (1996) A comparison of the *in vitro* fermentation of four foods by inocula from bovine rumen digesta and equine caecal digesta using the pressure transducer technique. *Animal Science* 62, 683 (abstract).

Lowman, R.S., Theodorou, M.K., Dhanoa, M.S., Hyslop, J.J. and Cuddeford, D. (1997) Evalution of an *in vitro* gas production technique for estimating the *in vivo* digestibility of equine feeds. In: *Proceedings of the 15th Equine Nutrition and Physiology Symposium.* Forth Worth, Texas, pp. 1–2.

MacCormack, J.A. and Bruce, J.M. (1991) The horse in winter – shelter and feeding. *Farm Buildings Progress* 105, 10–13.

Macheboeuf, D. and Jestin, M. (1998) Utilisation of the gas test method using horse faeces as a source of inoculation. In: *In Vitro Techniques for Measuring Nutrient Supply to Ruminants.* Occasional Publication No. 22. BSAS, Penicuik, Edinburgh.

Macheboeuf, D., Marangi, M., Poncet, C. and Martin-Rosset, W. (1995) Study of nitrogen digestion from different hays by the mobile nylon bag technique in horses. *Annales Zootechnic* 44, 219 (abstract).

Macheboeuf, D., Poncet, C., Jestin, M. and Martin-Rosset, W. (1996) Use of a mobile nylon bag technique with caecum fistulated horses as an alternative method for estimating pre-caecal and total tract nitrogen digestibilities of feedstuffs. In: *Proceedings of the 47th European Association of Animal Production Meeting.* Lillehammer, Norway, p. 296 (abstract).

Macheboeuf, D., Jestin, M., Andrieu, J. and Martin-Rosset, W. (1998) Prediction of the organic matter digestibility of forages in horses by the gas test method. In: *In Vitro Techniques for Measuring Nutrient Supply to Ruminants.* Occasional Publication No. 22. BSAS, Penicuik, Edinburgh.

Martin, R.G., McMeniman, N.P., Norton, B.W. and Dowsett, K.F. (1996) Utilisation of endogenous and dietary urea in the large intestine of the mature horse. *British Journal of Nutrition* 76, 373–386.

Martin-Rosset, W. and Doreau, M. (1984) Besoins et alimentation de la jument. In: Jarrige, R. and Martin-Rosset, W. (eds), *Le Cheval, Reproduction, Sélection, Alimentation, Exploitation.* INRA Publications, Versailles, France, pp. 355–370.

Martin-Rosset, W. and Dulphy, J.P. (1987) Digestibility interactions between forages and concentrates in horses: influence of feeding level–comparison with sheep. *Livestock Production Science* 17, 263–276.

Martin-Rosset, W., Boccard, R., Jussiaux, M., Robelin, J. and Trilland-Geyl, C. (1983) Croissance relative des différents tissus, organes et régions corporelle entre 12 et 30 mois chez le cheval de boucherie de différentes races lourdes. *Annales Zootechnie* 32, 153–174.

Martin-Rosset, W., Andrieu, J., Vermorel, M. and Dulphy, J.P. (1984) Valeur nutritive des aliments pour le cheval. In: Jarrige, R. and Martin-Rosset, W. (eds), *Le Cheval, Reproduction, Sélection, Alimentation, Exploitation.* INRA Publications, Versailles, France, pp. 208–238.

Martin-Rosset, W., Vermorel, N., Doreau, M., Tisserand, J.L. and Andrieu, J. (1994) The French horse feed evaluation systems and recommended allowances for energy and protein. *Livestock Production Science* 40, 37–56.

Martin-Rosset, W., Andrieu, J. and Jestin, M. (1996a) Prediction of the organic matter digestibility (OMD) of forages in horses from the chemical composition. In: *Proceedings of the 47th European Association of Animal Production Meeting.* Lillehammer, Norway, p. 295 (abstract).

Martin-Rosset, W. Andrieu, J. and Jestin, M. (1996b) Prediction of the organic matter digestibility (OMD) of forages in horses by pepsin-cellulase method. In: *Proceedings of the 47th European Association of Animal Production Meeting,* Lillehammer, Norway, p. 294 (abstract).

McBride, G.E., Christopherson, R.J. and Sauer, W. (1985) Metabolic rate and plasma thyroid hormone concentrations of mature horses in response to changes in ambient temperature. *Canadian Journal of Animal Science* 65, 375–382.

McLean, B.M.L., Hyslop, J.J., Longland, A.C. and Cuddeford, D. (1998) Physical processing of barley and its effects on intra-caecal pH and volatile fatty acid parameters in ponies offered barley based diets. In: *Proceedings of the 2nd Annual Conference of the European Society of Veterinary and Comparative Nutrition.* Vienna, Austria, 61 (abstract).

Memedeikin, V.G. (1990) The energy and nitrogen systems used in the USSR for horses. In: *Proceedings of the 41st European Association of Animal Production Meeting.* Toulouse, France, p. 382 (abstract).

Menke, K.H. and Steingass, H. (1988) Estimation of the energetic feed value obtained from chemical analysis and *in vitro* gas production using rumen fluid. *Animal Research and Development* 28, 7–55.

Meyer, H. (1983) Protein metabolism and protein requirements in horses. In: *Proceedings of the 4th International Symposium on Protein Metabolism and Nutrition.* Clermont-Ferrand, France, Publication No. 16, INRA, Paris.

Meyer, H. and Ahlswede, L. (1976) Intrauterine growth and the body composition of foals and the nutrient requirements of pregnant mares. *Ubersichten zur Tierernährung* 4, 263–292.

Miraglia, N. and Olivieri, O (1990) Statement and expression of the energy and nitrogen value of feedstuffs in Southern Europe. In: *Proceedings of the 41st European Association of Animal Production Meeting.* Toulouse, France, p. 390 (abstract).

Moore-Colyer, M.J.S., Hyslop, J.J., Longland, A.C. and Cuddeford, D. (1997a) Degradation of four dietary fibre sources by ponies as measured by the mobile bag technique. In: *Proceedings of the 15th Equine Nutrition and Physiology Symposium.* Fort Worth, Texas, pp. 118–119.

Moore-Colyer, M.J.S., Hyslop, J.J., Longland, A.C. and Cuddeford, D. (1997b) The degradation of organic matter and crude protein of four botanically diverse feedstuffs in the foregut of ponies as measured by the mobile bag technique. In: *Proceedings of the British Society of Animal Science,* Scarborough, p. 120.

Moore-Colyer, M.J.S., Longland, A.C., Hyslop, J.J. and Cuddeford, D. (1998) The degradation of protein and non-starch polysaccharides (NSP) from botanically diverse sources of dietary fibre by ponies as measured by the mobile bag technique. In: *In Vitro Techniques for Measuring Nutrient Supply to Ruminants.* Occasional Publication No. 22, BSAS, Penicuik, Edinburgh.

Morrison, F.B. (1937) *Feeds and Feeding. Handbook for the Student and Stockman,* 20th edn. Ithaca, New York.

National Research Council (1978) *Nutrient Requirements of Horses,* 4th revised edn. National Academy of Sciences, Washington, DC.

National Research Council (1989) *Nutrient Requirements of Horses,* 5th revised edn. National Academy of Sciences, Washington, DC.

Nedkvitne, J.J. (1976) *Forelesningar om Fôring av Hestar.* Agricultural University of Norway, pp. 24–26.

Neuhaus, U. (1959) Milch und Milchgewinnung von Pferdestuten. *Zeitschrift für Tierzüchtungsbiologie* 73, 370–392.

Olsson, N.A. and Ruudvere, A. (1955) The nutrition of the horse. *Nutrition Abstracts and Reviews* 25, 1–18.

Ørskov, E.R. and MacLeod, N.A. (1990) Dietary-induced thermogenesis and feed evaluation in ruminants. *Proceedings of the Nutrition Society* 49, 227–237.

Ørskov, E.R. and McDonald, I. (1979) The estimation of protein degradability in the rumen from incubation measurements weighted according to the rate of passage. *Journal of Agricultural Science, Cambridge* 92, 499–503.

Osuji, P.O., Gordon, J.F. and Webster, A.J.F. (1975) Energy exchanges associated with eating and rumination in sheep given grass diets of different physical forms. *British Journal of Nutrition* 34, 59–71.

Ott, E.A. and Asquith, R.L. (1986) Influence of level of feeding and nutrient content of the concentrate on growth and development of yearling horses. *Journal of Animal Science* 62, 290–299.

Ott, E.A., Asquith, R.L., Feaster, J.P. and Martin, F.G. (1979) Influence of protein level and quality on the growth and development of yearling foals. *Journal of Animal Science* 49, 620–628.

Pagan, J.D. (1994) Digestibility trials provide evaluation of feedstuffs. *Feedstuffs* 66, 14–15.

Pagan, J.D. and Hintz, H.F. (1986a) Equine energetics. 1. Relationship between bodyweight and energy requirements in horses. *Journal of Animal Science* 63, 815–821.

Pagan, J.D. and Hintz, H.F. (1986b) Equine energetics. II. Energy expenditure in horses during submaximal exercise. *Journal of Animal Science* 63, 822–830.

Pagan, J.D., Hintz, H.F. and Rounsaville, T.R. (1984) The digestible energy requirements of lactating pony mares. *Journal of Animal Science* 58, 1382–1387.

Pagan, J.D., Essén-Gustavsson, B., Lindholm, A. and Thornton, J. (1987) The effect of dietary energy source on exercise performance in standardbred horses. In: *Equine Exercising Physiology 2*. Granta Editions, Cambridge, UK, pp. 686–700.

Potter, G.D., Evans, J.W., Webb, G.W. and Webb, S.P. (1987) Digestible energy requirements of Belgian and Percheron horses. In: *Proceedings of the 10th Equine Nutrition and Physiology Symposium*. Fort Collins, Colorado, pp. 133–138.

Potter, G.D., Gibbs, P.G., Haley, R.G. and Klenshoj, C. (1992) Digestion of protein in the small and large intestine of equines fed mixed diets. *Pferdeheilkunde* September 1992, 140–143.

Prior, R.L., Hintz, H.F., Lowe, J.E. and Visek, W.D. (1974) Urea recycling and metabolism of ponies. *Journal of Animal Science* 38, 565–571.

Reitnour, C.M. and Salsbury, R.L. (1976) Utilisation of proteins by the equine species. *American Journal of Veterinary Research* 37, 1065–1067.

Robelin, J. (1979) Influence de la vitesse de croissance sur la composition du gain de poids des bovins; variations selon la race et le sexe. *Annales Zootechnie* 28, 209–218.

Schmidt-Nielsen, K. (1984) *Scaling: Why Animal Size is so Important*. Cambridge University Press, Cambridge, UK.

Schryver, H.F., Meakim, D.W., Lowe, J.E., Williams, J., Soderholm, L.V. and Hintz, H.F. (1987) Growth and calcium metabolism in horses fed varying levels of protein. *Equine Veterinary Journal* 19, 280–287.

Slade, L.M., Robinson, D.W. and Casey, K.E. (1970) Nitrogen metabolism in non-ruminant herbivores. 1 – The influence of non-protein nitrogen and protein quality on the nitrogen retention of adult mares. *Journal of Animal Science* 30, 753–760.

Slade, L.M., Bishop, R., Morris, J.G. and Robinson, D.W. (1971) Digestion and absorption of ^{15}N-labelled microbial protein in the large intestine of the horse. *British Veterinary Journal*, 127, xi–xiii.

Smolders, E.A.A. (1990) Evolution of the energy and nitrogen systems used in The Netherlands. In: *Proceedings of the 41st European Association of Animal Production Meeting*. Toulouse, France, p. 386 (abstract).

Staun, H. (1990) Energy and nitrogen systems used in northern countries for estimating and expressing value of feedstuffs in horses. In: *Proceedings of the 41st European Association of Animal Production Meeting*. Toulouse, France, p. 388 (abstract).

Stefansdottir, G.J. (1996) Degradation of fibre-based foodstuffs in the caecum of ponies using the *in situ* technique. MSc thesis, University College of Wales, Aberystwyth, UK.

Steffansdottir, G.J., Hyslop, J.J. and Cuddeford, D. (1996) The *in situ* degradation of four concentrate feeds in the caecum of ponies. *Animal Science* 62, 646 (abstract).

Stillions, M.C. and Nelson, W.E. (1972) Digestible energy during maintenance of the light horse. *Journal of Animal Science* 34, 981–982.

Theodorou, M.K., Williams, B.A., Dhanoa, M.S., McAllan, A.B. and France, J. (1994) A simple gas production method using a pressure transducer to determine the fermentation kinetics of ruminant feeds. *Animal Feed Science and Technology* 48, 185–197.

Thomas, D.P. and Fregin, G.F. (1981) Cardiorespiratory and metabolic response to treadmill exercise in the horse. *Pflügers Archiv* 385, 65–70.

Thompson, K.N., Jackson, S.G. and Rooney, J.R. (1988) The influence of high planes of nutrition on skeletal growth and development of weanling horses. *Journal of Animal Science* 66, 2459–2467.

Thonney, M.L., Touchberry, R.W., Goodrich, R.D. and Meiske, J.C. (1976) Intraspecies relationship between fasting heat production and body weight: a re-evaluation of $W^{.75}$. *Journal of Animal Science* 43, 692–704.

Tisserand, J. L. (1988) Nutrition of the non-ruminant herbivores – horses. *Livestock Production Science* 19, 279–288.

Tomlinson, A.L. (1997) Voluntary feed intake and apparent digestibility *in vivo* of a mature threshed grass hay and an assessment of the mobile bag technique to study the dynamics of fibre degradation in the digestive tract of ponies. MSc thesis, University College of Wales, Aberystwyth, UK.

Ullrey, D.E., Struthers, R.D., Hendricks, D.G. and Brent, B.E. (1966) Composition of mare's milk. *Journal of Animal Science* 25, 217–222.

Vander noot, G.W. and Trout, J.R. (1971) Prediction of digestible components of forages by equines. *Journal of Animal Science* 33, 38–41.

Vermorel, M. and Martin-Rosset, W. (1997) Concepts, scientific bases structures and validation of the French horse net energy system (UFC). *Livestock Production Science* 47, 261–275.

Vermorel, M. and Mormède, P. (1991) Energy cost of eating in ponies. In: Wenk, C. and Boessigner, M. (eds), *Energy Metabolism of Farm Animals*. European Association of Animal Production Publication No. 58, pp. 437–440.

Vermorel, M. and Vernet, J. (1991) Energy utilisation of digestion end-products for maintenance in ponies. In: Wenk, C. and Boessinger, M. (eds) *Energy Metabolism of Farm Animals*. European Association of Animal Production Publication No. 58, pp. 433–436.

Vermorel, M., Jarrige, R. and Martin-Rosset, W. (1984) Métabolisme et besoins énergétiques du cheval, le système des UFC. In: Jarrige, R. and Martin-Rosset, W. (eds), *Le Cheval, Reproduction, Sélection, Alimentation, Exploitation*. INRA Publications, Versailles, France. pp. 239–276.

Vernet, J., Vermorel, M. and Martin-Rosset, W. (1995) Energy cost of eating long hay, straw and pelleted food in sport horses. *Journal of Animal Science* 61, 581–588.

Webster, A.J.F., Osuji, P.O., White, F. and Ingram, J.F. (1975) The influence of food intake on portal blood flow and heat production in the digestive tract of sheep. *British Journal of Nutrition* 34, 125–139.

Wolff, E. (1888) Principes de l'alimentation rationnelle due cheval; nouvelles série d'expériences réalisées à la Station d'Hohenheim en 1885–1886. *Annales des*

Sciences Agronomique 5ème année 2, 336–339. Translation by M. Margottet of the original article published in *Landwirtschaftliche Jahrboek* 1887, 3, 49–131.

Wolff, E., Funke, W., Kreuzhage, C. and Kellner, O. (1877) Pferde Futterungsversuche. *Landwirtschaftliche Versuch, Stu* 20, 125–168.

Wooden, G.R., Knox, K.L. and Wild, C.L. (1970) Energy metabolism of light horses. *Journal of Animal Science* 30, 544–548.

Wysocki, A.A. and Baker, J.P. (1975) Utilisation of bacterial protein from the lower gut of the horse. In: *Proceedings of the 4th Equine Nutrition and Physiology Symposium*. Davis, California, pp. 21–43.

12

Prediction of Response to Nutrients by Ruminants Through Mathematical Modelling and Improved Feed Characterization

D.E. Beever, J. France and G. Alderman

The University of Reading, Department of Agriculture, Reading, Berkshire, UK

INTRODUCTION

Effective prediction of how animals will respond to changes in their diet, management system and environment is desirable in most systems of animal production, whilst in some instances it is an essential prerequisite. For example, in closely integrated production systems such as intensively housed pigs and poultry, the existence of small financial margins combined with a high throughput of animals requires that the system of production achieves high accuracy of prediction, with minimization of all possible errors. Establishment of the objectives, in terms of live weight gain, carcass composition or egg output, has to be reconciled before the most satisfactory nutritional and management regimes can be elucidated, and failure to match observed performance with predicted response can have major financial implications. In ruminant production, whilst the specifications in many aspects are less exacting, the need to meet stated objectives is also of crucial importance. Thus, whether the system is concerned with a herd of high yielding dairy cows in which the aim is to meet a pre-determined quota of milk production, or growing heifers fed on low-quality forages but with the objective to meet a satisfactory mating weight at a pre-determined time, it is clear that once the objectives have been set, the system of production which is adopted should be capable of meeting expectations.

This chapter will examine the potential of mathematical models to predict animal response in systems of ruminant livestock production. A resumé of the elementary principles involved in the construction and use of mathematical models will be given. A brief review of current empirically based factorial models for calculating energy and protein requirements will be presented, highlighting their usefulness but also their increasing limitations as more exacting demands are made of ruminant production systems. The opportunities to revise these models will be examined, the potential of mechanistic models as alternatives will be advanced, and current progress in the use of mechanistic models to predict nutrient supply and animal performance will be considered, indicating the evolutionary nature of the subject. From this it will be established that current descriptions of feedstuffs are inadequate, and the need to improve the description of both feeds and animals will be indicated, with some suggestions for future developments in these areas.

PRINCIPLES OF MATHEMATICAL MODELLING

Science is concerned with prediction. In order to predict, models or conceptual schemes of the real world are needed. Indeed, models are the most potent way of expressing our ideas or theories of how the world works and, as in most areas of science, the language of expression is mathematics. Some consideration of principles of mathematical modelling is therefore necessary in any treatise on the prediction of animal response. In a scientific investigation, the objectives of the study should determine what modelling approach, if any, is to be used. It is therefore important in the first instance to realize how the different types of model relate to one another and to the structure of the problem under investigation.

Fundamental to mathematical modelling in biology is the concept of organizational hierarchy; a typical scheme for the animal sciences is as follows:

Level	Description of level
i + 3	Collection of organisms (herd, flock)
i + 2	Organism (animal)
i + 1	Organ
i	Tissue
i − 1	Cell
i − 2	Organelle
i − 3	Macromolecule

This scheme may be continued in both directions and, for ease of exposition, the different levels are labelled i + 1, i, i − 1, etc. It can be used to highlight the essential differences between the two principal approaches to mathematical modelling used in science: empirical modelling and mechanistic modelling.

Empirical models are models in which observational data are used to suggest the approximating quantification, both structural and numerical. Thus an empirical model sets out principally to describe, and is based on observation and experiment and not necessarily on any pre-conceived biological theory. The empirical approach is one looking at a single level in the organizational hierarchy (e.g. whole animal) and deciding what form of equation (or set of equations) can be used as a mathematical model and fitting it to the data. The approach derives from the philosophy of empiricism and adheres to the methodology of statistics.

Mechanistic models, on the other hand, are theory led, and relate processes and responses at different hierarchical levels (e.g. whole animal and tissue biochemistry). A mechanistic model attempts to give an understanding of causation and is concerned with the mechanisms present in a system. It is constructed by looking at the system, dividing it into its key components and attempting to analyse the whole system in terms of these key components and their interactions with one another. Mechanistic modelling follows the traditional reductionist method of the physical and chemical sciences. Thus, the mechanistic modeller attempts to describe level i behaviour in terms of level i – 1 attributes, whereas the empirical modeller describes level i behaviour in terms of attributes at level i.

Consideration of dynamic, deterministic, mechanistic modelling is central to this review, and there is a mathematically standard way of representing such models called the rate:state formalism. The system under investigation is defined at time t by q state variables: X_1, X_2, ..., X_q. These variables represent properties or attributes of the system, such as microbial mass, quantity of substrate, etc. The model then comprises q first-order differential equations which describe how the state variables change with time:

$$dX_i/dt = f_i(X_1, X_2, ..., X_q; P); i = 1, 2, ..., q \qquad (12.1)$$

where P denotes a set of parameters, and the function f_i gives the rate of change of the state variable X_i. The function f_i comprises terms which represent the rates of processes (with dimensions of state variables per unit time), and these can all be calculated from the values of the state variable alone, together with the values of any parameters and constants. The rate:state equations (Equation 12.1), are not as restrictive as might first be thought, as any higher order differential equation can be written as, and many partial differential equations well approximated by, a series of first-order differential equations.

In this type of mathematical modelling, the rate:state equations are formed through direct application of the laws of science (e.g. the law of mass conservation, the first law of thermodynamics) or by application of a continuity equation derived from more fundamental scientific laws. The rate:state equations can sometimes be solved analytically. Most models are too complex, however, and only numerical solutions can be obtained. Further discussion of the principles of mathematical modelling can be found in France and Thornley (1984) and Forbes and France (1993).

UK RUMINANT FEED REQUIREMENT SYSTEMS

In the UK, the energy requirement scheme for ruminants is based on the metabolizable energy (ME) system first proposed by Blaxter (1962), adopted by the Agricultural Research Council (ARC, 1965), implemented by the Ministry of Agriculture, Fisheries and Food (MAFF, 1975) and revised by ARC (1980) and the Agriculture and Food Research Council (AFRC, 1990). Protein requirements were expressed as digestible crude protein by Evans (1960) and MAFF (1976), and remained the unit in use until 1992. A scheme based on total absorbed amino acid nitrogen (TAAN) entering the duodenum was proposed by ARC (1980) and revised by ARC (1984). This scheme was not adopted for use in the UK since it predicted lower crude protein dietary levels for dairy cows than were needed in practice. A revised scheme based on metabolizable protein (MP) was proposed by AFRC (1992) and has now been adopted for use alongside the revised ME system of AFRC (1990) and AFRC (1993).

Although ME is a nutritionally accepted entity for energy, limitations to its use include the methods used to estimate its content in feedstuffs and non-recognition of the composition of energy-yielding substrates. Furthermore, the empirical relationships used to predict the efficiencies of ME utilization for maintenance (k_m) and lactation (k_l) in dairy cows from the metabolizability of the diet (q_m) suggest that, over the range of diet metabolizabilities likely to be encountered with dairy rations, variations in k_m and k_l are small, less than those which might be expected from biochemical considerations, when the nutrient make-up of the ME is taken into account. Similarly, there is debate over the applicability of the equation used to predict the efficiency of energy utilization for fattening and growth (k_f) in growing animals, derived from calorimetry studies, and their discrepancy with those derived from comparative slaughter studies which continues to give cause for concern (Thomas *et al.*, 1988). For both growing and lactating animals, however, the major weakness of the ME system is its inability to predict the composition of the final product (meat or milk). With the dairy cow, although prediction of total energy balance is generally satisfactory, significant and often unpredictable changes in energy partition between lactation and body tissue synthesis have been observed (Thomas *et al.*, 1987). It is impossible to determine to what extent this phenomenon is due to the inability of the ME system to deal with energy–protein interactions, changes in nutrient intake or animal hormone levels.

Like the ME system, the metabolizable protein system of AFRC (1992) is based on a factorial approach aimed at matching net protein requirements with dietary crude protein intake. It recognizes the importance of ruminally degradable and undegradable protein but, apart from representation of an energy–protein interaction in the rumen by estimation of the fermentable ME content of the diet, it fails (like the ME system) to recognize the importance of interactions elsewhere in the animal, such as with energy supply, as in the Dutch DVE system (Subnel *et al.*, 1994). Reservations have been expressed about the accuracy or constancy of some of the values assumed in the MP system, and its inability to represent adequately important factors such as the effects of frequency of feeding and type of supplement, which may have a major bearing on aspects such as microbial protein synthesis and the efficiency of utilization of absorbed amino acids. The net ruminal supply of effective ruminally degradable protein (and thus the supply of undegradable dietary protein to the small intestine) is predicted to vary according to level of feeding and predicted solid outflow rate from the rumen. However, the efficiency of utilization of absorbed amino acids is varied only as a function of amino acid supply by the introduction of the term 'relative value' (RV), which is assigned a different and constant value for each synthetic process; growth, pregnancy or lactation. Variations in the composition of the absorbed amino acid supply and the availability of non-amino acid substrates are not taken into account.

In a re-analysis of the experiment of Whitelaw *et al.* (1986), MacRae *et al.* (1988) indicated that for three increments of duodenally infused casein the incremental response in net productive nitrogen (milk + body nitrogen) was linear (0.64 g g^{-1} of N supplied) and close to expectations as outlined by ARC (1980) and AFRC (1992), but the relationship between milk N synthesis and N intake was curvilinear, with partial efficiencies for the three casein levels of 0.45, 0.25 and 0.16. Again, this indicates that present systems cannot adequately predict the partition of extra nutrients between milk synthesis and body metabolism from a measure of nutrient supply to the dairy cow (see Chapter 6 for details). Close study of the protein system for growing animals has not occurred, but reservations raised by Beever *et al.* (1990) in relation to the observed efficiency of amino acid utilization being lower (0.53) than the value currently adopted (0.59) and the effect of nutrition on the protein content of tissue gain both merit further investigation.

A number of reasons can be suggested to support the contention that modification of current systems is likely to have only limited success in the search for more reliable predictive routines:

- the failure of both the energy (ME) and protein (MP) systems to represent energy–protein interactions adequately
- their inabilities to examine the response of animals to nutritional changes
- failure to predict product composition or nutrient partition (specifically in dairy cows)
- the non-recognition of the nutrient composition of ME
- the use of a number of debatable assumptions and constants

can all be advanced as valid reasons for investigating alternative approaches.

MECHANISTIC MODELS OF NUTRIENT SUPPLY AND ANIMAL PERFORMANCE

In principle, there are at least four distinct areas which need to be considered in reviewing the development of mechanistic models of nutrient supply and interaction, and animal performance and productivity, namely intake regulation, rumen function, dairy cow metabolism and metabolism in the growing animal.

Intake Regulation

It is apparent that the level of animal productivity will be influenced by the amount of feed consumed, which will be a function of both the composition of the diet and the physiological state of the animal. Both the rate at which feed is degraded in the rumen and the degree of physical fill and metabolism in the animal will influence dietary intake, whilst with feeds such as grass silage it is recognized that the end-products of the fermentation processes which occurred during ensiling may have an important regulatory effect on the control of dietary intake (Buchanan-Smith, 1982). To date, intake modelling has been largely empirical. For example, Rook (1991), through a series of statistical procedures, proposed a number of empirical equations based on a range of experimental data sets and concluded that without improved feed description, progress with empirical models to predict voluntary feed intake would be restricted.

By comparison, few mechanistic intake models have been developed. The most recent attempt is the model developed by Poppi *et al.* (1994) to examine physical and metabolic limits to intake (see Chapter 13 for details). It used a metabolism submodel, based on the simplified model of nutrient utilization and body composition in growing cattle by France *et al.* (1987), to investigate metabolic factors involved in intake regulation, these being genetic limits to protein deposition, heat dissipation and ATP degradation via substrate cycling. The metabolism submodel relied on a knowledge of the profile of absorbed nutrients and utilized kinetics of known metabolic transactions to predict protein and fat deposition, heat production and ATP balance. However, the degree of representation of metabolism is such that the model is inappropriate for low-quality diets where body tissue has to be mobilized. Intake may also be limited by physical means, and three factors were investigated, namely rate of intake, rumen turnover of plant cell wall and faecal output. The determination of rumen turnover utilized a calculation based on fractional digestion rate, extent of digestion for escape to occur and maximum rumen fill. The approach was to identify pathways (factors) that could limit intake (six identified), calculate the maximum intake possible for each pathway and in doing so identify the pathway with the lowest intake, which became the predicted intake. Seven diets were chosen for examination, ranging from low-digestibility grass to a predominantly cereal diet. In general, the model behaved satisfactorily predicting intakes in line with those reported in the literature. Two major points arose. Firstly, for some diets, more than one pathway may be approaching its maximum intake simultaneously. This means that removal of one of the constraining pathways by some experimental means may not lead to the expected increase in intake, and hence will result in incorrect rejection of the hypothesis that this pathway was involved in intake control. Secondly, ATP degradation featured as a limiting pathway for most diets and reflects the importance that balance of nutrients has on intake regulation.

Rumen Function

By contrast, a larger mechanistic modelling effort has occurred in relation to the processes of rumen function (see Chapter 13 for details). Stimulated by the work of Baldwin *et al.* (1970), early efforts by Baldwin *et al.* (1977), Black *et al.* (1980/81) and France *et al.* (1982) were confined to attempts to simulate rumen function in adult sheep. Each of these models required an improved description of the feedstuffs, but parallel efforts by Reichl and Baldwin (1975) to provide a comprehensive description of the interactions of different microbial species were rejected on the grounds of complexity, and in all cases microbial biomass was represented as a single pool. In a series of tests of the model by Black *et al.* (1980/81), Beever *et al.* (1980/81) demonstrated the logical behaviour of the model but identified areas where its ability to predict reality was limited. As a research tool, however, this model was useful. It demonstrated the importance of ATP for maintenance of the microbial population, and illustrated the consequence, in terms of the yield of end-products of digestion, of an asynchrony between nutrient availability and microbial growth. The major contribution of France *et al.* (1982) was to promote a mathematical refinement based on the rate:state formalism of the earlier model by Black *et al.* (1980/81). A series of differential equations were proposed to describe the nine state variables, and improved representations of both microbial growth and lysis within the rumen were provided. Once again, this model was based on data from mature sheep and, when compared against experimental data obtained by Beever *et al.* (1981) for forages of contrasting physical form, gave reasonable qualitative agreement. To check model behaviour, France *et al.* (1982) calculated total carbon and nitrogen balance and found satisfactory agreement between the outputs and the respective inputs, but more detailed examination of the model revealed that the predicted mass of microbial material in the rumen was less than expected.

Whilst these early models made valuable contributions towards the overall objective of predicting nutrient supply, it was apparent that none of them were totally satisfactory in achieving this. Baldwin *et al.* (1987b) identified a number of weaknesses, in particular that none of the models attempted to predict nutrient supply in dairy cows. The model of the rumen of the dairy cow by Baldwin *et al.* (1987b) was part of a larger initiative to develop a model capable of examining whole lactational performance, and to achieve this objective three models were proposed (Baldwin *et al.*, 1987a,b,c) (see Chapter 14 for details). The model of rumen function was based on the previous efforts of France *et al.* (1982), but with a more comprehensive description of the chemical and physical characteristics of the diet. Both large and small particles and a water pool were explicitly represented, with ingested nutrients capable of entering any of these pools depending upon the proportion of large:small particles in the feed and the solubility of dietary nutrients. Large particles were assumed to be converted to small particles at a rate related to the extent of rumination, small particles were assumed to either pass out of the rumen or enter the water pool as a result of hydrolysis, whilst soluble nutrients were assumed to either undergo fermentation or pass out of the rumen. Only one microbial pool was represented within the model but microbial affinities were assigned for specific nutrients to distinguish between microbes utilizing starch or fibre, in order to control interactions between specific nutrients.

Russell *et al.* (1992) constructed a static, deterministic model of the rumen, using mostly empirical and some mechanistic elements, as part of the Cornell Net Protein and Carbohydrate model (CNCPS) (see Chapters 7 and 8 for details). They assumed two microbial groups, non-structural carbohydrate (NSC) and structural carbohydrate (SC) fermenting rumen bacteria, whose rate of growth was determined primarily by the amounts of feed carbohydrate degraded whilst feed was retained in the rumen. Growth of NSC bacteria was not limited by ammonia supply in the rumen, but SC bacteria were dependent upon peptides for 0.66 of their total N supply for maximum growth. Protozoa were not identified as a separate group in the rumen, but predicted microbial growth of both SC and NSC bacteria was reduced by 0.2 to account for protozoal depredation. Microbial growth was also limited when effective neutral detergent fibre (NDF) content (Mertens, 1985) fell below 20%, based on predicted reductions in rumen pH value. Microbial growth was not influenced by microbial mass or solid or liquid outflow rate, except in so far as these affected the amounts of carbohydrate degraded by the rumen bacteria.

The most recent and farthest developed model of rumen function is that by Dijkstra and co-workers (Dijkstra *et al.*, 1992; Neal *et al.*, 1992; Dijkstra, 1994), who built on the achievements of Baldwin *et al.* (1987b) primarily by improving microbial dynamics through distinguishing microbial categories, with specific emphasis on protozoa (see Chapter 13 for details). The model is driven by continuous inputs of nutrients, and currently consists of 17 state variables, which represent the nitrogen, carbohydrate, fatty acid and microbial pools in the rumen. Several protozoal characteristics are represented, including: preference for utilization of starch and sugars compared with fibre and of insoluble compared with soluble protein; engulfment and storage of starch; no utilization of ammonia to synthesize amino acids; engulfment and digestion of bacteria and protozoa; selective retention within the rumen; and death and lysis related to nutrient availability. The model was tested thoroughly on a range of diets and generally provided reasonable predictions, particularly of nutrient flows, with the exception of patterns of volatile fatty acid (VFA) production, i.e. acetate:propionate:butyrate ratios.

Dairy Cow Metabolism

Substantial mechanistic modelling effort has also occurred in relation to post-digestive metabolism. Significant amongst attempts for the dairy cow are those by Baldwin and co-workers and by Danfær (1990) (see Chapter 14 for details). In the model by Baldwin *et al.* (1987a), based on an earlier biochemical model of the lactating cow devised by Smith (1970), representation of hormonal regulation of nutrient utilization was attempted whilst recognizing that this aspect of metabolism is still not fully understood. Predicted glucose concentration compared with a standard glucose concentration was used as an index of hormonal activity, with catabolic hormones affecting the maximum reaction rate (V_{max}) and anabolic hormones affecting the reaction affinity constant (K). Protein metabolism in the viscera and in the lean body were represented separately, along with adipose tissue as the other major body tissue, and the mammary gland. The distinction between the body protein reserves was included to represent phenomena known to occur within the cow, especially in early lactation where visceral tissues will proliferate (Lindsay, 1993) whilst lean body will change much less dramatically (Gibb *et al.*, 1992) and will almost certainly undergo some depletion before subsequent repletion in later lactation.

The behaviour of both the metabolism and rumen models by Baldwin *et al.* (1987a,b) was sufficiently encouraging for Baldwin *et al.* (1987c) to produce a combined model of nutrient digestion and utilization. In fact, two combined models were produced, the first requiring a small integration interval and suitable for simulating the within-day dynamics of nutrient supply and metabolism. By inflating pool sizes and decreasing fast rate constants to overcome stiffness (France *et al.*, 1992), this model subsequently was modified so as to require an integration interval of 1 day and used to simulate full lactation cycles. In examining the effect of plane of nutrition and dietary protein content on milk output between weeks 2 and 26 of lactation, Baldwin *et al.* (1987c) predicted a lower and early peak yield on the low plane of nutrition and, when dietary energy and protein intakes were changed at week 12, the reduction in milk yield was greatest with the imposed reduction in energy intake (Fig. 12.1).

In a subsequent analysis, energy balance was examined in early and mid-lactation cows by considering a 50:50 forage:concentrate diet at different flat rates of feeding plus a standard allocation per unit of milk secreted. In early lactation, reducing flat rate feeding from 8 to 5 kg day⁻¹ caused a

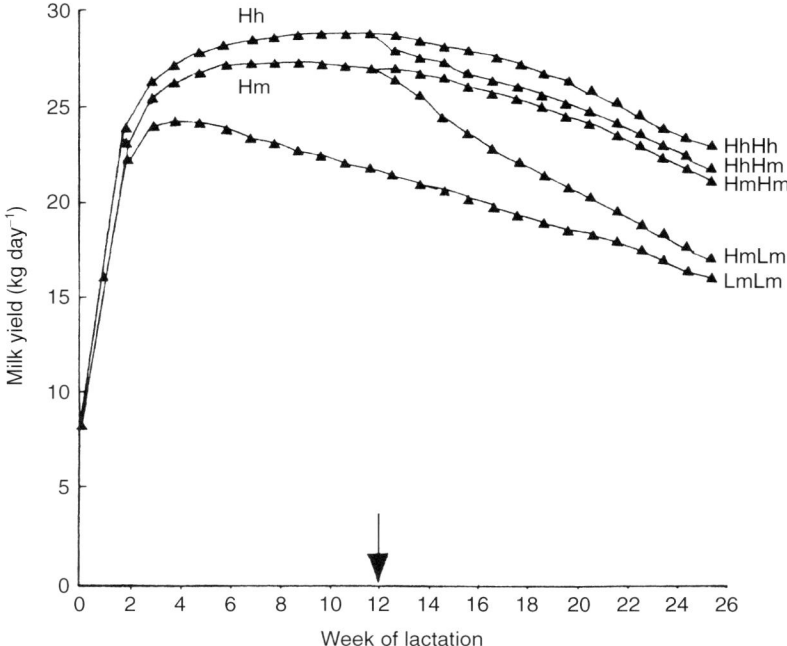

Fig. 12.1. Predicted effect of different feeding strategies upon lactational performance (Baldwin *et al.*, 1987c). L and H refer to low and high plane of nutrition, and m and h refer to medium and high protein respectively. ↓ Indicates diet change at week 12.

significant increase in body energy loss (23.2 MJ day^{-1}) with only a marginal
reduction in milk energy secretion. Partial efficiencies of milk secretion
were calculated to be 0.75 and 0.78 MJ of milk MJ^{-1} energy available above
maintenance, with calculated maintenance costs of 0.73 and 0.70 MJ day^{-1}
kg$^{-0.75}$ and efficiencies of ME utilization for maintenance of 0.78 and 0.81.
However, model behaviour with respect to the effect of high-concentrate feeding on milk fat secretion was less satisfactory. Changing from a high
forage-based diet which gave a milk fat content of 35.9 g kg^{-1} to a high
concentrate-based diet reduced fat content by only 7.0 g kg^{-1}, less than
expected in light of the severity of the imposed dietary change. As expected,
the predicted ratio of absorbed acetate (+ butyrate) to propionate declined
from 3.5 to 2.3 with the introduction of high-concentrate feeding, whilst
potential glucose supply increased from 7.6 to 9.6 mol day^{-1}. In contrast, the
supply of C$_2$ units as potential precursors for milk fat synthesis only fell from
55 to 50 mol day^{-1}, which may not have been sufficient to cause the expected
decline in milk fat synthesis and content. The VFA yields predicted by Baldwin
et al. (1987b) did not concur with observed values in every dietary situation
examined, possibly suggesting that representation of hexose disposal in the
rumen or stoichiometric yields of rumen VFAs used in the model are not totally
adequate. On the other hand, to suggest that the problem resides only with
prediction of supply could be misleading, and the representation of factors
controlling the partition of fat synthesis between adipose tissue and the
mammary gland should also be re-assessed.

The model of dairy cow metabolism by Danfær (1990) is at a similar level
of aggregation to that adopted by Baldwin and co-workers, but with a more
distinct and detailed representation of specific tissues and individual reactions.
The model has nine identifiable compartments (rumen, intestinal lumen, intestinal wall, liver, peripheral blood plus extracellular fluids, as well as mammary,
adipose, muscle and other tissues), has 77 state variables, and contains over
1500 equations. Full testing of the model has not been possible, and its present
size may prove to be a major weakness. Danfær (1990) compared high-starch
and high-protein diets using the model (Table 12.1).

Despite different protein intakes on the two diets, the yield of microbial
protein was highest on the starch diet, such that amino acid absorption was
similar for both diets. The starch diet, however, gave a substantially higher net
absorption of propionate and it was estimated that 0.77 of glucose irreversible
loss in the body was derived from propionate. In contrast, propionate
accounted for only 0.61 of glucose flux on the high-protein diet, with a
substantial amount being derived from gluconeogenic amino acids. The
starch diet promoted a higher milk yield but a much reduced fat content and
net synthesis of fat, which were more in line with expectations than the
simulations of Baldwin *et al.* (1987c).

Furthermore, the high-protein diet gave a higher milk protein content and
a similar milk protein yield to the high-starch diet, despite the suggestion of
enhanced catabolism of absorbed amino acids for the synthesis of glucose on
this diet. Milk energy outputs were calculated from the simulated data using
the equation of Tyrrell and Reid (1965), assuming that milk from both diets

Table 12.1. Effect of high-starch and high-protein diets on energy and protein utilization by lactating dairy cows as predicted by Danfær (1990).

	High starch	High protein
Daily intakes (kg day^{-1})		
Dry matter	18.6	18.6
Soluble carbohydrates	0.63	4.02
Starch	6.44	0.82
Cell walls	7.38	7.37
Protein	2.31	3.59
Absorption (kg day^{-1}):		
Propionate	2.59	1.82
Glucose	0.15	0.14
Amino acids	2.33	2.50
Milk output (kg day^{-1}):		
Total	33.7	29.9
Fat	1.08	1.28
Protein	0.98	1.03
Energy (MJ day^{-1})	91	98

had a mean lactose content of 47 g kg^{-1}. Milk energy content was 21% higher on the high-protein diet though milk energy yield was only 8% higher, a difference attributable to the enhanced mobilization of tissue observed on the high-protein diet.

Metabolism in the Growing Animal

Mechanistic modelling effort in relation to metabolism in the growing ruminant has focused primarily on sheep rather than cattle, with significant attempts having been made by Gill and co-workers, Sainz and Wolff (1990) and Gerrits *et al.* (1997a,b). In order to reconcile some of the inadequacies of the ME system discussed earlier, Gill *et al.* (1984) undertook the development of a mechanistic model to simulate the metabolism of absorbed energy-yielding nutrients in young sheep which encompassed nutrient interactions and the metabolic fate of individual nutrients. The model, which had 12 state variables, utilized standard expressions from enzyme kinetics (principally of the Michaelis–Menten form) for the flux equations and, whilst hormonal regulation of metabolism was not specifically represented, a number of modifiers to either enhance or inhibit individual reactions were included in these equations. Model input was based on six absorbed nutrients, namely the three principal VFAs, amino acids, glucose and lipid. In addition to these six nutrients, the model included ATP, reduced nicotinamide adenine dinucleotide phosphate (NADPH), triglyceride, body and wool protein and glycogen as important body pools. In a series of simulations, Gill *et al.* (1984) showed that increasing the protein energy content from 0.10 to 0.25 and glucose energy from 0 to 0.05 of total absorbed energy, whilst the lipid contribution and VFA proportions (but not VFA energy) were held constant, gave marked effects with respect to estimates of efficiency of energy utilization. Simulation at two levels of energy intake illustrated that the efficiency of energy utilization declined with increased energy intake (Fig. 12.2), challenging the concept of a fixed efficiency at all feeding levels, and showed that the estimates of efficiency were significantly affected by the composition of the absorbed energy.

A consistent positive response was established with respect to increased glucose supply, whereas, with increased protein, the effect on efficiency was positive initially but subsequently became negative, and appeared to be influenced by the level of glucose supplied. In these simulations, it is possible that the highest level of protein supplied exceeded the animal's protein requirement, with catabolism of the surplus protein contributing to the reduction in efficiency, whilst the interaction with glucose may be related to gluconeogensis from amino acids which, in some situations, could be of quantitative significance.

Subsequently, Black *et al.* (1987a) used the model to examine the control of acetate utilization in growing sheep. By varying the nature of the end-products of digestion, values for the proportion of additional acetate energy which was retained in the body, k_{Ac}, were obtained (Table 12.2). At both levels of glucose supply, the efficiency of acetate utilization increased in response to increased protein supply, whilst a marked response to increased glucose supply was observed at each level of protein. Those diets which gave low values of k_{Ac} were associated with a considerable flux of ATP through the

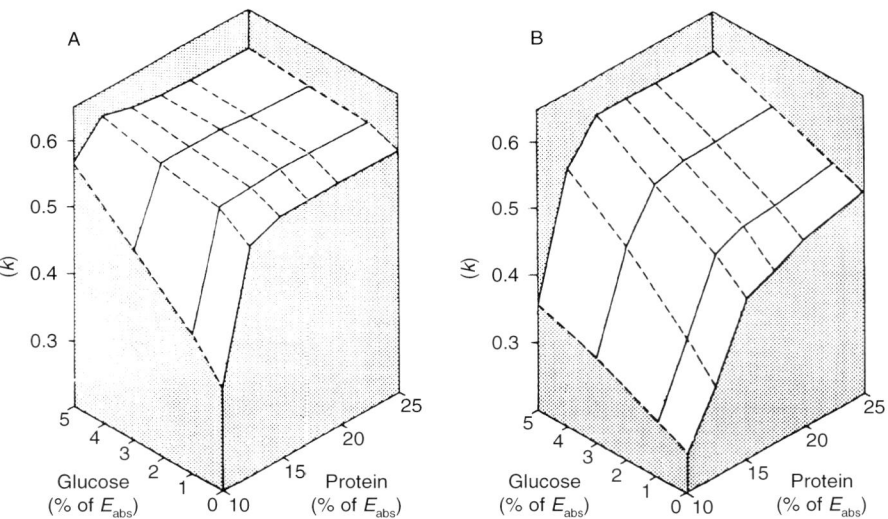

Fig. 12.2. Simulated response in efficiency of energy utilization for growth (k) to protein and glucose (both as % of absorbed energy) increasing from 10 to 25% and 0 to 5%, respectively, at ME intakes increasing from (A) 8.7 to 9.6 and (B) 9.6 to 10.5 MJ day⁻¹ (Gill *et al.*, 1984).

Table 12.2. The effect of variations in the composition of the end-products of digestion on efficiency of utilization of added acetate (2 mol day^{-1}) as predicted by Black *et al.* (1987a).

Glucose level (%)[a]	Protein level (%)[a]			
	10.0	12.5	15.0	20.0
1.0	0.16	0.26	0.36	0.43
5.0	0.35	0.47	0.55	0.56

[a]Calorific value of absorbed protein (glucose) as a percentage of total absorbed energy in the non-supplemented (acetate) diet.

degradation pathway (i.e. substrate cycles) related to an inhibition in acetyl coenzyme A (acetyl-CoA) utilization and an increase in its oxidation.

Subsequent examination of the model revealed that the response to glucose described above was due primarily to the increased provision of NADPH, which is essential for the synthesis of fatty acids from acetyl-CoA. In contrast, it appeared that the effect of protein on acetate utilization was not related directly to NADPH production, contradicting the suggestion of MacRae *et al.* (1985) that in such situations NADPH production is enhanced through an increased availability of gluconeogenic amino acids. To investigate further the effect of protein, Black *et al.* (1987b) simulated the consequence of increased absorbed amino acid supply on two basal diets supplying 1 or 5% of the total absorbed energy as glucose. The first level of amino acid supplementation gave values of k_f, the efficiency of energy utilization for fattening and growth, of 0.23 and 0.003 for the 1 and 5% glucose diets respectively (Fig. 12.3a).

These low values were due primarily to an increased requirement of ATP for protein synthesis, which showed the largest incremental response for all levels of amino acid supply. Consequently, fat deposition declined on both diets but thereafter increased substantially (Fig. 12.3c), whilst protein deposition showed only modest increases at the two highest levels of amino acid addition (Fig. 12.3b), and values of k_f of 0.37 and 0.46 were obtained for the low- and high-glucose diets respectively (Fig. 12.3a). Furthermore, at the lowest amino acid increment, it was predicted that 63% was used for protein synthesis, with 31% oxidized and 6% converted to glucose. In contrast, for the two highest amino acid increments, the corresponding values were 11, 73 and 16%, indicating that amino acid supply was in excess of amino acid requirements.

The model of lamb metabolism and growth by Sainz and Wolff (1990) was undertaken as part of a study of the mechanisms of action of growth promotants. This model, developed from the earlier model of Gill *et al.* (1984) and incorporating concepts advanced by Baldwin and Black (1979) and Baldwin *et al.* (1987a), related tissue growth to DNA accretion and protein turnover, and state variables include circulating amino acids, glucose, lipids and acetate, four protein pools (carcass, viscera, other tissues and wool) and storage triacylglycerol. Model inputs were the principal absorbed nutrients (amino acids, acetate, propionate, etc.). Flux equations were mainly of

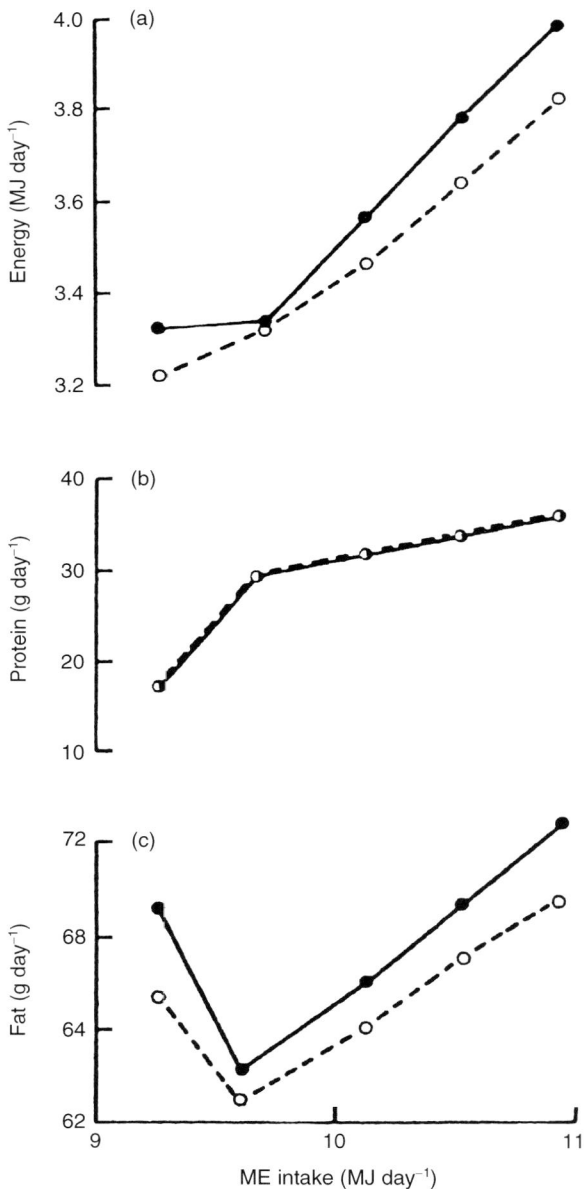

Fig. 12.3. Predicted effect on retention of (a) energy, (b) protein and (c) fat of increasing ME intake by adding amino acids to diets providing 1% (○ —○) or 5% (● —●) glucose (Black *et al.*, 1987b).

the Michaelis–Menten form, allowing for nutrient utilization patterns to be determined by relative tissue affinities for substrates (K), enzymic capacities (V_{max}) and substrate concentrations. Protein degradation rates were defined as first-order with respect to protein. The model adequately simulated growth from 20 to 40 kg of empty body weight, and changes in nutrient input yielded reasonable energy balance response patterns, although efficiencies were greater than those observed experimentally. Variations in VFA absorption patterns were accommodated, and predicted nitrogen retention was approximated with experimental observations. The model also responded to changes in dietary protein level, with body fat accumulation varying inversely with amino acid absorption. Thus the model was found to perform adequately for the purpose of examining mechanisms responsible for alteration of growth and body composition.

As part of a research programme to examine the cause of low efficiencies of nutrient utilization in forage-fed animals, Gill *et al.* (1989a) constructed a mathematical integration of whole-body protein synthesis and degradation in growing lambs. This model was based on protein metabolism in ten individual tissues, namely adipose, central nervous system (CNS), gastrointestinal tract (GIT), heart, kidney, liver, muscle, pancreatic and salivary glands (PSG), reticulo-endothelial system (RES) and skin, together with a blood pool of amino acids. The model was used to examine whole-body protein synthesis at different rates of growth and to compare the relative contributions of each tissue. The GIT (25–26%), skin (23–26%) and muscle (21–26%) had the highest contributions to total protein synthesis, followed by the liver (13–14%), RES (6–7%) and PSG (3–6%), while adipose, CNS, heart and kidney together contributed less than 5%. These values accord, as far as can be ascertained, with experimental findings. Subsequently, Gill *et al.* (1989b) extended the mathematical representation to take account of the energy costs of protein turnover and the transport of sodium and potassium ions.

This model was then used to examine the relative contributions of these processes to ATP expenditure at two different growth rates. Protein turnover was found to account for 19% of whole-body ATP expenditure at both growth rates examined, with the GIT accounting for 25–27%, muscle for 21–26%, skin for 23–26% and liver for 13% of total protein turnover energy costs. The contribution of Na^+,K^+-transport increased from 18 to 23% of whole-body heat production as growth rate increased, with the GIT accounting for 39 and 50%, muscle for 17 and 10% and liver for 18 and 23% of total Na^+,K^+-transport costs at low and high nutrient inputs, respectively. Thus, protein turnover accounted for 19% of the increment in ATP expenditure due to the increased nutrient input at the higher growth rate, while Na^+,K^+-transport accounted for 39%, fat turnover and accretion accounted for 25%, leaving 17% of the ATP increment unaccounted for.

Gerrits *et al.* (1997a) developed a mechanistic model integrating energy and protein metabolism in the pre-ruminant calves of 80 to 240 kg live weight (see Chapter 15 for details). It is based on two calf feeding experiments specifically designed for the purpose, which used six protein levels and two energy levels (Gerrits *et al.*, 1996). The model simulates the partitioning of nutrients from ingestion through intermediary metabolism growth, consisting of accretions of protein, fat, ash and water. The model contains ten state variables, comprising fatty acids, glucose, acetyl-CoA and amino acids as metabolite pools, and fat, ash and protein in muscle, hide, bone and viscera as body constituent pools. The turnover of fat and protein is represented. The model also includes the facility to check dietary amino acid imbalance and can be used to predict amino acid requirements on a theoretical basis.

Gerrits *et al.* (1997b) carried out sensitivity and behavioural analysis of their model and tests against independent data. Simulation of two other calf feeding experiments showed that rates of gain of live weight, protein and fat were predicted satisfactorily. The model is clearly sensitive to 25% changes in the kinetic parameters describing muscle protein synthesis and amino acid oxidation. Comparison of experimentally derived amino acid requirements with model predictions was satisfactory for calves of approximately 90 kg live weight, but lower requirements for lysine and methionine + cystine were suggested at about 230 kg live weight. This indicates that more emphasis needs to be placed on the inevitable oxidative losses of amino acids included in the model as published.

IMPROVED FEED CHARACTERIZATION

Even without changes in the current feeding systems, designed primarily to meet the animals energy and protein requirements, it is accepted that present descriptions of feedstuffs are insufficient (see Chapter 2 for details). Also, future use of the polyester bag procedure (Ørskov and McDonald 1979), an *in situ* technique used extensively to determine ruminal degradation, to describe feedstuffs routinely is open to question in light of the input of animal and labour resources required. Furthermore, the data obtained from the procedure need to be interpreted with caution, especially as they are claimed to measure events rather than describe the components required to predict events. In this context, current interest in devising laboratory methods to predict polyester bag results as a true reflection of *in vivo* events, without adequate description of those events, needs re-addressing (Beever, 1993). The development of the gas pressure transducer technique (Theodorou *et al.*, 1994) for assessing rates of nutrient degradation by rumen liquor offers promise in supplying kinetic data for rumen models.

Some improvements in feed characterization at both the research and on-farm level are occurring within the UK, but progress is slow. The proportion of conserved feeds which are now chemically analysed prior to feeding is increasing, as well as the range of analyses which are undertaken. The situation with concentrates, however, is worse, with the principal feed description available for compounds being only the compounders' statutory declarations of chemical composition, although, increasingly, formulation details are also disclosed. However, this requires the use of 'straights' composition data using historic and out-of-date feed composition tables (MAFF, 1992; AFRC, 1993). Nevertheless, near infrared reflectance spectrometry (NIRS) (Shenk *et al.*, 1981; Baker and Barnes, 1990; Park *et al.*, 1997) and other techniques are emerging, but NIRS measurements have to be calibrated using large collections of stored feedstuff samples with an associated database of their chemical composition or animal measurements, such as digestibility or intake.

Any new system of feed characterization must be rapid and relatively inexpensive. In this context, NIRS has much to offer. However, before such exciting technologies can be harnessed properly, it is necessary to reconcile what the various principal requirements are. It is our belief that all future systems, whether for dairy or beef cattle, need to be nutrient based. Consequently, both comprehensive carbohydrate and protein characterization will be necessary, with due recognition to all significant nutritional entities which are likely to have different metabolic fates. Sugar, starch and fibre represent minimal requirements for carbohydrates. Sugar probably can be described adequately as one entity, which will be completely degraded within the rumen. In contrast, descriptions of starch and fibre will need to recognize the heterogenous nature of both these chemical entities with respect to potential ruminal degradability and their susceptibility to microbial degradation (i.e. maximum extent and rate of degradation). Pectins are anomalous, in that, whilst they are associated with the cell wall structures, they are degraded rapidly in the rumen, giving a VFA pattern different from that of sugars and starches. The crude protein fraction of most feeds can be largely accounted for by the contents of true protein, peptides and amino acids, and ammonia. Clearly, the relative proportions of these three fractions can have a major bearing on the processes of ruminal degradation, whilst a further distinction of true protein in terms of its maximum extent and rate of degradation will be necessary. In some feeds, such as ensiled forages and other partially fermented substrates, there may be a need to recognize the occurrence of amines and other related products, primarily for the negative effects they may have on microbial activity and feed intake. Hopefully, most of these requirements will be met in the long term by the use of appropriate chemical analysis.

Dewhurst and colleagues, in recognizing that ME was not an adequate description of the energy value of feeds for ruminants, proposed a model to predict energy yields of feedstuffs. Originally (Dewhurst *et al.*, 1986), the model was based in part on Weende proximate analysis but, subsequently, Dewhurst and Webster (1989) drew attention to four specific feed fractions, namely quickly fermentable energy (QFE), slowly fermentable energy (SFE), unfermentable digestible energy and unfermentable indigestible energy, and suggested ways in which such entities could be quantified in feedstuffs. Furthermore, in constructing the model, the authors recognized the importance of rate and extent of degradation in the rumen. Dewhurst and Webster (1989) provided evidence of the behaviour of their model by comparing predicted ruminal degradation of SFE with the observed apparent degradability of NDF for two diets comprising nutritionally improved straw and wheat (70:30 and 30:70) offered at two intake levels to sheep. The results are summarized in Table 12.3 where it can be seen that on the high-straw diet, predicted and observed results were in close agreement, whereas on the high-wheat diet, overpredictions for the digestion of SFE were obtained. This led the authors to suggest that the model needed to be

Table 12.3.   Predicted degradation of slowly fermented energy (SFE) and observed degradation of neutral detergent fibre (NDF) for two contrasting diets comprising wheat (W) and nutritionally improved straw (NIS) (taken from Webster, 1989).

	NIS:W 70:30		NIS:W 30:70	
Feeding level	Low	High	Low	High
Degradation (g kg^{-1})				
SFE (predicted)	632	558	716	681
NDF (observed)	670	594	621	521

refined to take account of possible negative effects of QFE on the degradation of SFE.

Whilst the initiative taken by Dewhurst and colleagues is to be welcomed, the methods used to describe the defined feed categories quantitatively is open to debate. QFE is assumed to comprise sugars and α-glycans and, whilst these can be quantified rapidly, the assumption that all QFE is degraded quickly and completely in the rumen fails to take account of any possible differences due to the nature of the starch in the diet (e.g. barley versus maize starch). The degradability of SFE is calculated from the proportion of potentially degradable β-glycans and the rate of degradation of β-glycans, both of which are predicted from an estimate of the acid detergent lignin (ADL) content of the diet. Similarly, undegradable indigestible energy contents are predicted from estimates, either of ADL or acid-insoluble ash content. The robustness of these approaches to deal with the range of feedstuffs used in the ruminant livestock industry has not yet been examined sufficiently. Despite this and the tendency of the model to overpredict the true ME content of raw materials and some compound feeds (thought to be related primarily to the inhibitory effect of sugars and α-glycans on the rate and extent of degradation of β-glycans), the model provides a basis for future developments.

A similar approach was used by Sniffen *et al.* (1992) in the development of the CNCPS rumen model referred to earlier (see Chapters 7 and 8 for details). The analytical scheme of Goering and Van Soest (1970) for NSC carbohydrates is used for the SFE fractions, whilst QFE consists of starch measurements, with sugars estimated by difference from the estimate of cell contents (1000 − NDF) minus starch, protein, fat and minerals. This leads to errors when any significant amounts of fermentation acids or pectins are present in the feeds. Protein fractions are determined as by Van Soest *et al.* (1981). These are then combined with literature values for rates of degradation of the identified feed fractions, rather than predicted from feed composition data.

In justifying their approach, Dewhurst and Webster (1989) stated three important criteria which are of relevance to those engaged in developing improved feedstuff characterization as a means of providing more accurate prediction of animal performance.

These authors stated that any attempt to improve feedstuff characterization must:

- be based on robust measurements suitable for routine use in the advisory and feedstuffs industry;
- adequately describe the most important processes of digestion and metabolism, whilst allowing for further developments; and
- provide a better prediction of nutritive value, and ultimately animal performance, than is currently available.

CONCLUSIONS

This chapter has, through consideration of current systems for calculating the energy and protein requirements of ruminants, established that appropriate schemes for predicting animal response to nutrients in systems of ruminant livestock production are not presently available within the UK. Modification and revision of these empirically based factorial feeding systems are unlikely to yield successful predictions of animal production responses to changes in nutrient intake. The need to consider the development of mechanistic mathematical models which describe the interactions of individual nutrients is proposed. Present mechanistic modelling efforts in areas associated with intake regulation and with metabolism within the rumen and body tissues of both growing and lactating animals are reviewed, highlighting where current limitations exist. Finally, the need to improve feedstuff characterization is suggested and illustrated using a model for improving the prediction of the true metabolizable energy content of diets for ruminants.

REFERENCES

Agricultural and Food Research Council (1990) Technical Committee on Responses to Nutrients, Report No. 5. Nutritive requirements of ruminant animals: energy. *Nutrition Abstracts and Reviews Series B* 60, 729–804.

Agricultural and Food Research Council (1992) Technical Committee on Responses to Nutrients, Report No. 9. Nutrient requirements of ruminant animals: protein. *Nutrition Abstracts and Reviews Series B* 62, 787–835.

Agricultural and Food Research Council (1993) *Energy and Protein Requirements of Ruminants.* CAB International, Wallingford, UK.

Agricultural Research Council (1965) *The Nutrient Requirements of Farm Livestock, No. 2 Ruminants.* Agricultural Research Council, London.

Agricultural Research Council (1980) *The Nutrient Requirements of Ruminant Livestock.* Commonwealth Agricultural Bureaux, Farnham Royal, Slough, UK.

Agricultural Research Council (1984) *The Nutrient Requirements of Farm Livestock, Supplement No.1.* Commonwealth Agricultural Bureaux, Farnham Royal, Slough, UK.

Baker, C.W. and Barnes, R. (1990) The application of near infra-red spectrometry to forage evaluation in the Agricultural Development and Advisory Service. In:

Wiseman, J. and Cole, D.J.A. (eds), *Feedstuff Evaluation*. Butterworths, London, pp. 337–351.

Baldwin, R.L. and Black, J.L. (1979) Simulation of the effects of nutritional and physiological status on the growth of mammalian tissue: description and evaluation of a computer program. *Animal Research Laboratory Technical Paper No. 6*. Commonwealth Scientific and Industrial Research Organisation, Melbourne, Australia.

Baldwin, R.L., Lucas, H.L. and Cabrera, R. (1970) Energetic relationships in the formation and utilization of fermentation end products. In: Phillipson, A.T. (ed.), *Physiology of Digestion and Metabolism in the Ruminant*. Oriel Press, Newcastle-upon-Tyne, UK, pp. 319–334

Baldwin, R.L., Koong. L.J. and Ulyatt, M.J. (1977) A dynamic model of ruminant digestion for evaluation of factors affecting nutritive value. *Agricultural Systems* 2, 255–288.

Baldwin, R.L., France, J. and Gill, M. (1987a) Metabolism of the lactating cow. I. Animal elements of a mechanistic model. *Journal of Dairy Research* 54, 77–105.

Baldwin, R.L., Thornley, J.H.M. and Beever, D.E. (1987b) Metabolism of the lactating cow. II. Digestive elements of a mechanistic model. *Journal of Dairy Research* 54, 107–131.

Baldwin, R.L., France, J., Beever, D.E., Gill, M. and Thornley, J.H.M. (1987c) Metabolism of the lactating cow. III. Properties of mechanistic models suitable for evaluation of energetic relationships and factors involved in the partition of nutrients. *Journal of Dairy Research* 54, 133–145.

Beever, D.E. (1993) Characterisation of forages. In: Garnsworthy, P.C. and Cole, D.J.A. (eds), *Recent Advances in Animal Nutrition – 1993, Proceedings of the 27th Nottingham Feed Manufacturers Conference*. Nottingham University Press, Nottingham, UK, pp. 3–18.

Beever, D.E., Black, J.L. and Faichney, G.J. (1980/81) Simulation of the effects of rumen functions on the flow of nutrients from the stomach of sheep. Part 2. Assessment of computer predictions. *Agricultural Systems* 6, 221–241.

Beever, D.E., Osbourn, D.F., Cammell, S.B. and Terry, R.A. (1981) The effect of grinding and pelleting on the digestion of Italian ryegrass and timothy by sheep. *British Journal of Nutrition* 46, 357–370.

Beever, D.E., Gill, M., Dawson, J.M. and Buttery P.J. (1990) The effect of fishmeal on the digestion of grass silage by growing cattle. *British Journal of Nutrition* 63, 489–502.

Black, J.L., Beever, D.E., Faichney, G.J., Howarth, B.R. and Graham, N.McC. (1980/81) Simulation of the effects of rumen function on the flow of nutrients from the stomach of sheep. Part 1. Description of a computer model. *Agricultural Systems* 6, 195–219.

Black, J.L., Gill, M., Beever, D.E., Thornley, J.H.M. and Oldham. J.D. (1987a) Simulation of the metabolism of absorbed energy yielding nutrients in young sheep: efficiency of utilization of acetate. *Journal of Nutrition* 117, 105–115.

Black J.L., Gill, M., Thornley, J.H.M., Beever, D.E. and Oldham, J.D. (1987b) Simulation of the metabolism of absorbed energy-yielding nutrients in young sheep: efficiency of utilization of lipid and amino acid. *Journal of Nutrition* 117, 116–128.

Blaxter, K.L. (1962) *The Energy Metabolism of Ruminants*. Hutchinson, London.

Buchanan-Smith, J.B. (1982) Voluntary intake in ruminants as affected by silage extracts and amines in particular. In: Thomson, D.J., Beever, D.E. and Gunn, R.G. (eds), *Forage Protein in Ruminant Animal Production, British Society of Animal Production Occasional Publication No. 6*. British Society of Animal Production, Thames Ditton, UK, pp. 180–182.

Danfær, A. (1990) A dynamic model of nutrient digestion and metabolism in lactating dairy cows. PhD Thesis, National Institute of Animal Science, Report No. 671, Foulum, Denmark.

Dewhurst, R.J. and Webster, A.J.F. (1989) Development of a practical, deterministic model for the prediction of true metabolizable energy in forages and compound feeds. In: Van der Honing, Y. and Close, W.H. (eds), *Energy Metabolism of Farm Animals*. European Association of Animal Production Publication No. 43. Pudoc, Wageningen, The Netherlands, pp. 223–230

Dewhurst, R.J., Webster, A.J.F., Wainman, F.W. and Dewey, P.J.S. (1986) Prediction of the true metabolizable energy concentration in forages for ruminants. *Animal Production* 43, 183–194.

Dijkstra, J. (1994) Simulation of the dynamics of protozoa in the rumen. *British Journal of Nutrition* 72, 677–699.

Dijkstra, J., Neal, H.D.StC., Beever, D.E. and France, J. (1992) Simulation of nutrient digestion, absorption and outflow in the rumen: model description. *Journal of Nutrition* 122, 2239–2256.

Evans, R.E. (1960) *Rations for Livestock, Bulletin 48*. HMSO, London.

Forbes, J.M. and France, J. (1993) *Quantitative Aspects of Ruminant Digestion and Metabolism*. CAB International, Wallingford, UK.

France, J. and Thornley, J.H.M. (1984) *Mathematical Models in Agriculture*. Butterworths, London.

France, J., Thornley, J.H.M. and Beever, D.E. (1982) A mathematical model of the rumen. *Journal of Agricultural Science, Cambridge* 99, 343–353.

France, J., Gill, M., Thornley, J.H.M. and England, P. (1987) A model of nutrient utilisation and body composition in beef cattle. *Animal Production* 44, 371–385.

France, J., Thornley, J.H.M., Baldwin, R.L. and Crist, K.A. (1992) On solving stiff equations with reference to simulating ruminant metabolism. *Journal of Theoretical Biology* 156, 525–539.

Gerrits, W.J.J., Tolman, G.H., Schrama, J.W., Tamminga, S., Bosch, M.W. and Verstegen, M.W.A. (1996) Effect of protein and protein-free energy intake on protein and fat deposition rates in pre-ruminant calves of 80–240 kg live weight. *Journal of Animal Science* 74, 2129–2139.

Gerrits, W.J.J., Dijkstra, J. and France, J. (1997a) Description of a model integrating protein and energy metabolism in pre-ruminant calves. *Journal of Nutrition* 127, 1229–1242.

Gerrits, W.J.J., France, J., Dijkstra, J., Bosch, M.W., Tolman, G.H. and Tamminga, S. (1997b) Evaluation of a model integrating protein and energy metabolism in pre-ruminant calves. *Journal of Nutrition* 127, 1243–1252.

Gibb, M.J., Ivings, W.E., Dhanoa, M.S. and Sutton, J.D. (1992) Changes in body components of autumn-calving Holstein–Friesian cows over the first 29 weeks of lactation. *Animal Production* 55, 339–360.

Gill, M., Thornley, J.H.M., Black, J.L. Oldham, J.D. and Beever, D.E. (1984) Simulation of the metabolism of absorbed energy-yielding nutrients in young sheep. *British Journal of Nutrition* 52, 621–649.

Gill, M., France, J., Summers, M., McBride, B.W. and Milligan, L.P. (1989a) Mathematical integration of protein metabolism in growing lambs. *Journal of Nutrition* 119, 1269–1286.

Gill, M., France, J., Summers, M., McBride, B.W. and Milligan, L.P. (1989b) Simulation of the energy costs associated with protein turnover and Na^+, K^+-transport in growing lambs. *Journal of Nutrition* 119, 1287–1299.

Goering, H.K. and Van Soest, P.J. (1970) *Forage Fiber Analysis.* Agricultural Handbook No.379, Agricultural Research Service, USDA, Washington, DC.

Lindsay, D.B. (1993) Metabolism in the portal-drained viscera. In: Forbes, J.M. and France, J. (eds) *Quantitative Aspects of Ruminant Digestion and Metabolism.* CAB International, Wallingford, UK, pp. 267–289.

MacRae, J.C., Dewey, P.J.S., Brewer, A.C., Brown, D.S. and Walker, A. (1985) The efficiency of utilization of metabolizable energy and apparent absorption of amino acids in sheep given spring- and autumn-harvested dried grass. *British Journal of Nutrition* 54, 197–209.

MacRae, J.C., Buttery, P.J. and Beever, D.E. (1988) Nutrient interactions in the dairy cow. In: Garnsworthy, P.C. (ed.), *Nutrition and Lactation in the Dairy Cow.* Butterworths, London, pp. 55–75.

Mertens, D.R. (1985) Effect of fiber on feed quality for dairy cows. In: *Proceedings of the 48th Minnesota Nutrition Conference.* St. Paul, Minnesota, p. 209.

Ministry of Agriculture, Fisheries and Food (1975) *Energy Allowances and Feeding Systems for Ruminants, Technical Bulletin No. 33.* HMSO, London.

Ministry of Agriculture, Fisheries and Food (1976) *Nutrient Allowances and Composition of Feedingstuffs for Ruminants.* Booklet LGR21, MAFF Publications, Pinner, UK.

Ministry of Agriculture, Fisheries and Food (1992) *Feed Composition – UK Tables of Feed Composition and Nutritive Value for Ruminants,* 2nd edn, Ministry of Agriculture, Fisheries and Food Standing Committee on Tables of Feed Composition, Chalcombe Publications Canterbury, UK.

Neal, H.D.StC., Dijkstra, J. and Gill, M. (1992) Simulation of nutrient digestion, absorption and outflow in the rumen: model evaluation. *Journal of Nutrition* 122, 2257–2272.

Ørskov, E.R. and McDonald, I. (1979) The estimation of protein degradability in the rumen from incubation measurements weighted according to rate of passage. *Journal of Agricultural Science, Cambridge* 92, 499–503.

Park, R.S., Agnew, R.E., Gordon, F.J. and Steen, R.W.J. (1997) The potential of near infra-red reflectance spectroscopy (NIRS) to evaluate digestibility and degradability parameters of undried silage. In: *Proceedings of the British Society of Animal Science,* Scarborough, p. 205.

Poppi, D.P., Gill, M. and France J. (1994) Integration of theories of intake regulation in growing ruminants. *Journal of Theoretical Biology* 167, 129–145.

Reichl, J.R. and Baldwin, R.L. (1975) Rumen modelling: rumen input–output balance models. *Journal of Dairy Science* 58, 879–890.

Rook, A.J. (1991) On behalf of the AFRC Technical Committee on Responses to Nutrients, Report No. 8. The Voluntary Intake of Cattle. *Nutrition Abstracts and Reviews Series B* 61, 815–823.

Russell, J.B., O'Connor, J.D., Fox D.G., Van Soest, P.J. and Sniffen, C.J., (1992) A net carbohydrate and protein system for evaluating cattle diets: I Ruminant fermentation. *Journal of Animal Science* 70, 3551–3561.

Sainz, R.D. and Wolff, J.E. (1990) Development of a dynamic, mechanistic model of lamb metabolism and growth. *Animal Production* 51, 535–549.

Shenk, J.S., Landa, I., Hoover, M.R. and Westerhaus, M.O. (1981) Description and evaluation of a near infra-red reflectance spectro-computer for forage and grain analysis. *Crop Science* 21, 355–358.

Smith, N.E. (1970) Modelling studies of ruminant metabolism. PhD thesis, University of California, Davis, California.

Sniffen, C.J., O'Connor, J.D., Van Soest, P.J., Fox D.G. and Russell, J.B., (1992) A net carbohydrate and protein system for evaluating cattle diets: II Carbohydrate and protein availability. *Journal of Animal Science* 70, 3562–3577.

Subnel, A.P.J., Meijer, R.G.M., Van Straalen, W.M. and Tamminga, S. (1994) Efficiency of milk protein production in the DVE protein system. *Livestock Production Science* 40, 215–224.

Theodorou, M.K., Williams, B.A., Dhanoa, M.S., McAllan, A.B. and France, J. (1994) A simple gas production method using a pressure transducer to determine the fermentation kinetics of ruminant feeds. *Animal Feed Science and Technology* 48, 185–197.

Thomas, C., Gibbs, B.G., Beever, D.E. and Thurnham, B.R. (1988) The effect of date of cut and barley substitution on gain and on the efficiency of utilization of grass silage by growing cattle. 1. Gains in live weight and its components. *British Journal of Nutrition* 60, 297–306.

Thomas, P.C., Chamberlain, D.G., Martin, P.A. and Robertson, S. (1987) Dietary energy intake and milk yield and composition in dairy cows. In: Moe, P.W., Tyrell, H.F. and Reynolds, P.J. (eds), *Energy and Metabolism of Farm Animals*. European Association of Animal Production Publication No. 32, Rowman Littlefield, New Jersey, pp. l88–191.

Tyrrell, H.F. and Reid, J.T. (1965) Prediction of the energy value of cow's milk. *Journal of Dairy Science* 48, 1215–1223.

Van Soest, P.J., Sniffen, C.J., Mertens, D.R., Fox, D.G., Robinson, P.H. and Krishnamorthy, U.C. (1981) A net protein system for cattle: the rumen sub-model for nitrogen. In: Owen, F.N. (ed.), *Protein Requirements for Cattle: Proceedings of an International Symposium MP 109*. Division of Agriculture, Oklahoma State University, Stillwater, Oklahoma, pp. 265–279.

Whitelaw, F.G., Milne, J.S., Ørskov, E.R. and Smith, J.S. (1986) The nitrogen and energy metabolism of lactating cows given abomasal infusions of casein. *British Journal of Nutrition* 55, 537–556.

13 Analyses of Modelling Whole-rumen Function

J. Dijkstra[1] and A. Bannink[2]

[1]Animal Nutrition Group, Wageningen Institute of Animal Sciences, Wageningen Agricultural University, Wageningen; [2]DLO-Institute for Animal Science and Health, Department of Ruminant Nutrition, Lelystad, The Netherlands

INTRODUCTION

The particular ability of ruminants to utilize highly fibrous diets, which are of low nutritional value for simple-stomached animals and men, contributes to their provision for both the nutritional and economic requirements of the human population. Systems of ruminant production vary widely as influenced by local conditions and, consequently, so do key research interests. The growing human population in tropical countries leads to increasing demands for animal products, whereas the availability of land for production of feed for farm animals is reduced because of the pressure on land use for other activities (Preston and Leng, 1987). On the other hand, in western areas, where animal industry has developed intensive systems, measures have been or will be taken to reduce the negative impact of ruminant production on the environment (Tamminga, 1996).

The major nutrients available for absorption, which arise from the microbial activities in the rumen, are volatile fatty acids (VFAs) and amino acids in microbial protein. Optimization of the rumen fermentation processes in order to maximize the supply of available nutrients and to reduce the formation of waste products has been a key area of research in ruminant nutrition. The increase in knowledge on rumen function facilitated the development of mechanistic modelling as a research tool in this area. Stimulated by the work of Baldwin *et al.* (1970), several models of whole-rumen function have been developed which integrate knowledge on various aspects of the processes in the rumen. These models do not necessarily share common objectives, and the evaluation of models generally depends on an appraisal of the total effort in relation to the objectives of the modelling exercise (France and Thornley, 1984; Baldwin, 1995; Dijkstra and France, 1995). The aim of this chapter is to describe objectives and applications of mechanistic rumen models and their contribution to the advancement of ruminant production, illustrated by a few selected examples of rumen models. This chapter does not intend to compare rumen models, because recent papers already describe comparative aspects of whole-rumen function (Dijkstra and France, 1996), diet-specific input parameters (Bannink *et al.*, 1997a), extra-microbial and microbial processes (Bannink and De Visser, 1997), microbial protein synthesis (Dijkstra *et al.*, 1998a), the optimization of rumen processes using models (Sauvant *et al.*, 1995) and the predictive ability of models in comparison with each other (Kohn *et al.*, 1995; Bannink *et al.*, 1997c).

BRIEF HISTORY OF RUMEN MODELS

Models of Rumen Function in Sheep

Much of the development of mechanistic mathematical models of whole-rumen function is due to the earlier work of Baldwin *et al.* (1970). They integrated knowledge available at that time on energy and N metabolism within the rumen in order to predict the formation and utilization of fermentation end-products in sheep. This model comprises a system of first-order differential equations, and all the fluxes represented are assumed to obey mass–action kinetics. These early efforts obviously were limited by the availability of both research data and computer power. Reichl and Baldwin (1975) developed a linear programming model to evaluate competitive relationships among eight rumen microbial groups as affected by protein and carbohydrate availability. However, considerable simplifications of the rumen microbial groups had to be made when solving the model and, in their evaluation, Reichl and Baldwin (1976) concluded that the concepts represented in the model were not sufficient. In another model of rumen processes in sheep (Baldwin *et al.*, 1977), the objectives were to identify areas of inadequate knowledge and to test hypotheses regarding factors likely to affect nutritive value. The simulation results stimulated further experimental research into VFA stoichiometry and rumination patterns (Murphy *et al.*, 1982, 1983). An update of the Baldwin *et al.* (1977) model was described by Murphy *et al.* (1986) and includes more extensive representation of fibre present in particles of different size and of the production of VFAs in the rumen.

Black *et al.* (1981) judged the previously mentioned rumen models too complex and detailed to be readily compatible with an existing sheep growth model. They developed a model of rumen fermentation with the objective of predicting nutrient flows from the stomach. Some processes are represented using existing empirical equations, whilst a number of microbial aspects are represented in a mechanistic way. Black *et al.* (1981) acknowledged that integration of the difference equations that constitute their model was not satisfactory from a mathematical point of view, and suggested that the model structure could be improved by conforming to the standard rate:state formalism. This mathematical refinement based on the rate:state formalism was promoted by France *et al.* (1982), attempting to provide a mechanistic description and quantitative understanding of the main processes occurring in the rumen of sheep. Most of the rumen models developed subsequently have their mathematical basis in the France *et al.* (1982) model. A major departure from all previous approaches was that, in addition to the mass–action kinetics, Michaelis–Menten (saturation) kinetics were used to describe fluxes. For the first time, observations with pulsed dietary inputs were compared with predictions. With constant feed intake, model behaviour was satisfactory, but considerable discrepancies were found between experimental data and model output with pulsed feed intake.

Models of Rumen Function in Cattle

Baldwin *et al.* (1987b) were the first to develop a model of rumen function in dairy cattle. Their objective was to predict the rate and pattern of nutrient absorption in the dairy cow. The limited evaluation against *in vivo* data in high producing dairy cattle was satisfactory for most diets, with the noticeable exception of the VFA production rates. Some of the suggestions to improve the model, e.g. depressed microbial activity at reduced pH, were addressed in an update of this model (Argyle and Baldwin, 1988; Baldwin, 1995). Following the work of Baldwin *et al.* (1987a,b), Danfær (1990) also dealt with the issue of a mechanistic model of dairy cow metabolism. Rumen function was represented using distinct components of microbial mass as state variables in the model, which differed from all previous approaches. Evaluation of this model was not independent, as experimental data used in part to develop the model were used subsequently in its evaluation.

The Cornell net carbohydrate and protein system was developed to provide estimates of nutrients available for absorption in cattle, and is claimed to serve as a research tool as well as to serve in practical ration balancing. Russell *et al.* (1992) and Sniffen *et al.* (1992) describe its rumen submodel. This submodel consists of static equations, representing degradation and passage of several feed fractions, and equations to represent growth and outflow of microorganisms. A first update of the rumen submodel contains equations to predict individual amino acid flows (O'Connor *et al.*, 1993). A later update was published by Pitt *et al.* (1996) and addresses the prediction of pH and VFA in the rumen but, as before, VFA molar proportions were not predicted accurately. In addition, problems were encountered with the prediction of starch degradation.

Dijkstra *et al.* (1992) built on the achievements of Baldwin *et al.* (1987b) primarily by improving microbial dynamics. The objective of their modelling exercise was to examine the effect of supplementation of forage diets on the profile of nutrients available for absorption in cattle. Specific areas of improvement of representation included the combination of microbial substrate preference, differential outflow and microbial composition, and recycling of microbial matter. This model was tested thoroughly on a range of diets and generally provided reasonable predictions, with again the main exception of VFA molar proportions (Neal *et al.*, 1992). It was also shown that the representation of protozoan activities needed more attention, and this was dealt with subsequently in the model of Dijkstra (1994a), in which the objective was to provide a quantitative understanding of protozoan dynamics. This mathematical integration of dietary, bacterial and protozoal factors improved understanding of determinants involved in the contribution of protozoa to fibre degradation (Dijkstra and Tamminga, 1995) and microbial N recycling (Dijkstra *et al.*, 1998b) in the rumen.

Lescoat and Sauvant (1995) developed another rumen model in order to predict duodenal amino acid flux and VFA production in dairy cattle. Unlike all other rumen models in cattle, in their model virtually all fluxes are in mass–action form. The model was evaluated against data obtained from goats fed a range of diets at maintenance level, and the duodenal fluxes of lysine and methionine evaluated using data from beef and dairy cattle. Whilst lysine and methionine fluxes were predicted satisfactorily, a large bias was reported for the predicted VFA concentration in rumen fluid, as well as for ruminal ammonia concentration and microbial efficiency.

Dijkstra *et al.* (1996a,b) constructed a simple mechanistic model of digestion of sugarcane in cattle, to be applied in the search for suitable supplements. The aggregation level and mathematical approach of this model were based on the model of France *et al.* (1982). The development and evaluation of this model will be discussed later in this chapter.

OBJECTIVES IN RUMEN MODELLING

Modelling is a central and integral part of the scientific method, as models provide scientists with representations that they can use (for a review of modelling and methodology in animal sciences, see Dijkstra and France, 1995). All applications of rumen models mentioned in the previous section are compatible with the general goal of scientists working on rumen function, which is to increase our understanding and control of the rumen fermentation processes. In general, objectives in rumen modelling can be classified into objectives in applied (management) modelling and objectives in research modelling. Whilst there is no absolute difference between a research model and an applied model, there is a very definite difference in the process of their evaluation (France and Thornley, 1984). With a research model, hypotheses or concepts can be evaluated for (in)adequacy in an integrated framework, and the model need not necessarily describe accurately what happens. On the other hand, an applied model must give predictions that are clearly better than existing practice, if it is to be applied successfully

To illustrate the contribution of rumen models to the advancement of ruminant science, in the following sections three different types of modelling objectives will be described using selected examples of rumen models. Firstly, the integration of knowledge on individual components of rumen function by rumen modelling will be described. Next, the role of rumen models in research programmes will be depicted. Finally, the application of rumen models in management will be described.

INTEGRATION OF KNOWLEDGE ON INDIVIDUAL ELEMENTS

The rumen can be viewed as consisting of several components which together determine its functioning and the type and level of substrates available for absorption. Basic elements of the rumen ecosystem include the energy-yielding substrates, the microorganisms and the end-products of fermentation. Considerable research on various aspects of rumen fermentation processes has accumulated over time. However, concentrating research on individual components of the rumen fermentation, rather than on its integration, has resulted in insufficient information on many of the important mechanisms that link the individual components (Gill *et al.*, 1989). Mechanistic modelling is a proper tool to achieve such integration. In mechanistic modelling, a biological problem is examined by looking downwards on the organizational hierarchy (reduction and analysis), with the synthetic process of modelling to reach the original level again (integration) (France and Thornley, 1984; Dijkstra and France, 1995). Thus, models can provide a means of bringing together the knowledge about several components of a system, to give a coherent view of the behaviour of the whole system and increase understanding.

Integration of Knowledge to Predict Responses Measured Frequently

Among the most important and frequently measured features of rumen functioning is the supply of microbial protein to the duodenum. The prediction of microbial protein supply is based on knowledge of individual components of rumen function. These components include the degradation of substrates in the rumen, the efficiency of growth of rumen microorganisms, the metabolism of individual microbial species, etc. Considerable research evidence on each of these components is available. For example, databases containing information on *in situ* protein and carbohydrate degradation of various feeds (Tamminga *et al.*, 1990) and extensive knowledge on the rumen microbial ecosystem (Hobson, 1988; Williams and Coleman, 1992; Hobson and Stewart, 1997) are available. In integrating knowledge about the components, one of the tasks of the modeller is to extract from this wealth of information the concepts and data required for the specific objective of the model. Not surprisingly, a range of models exists to predict microbial protein supply, incorporating and representing to a highly variable extent the individual components which affect microbial protein supply (reviewed by Dijkstra *et al.*, 1998a).

Even when just one specific component of rumen function is to be included in a model, its representation can vary widely. An illustration of this is the representation of the effect of pH upon fibre degradation and microbial metabolism. Since the products of fibre hydrolysis are a major source of energy for microbial growth in the rumen, proper representation of fibre degradation is vital to predict microbial protein supply satisfactorily. Many *in vitro* experiments have quantified the effect of pH on fibre degradation (Terry *et al.*, 1969; Stewart, 1977; Hoover *et al.*, 1984; Mould and Ørskov, 1984; Shriver *et al.*, 1986; Grant and Mertens, 1992a,b). *In vivo*, the pH below which fibre degradation is reduced is approximately 6.3 (reviewed by Erdman, 1988). Based on these *in vitro* and *in vivo* data, the representations of the effect of pH upon fibre degradation adopted in rumen models differ widely (Fig. 13.1). Argyle and Baldwin (1988) assumed no inhibition above pH 6.2, a linear

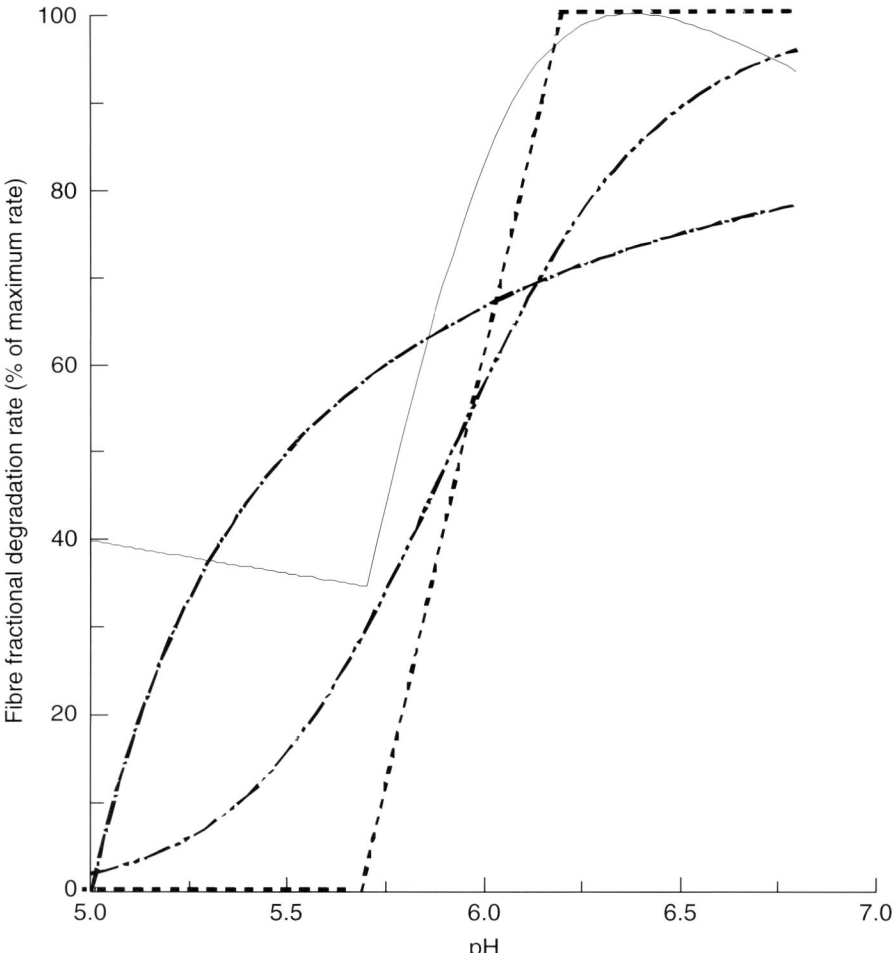

Fig. 13.1. The simulated effect of pH on fibre fractional degradation rate according to the model of Argyle and Baldwin (1988) (------), Dijkstra *et al.* (1992) (–··–··–), Lescoat and Sauvant (1995) (–·–·–·) and Pitt *et al.* (1996) (———). Maize silage (fractional fibre degradation rate 0.07 h^{-1}) is taken for the Pitt *et al.* (1996) Cornell model update.

decrease in fibre degradation rate between pH 5.7 and 6.2, and zero fibre degradation when the pH drops below 5.7. Dijkstra *et al.* (1992) included a sigmoidal relationship between fibre degradation rate and pH, with an affinity constant (the pH where fibre degradation rate is 50% of the maximum rate) of 6.0 and a steepness parameter (determining the sigmoidicity of the curve) of 22.9. Lescoat and Sauvant (1995) adopted a saturation-type relationship between fibre degradation rate and pH, with an affinity constant of 5.5. Pitt *et al.* (1996) applied a non-linear relationship to calculate the decrease in microbial yield as pH declines, and adjusted the fractional degradation rate of fibre accordingly. In this approach, the change in fibre fractional degradation rate due to change in pH depends upon the maximum rate of degradation, the latter being a feed characteristic and an input to the model. At a pH of 5.7, however, the lowest fractional degradation rate will have been achieved, which invariably equals $0.025\ \text{h}^{-1}$, independently of the maximum fibre degradation rate of the feedstuff. In comparing these approaches with the Pitt *et al.* (1996) approach, Fig. 13.1 includes maize silage (potential fractional degradation of the fibre equals $0.07\ \text{h}^{-1}$) as an example of the relationship between pH and fibre degradation rate.

At a pH of 6.7, the predicted fibre fractional degradation rate is at least 90% of the maximum rate, except for the Lescoat and Sauvant (1995) approach (77% of maximum rate) (Fig. 13.1). At a pH of 5.7, this rate has decreased slightly to 57% of the maximum rate in the approach of Lescoat and Sauvant (1995), but reduced to zero in that of Argyle and Baldwin (1988), with the other approaches having intermediate values. Below a pH of 5.7, there is a further rapid decrease in fractional degradation rate according to Dijkstra *et al.* (1992) and Lescoat and Sauvant (1995). In sharp contrast, Pitt *et al.* (1996) assume another increase in fibre degradation rate when the pH is decreased below 5.7. In the latter approach, separate equations are employed to relate fibrolytic microbial growth and fibre degradation to rumen pH, giving rise to unrealistic behaviour. For example, with pH values below 5.7, these equations predict fibrolytic microbial growth to be zero whilst fibre degradation continues. This example thus illustrates the importance of the formulation of hypotheses, of the choice of system components and of the identification of relationships, as well as the inadequacies that might arise from improper representation of the biological processes.

Integration of Knowledge to Predict Responses Difficult to Measure

The reductionist approach in mechanistic modelling enables the integration of components of a system and provides a powerful tool to improve understanding and predict the behaviour of the system. This can be of particular significance when it is difficult or impossible to measure experimentally a certain system element, or to measure elements simultaneously. In these situations, modelling allows quantitative description and examination of the relevant interactions and of the importance of such processes within the system.

An example in rumen function concerns the recycling of microbial N within the rumen. Rumen protozoa have a major impact upon microbial N recycling within the rumen through engulfment of microorganisms and autolysis. *In vitro*, bacterial protein breakdown is reduced by some 90% upon removal of protozoa (Wallace and McPherson, 1987). Defaunation of the rumen invariably increases the efficiency of microbial protein synthesis *in vivo* (review in Jouany *et al.*, 1988). Isotope tracer methods have been employed in a limited number of experiments to quantify microbial N recycling directly in the rumen. The estimates of microbial N recycling thus obtained vary widely, from 20 to 90% of gross microbial N synthesis (reviewed by Firkins, 1996). Problems using isotope tracer methods include unrepresentative labelling of microbial samples, inability to measure all fluxes involved, and difficult biological interpretation of the measured pools and fluxes in the complex rumen ecosystem.

An alternative way of examining microbial N recycling within the rumen employs modelling, in which hypotheses about the processes contributing to recycling and the relevant components and transactions between components are integrated and represented mathematically. Dijkstra (1994a) developed a rumen model with emphasis on protozoal metabolism, to provide a quantitative understanding of the protozoal dynamics. This model has been applied to examine the effects of dietary factors and protozoal activities on microbial N recycling in the rumen of cattle (Dijkstra *et al.*, 1998b). The model comprises 19 state variables, representing carbohydrate entities (fibre, starch and sugars), non-microbial N-containing entities (protein and ammonia), fatty acid-containing entities (lipid and VFAs) and microbial entities (bacteria and protozoa). Several protozoal characteristics are represented, including preference for utilization of starch and sugars, engulfment of microorganisms, and selective retention within the rumen. In the model, engulfment of microorganisms by protozoa and lysis of protozoal cells contribute to recycling of microbial N within the rumen. In the steady state, recycling (% of gross microbial N synthesis) represents the proportion of microbial N synthesized but not washed out of the rumen.

Simulation results for a number of all-roughage diets (hay, grass and maize silages) at two levels of daily dry matter intake (DMI) are presented in Table 13.1 (for details of diets, see Dijkstra *et al.*, 1998b). The high starch content and low N content of maize silage, compared with grass and hay, results in a high, simulated high proliferation of protozoa. Because of this high protozoal activity on the maize silage diets, microbial recycling within the rumen is much higher than on the hay or grass diets. An increased DMI is expected to reduce recycling, because of increased fractional rates of passage. However, this intake effect is least pronounced with the maize silage diet. Firkins *et al.* (1992) estimated microbial N recycling via the ammonia and non-ammonia pools using compartmental analysis. Their estimates of microbial N recycling are much higher than those obtained by methods which determine recycling through the ammonia pool only, and they concluded that their figures represented ruminal conditions more accurately. However, maize silage was the major component of the diet used in Firkins *et al.* (1992) and the results of the integrated model (Table 13.1) indicate that at least part of the higher microbial N recycling estimated by Firkins *et al.* (1992) is related to the diet used. The same model has been applied to increase understanding of the contribution of protozoa and bacteria to fibre degradation (Dijkstra and Tamminga, 1995), for which no *in vivo* measurements have been published. In conclusion, integration of concepts and data by mechanistic modelling allows an improved understanding of observed responses and a prediction of responses that are difficult to measure experimentally.

ROLE OF RUMEN MODELS IN RESEARCH PROGRAMMES

The development of a mechanistic research model essentially does not differ from the experimental approach (France and Thornley, 1984; Baldwin, 1995; Dijkstra and France, 1995). The first step in the application of scientific precepts to a problem is to identify objectives. Next, appropriate information is collated in order to generate theories and hypotheses that subsequently are tested against observations. In experimental research, it is virtually impossible to measure all of the system variables of the system of interest, and only a limited set of system variables will be available. This restriction hampers the simultaneous testing of all the mechanisms involved in an integrated system. On the other hand, in the modelling approach, assumptions and hypotheses are defined explicitly, which allows testing of the hypotheses in an integrative manner. However, individual relationships between model variables can be subject to considerable uncertainty. The testing of such a relationship with data from selected experiments will be a test of that specific relationship only, and can hardly be considered a test of the model as the integrated representation of the mechanisms involved. In other words, a successful test of a single relationship will not necessarily lend support to the whole model. Much more

Table 13.1. Simulated protozoal and total microbial biomass, gross microbial N synthesis and microbial N recycling in the rumen of cattle fed all hay, grass and maize silage diets at 9.2 and 17.1 kg DMI day^{-1} (for details of simulations and diets, see Dijkstra *et al.*, 1998b).

	Grass DMI (kg day^{-1})		Hay DMI (kg day^{-1})		Maize silage DMI (kg day^{-1})	
	9.2	17.1	9.2	17.1	9.2	17.1
Rumen protozoal N mass (g N)	23.7	25.2	22.6	25.3	46.9	46.6
Rumen microbial N mass (g N)	125.5	249.5	132.7	248.0	88.5	150.3
Gross microbial N synthesis (g N day^{-1})	213.5	393.1	220.8	408.7	178.6	352.1
Microbial N recycling (% of gross synthesis)	49.1	37.9	46.8	36.3	68.3	66.5

will be learned from testing whole model behaviour and attempting to evaluate the usefulness and realism of the mechanisms described. In this way, mechanistic modelling can be a useful tool in a research programme, with the experimental and modelling objectives highly interrelated. To illustrate the contribution of modelling to increase knowledge of rumen fermentation processes, recent results of modelling VFA production will be discussed.

VFA Production and VFA Molar Proportions

The major VFAs produced in the rumen are acetic acid, propionic acid and butyric acid. The absorbed VFAs make a large contribution to the total amount of absorbed energy in the ruminant. However, extant models predict the relative proportions of individual VFAs in the rumen of cattle inaccurately (Baldwin *et al.*, 1987b; Neal *et al.*, 1992; Kohn *et al.*, 1995; Pitt *et al.*, 1996; Bannink *et al.*, 1997b,c). Dietary factors have a large effect on the proportion in which each of these acids is produced in the rumen (Murphy *et al.*, 1982; Baldwin, 1995). Since *in vivo* quantification of VFA production is difficult and expensive, only VFA concentrations in rumen fluid are measured routinely in rumen digestion trials. Molar proportions of VFAs in rumen fluid may be an unreliable estimate of the relative proportion of production rates where the fractional absorption rates of individual VFAs differ. In the last decade, data have become available on the measured rate of net (arterial minus venous) VFA appearance in the portal blood of lactating cows (Lomax and Baird, 1983; Huntington *et al.*, 1985; Reynolds and Huntington, 1988; De Visser *et al.*, 1997). However, these measurements again do not necessarily give an accurate indication of VFA production rates in the rumen. First, additional VFA appearance in the portal blood occurs because of VFA production in and absorption from the large intestine. Second, during the absorption of VFAs through the gastrointestinal tract wall, a variable part of the VFA is metabolized (for a review, see Bergman, 1990). Thus, integration is required of the processes controlling the production of individual VFAs, the absorption of VFAs from the rumen, and the metabolism of VFAs in the rumen wall.

One of the few examples of identifying the stoichiometry of rumen fermentation by a modelling approach is the work of Koong *et al.* (1975). This model requires observed quantities of five types of fermented substrate as an input and describes their conversion into either a single type of microbial mass or four types of VFAs by static relationships. A single VFA coefficient was defined for the conversion of each type of substrate into each type of VFA. By fitting predicted against observed molar proportions of each individual VFA in rumen fluid, Murphy *et al.* (1982) applied this model to estimate two distinct sets of coefficient values, one for roughage diets and one for concentrate diets, based on a large data set of substrate digestion and molar proportions of VFAs in rumen fluid of sheep and beef cattle. In evaluating the coefficient estimates obtained, they demonstrated an accurate prediction, with more than 90% of the observed variance explained, mainly because of high leverage of the much higher production rates measured in cattle as opposed to those measured in sheep. These coefficient estimates were applied in the construction of mechanistic models of whole rumen function in dairy cattle. Model evaluations however, showed that none of the models predicted VFA molar proportions accurately (Baldwin *et al.*, 1987b; Neal *et al.*, 1992; Kohn *et al.*, 1995; Bannink *et al.*, 1997b,c). Argyle and Baldwin (1988) adapted the values of the coefficients related to soluble carbohydrate and starch fermentation to account for the effect of rumen pH on fermentation stoichiometry as established *in vitro*. In evaluating these adaptations, Bannink *et al.* (1997b) yet again did not obtain an improved VFA prediction. Despite this inaccuracy, the coefficient values derived by Murphy *et al.* (1982) are thought to be the best estimates available at present (Baldwin, 1995).

Several hypotheses have been postulated to explain the inaccuracy of VFA predictions. Some of these hypotheses have been evaluated by Bannink *et al.* (1997b) by changing elements of the rumen model of Dijkstra (1994a). The alternatives evaluated were: (i) representation of discontinuous feed intake patterns; (ii) separation between rumen substrate pools of roughage as opposed to concentrate origin; (iii) changing VFA absorption kinetics; (iv) changing the VFA coefficients; and (v) representation of particle dynamics. Only the change of VFA absorption kinetics and VFA coefficients appeared to have an effect large enough to account for the prediction error of molar proportions of VFA; however, prediction error remained large (Fig. 13.2). It was concluded that, given a proper prediction of nutrient digestion in the rumen, only VFA absorption kinetics and VFA coefficients remain likely candidates to cause inaccurate VFA prediction.

In view of these simulation results, it was hypothesized further that the VFA coefficients currently applied are inaccurate, because Murphy *et al.* (1982) used fermentation data of sheep and steers fed at or somewhat above main-tenance level, which might be not representative of lactating cows with moderate to high feed intake levels. Hence, the method of Murphy *et al.* (1982) was repeated (results not published) using a large data set obtained in lactating dairy cows selected from published results and from our own laboratories. However, before deriving a new set of VFA coefficients from data with lactating cows, it seemed necessary to test the potential of accurate coefficient estimation with this model. A data set was simulated using a set of arbitrarily chosen, but physiologically meaningful, values of VFA coefficients. Next, it was determined whether the method was capable of re-estimating the coefficient values from the simulated data. Typical results obtained with an arbitrary set of VFA coefficients are shown in Fig. 13.3. These results demon-strate that the method is capable of reproducing the original set of coefficient values, except for the small coefficient values of valeric acid production. Moreover, the coefficient estimates for the production of propionic acid and butyric acid from fermented protein were more prone to error. This susceptibility may not be due to the data set used, as suggested by Murphy *et al.* (1982), but may be inherent in the statistical procedure and rumen model. The coefficients that could be estimated accurately were estimated with an error that remained below 1% of the coefficient value in general. These findings fully correspond to the results and conclusions of Murphy *et al.*

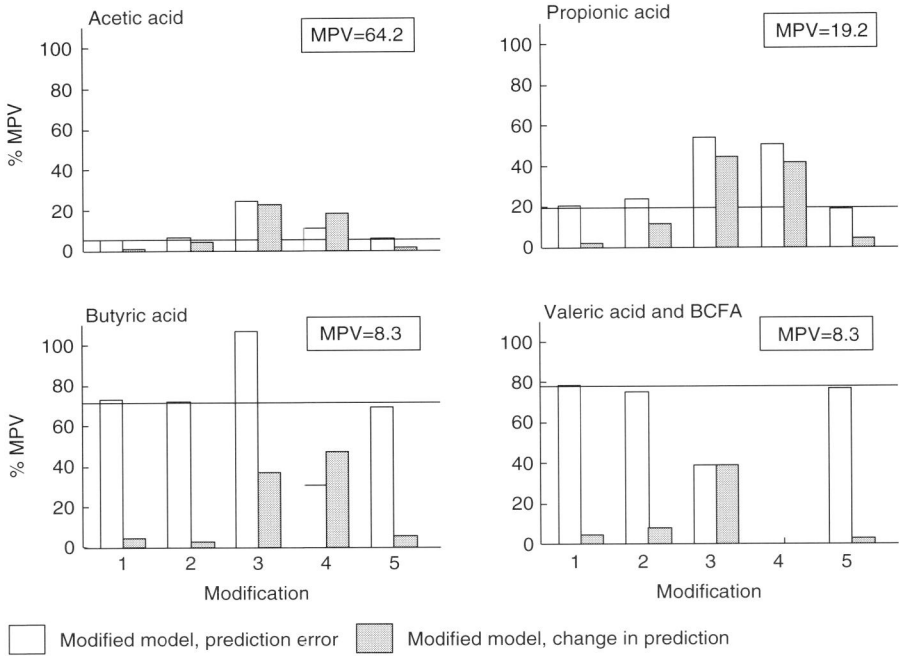

Fig. 13.2. Change (▨) and error (▢) of predicted molar proportions of rumen VFAs (both expressed as a percentage of the mean predicted value (MPV) with the original unmodified model that is indicated in the boxes) with a modified representation of five components in the model of whole-rumen function of Dijkstra (1994a). The model was modified in terms of feed intake pattern (1), distinction between roughage and concentrate substrate (2), VFA absorption kinetics (3), VFA coefficients (4) and representation of particle dynamics (5). Solid lines indicate the prediction error of the unmodified model. See Bannink *et al.* (1997b) for further details.

(1982), and were confirmed when other sets of coefficient values were used, or regression parameters (Koong *et al.*, 1975) were adapted, or with statistical weighing of data (results not shown). The average of coefficient estimates obtained from multiple runs, with a different small subset of data omitted from the total data set in each run (jack-knife procedure; Murphy *et al.*, 1982), reduced the estimation errors of all coefficients substantially to less than 0.5% of the coefficient value.

Having established that the method is capable, in principle, of accurate estimation of the full set of significant coefficient values with an ideal representation of rumen fermentation by the model, the method was applied to the data set of lactating dairy cattle described previously. It appeared that particularly the observed variation in molar proportions of acetic acid and butyric acid could not be explained well by the newly fitted coefficients (Fig. 13.4). A comparison of the molar proportions of VFAs predicted using the

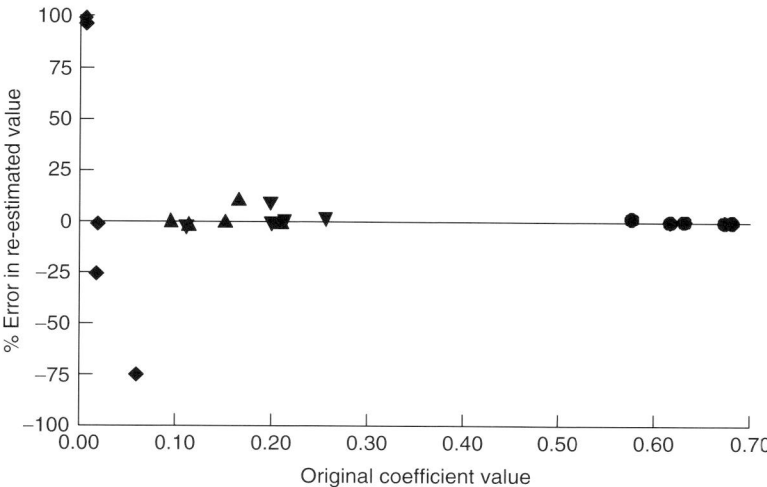

Fig. 13.3. Typical example of errors in estimated coefficient values with the method of Murphy *et al.* (1982). Coefficient values (of acetic acid production (●), of propionic acid production (▲), of butyric acid production (▼) and of valeric and branched chain fatty acid production (◆)) were estimated from a data set simulated from an arbitrarily chosen set of original coefficient values (values on horizontal axis).

newly fitted coefficient estimates with the molar proportions predicted with the coefficient values published by Murphy *et al.* (1982) shows a clear systematic shift. These findings indicate that the static model used by Murphy *et al.* (1982) is an oversimplified representation of some aspects of ruminal VFA production. The results also confirm the suggestions previously made (Dijkstra, 1994b) that additional factors related to the redox balance in the rumen need to be incorporated in order to improve the explanation of the molar proportions of rumen VFAs. These factors include the differential VFA absorption rates, and provisions for the effect of amount and rate of substrate degradation, rumen fluid pH and microbial species preferences on the type of VFA produced.

Alternative approaches to improve the prediction of molar VFA proportions were taken by Kohn and Boston (1995) and Pitt *et al.* (1996). Kohn and Boston (1995) described rumen fermentation at a lower aggregation level through the representation of thermodynamic principles of the fermentation process. This approach incorporates concepts not previously used in rumen modelling. However, the model and its evaluation results have not been fully published yet. Pitt *et al.* (1996) included lactic acid production and metabolism by rumen microorganisms and the effect of pH on these processes. The regression equations developed indicate no lactate production from fermentation of fibre, and increased lactate production in response to decreased pH of rumen fluid. The partial conversion of lactate into VFA is represented using piece-wise, pH-related, regression equations. However, their evaluation revealed that this incorporation of lactic acid dynamics, in comparison with the Murphy *et al.* (1982) coefficients, did not improve

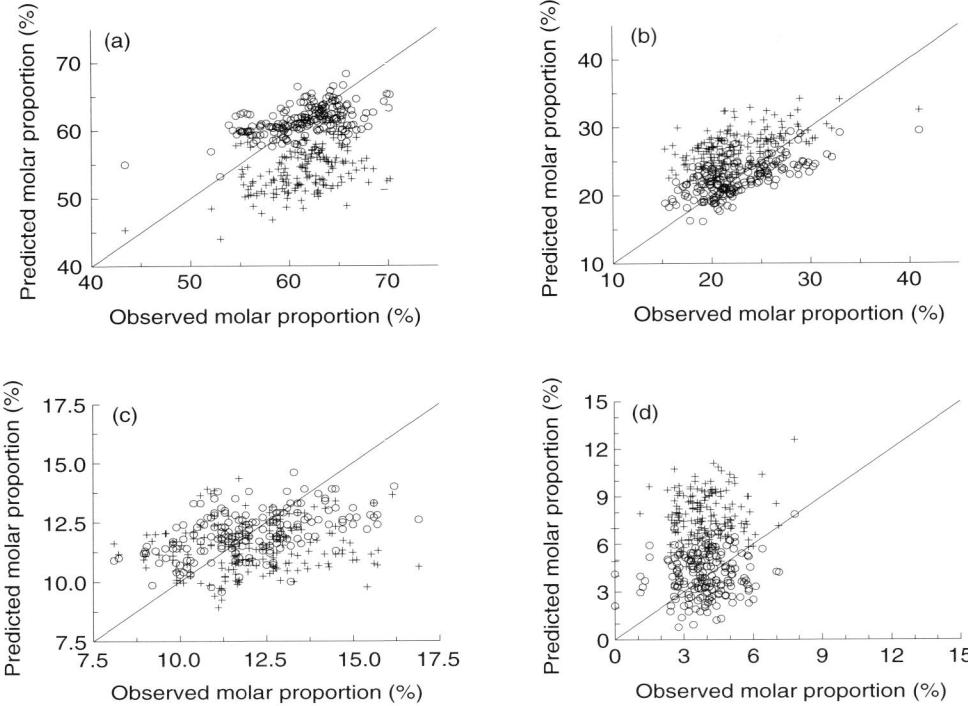

Fig. 13.4. Observed against predicted molar proportions of acetic acid (a), propionic acid (b), butyric acid (c) and valeric acid (d) with the newly fitted coefficient estimates obtained from this data set (O) and with the coefficient estimates obtained by Murphy *et al.* (1982) (+).

the accuracy of the prediction of VFA molar proportions. Remarkably, this evaluation showed that the predictions of molar proportions of, particularly, propionic and butyric acid were sensitive to dietary variations in the evaluation studies when the Murphy *et al.* (1982) coefficients were applied, but insensitive when the coefficients including lactic acid dynamics were applied. Since all the involved relationships were derived from *in vitro* results, it was inherently assumed that *in vitro* VFA and lactic acid production correspond to that occurring *in vivo*. Given the inaccuracy and insensitivity to dietary changes, this modelling exercise indicates that *in vitro* data might not be suitable to estimate the VFA molar proportions in a rumen model. Consequently, experiments are required with data on VFA that are obtained using corresponding diets or substrates *in vitro* and *in vivo*.

At present, it is not clear which approach is most promising in order to improve the prediction of molar proportions of VFA: either the incorporation of additional explaining factors in the empirical approach of Murphy *et al.* (1982), or following an approach such as Kohn and Boston (1995) to identify the mechanisms which govern the partition of substrate into individual VFAs at a lower level of aggregation. Regardless of the approach that will be chosen, the additional problem that remains to be solved is the identification of the relationship between the rumen concentration of VFAs and their rate of production, absorption and metabolism during absorption and portal appearance. In order to identify the stoichiometry of rumen fermentation and VFA production from indirect measures such as rumen VFA concentrations or rates of appearance of VFAs in portal blood, modelling techniques have been and will be helpful in creating a better understanding and in clarifying the assumptions made in order to interpret experimental results.

APPLIED WHOLE-RUMEN MODELS

Prediction of live weight gain and milk yield, and meat and milk composition from specified diets is desired to optimize farm management. Rumen models are required as the first stage in models to predict ruminant production, because the amount and type of nutrients available for absorption, and subsequently available to the tissues, generally differ largely from the profile of nutrients in the feed. These differences result from the metabolic activities of rumen microorganisms. At present, most applied models for predicting nutrient availability and ruminant production are based on empirical relationships. Empirical models usually are related directly to observational data and are simple and thereby easily adopted in practice. However, these models are not applicable outside the range of conditions covered by the data used to derive the equations. Mechanistic models allow the compartmentalization of the system, which potentially enables the use of experimental data of individual components of the system and can be applied over a wider range of conditions. Therefore, mechanistic models will probably form the basis of future feed evaluation systems, given that understanding of biological processes advances and computer capacity increases (e.g. Gill *et al.*, 1989; Beever and Cottrill, 1994; Baldwin, 1995; Hanigan *et al.*, 1997; Dijkstra *et al.*, 1998a).

Progress towards a mechanistic prediction of nutrient availability has been encouraging, but some major limitations have to be overcome. In a comparative evaluation of extant rumen models, Dijkstra and France (1996) identified the critical issues. These include the representation of discontinuous feeding regimes and the outflow of material from the rumen, the microbial distribution within the rumen, the interactions between microorganisms, the amino acid composition of undegraded feed and microbial matter, and the factors determining the VFA profile. Another topic is that of input parameters required to run a model. Bannink *et al.* (1997a) investigated the impact of diet-specific input parameters on simulated rumen function using the rumen models of Baldwin *et al.* (1987b), Danfær (1990) and Dijkstra *et al.* (1992). In addition to the major driving variables (feed intake and feed composition), each model requires a specific selection of diet-specific input parameters, including comminution rate, pH and carbohydrate partitioning parameters. These diet-specific parameters often cannot be readily derived from standard rumen measurements. Bannink *et al.* (1997a) concluded that these parameters have a large impact on the predicted availability of nutrients for absorption. Thus, it is clear that, in order to apply rumen models as a management tool, improved representation of specific aspects of rumen function is required, as well as further development of versions using parameter inputs that can be derived easily from standard rumen measurements.

Simulation of Nutrient Availability in an Applied Model

Although the accuracy of model predictions of nutrient availability on a range of diets is not ideal, mechanistic models have been applied successfully in specific dietary situations. An illustration of such an applied model is the mechanistic model of digestion and absorption of nutrients in cattle fed sugar-cane-based diets, developed by Dijkstra *et al.* (1996a). Sugarcane is an important forage in a number of tropical countries, but milk production of dairy cattle fed unsupplemented sugarcane generally is low (Leng and Preston, 1988). This sugarcane model has been employed in the search for suitable feed supplements to improve milk production, with a view to eliminate unnecessary feeding trials. The model simulates the ruminal degradation of feed, the synthesis of microbial protein and the production of VFAs in the rumen, as well as the intestinal digestion and absorption of nutrients. Next, the availability of glucose and its major precursor (propionic acid), of amino acids (from microbial and undegraded feed protein) and of long chain fatty acids is predicted. The availability of energy for metabolic processes in the cow is also predicted, calculated from glucose, amino acids, long chain fatty acid and VFA availability and their respective energy contents (see Dijkstra *et al.*, 1996b). Finally, the potential production of milk, given a fixed milk composition, is calculated based on the amount of available glucose and propionic acid, amino acids, long chain fatty acids and energy. Comparison of the potential production allows the identification of the most limiting group of nutrients.

The potential milk production of cattle according to simulated nutrient availability on sugarcane-based diets is presented in Fig. 13.5. The basal diet in these simulations consisted of sugarcane with urea (40 g of urea kg^{-1} of DM). The simulated level of absorbed amino acids with the basal diet was low in comparison with the level of absorbed energy and glucogenic substrates; this low amino acid availability was the factor which most limited milk production. *Leucaena* addition to this basal diet improves the availability of amino acids, and the availability of long chain fatty acids and energy becomes the limiting factor. Many long chain fatty acids can be synthesized from other substrates (including acetic acid), and the long chain fatty acid availability simulated by the model may not be the true limiting factor. However, the theoretical efficiency of direct incorporation of long chain fatty acids into milk lipids is much higher than the conversion of, for example, acetic acid into long chain fatty acids (Baldwin, 1995). Thus, such a shortage of long chain fatty acids will reduce the energetic efficiency of milk synthesis. The predicted potential milk production according to the most limiting nutrients available agrees well with

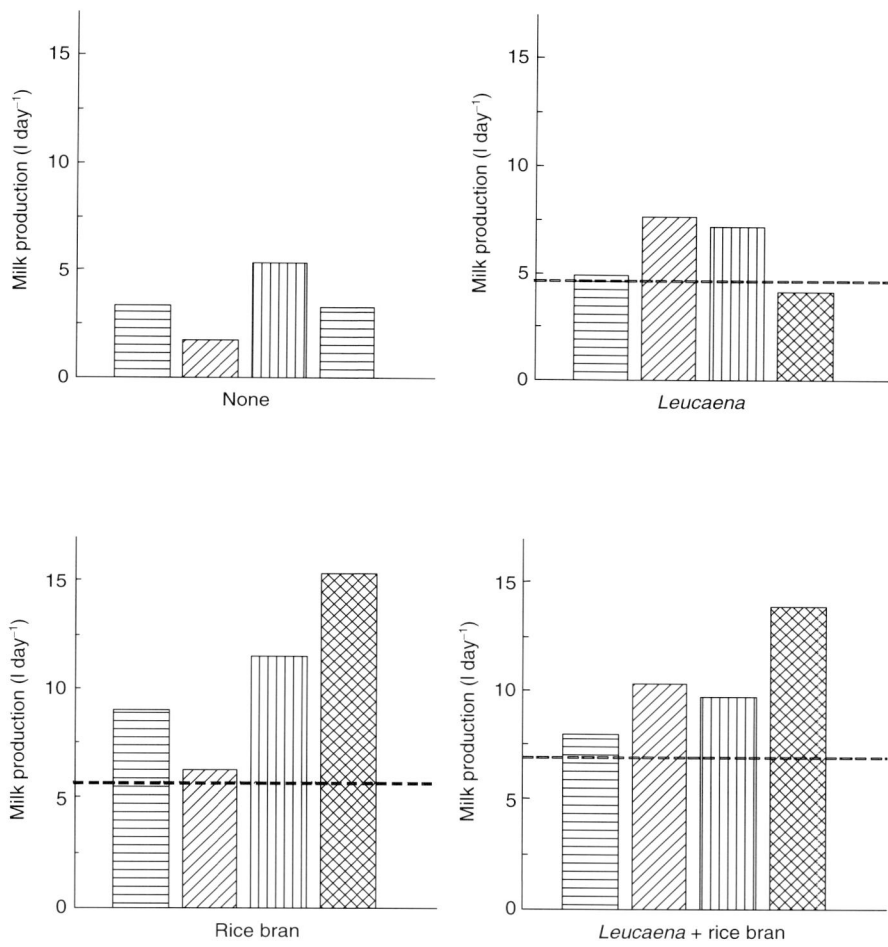

Fig. 13.5. Potential milk production according to simulated availability of energy (≡), amino acids (▨), glucose and propionate (▥) and long chain fatty acids (▩) on unsupplemented basal diet (sugarcane and 40 g urea kg⁻¹ of sugarcane DM), basal diet supplemented with *Leucaena*, basal diet supplemented with rice bran, and basal diet supplemented with *Leucaena* and rice bran. Simulations were done with the model described by Dijkstra *et al.* (1996a). The dashed line represents average milk production observed with Brown Swiss × Zebu cattle (Alvarez and Preston, 1976; Alvarez *et al.*, 1978). DMI of the basal diet is 7.5 kg day⁻¹, that for the supplemented diets was as reported by Alvarez and Preston (1976) and Alvarez *et al.* (1978).

observed milk production of Brown Swiss × Zebu cattle fed this sugarcane plus *Leucaena* diet (mean 4.6 l day⁻¹; Alvarez and Preston, 1976; Alvarez *et al.*, 1978). Rice bran supplementation of the basal diet will improve the predicted availability of energy and long chain fatty acids, but then amino acids again become most limiting (Fig. 13.5). The potential milk production according to amino acid availability is only slightly higher than the observed milk production (mean 5.5 l day^{-1}; Alvarez and Preston, 1976; Alvarez *et al.*, 1978). Finally, the supplementation with both *Leucaena* and rice bran improved the potential milk yield further, with energy availability limiting the milk production. Potential milk production predicted by the model was again slightly higher than that observed experimentally (mean 6.7 l day^{-1}; Alvarez and Preston, 1976; Alvarez *et al.*, 1978). These examples show that the model is capable of indicating those nutrients that can limit milk production, and that the model may be applied to search for supplements and to optimize supplementation levels.

Application of Models to Reduce Excretion of Wastes

The animal industry may contribute to environmental problems, particularly in intensive systems. In a number of countries, measures have been taken to reduce the negative impact of ruminant production on the environment (Tamminga, 1996). Nutrition management of farm animals has been used as a tool to maximize the output of useful products (milk, meat), and now also may become a tool to help control environmental pollution from animal production.

A particular problem in intensive ruminant production systems is the surplus of N. For example, under Dutch feeding regimens, with some 55% of the total diet fed being home grown (grass, maize silage, etc.), 75–85% of the N ingested by dairy cows is excreted in faeces and urine. Important N losses arise from rumen fermentation processes. These losses include the degradation of protein in the rumen and subsequent excretion of N in urine, and the incorporation of feed protein N into rumen microbial nucleic acids which cannot be utilized in intermediary metabolism (Tamminga, 1992). Reductions in rumen N losses are possible by reducing the dietary N level or reducing the degradation of feed protein. However, such reductions may decrease the ruminal degradation of other dietary ingredients, particularly when feed intake is high (Oldham, 1984). Another alternative to reduce rumen N losses is to supplement the diet with particular supplements aimed at an improved balance between energy and N availability in the rumen (Van Vuuren and Meijs, 1987). Current protein evaluation systems may provide information on the balance between N availability in the rumen and N incorporation into microbial matter, to give an indication of N losses in the rumen. However, these systems, including the Cornell system (Russell *et al.*, 1992; Sniffen *et al.*, 1992), do not incorporate some of the important interactions between N availability and DM degradation in the rumen. This is one of the major reasons why present protein evaluation systems are not sufficiently accurate to be used as a tool for the reduction of N losses from the rumen, as reviewed by Dijkstra *et al.* (1998a). In that review, an evaluation of different mathematical approaches to predict microbial protein supply is presented, and the importance of a sound representation of the needs of the rumen microbes for energy sources (primarily carbohydrates) and for N sources (peptides, amino acids and/or ammonia) and their interactions is emphasized. However, the structure of current protein evaluation systems does not allow useful incorporation of existing and new knowledge of rumen microbial metabolism (Dijkstra *et al.*, 1998a). In particular, in these systems, no representation of the differences in the amount of carbohydrates required for protein synthesis with ammonia or with pre-formed amino acids is included, the net efficiency of capture of degraded N by microbes cannot exceed 100% (usually between 80 and 100%), and the curvilinear response of microbial protein synthesis to the supply of N sources relative to that of carbohydrates is ignored (see also the review on N conversion of McAllan *et al.*, 1987).

A number of mechanistic rumen models do represent N transactions in the rumen and interactions with other feed ingredients. An illustration of the potential of mechanistic models for providing quantitative information on N losses in the rumen is given in Table 13.2, using the sugarcane model described before (Dijkstra *et al.*, 1996a) and various urea supplementation levels. Because of the very low N content of sugarcane, urea is usually added in practice to improve the N availability and consequently the DM degradation in the rumen. The model can be employed to examine the useful incorporation of N into microbial protein and the N fluxes to and from the rumen. Without urea addition, DM degradation and microbial activity are low, and predicted total non-ammonia nitrogen (NAN) outflow (consisting mainly of microbial NAN) is only 39.6 g of N day^{-1}. Note that because of endogenous urea input, this total NAN outflow clearly exceeds the dietary N intake (22.6 g of N day^{-1}), a result which would have been difficult to predict using the current protein evaluation systems. The ammonia concentration of rumen fluid is very low and only a small quantity of ammonia N is absorbed or washed out from the rumen (0.8 g day^{-1}). A much larger quantity of urea N is transferred into the rumen via saliva and across the rumen wall (predicted to be 17.8 g day^{-1}). This urea N originates mainly from protein (absorbed protein and body protein) deaminated in intermediary metabolism. Hence, there is a predicted net N influx (17.0 g day^{-1}) and the contribution of rumen N losses to N

Table 13.2. Simulated dietary N intake, duodenal NAN outflow, urea N transfer into the rumen, ammonia N absorption and outflow, and net N efflux from the rumen on sugarcane diets supplemented with urea fed to Holstein × Zebu cattle (DMI 7.5 kg day^{-1}) (for further details of simulations and diets, see Dijkstra *et al.*, 1996a,b).

	Urea supplementation (g kg^{-1} sugarcane DM)		
	0	20	40
N intake (g day^{-1})	22.6	88.9	152.8
Duodenal NAN outflow (g day^{-1})	39.6	75.3	75.5
Urea N transfer to the rumen (g day^{-1})	17.8	37.2	39.5
Ammonia N outflow and absorption from the rumen (g day^{-1})	0.8	50.8	116.8
Rumen net N efflux (g day^{-1})	−17.0	13.6	77.3

excretion in the urine will be low. The optimum urea supplementation level with respect to substrate degradation and microbial protein synthesis is between 20 and 40 g kg^{-} of sugarcane DM (Dijkstra *et al.*, 1996b). The simulations (Table 13.2) clearly show that increasingly larger amounts of urea N added to the feed are absorbed or washed out from the rumen (net N efflux increases from 13.6 to 77.3 g of N day^{-1} when urea supplementation is increased from 20 to 40 g kg^{-1} of DM), increasing the loss of N in urine. Thus, the model provides quantitative information on N fluxes and potential N deposition from rumen processes. With these particular diets, a steady state has been assumed, since a slow and fairly continuous ingestion of sugarcane is observed in practice. For other diets, the synchronization between carbohydrate and N release in the rumen may significantly affect the N utilization and efflux from the rumen and may require non-steady-state models (Dijkstra *et al.*, 1998a).

CONCLUSIONS

Stimulated by the work of Baldwin *et al.* (1970), several models of whole-rumen function have been developed which integrate knowledge on various aspects of the processes in the rumen. These models have been applied to increase our knowledge of fermentation processes in the rumen, to aid in the design and interpretation of experiments and to predict nutrient supply and excretion of waste. Successful application of rumen models should lead to increasing confidence and widespread use of these models, not only in research programmes but also in farm management.

REFERENCES

Alvarez, F.J. and Preston, T.R. (1976) *Leucaena leucocephala* as protein supplement for dual purpose milk and weaned calf production on sugarcane based ratios. *Tropical Animal Production* 1, 112–118.

Alvarez, F.J., Wilson, A. and Preston, T.R. (1978) *Leucaena leucocephala* as protein supplement for dual purpose milk and weaned calf production on sugarcane based diets: comparisons with rice polishings. *Tropical Animal Production* 3, 51–55.

Argyle, J.L. and Baldwin, R.L. (1988) Modelling of rumen water kinetics and effects of rumen pH changes. *Journal of Dairy Science* 71, 1178–1188.

Baldwin, R.L. (1995) *Modelling Ruminant Digestion and Metabolism.* Chapman & Hall, London.

Baldwin, R.L., Lucas, H.L. and Cabrera, R. (1970) Energetic relationships in the formation and utilization of fermentation end-products. In: Phillipson, A.T., Annison, E.F., Armstrong. D.G., Balch, C.C., Comline, R.S., Hardy, R.S., Hobson, P.N. and Keynes, R.D. (eds), *Physiology of Digestion and Metabolism in the Ruminant.* Oriel Press, Newcastle-upon-Tyne, UK, pp. 319–334.

Baldwin, R.L., Koong, L.J. and Ulyatt, M.J. (1977) A dynamic model of ruminant digestion for evaluation of factors affecting nutritive value. *Agricultural Systems* 2, 255–288.

Baldwin, R.L., France, J. and Gill, M. (1987a) Metabolism of the lactating cow. I. Animal elements of a mechanistic model. *Journal of Dairy Research* 54, 77–105.

Baldwin, R.L., Thornley, J.H.M. and Beever, D.E. (1987b) Metabolism of the lactating cow. II. Digestive elements of a mechanistic model. *Journal of Dairy Research* 54, 107–131.

Bannink, A. and De Visser, H. (1997) Comparison of mechanistic rumen models on mathematical formulation of extramicrobial and microbial processes. *Journal of Dairy Science* 80, 1296–1314.

Bannink, A., De Visser, H., Dijkstra, J. and France, J. (1997a) Impact of diet-specific input parameters on simulated rumen function. *Journal of Theoretical Biology* 184, 371–384.

Bannink, A., De Visser, H., Klop, A., Dijkstra, J. and France, J. (1997b) Causes of inaccurate prediction of volatile fatty acids by simulation of rumen function in lactating cows. *Journal of Theoretical Biology* 189, 353–366.

Bannink, A., De Visser, H. and Van Vuuren, A.M. (1997c) Comparison and evaluation of mechanistic rumen models. *British Journal of Nutrition* 78, 563–581.

Beever, D.E. and Cottrill, B.R. (1994) Protein systems for feeding ruminant livestock: a European assessment. *Journal of Dairy Science* 77, 2031–2043.

Bergman, E.N. (1990) Energy contribution of volatile fatty acids from the gastro-intestinal tract in various species. *Physiological Reviews* 70, 567–590.

Black, J.L., Beever, D.E., Faichney, G.J., Howarth, B.R. and Graham, N.McC. (1981) Simulation of the effects of rumen function on the flow of nutrients from the stomach of sheep: part 1 – description of a computer program. *Agricultural Systems* 6, 195–219.

Danfær, A. (1990) A dynamic model of nutrient digestion and metabolism in lactating dairy cows. PhD thesis, National Institute of Animal Science, Foulum, Denmark.

De Visser, H., Valk, H., Klop, A., Van der Meulen, J., Bakker, J.G.M. and Huntington, G.B. (1997) Nutrient fluxes in splanchnic tissue in dairy cows: influence of grass quality. *Journal of Dairy Science* 80, 1666–1673.

Dijkstra, J. (1994a) Simulation of the dynamics of protozoa in the rumen. *British Journal of Nutrition* 72, 679–699.

Dijkstra, J. (1994b) Production and absorption of volatile fatty acids in the rumen. *Livestock Production Science* 39, 61–69.

Dijkstra, J. and France, J. (1995) Modelling and methodology in animal science. In: Danfær, A. and Lescoat, P. (eds), *Proceedings of the Fourth International Workshop on Modelling Nutrient Utilisation in Farm Animals*. National Institute of Animal Science, Foulum, Denmark, pp. 9–18.

Dijkstra, J. and France, J. (1996) A comparative evaluation of models of whole rumen function. *Annales de Zootechnie* 45 (Suppl. 1), 175–192.

Dijkstra, J. and Tamminga, S. (1995) Simulation of the effects of diet on the contribution of rumen protozoa to degradation of fibre in the rumen. *British Journal of Nutrition* 74, 617–634.

Dijkstra, J., Neal, H.D.St.C., Beever, D.E. and France, J. (1992) Simulation of nutrient digestion, absorption and outflow in the rumen: model description. *Journal of Nutrition* 122, 2239–2256.

Dijkstra, J., France, J., Neal, H.D.St.C., Assis, A.G., Aroeira, L.J.M. and Campos, O.F. (1996a) Simulation of digestion in cattle fed sugar cane: model development. *Journal of Agricultural Science, Cambridge* 127, 231–246.

Dijkstra, J., France, J., Assis, A.G., Neal, H.D.St.C., Campos, O.F. and Aroeira, L.J.M. (1996b) Simulation of digestion in cattle fed sugar cane: prediction of nutrient supply for milk production with locally available supplements. *Journal of Agricultural Science, Cambridge* 127, 247–260.

Dijkstra, J., France, J. and Davies, D.R. (1998a) Different mathematical approaches to estimating microbial protein supply in ruminants. *Journal of Dairy Science* 81, 3370–3384.

Dijkstra, J., France, J. and Tamminga, S. (1998b) Quantification of the recycling of microbial nitrogen in the rumen using a mechanistic model of rumen fermentation processes. *Journal of Agricultural Science, Cambridge* 130, 81–94.

Erdman, R.A. (1988) Dietary buffering requirements of the lactating dairy cow: a review. *Journal of Dairy Science* 71, 3246–3266.

Firkins, J.L. (1996) Maximizing microbial protein synthesis in the rumen. *Journal of Nutrition* 126, 1347S–1354S.

Firkins, J.L., Weiss, W.P. and Piwonka, E.J. (1992) Quantification of intraruminal recycling of microbial nitrogen using nitrogen-15. *Journal of Animal Science* 70, 3223–3233.

France, J. and Thornley, J.H.M. (1984) *Mathematical Models in Agriculture.* Butterworths, London.

France, J., Thornley, J.H.M. and Beever, D.E. (1982) A mathematical model of the rumen. *Journal of Agricultural Science, Cambridge* 99, 343–353.

Gill, M., Beever, D.E. and France, J. (1989) Biochemical bases needed for the mathematical representation of whole animal metabolism. *Nutrition Research Reviews* 2, 181–200.

Grant, R.J. and Mertens, D.R. (1992a) Development of buffer systems for pH control and evaluation of pH effects on fiber digestion *in vitro. Journal of Dairy Science* 75, 1581–1587.

Grant, R.J. and Mertens, D.R. (1992b) Influence of buffer pH and raw corn starch addition on *in vitro* fiber digestion kinetics. *Journal of Dairy Science* 75, 2762–2768.

Hanigan, M.D., Dijkstra, J., Gerrits, W.J.J. and France, J. (1997) Modelling post-absorptive protein and amino acid metabolism in the ruminant. *Proceedings of the Nutrition Society* 56, 631–643.

Hobson, P.N. (ed.) (1988) *The Rumen Microbial Ecosystem.* Elsevier Science Publishers, London.

Hobson, P.N. and Stewart, C.S. (eds) (1997) *The Rumen Microbial Ecosystem*, 2nd edn. Blackie Academic and Professional, London.

Hoover, W.H., Kincaid, C.R., Varga, G.A., Thayne, W.V. and Junkins, L.L., Jr (1984) Effects of solid and liquid flows on fermentation in continuous cultures. IV. pH and dilution rate. *Journal of Animal Science* 58, 692–699.

Huntington, G.B. and Tyrrell, H.F. (1985) Oxygen consumption by portal-drained viscera of cattle: comparison of analytical techniques and relationship to whole body oxygen consumption. *Journal of Dairy Science* 68, 2727–2731.

Jouany, J.P., Demeyer, D.I. and Grain, J. (1988) Effect of defaunating the rumen. *Animal Feed Science and Technology* 21, 229–265.

Kohn, R.A. and Boston, R.C. (1995) The application of thermodynamic principles to understand changes in rumen metabolism. *Journal of Dairy Science* 78 (Suppl. 1), 445.

Kohn, R.A., Boston, R.C., Ferguson, J.D. and Chalupa, W. (1995) The integration and comparison of dairy cow models. In: Danfær, A. and Lescoat, P. (eds), *Proceedings of the Fourth International Workshop on Modelling Nutrient Utilisation in Farm Animals.* National Institute of Animal Science, Foulum, Denmark, pp. 117–128.

Koong, L.J., Baldwin, R.L., Ulyatt, M.J. and Charlesworth, T.J. (1975) Iterative computation of metabolic flux and stoichiometric parameters for alternate pathways in rumen fermentation. *Computer Programs in Biomedicine* 4, 209–213.

Leng, R.A. and Preston, T.R. (1988) Constraints to the efficient utilization of sugarcane and its byproducts as diets for production of large ruminants. In: Sansouay, R., Aarts, G. and Preston, T.R. (eds), *Sugarcane as Feed*. FAO, Rome, pp. 284–309.

Lescoat, P. and Sauvant, D. (1995) Development of a mechanistic model for rumen digestion validated using the duodenal flux of amino acids. *Reproduction, Nutrition, Developpement* 35, 45–70.

Lomax, M.A. and Baird, G.D. (1983) Blood flow and nutrient exchange across the liver and gut of the dairy cow. Effects of lactation and fasting. *British Journal of Nutrition* 49, 481–496.

McAllan, A.B., Siddons, R.C. and Beever, D.E. (1987) The efficiency of conversion of degraded nitrogen to microbial protein in the rumen of cattle and sheep. In: Jarrige, R. and Alderman, G. (eds), *Agriculture. Feed Evaluation and Protein Requirement Systems for Ruminants*. Commission of the European Communities, Luxembourg, pp. 111–128.

Mould, F.L. and Ørskov, E.R. (1984) Manipulation of rumen fluid pH and its influence on cellulolysis *in sacco*, dry matter degradation and the rumen microflora of sheep offered either hay or concentrate. *Animal Feed Science and Technology* 10, 1–11.

Murphy, M.R., Baldwin, R.L. and Koong, L.J. (1982) Estimation of stoichiometric parameters for rumen fermentation of roughage and concentrate diets. *Journal of Animal Science* 55, 411–421.

Murphy, M.R., Baldwin, R.L., Ulyatt, M.J. and Koong, L.J. (1983) A quantitative analysis of rumination patterns. *Journal of Animal Science* 56, 1236–1240.

Murphy, M.R., Baldwin, R.L. and Ulyatt, M.J. (1986) An update of a dynamic model of rumen fermentation. *Journal of Animal Science* 62, 1412–1422.

Neal, H.D.St.C., Dijkstra, J. and Gill, M. (1992) Simulation of nutrient digestion, absorption and outflow in the rumen: model evaluation. *Journal of Nutrition* 122, 2257–2272.

O'Connor, J.D., Sniffen, C.J., Fox, D.G. and Chalupa, W. (1993) A net carbohydrate and protein system for evaluating cattle diets: IV. Predicting amino acid adequacy. *Journal of Animal Science* 70, 3551–3561.

Oldham, J.D. (1984) Protein–energy interrelationships in dairy cows. *Journal of Dairy Science* 67, 1090–1114.

Pitt, R.E., Van Kessel, J.S., Fox, D.G., Pell, A.N., Barry, M.C. and Van Soest, P.J. (1996) Prediction of ruminal volatile fatty acids and pH within the net carbohydrate and protein system. *Journal of Animal Science* 74, 226–244.

Preston, T.R. and Leng, R.A. (1987) *Matching Ruminant Production Systems with Available Resources in the Tropics and Sub-Tropics*. Penambul Books, Armidale, Australia.

Reichl, J.R. and Baldwin, R.L. (1975) Rumen modelling: rumen input–output balance models. *Journal of Dairy Science* 58, 879–890.

Reichl, J.R. and Baldwin, R.L. (1976) A rumen linear programming model for evaluation of concepts of rumen microbial function. *Journal of Dairy Science* 59, 439–454.

Reynolds, P.J. and Huntington, G.B. (1988) Net portal absorption of volatile fatty acids and L(+)-lactate by lactating Holstein cows. *Journal of Dairy Science* 71, 124–133.

Russell, J.B., O'Connor, J.D., Fox, D.G., Van Soest, P.J. and Sniffen, C.J. (1992) A net carbohydrate and protein system for evaluating cattle diets: I. Ruminal fermentation. *Journal of Animal Science* 70, 3551–3561.

Sauvant, D., Dijkstra, J. and Mertens, D. (1995) Optimization of ruminal digestion: a modelling approach. In: Journet, M., Grenet, E., Farce, M.-H., Thériez, M. and Demarquilly, C. (eds), *Recent Developments in the Nutrition of Herbivores*. INRA Editions, Paris, pp. 143–165.

Shriver, B.J., Hoover, W.H., Sargent, J.P., Crawford, R.J. Jr and Thayne, W.V. (1986) Fermentation of a high concentrate diet as affected by ruminal pH and digesta flow. *Journal of Dairy Science* 69, 413–419.

Sniffen, C.J., O'Connor, J.D., Van Soest, P.J., Fox, D.G. and Russell, J.B. (1992) A net carbohydrate and protein system for evaluating cattle diets: II. Carbohydrate and protein availability. *Journal of Animal Science* 70, 3562–3577.

Stewart, C.S. (1977) Factors affecting the cellulolytic activity of rumen contents. *Applied and Environmental Microbiology* 33, 497–502.

Tamminga, S. (1992) Nutrition management of dairy cows as a contribution to pollution control. *Journal of Dairy Science* 75, 345–357.

Tamminga, S. (1996) A review on environmental impacts of nutritional strategies in ruminants. *Journal of Animal Science* 74, 3112–3124.

Tamminga, S., Van Vuuren, A.M., Van der Koelen, C.J., Ketelaar, R.S. and Van der Togt, P.L. (1990) Ruminal behaviour of structural carbohydrates, non-structural carbohydrates and crude protein from concentrate ingredients in dairy cows. *Netherlands Journal of Agricultural Science* 38, 513–526.

Terry, R.A., Tilley, J.M.A. and Outen, G.E. (1969) Effect of pH on cellulose digestion under *in vivo* conditions. *Journal of the Science of Food and Agriculture* 20, 317–320.

Van Vuuren, A.M. and Meijs, J.A.C. (1987) Effect of herbage composition and supplement feeding on the excretion of nitrogen in dung and urine by grazing dairy cows. In: Van der Meer, N.G., Unwin, R.J., Van Dijk, T.A. and Emmink, G.C. (eds), *Animal Manure on Grassland and Fodder Crops*. Martinus Nijhoff, Dordrecht, The Netherlands, pp. 17–28.

Wallace, R.J. and McPherson, C.A. (1987) Factors affecting breakdown of bacterial protein in rumen fluid. *British Journal of Nutrition* 58, 313–323.

Williams, A.G. and Coleman, G.S. (1992) *The Rumen Protozoa*. Springer-Verlag, New York.

14 Modelling the Lactating Dairy Cow

R.L. Baldwin and K.C. Donovan

Department of Animal Science, University of California, Davis, California, USA

INTRODUCTION

A dynamic, mechanistic model of digestion and metabolism within the lactating dairy cow will be used to illustrate the evolution of a model through an interplay of modelling and experimental research. Further, the model will be used to illustrate the potential utility of the use of dynamic, mechanistic models in support of management decision-making in animal agriculture. Mathematical modelling techniques assist researchers in a better understanding of animals and animal production systems. Models are constructed with the intention of evaluating hypotheses and advancing science. After model construction and evaluation, proper experimentation increases the usefulness of the data gathered both qualitatively and quantitatively. When differential equations that describe changes of state variables and constants used to implement either Michaelis–Menten enzyme or mass–action kinetics used in the model are valid, simulated results will reflect reality. If not, the results of simulations reveal inadequacies in our current understanding of the animal system.

BACKGROUND AND DEFINITIONS

Models can be classified in a number of different ways. The classification we prefer in accord with our objective follows Fig. 14.1.

In this context, dynamic models are made up of differential equations of the form $dx/dt = F(1) + F(2) ..... F(i)$ and are solved over time using a numerical integration technique. Models made up of equations of this form allow users to trace the behaviour of a system through time such that simulations accommodate both the quantitative and dynamic domains. This is important in modelling animal systems since animals change over time, and

Dynamic	vs	Static
Deterministic	vs	Stochastic
Mechanistic	vs	Empirical

Fig. 14.1.   Mocel classification (adapted from Thornley and France, 1984).

often past or current management decisions influence subsequent function. Static models are usually algebraic in form, contain no time-dependent variables and are solved once for a given set of conditions specified as input. Static models have proven to be quite useful over the past 100 years but have a number of severe limitations, which have been discussed elsewhere (Thornley and France, 1984; Robson and Poppi, 1990; Forbes and France, 1993; Baldwin, 1995; and others).

Deterministic models yield a single answer for each simulation run since all parameters are entered as exact values. The simulation outputs are nominally considered to represent the average animal of the population. The term stochastic is usually taken to reflect either uncertainty regarding cause or effect relationships within the system or true sources of variance in a population of animals. In both cases, a model becomes stochastic when parameters are specified as a mean value ± a standard deviation, and the value used in the model is allowed to vary randomly in sequential runs where ranges specified by the standard deviation and a random number generator define the actual parameter values used. Stochastic models must be run a number of times in order to obtain an estimate of the population mean. Such models also yield estimates of variance within the simulated population. Introduction of stochastic elements, which reflect a lack of understanding of cause and effect relationships within the system, is totally unacceptable given our research objectives. On the other hand, given a basically sound deterministic model and specific knowledge that one or more specific (genetic) traits are major causes of variation in responses among animals, it can be quite instructive to introduce these as stochastic elements into the model and ask the question: how much of the total variance in a population can be explained on the basis of these one or two variable traits?

The term empirical usually applies to regression equations which fit to a specific data set and are used to describe relationships among two or more variables observed in the animal. The only constraint imposed upon empirical equations is that the equations represent a statistical best fit of the data set. Coefficients in the equations need not and most often do not imply anything about underlying relationships. Empirical equations must be applied carefully to ensure that they are not used to simulate situations not defined by the data set upon which they are based.

Mechanistic, sometimes called theoretical, equations are, by definition, not derived from a particular set of data but rather from concepts regarding the fundamental nature of the system. The concepts upon which mechanistic equations and, ideally, parameter values used to implement the equations are based arise from studies of lower level functions conducted at the organ, tissue or cell levels. Well-conceived models based upon mechanistic equations should have explanatory power and apply generally (for more detailed discussions, see France and Thornley, 1984; Baldwin, 1995).

Our current model of digestion and metabolism of a lactating dairy cow is a dynamic, deterministic and mechanistic model. Dynamic, mechanistic models have several properties which enhance our ability to anticipate effects of previous and current planes of nutrition upon current and future performance, and have explanatory power that has helped advance our understanding of ruminant digestion and metabolism.

THE LACTATING COW MODEL

Model Description

A block diagram of our basic model of the metabolism of a lactating cow is presented in Fig. 14.2. The model looks quite complex at first glance because of interactions among nutrients and pools, but in truth is quite simple as only 12 state variables are represented in the model. Of these, blood acetate, amino acids, lipids (as fatty acids) and glucose; body and visceral protein; storage triacylglyceride; and milk yield are presented in Fig. 14.2. Urea, the lactation hormone complex, udder enzymatic capacity and milk in the udder are the additional state variables.

Initial formulation of the model required an extensive survey of the literature to ascertain relative organ weights and energy expenditures, blood and tissue metabolite levels under varying conditions, rates of nutrient turnover and oxidation in intact animals and individual tissues, arteriovenous differences for specific organs, and many more data (Baldwin, 1995). In all modelling studies, problems are encountered with the availability of data required for unique definition of equation forms and parameter values. Some of these limitations compromise the goal stated above that all concepts and numerical inputs to mechanistic models be firmly based upon experimental data. We are constantly confronted with this problem even though our experimental programme for the past 25 years has been dedicated to the collection of required data. Numerous examples of cases where our mechanistic models were found to be inadequate at either the conceptual or numerical input levels, or both, and experiments undertaken to alleviate these problems were discussed by Baldwin (1995). Only one example of approaches undertaken to correct these problems will be presented here. Note in the block diagram in Fig. 14.3 that four equations determine milk synthesis (one arrow equals one equation):

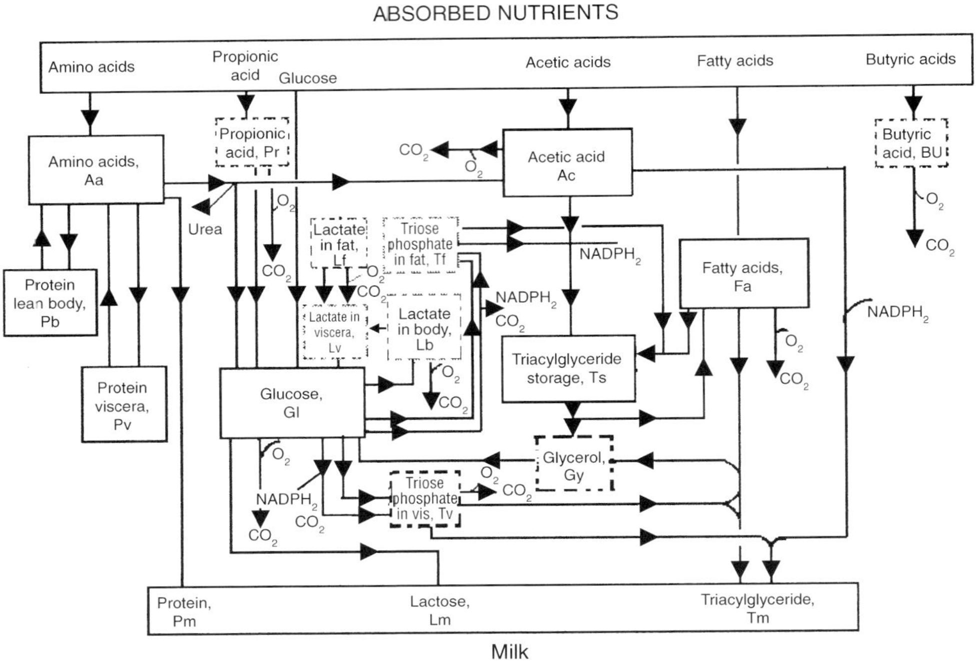

ABSORBED NUTRIENTS

Fig. 14.2. Diagrammatic representation of parts of the metabolic element of the lactating dairy cow model. Boxes enclosed by solid lines indicate state variables. Boxes enclosed by broken lines indicate metabolites explicitly considered but not represented as state variables. Arrows indicate fluxes (adapted from Baldwin, 1995).

The equation for milk protein (Pm) synthesis is

$$U_{Aa,Pm} = V_{Aa,Pm} (UENZ)/(1.0 + k_{Aa,Pm}/cAa) \tag{14.1}$$

where $U_{Aa,Pm}$ is the rate of amino acid (Aa) incorporation into milk protein in mol day^{-1}, $V_{Aa,Pm}$ is the scalar used to convert udder enzyme (UENZ) to the current mammary gland enzymatic capacity for protein synthesis to the maximal capacity analogous to the V_{max} of a Michaelis–Menten equation for the reaction in mol day^{-1}, $k_{Aa,Pm}$ is the apparent affinity (mol day^{-1}) of the udder for amino acid incorporation into protein, and cAa is the current concentration of amino acid in blood in mol l^{-1};

The equations for incorporation of blood lipids (FA) and acetate (Ac) into milk fat (Tm) in mol day^{-1} are

$$U_{FA,Tm} = V_{FA,Tm} (UENZ)/(1.0 + k_{FA,Tm}/cFA + k1_{FA,Tm}/cGl) \tag{14.2}$$

$$U_{Ac,Tm} = V_{Ac,Tm} (UENZ)/(1.0 + k_{Ac,Tm}/cAc + k1_{Ac,Tm}/cGl) \tag{14.3}$$

where $V_{FA,Tm}$ and $V_{Ac,Tm}$ are scalars for UENZ, $k_{FA,Tm}$ and $k_{Ac,Tm}$ are apparent udder affinities for blood lipids and acetate (mol l^{-1}), $k1_{FA,Tm}$ and $k1_{Ac,Tm}$ are apparent affinities (mol l^{-1}) of the udder for glucose use for esterification of FA and the use of glucose as a source of NADPH$_2$ for fatty acid formation from acetate and esterification of resulting fatty acids, respectively.

The equation for lactose (Lm) synthesis is

$$U_{Gl,Lm} = V_{Gl,Lm} (\text{UENZ})/(1.0 + k_{Gl,Lm}/c\text{Gl} + k1_{Gl,Lm}/c\text{Aa}) \tag{14.4}$$

where $U_{Gl,Lm}$ is the rate of utilization of glucose (Gl) for lactose (Lm) synthesis in mol day^{-1}, $V_{Gl,Lm}$ is a scalar, $k_{Gl,Lm}$ is the affinity (mol l^{-1}) of the udder for glucose use in lactose synthesis and $k1_{Gl,Lm}$ is the affinity (mol l^{-1}) of the udder for amino acid use for α-lactalbumin synthesis which is, in turn, essential to lactose synthesis. These four equations are clearly highly aggregated, as is appropriate for animal level models. Formulation and parameterization of these equations was not at all straightforward or possible on the basis of whole-animal input–output data. In order to help resolve this dilemma in the development of the mammary element of the whole-animal model, two approaches were utilized. The first was to conduct extensive studies of mammary tissue metabolism using mainly *in vitro* radiotracer techniques and *in vivo* arteriovenous difference studies across the udder in order to establish appropriate input–output relationships and resolve such issues as pathways of nutrient utilization under various conditions and, most importantly, interactions among nutrients. The second approach has been to utilize these and other (enzyme, metabolite, etc.) data to formulate detailed models of tissue metabolism which can then be used to identify dominant features of tissue metabolism which must be captured in equations used at the animal level and to help parameterize these highly aggregated representations. Initially, Equation 14.1 was considered adequate to the simulation of milk protein synthesis. Analyses conducted using the mammary model, and attendant data used to construct the detailed model, indicated that changes in the availabilities of amino acids, in general, were required for correct simulations of the effects of specific diets upon protein concentrations in milk. These observations led to revision of the whole-animal model (see comments on AaMolly discussed below) In other words, detailed analyses led to revision of Equation 14.1. The same argument was found to apply to lactose synthesis in the mammary model, and appropriate changes were substituted for Equation 14.4 in the whole-animal model. In contrast, detailed analyses of metabolic patterns using the mammary model indicated that interactions represented in Equations 14.2 and 14.3 were adequate to simulations of whole-animal metabolism.

A block diagram of our current model of cow mammary gland metabolism is presented in Fig. 14.3 by way of illustration. Abbreviations used in the block diagram are defined in Table 14.1.

The *in vitro* data summarized in Figs 14.4–14.7 and Table 14.2 are presented to illustrate the utility of such data in modelling tissue metabolism. Figures 14.4 and 14.5 illustrate a common result in that most tissues exhibit saturation kinetics *in vitro*. Such observations clearly suggest that Michaelis–Menten (saturation kinetic) type equations are appropriate to represent most tissue level transactions even though the concentrations required to saturate the system may be considerably above the concentrations normally encountered under normal physiological conditions. The data

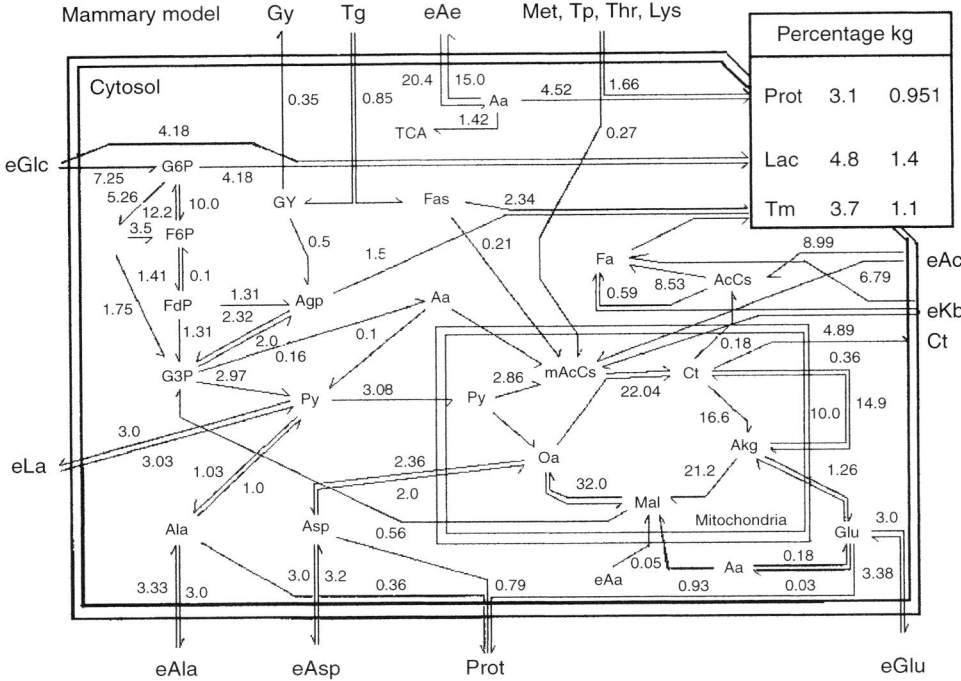

Fig. 14.3. Block diagram of the cow mammary gland model. Solid double lines enclose cytosolic and mitochondrial spaces. Pools are identified by two-, three- or four-letter codes. Arrows between pools indicate fluxes, and numbers associated with the arrows are the fluxes at steady-state concentrations (M day⁻¹). Extracellular space is that area outside of the cytosolic space. Pool sizes of extracellular nutrients do not change throughout a particular simulation, while pool sizes of intracellular nutrients and products do change dependent upon the relevant inputs and outputs (fluxes) to the pool. Abbreviations other than those defined in standard biochemical tests are milk fat (Tm), cytosolic acetyl-CoA (AcCs), mitochondrial acetyl-CoA (mAcCs) (adapted from Baldwin, 1995).

presented in Figs 14.4 and 14.5 indicate a feature of experimental design which is essential to gaining insights into critical tissue metabolic properties. This is the use of radioisotope tracers labelled in specific positions. The difference in the patterns of lactate-1-14C and lactate-2-14C oxidation in Fig. 14.4 reflects the fact that most of the conversion of lactate-1-14C to CO₂ is due to the action of pyruvate dehydrogenase, while most of the oxidation of lactate-2-14C reflects the conversion of lactate to acetyl-CoA which subsequently is oxidized via the tricarboxylic acid cycle. The difference between rates of oxidation of the two radiolabelled carbons of lactate also indicates how much acetyl-CoA formed from lactate via the pyruvate dehydrogenase reaction is metabolized via pathways other than the tricarboxylic acid pathway.

Similarly, the differences between glucose-1-14C and glucose-2-14C oxidation depicted in Fig. 14.5 provide insight regarding the relative use of glucose-6-P via the pentose phosphate and Embden–Meyerhof pathways and,

Table 14.1. Extracellular and intracellular concentrations of metabolites adopted to simulate reference state[a].

Extracellular metabolite code[b]	Chemical name	Intracellular metabolite code[c]	Chemical name
eAe	Amino acids	Aa	Amino acids
Ac	Acetate	AcCs	Cytosolic acetyl-CoA
eAla	Alanine	Ad	ADP
eAsp	Aspartate + asparagine	Agp	α-Glycerol phosphate
Glc	Glucose	Akg	α-Ketoglutarate
eGlu	Glutamate	Ala	Alanine
Kb	Ketone bodies	Asp	Aspartate
La	Lactate	At	ATP
eLys	Lysine	Cd	Carbon dioxide
eMet	Methionine	Cs	Cytosolic CoA
Tg	TAG + NEFA + glycerol	Ct	Citrate
eThr	Threonine	F6P	Fructose-6-phosphate
TP	Tyrosine + phenylalanine	FaCs	Fatty acyl-CoA
		Fd	FAD
		Fdh	FADH$_2$
		Fbp	Fructose-1,6-bisphosphate
		G3P	Glycerol-3-phosphate
		G6P	Glucose-6-phosphate
		Glu	Glutamate
		Gy	Glycerol
		mAcCs	Mitochondrial acetyl-CoA
		Mal	Malate
		mCs	Mitochondrial CoA
		Nd	NAD
		Ndh	NADH
		NH$_3$	Ammonia
		Np	NADP
		Nph	NADPH
		O$_2$	Oxygen
		Oa	Oxaloacetate
		Py	Pyruvate
		Ru	Ribulose-5-phosphate

[a]Values were derived from blood metabolite data presented by Smith (1970), Miller *et al.* (1991), and Hanigan *et al.* (1992).
[b]The prefix e indicates concentration in the extracellular as distinct from the intracellular space; TAG = triacylglycerol; and NEFA = non-esterified fatty acids.
[c]Derived from data summarized by Baldwin and Yang (1974).

in this case, the effect of acetate availability upon the contributions of these two alternative pathways. These types of data can be evaluated quantitatively and objectively using tissue level models, and are very helpful or even essential in formulating equations for animal level models.

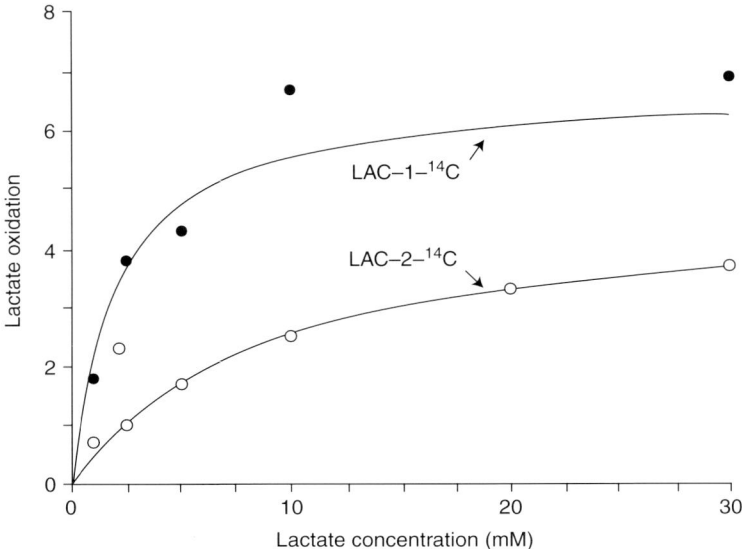

Fig. 14.4. Oxidation of [1-^{14}C]lactate (LAC-1-^{14}C) and [2-^{14}C]lactate (LAC-2-^{14}C) in the presence of 2 mM glucose and 2 mM acetate in cow mammary tissue (adapted from Forsberg *et al.*, 1985b).

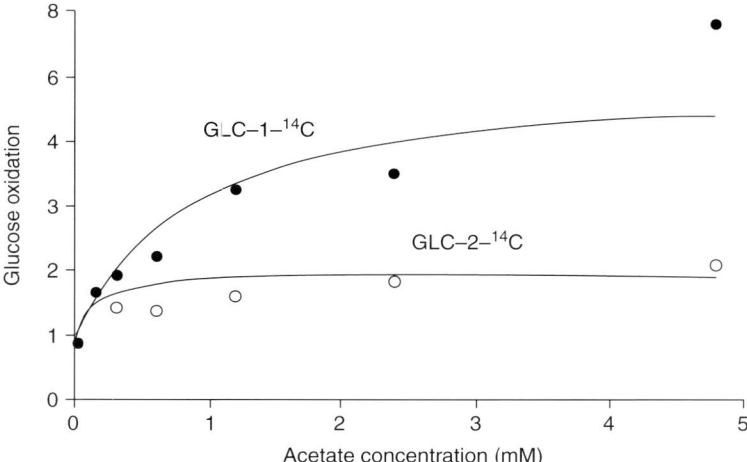

Fig. 14.5. Effects of acetate concentrations upon oxidation rates of [1-^{14}C]glucose (GLC-1-^{14}C) and [2-^{14}C]glucose (GLC-2-^{14}C) in cow mammary tissue (adapted from Forsberg *et al.*, 1985a).

Figures 14.6 and 14.7 illustrate the importance of considering nutrient interactions in tissue level models.

The data in Fig. 14.6 clearly indicate that rates of lactate conversion to fatty acids in cow mammary tissue are linear functions of lactate concentrations up to supraphysiological concentrations. However, at physiological

Table 14.2. Effect of acetate on apparent kinetic parameters for [1-^{14}C]glucose use.

Product	Acetate (mM)	K_{glc}(mM)[a]	V_{max}[b]
Lactose	2	3.4	2.3
	8	1.9	1.9
Glycerol	2	1.6	1.2
	8	0.8	0.6
Citrate	2	8.1	1.3
	8	0.9	0.4

[a]Apparent affinity for conversion of glucose to product.
[b]Apparent maximum capacities for conversion of glucose-1-^{14}C to product expressed as μg of atoms tracer converted $(g \times h)^{-1}$.

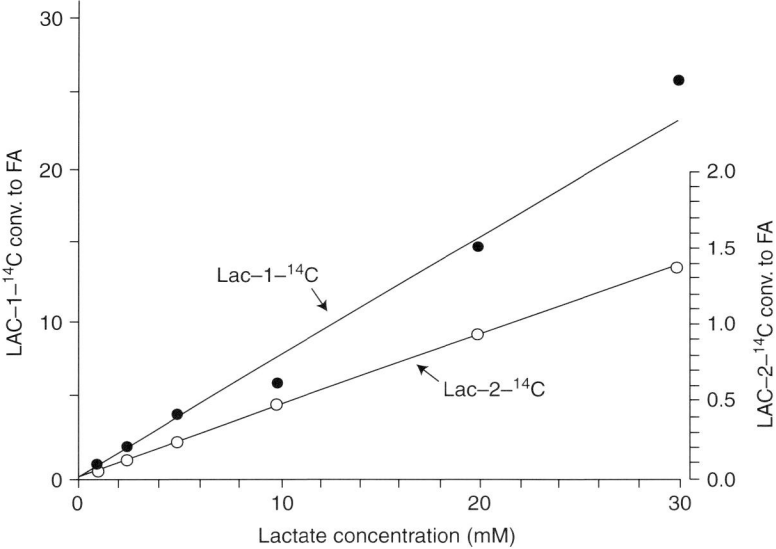

Fig. 14.6. Conversion of lactate carbons 1 and 2 (Lac-1-^{14}C and Lac-2-^{14}C) to fatty acids (FA) in the presence of 2 mM glucose and 2 mM acetate (adapted from Forsberg *et al.*, 1985b).

concentrations of acetate, lipogenesis from lactate is strongly inhibited due to the negative feedback of acetyl-CoA on the pyruvate dehydrogenase reaction (Fig. 14.7). Several additional interactions are summarized in Table 14.2. These interactions must be accounted for in tissue level models and, when found to be quantitatively important under physiological conditions, must be captured in equations forms adopted for use in whole-animal models.

We have found in the course of development of mechanistic models of animal metabolism that the first basic step in formulation of a model should be definition of a standard or reference condition for input–output relationships. Usually, the reference conditions are set for a perceived average tissue or cow.

This is very useful in initial parameterization of the model and calculations of numerical values within the model. Also, when the model is solved for a steady-state (constant inputs, concentrations, etc.), the reference conditions are very helpful in identifying errors in coding. The reference condition set for inputs and outputs for our mammary gland model and mammary elements in the lactating cow model are presented in Table 14.3.

This reference condition is an assumption, but it is a useful one, as noted above and discussed in detail by Baldwin (1995). If the model is well

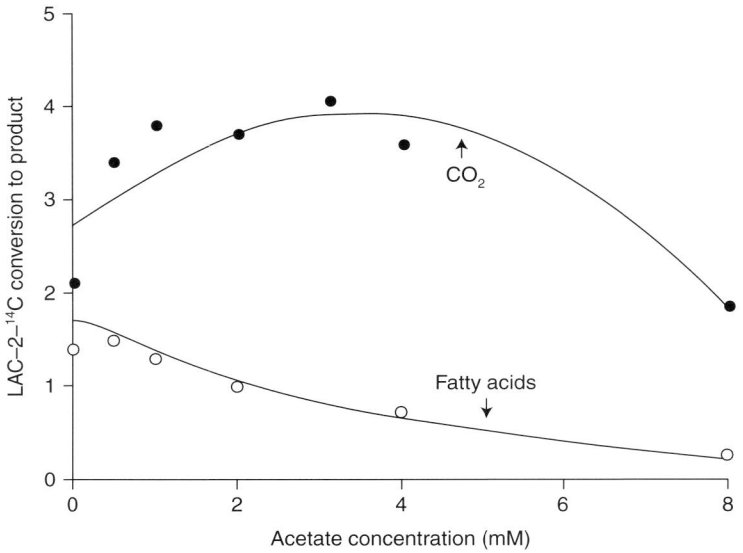

Fig. 14.7. Effect of acetate on conversion of [2-^{14}C]lactate to fatty acids and carbon dioxide in cow mammary tissue. The respective glucose and lactate concentrations were 2 and 30 mM (adapted from Forsberg *et al.*, 1985b).

Table 14.3. Uptake–output balance for mammary glands of a model cow producing 30 kg of milk per day.

Nutrient	Uptake day^{-1}		Milk component	Output day^{-1}	
	Moles	MJ		Moles	MJ
Glucose	11.4	32.1	Lactose	4.10	23.2
Amino acids	8.3	21.6	Protein	7.86	20.4
Acetate	13.0	11.4	TAG	4.80	45.6
βHBA	0.94	1.9	Citrate	0.34	0.7
Stearate	3.09	35.0			
Glycerol	−0.83	−1.4			

Mammary efficiency = MJ output per MJ uptake = 0.89 (from Baldwin, 1995).
βHBA = β-hydroxybutyrate; TAG = triglyceride.

conceived and approaches being capable of simulating reality, responses in blood nutrient concentrations as influenced by alternative diets, feed intakes, stages of lactation, etc. should occur in the animal and cause expected (observed) differences in nutrient availabilities to the udder and the composition and yield of milk.

The challenge which a mechanistic model must face is illustrated further in Fig. 14.8 which is a plot of radiolabelled fatty acid oxidation rates as a function of fatty acid concentrations in blood. Notice the variance about the statistical best-fit curve. In particular, notice that at a fatty acid concentration of approximately 40 mg l^{-1}, rates of fatty acid oxidation range from 0.12 to 0.24 mg (min × kg)$^{-1}$. A mechanistic model must simulate and help explain this variance which is largely due to the availability of other oxidizable nutrients.

BEHAVIOUR OF THE MODEL

Alternative Feeding Strategies

In our introduction, we emphasized that we prefer dynamic models to the static models in current use for estimating nutrient requirements of animals because 'animals change over time and often past or current management decisions influence subsequent function'. Broster and Broster (1984) summarized results of a large number of full lactation studies with lactating dairy cows subjected to alternative feeding strategies. A summary figure from their analysis is presented in Fig. 14.9 to illustrate the effects of feeding strategy on current productivity and subsequent performance. A clear-cut carry-over

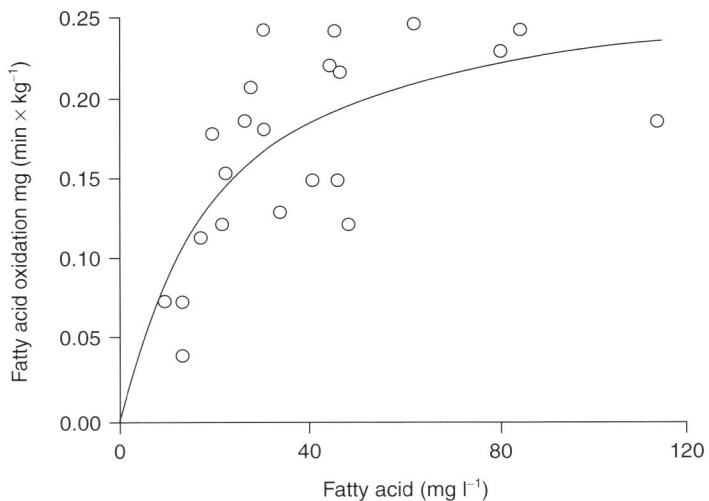

Fig. 14.8. Representation of the effect of plasma free fatty acid concentration and rates of fatty acid oxidation (from Baldwin, 1995).

or residual effect of high rates of feeding early in lactation upon subsequent performance is evident in Fig. 14.9. It should be clear that to be acceptable, a dynamic cow model should exhibit these responses. A modelling analysis of effects of feeding cows at low (L_) and high (H_) energy intakes and medium (_M) and high (_H) protein rations upon lactation performance is presented in Fig. 14.10.

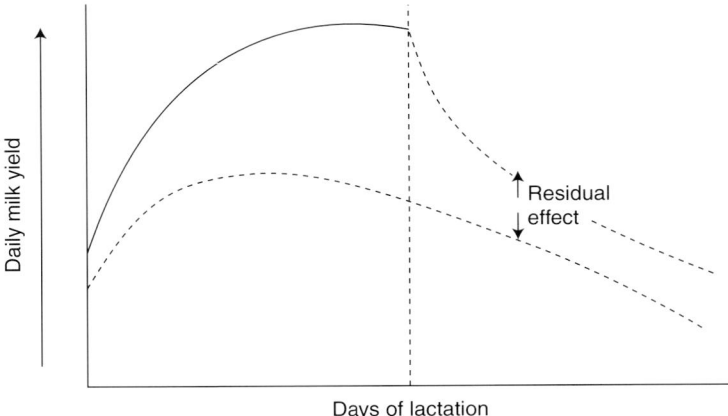

Fig. 14.9. Long-term effects of high (——) and low (---) planes of nutrition on milk production of dairy cows (adapted from Broster and Broster, 1984).

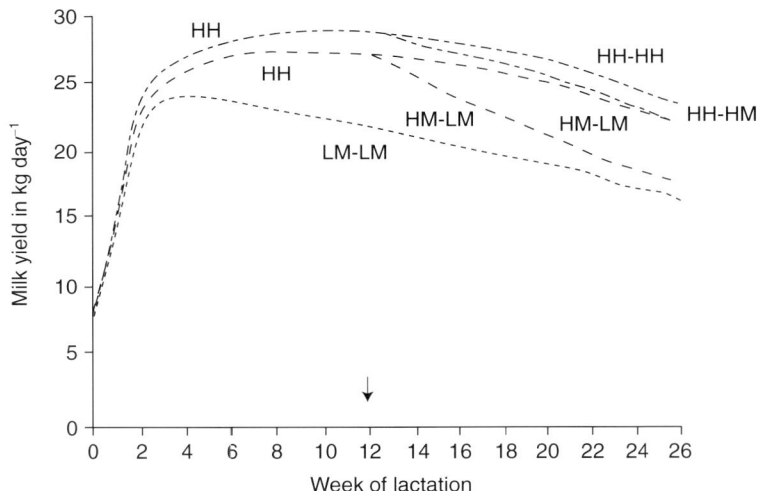

Fig. 14.10. Simulated effects of different feeding strategies upon lactation performance. L_ indicates a feeding rate of 5 kg day^{-1} plus 1 kg feed per 3 kg milk averaged over the previous 3 weeks. H_ indicates a feeding rate of 8 kg day^{-1} plus 1 kg feed per 3 kg average daily milk yield. _M indicates the standard forage:concentrate (50:50) ration of 15% crude protein. _H indicates that the standard ration was adjusted to 18% crude protein with fish meal. Simulated changeovers of diet and feeding strategy occurred at week 12 of lactation (from Baldwin, 1995).

The model does exhibit the required characteristics. In fact, simulated responses were very close to observed values (Baldwin *et al.*, 1987). Another attribute which a mechanistic model should have is the capability of evaluating hypotheses for probable adequacy as explanations of observed responses. Responses to changes in diet and intake in terms of estimates of ME and balances among products of digestion are appropriate. The model has the capacity to simulate the effects of previous and current nutritional management strategies upon performance in dairy cattle. The simulated and observed results of such an evaluation are presented in Fig. 14.11.

Several hypotheses considered as possible mechanisms for explaining responses of lactating cows to administration of recombinant bovine somatotropin (rBST) were evaluated. The only hypothesis adequate to explain the response was that rBST acted directly or indirectly to enhance the metabolic capacity of the udder. The simulated response is presented in Fig. 14.11.

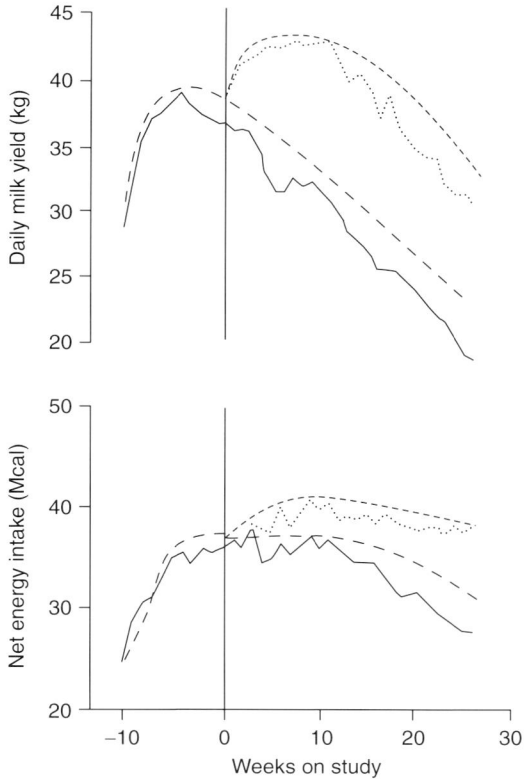

Fig. 14.11. Comparison of effects of growth hormone administration *in vivo* and lactation hormone (rBST) administration in a model. Solid lines indicate the observed milk yields and energy intakes of control cows. Dotted lines indicate the observed milk yields and energy intakes of cows treated with rBST. Dashed lines indicate simulated responses in daily milk yields and energy intakes.

The lactating cow model has been subjected to a large number of challenges and generally has survived well, although both experimental and modelling efforts continue to improve critical elements which misbehave under some circumstances. This is to be expected since research also continues to improve upon the static, largely, empirical models in current use for estimating human and animal nutrient requirements. On the other hand, we continue to gain confidence in the model and have started to evaluate the model in terms of potential use in animal agriculture to support management decision-making. Results of a study of a model to predict milk production and profitability of several alternative diets and feeding strategies in 100 cow herds are presented in Table 14.4.

In these simulations, a stochastic variable defining the genetic capacities for milk production of cows in the herd was introduced as a mean value ± a standard deviation. This was done to evaluate the effects of diet and feeding strategy for each cow in the herd rather than the average cow. In this fashion, feeding cows of low genetic merit high- and medium-quality diets (diet 1 and 2) and feeding cows of high genetic merit medium- and poor-quality diets (diet 1 and 3) upon herd performance and profitability can be evaluated. The effects of alternative feeding strategies (all diets and feeding strategies simulated are used in practice) upon overall herd profitability were assessed. Two trends are evident in the simulation outputs. Starting lactations with diet 2 instead of diet 1 resulted in higher averages for milk production and profit (Table 14.4). The increased protein input during early lactation to cows of above average genetic potential paid dividends. The second trend evident was that the longer a more nutrient-dense diet was fed, the greater the resulting increases in performance and profit. With slightly different feed costs, the ranking of the feeding strategies in terms of profit could be affected. Effects of such differences produced using conventional least-cost ration formulations and/or differing right hand side constraints in least-cost ration formulations can be evaluated similarly using the model.

This limited evaluation demonstrates the value of the use of dynamic models in evaluating alternative feeding strategies and enables researchers or managers to test these in the real world. As there are an infinite number of feeding strategies, this approach could be used to identify those that are most optimal in terms of profit or other criteria such as the efficiency of nitrogen utilization or nitrogen excretion which may be important from the environmental point of view.

Amino Acid Metabolism

Recently, the effects of amino acid availability upon casein, α-lactalbumin synthesis, stoichiometry and degradation rates of amino acids, fractions of amino acids entering the circulation and infusion of or feeding of rumen-protected sulphur amino acids (cysteine and methionine), lysine and phenylalanine plus tyrosine were introduced into the lactating cow model. This enabled

Table 14.4. Evaluation of alternative herd feeding strategies[1].

Strategy	Diet and strategy	Milk (kg)	FCM (lb)	DMI (kg)	EBW[2] (kg)	Profit[3] ($)
1	1	9,044[b]	19,648[a]	6,295[c]	690[bc]	1,888[c]
2	2	10,270[g]	20,007[a]	6,619[e]	701[f]	2,179[i]
3	3	8,324[e]	18,155[c]	6,069[d]	585[g]	1,708[h]
4	1→3 T = 98	8,502[d]	18,776[d]	6,123[d]	598[d]	1,766[g]
5	163 M = 20	9,015[b]	19,580[a]	6,285[c]	689[b]	1,882[c]
6	261 T = 98	9,899[ac]	19,761[a]	6,447[a]	694[c]	2,061[e]
7	2→1→3 T = 98 T = 210	9,829[a]	19,654[a]	6,455[a]	653[a]	2,022[a]
8	2→1 M = 29.5	10,079[c]	19,709[a]	6,483[a]	690[b]	2,035[ad]
9	2→1→3 M = 29.5 M = 20	10,095[c]	19,747[a]	6,483[a]	689[b]	2,031[ae]
Pooled standard error		63	155	21	1	14

[1]The first three strategies involved feeding the simulated herd of 100 cows diets 1, 2 or 3 for the whole lactation. Diets 1 and 2 were 60% concentrate diets with crude protein contents of 15 and 18%, respectively. Diet 3 was 30% concentrate with a crude protein content of 15%. Respective prices of the three diets were 0.123, 0.138 and 0.11 $ kg^{-1}. Strategy 4 involved feeding diet 1 for 98 days and diet 3 for the remainder of lactation. Strategy 5 involved starting lactation with diet 1 and switching to diet 3 after milk production decreased to 20 kg day^{-1}. Strategy 6 simulated feeding individual cows diet 2 until day 98 of lactation and diet 1 thereafter. Strategy 7 involved changing from diet 2 to diet 1 at day 98 of lactation and diet 3 at day 210 of lactation. Strategy 8 involved changing from diet 2 to diet 1 when milk production dropped below 29.5 kg day^{-1}. Strategy 9 involved a shift from diet 2 to diet 1 when milk production dropped below 29.5 kg day^{-1} and to diet 3 after milk production decreased to 20 kg day^{-1}. Values with differing superscripts within a column were significantly different from one another according to the Student–Newman–Kuhl test ($P < 0.05$). Strategy 7 was arbitrarily assigned the superscript 'a' for the purposes of a common reference point.
[2]Empty body weight at the end of lactation.
[3]Profit over feed costs.

simulations of experiments in which rumen-protected amino acids were fed or proteins were infused into the abomasum. The model inputs required for this revision were the amino acid contents of microbial, milk, body and visceral proteins and α-lactalbumin. These changes allowed use of the model to simulate body and visceral protein turnover, milk component synthesis and amino acid degradation.

Synthesis of body and visceral proteins, casein and α-lactalbumin were incorporated, with provisions that any of the amino acid pools can limit the synthesis of individual proteins explicitly. Amino acids are degraded to form urea, glucose and acetate, which in turn are used as carbon, nitrogen and energy sources. Representation of amino acid entry and degradation rates enabled computation of the effects of diet upon the mix of amino acids in the circulation and, as a result, protein synthesis and calculations of stoichiometric relationships in amino acid degradation. Equations for estimates of glucose formation, acetyl-CoA yields, $NADH_2$ yields, $FADH_2$ yields, O_2 uptake, net adenine nucleotide metabolism and CO_2 production from degradation of amino acid pools are calculated within the model based upon established pathways of amino acid metabolism.

The simulation of effects of amino acid and protein supplementation upon animal performance are illustrated in Tables 14.5 and 14.6. Table 14.5 presents simulated responses to supplementation of the reference diet and a corn gluten meal-based diet with sulphur amino acids (SAa), lysine (Lys), SAa plus Lys and with casein per abomasal infusion on day 84 of lactation. In Table 14.6, full lactation data for the same treatments are presented. Responses to supplementation with SAa alone or with Lys alone were relatively minor because both are very close to limiting when the reference diet is fed. When the concentration of one of those amino acids in blood was increased by supplementation, the other amino acid became limiting. Effects upon daily milk (DMILK) and total volume milk (TVMLK) and milk protein (PPM) were significant but relatively minor. When the availabilities of both SAa and Lys were increased, milk production at day 84 increased 5.8%, daily milk protein yield increased 10% and amino acids other than SAa and Lys became limiting (Table 14.5).

When a corn-based diet with corn gluten meal as the protein supplement was input to the model, Lys was limiting such that Lys supplementation resulted in a 7.9% increase in predicted daily milk production and a 8.5% increase in daily milk protein yield at day 84 of lactation (Table 14.5). Supplementation of the corn-based diet with Lys and SAa resulted in a 30% increase in milk production at day 84 of lactation and increased total milk and protein yields. At the rates of Lys and SAa supplementation specified in these simulations, amino acid availability in general became limiting (see Tables 14.5 and 14.6).

The model responses to SAa and Lys supplementation are somewhat higher than those reported by Clark (1975) and Polan *et al.* (1991), but their experimental periods were shorter and the rates of SAa and Lys supplementation in the simulations were higher than those used in the cited experiments. The responses to abomasal infusion of casein were greater than those reported by Whitelaw *et al.* (1986). Feed intake responses that were used in the simulations were greater than those observed, leading to enhancements of the simulated responses.

Table 14.5. Effects of base diets and supplements on model outputs on day 84 of simulated lactations[a].

	DMILK (kg day⁻¹)	PPM (%)	PTM (%)	FDDMIN (kg day⁻¹)	EBW (kg)	cTAa (M × 10⁻³)	cAa (M × 10⁻³)	cSAa (M × 10⁻⁵)	cLys (M × 10⁻⁴)	cPT (M × 10⁻⁴)	Pm lim. Aa
Reference	30.9	3.24	4.4	21.7	578	2.2	1.97	6.24	1.01	1.29	SAa
+SAa	30.9	3.34	4.4	21.7	578	2.1	1.77	10.5	0.97	1.24	Lys
+Lys	30.8	3.27	4.4	21.7	578	2.2	1.80	6.23	2.39	1.29	SAa
+SAa + Lys	32.7	3.37	4.2	22.1	578	1.9	1.46	10.1	2.19	1.17	Aa
+Casein	36.0	3.24	3.8	22.8	585	2.9	2.6	7.16	1.32	1.50	SAa
Corn gluten meal	25.4	3.29	4.9	20.3	574	2.6	2.27	6.5	0.84	1.60	Lys
+SAa	25.4	3.29	4.9	20.3	574	2.5	2.15	11.2	0.84	1.60	Lys
+Lys	27.4	3.31	4.9	20.3	576	2.5	2.15	6.3	0.88	1.60	Lys
+SAa + Lys	33.0	3.37	4.1	22.2	579	1.9	1.47	9.8	1.68	1.35	Aa
+Casein	33.4	3.31	4.1	22.3	584	3.14	2.79	7.20	1.02	1.77	SAa

[a]Values presented are outputs simulated for day 84 of a simulated full lactation when the default diet was not supplemented or supplemented with SAa (0.1 mol day⁻¹), Lys (0.3 mol day⁻¹), SAa plus Lys or with casein (200 g day⁻¹) per abomasum. In the second series of runs, corn gluten meal was the primary protein source with no supplement or supplemented with SAa, Lys, SAa plus Lys or with casein per abomasum. DMILK, daily milk yield; PPM, percentage of protein in milk; PTM, percentage of milk fat; FDDMIN, feed dry matter intake; EBW, empty body weight; cTAa, total concentrations of amino acids in blood; cAa, concentrations of amino acids in blood not represented explicitly; cSAa, concentrations of sulphur amino acids in blood; cLys, concentration of lysine in blood; cPT, concentration of phenylalanine plus tyrosine in blood; and Pm lim. Aa, the amino acid limiting to milk protein synthesis.

Table 14.6. Effects of base diets and supplements on model outputs after a 305 day simulated lactation[a].

	TVMLK (kg)	TMLKPm (kg)	TMLKTm (kg)	TDMIN (kg)	EBW (kg)	WtB (kg)	WtV (kg)	WtFAT (kg)
Reference	7313	236	295	5769	636	436	94.3	105.1
+SAa	7388	252	295	5798	637	439	94.6	103.6
+Lys	7298	236	297	5768	636	436	94.3	105.8
+SAa + Lys	8235	273	298	6028	651	440	95.7	116.0
+Casein	8507	274	298	6079	662	439	95.4	127.4
Corn gluten meal	6492	212	294	5551	701	436	94.1	96.7
+SAa	6504	213	294	5556	628	437	94.0	96.6
+Lys	6854	223	295	5651	631	436	94.2	100.8
+SAa + Lys	8357	278	298	6059	653	440	95.7	117.2
+Casein	8189	265	297	6007	659	439	95.3	125.0

[a]Values presented are outputs simulated for a simulated full lactation when the default diet was not supplemented or supplemented with SAa (0.1 mol day[-1]), Lys (0.3 mol day[-1]), SAa plus Lys or with casein (200 g day[-1]) per abomasum. In the second series of runs, corn gluten meal was the primary protein source with no supplement or supplemented with SAa, Lys, SAa plus Lys or with casein per abomasum. TVMLK, total milk yield ; TMLKPm, total milk protein yield; TMLKTm, total milk fat yield; TDMIN, total dry matter intake; EBW, empty body weight; WtB, body weight; WtV, viscera weight; and WtFAT, fat weight.

CONCLUSIONS

Dynamic, mechanistic lactating dairy cow models allow the generation of data for evaluations of animal performance and managerial decisions. A large number of variables can be examined to give direction to decision-making by scientists and managers. Current practitioners may address these issues intuitively or through the use of static equation systems, but both approaches are, at most, semi-quantitative when used in evaluations of lactation and economic performance. In our view, dynamic models are absolutely essential to economic evaluations of risks associated with current decisions because these clearly influence subsequent performance. When a proven, dynamic, mechanistic model becomes available to practitioners to generate data such as those presented in Table 14.4 and, further, to enable cause and effect analyses of underlying reasons for observed responses, we will have created a superior instrument for both the teaching and application of our science. This is the challenge we pose, to ourselves and others who share this vision.

For a number of years, milk protein production has been understood to be limited by the intestinal supply of particular amino acids, most often methionine and lysine (Clark, 1975). Although several systems have been proposed for the calculation of individual amino acid requirements and their dietary balance (Evans and Patterson, 1985; Madsen, 1985; Mertens, 1985), the provisions in this model for amino acid metabolism constitute the first dynamic model of lactational responses to potentially limiting amino acids.

Experimental approaches used in amino acid nutrition studies of dairy cows can now be simulated and analysed in a manner that has not previously been realized. As new data on amino acid metabolism become available, those data can be utilized for parameter estimation in an attempt to improve the accuracy of predicted milk protein responses to amino acid supplementation.

REFERENCES

Baldwin, R.L. (1995) *Modeling Digestion and Metabolism.* Chapman & Hall, London.

Baldwin, R.L. and Yang, Y.T. (1974) Enzymatic and metabolic changes in the development of lactation. In: Larson, B.L. and Smith, V.R. (eds), *Lactation: A Comprehensive Treatise.* Academic Press, New York, pp. 349–411.

Baldwin, R.L., France, J. and Gill, M. (1987) Metabolism of the lactating cow I. Animal elements of a mechanistic model. *Journal of Dairy Research* 54, 77–105.

Broster W.H. and Broster, V.J. (1984) Reviews of the progress of dairy science: long term effects of plane of nutrition on performance of the dairy cow. *Journal of Dairy Research* 51, 149–163.

Clark, J.H. (1975) Lactational response to postruminal administration of proteins and amino acids. *Journal of Dairy Science* 58, 1178–1197.

Evans, E.H. and Patterson, R.J. (1985) Use of dynamic modelling seen as a good way to formulate crude protein, amino acid requirements for cattle diets. *Feedstuffs* 57, 24–27.

Forbes, J.M. and France, J. (1993) *Quantitative Aspects of Ruminant Digestion and Metabolism.* CAB International, Wallingford, UK.

Forsberg, N.E., Baldwin, R.L. and Smith, N.E. (1985a) Roles of glucose and its interactions with acetate in maintenance and biosynthesis in bovine mammary tissue. *Journal of Dairy Science* 68, 2544–2549.

Forsberg, N.E., Baldwin, R.L. and Smith, N.E. (1985b) Roles of lactate and its interactions with acetate in maintenance and biosynthesis in bovine mammary tissue. *Journal of Dairy Science* 68, 2550–2556.

France, J. and Thornley, J.H.M. (1984) *Mathematical Models in Agriculture.* Butterworth, London.

Hanigan, M.D., Calvert, C.C., DePeters, E.J., Reis, B.L. and Baldwin, R.L. (1992) Kinetics of amino acid extraction by lactating mammary glands in control of sometribove-treated Holstein cows. *Journal of Dairy Science* 75, 161–173.

Madsen, J. (1985) The basis for the proposed Nordic protein evaluation system for ruminants. The AAT-PBV system. *Acta Agriculturae Scandanavicaa, Supplement* 25, 9–20.

Mertens, D.R. (1985) Factors influencing feed intake in lactating cows. From theory to application using neutral detergent fiber. In: *Proceedings of the Georgia Nutrition Conference.* University of Georgia, Atlanta, Georgia, pp. 1–18.

Miller, P.S., Reis, B.L., Calvert, C.C., DePeters, E.J. and Baldwin, R.L. (1991) Patterns of nutrient uptake by the mammary glands of lactating dairy cows. *Journal of Dairy Science* 74, 3791–3799.

Polan, C.E., Cummins, K.A., Sniffen, C.J., Muscato, T.V, Vicini, J.L., Crooker, B.A., Clark, J.H., Johnson, D.G., Otterby, D.E., Guillaume, B., Muller, L.D., Varga, G.A., Murray, R.A. and Peirce-Sandner, S.B. (1991) Responses of dairy cows to supplemental rumen-protected forms of methionine and lysine. *Journal of Dairy Science* 74, 2997–3013.

Robson, A.B. and Poppi, D.P. (eds) (1990) *Third International Workshop on Modelling Ruminant Digestion and Metabolism in Farm Animals*. Lincoln University, Canterbury, New Zealand.

Smith, N.E. (1970) Quantitative simulation analyses of ruminant metabolic functions: basal; lactation; milk fat depression. PhD dissertation, University of California, Davis, California.

Thornley, J.H.M. and. France, J. (1984) Role of modeling in animal production research and extension work. In: Baldwin, R.L. and Bywater, A.C. (eds), *Modeling Ruminant Digestion and Metabolism*. Department of Animal Science, University of California, Davis, California, pp. 4–9.

Whitelaw, F.G., Milne, J.S., Orskov, E.R. and Smith, J.S. (1986) The nitrogen and energy metabolism of lactating cows given abomasal infusions of casein. *British Journal of Nutrition* 55, 537–556.

15

Modelling Growth and Wool Production in Ruminants

W.J.J. Gerrits[1,2] and J. Dijkstra[1]

[1]Animal Nutrition Group, Wageningen Institute of Animal Sciences, Wageningen Agricultural University; [2]TNO Nutrition and Food Research Institute, Department of Human and Animal Nutrition, Wageningen, The Netherlands

INTRODUCTION

Modelling the growth of ruminants has received much attention throughout the years, both in the field of practical animal nutrition and in research. For practical animal nutritionists, the possibility to manipulate growth rates and body composition by means of nutrition has been sufficient reason to develop empirical models. These models predict body fatness from energy intake and growth rate (Lofgreen and Garrett, 1968; Fox and Black, 1984) or predict energy and protein utilization from feed intake and composition (Graham *et al.*, 1976). Factors other than nutrition, such as sex and breed, have led to the development of such relationships for individual breeds or sexes separately (Robelin, 1979; Korver *et al.*, 1988; Keele *et al.*, 1992). For research purposes, several nutrient partitioning models have been developed for sheep (Gill *et al.*, 1984; Sainz and Wolff, 1990) and for beef cattle (Oltjen *et al.*, 1986; France *et al.*, 1987; Di Marco *et al.*, 1989). These models were driven mostly by absorbed nutrients. However, the amount of nutrients available for absorption depends on voluntary intake, rumen fermentation and digestive processes.

Several reviews have discussed and/or compared the performance of models of ruminant growth, each with their own objective (Oltjen, 1989; Arnold and Bennett, 1991a,b; Schmidely, 1996; Hanigan *et al.*, 1997). This chapter focuses on several aspects of modelling growth in ruminant animals, following the nutrients from intake through the gastrointestinal tract to their deposition in body tissues. Emphasis is on dynamic models and on post-absorptive nutrient partitioning. Several aspects of modelling post-absorptive nutrient partitioning will be discussed: (i) choice of body components; (ii) the need to move towards nutrient-based systems; (iii) (dis)advantages of the conventional maintenance and production approach versus the representation of biological processes; and (iv) representation of mechanisms regulating nutrient partitioning.

Wool growth is included in some of the existing growth models (Blackburn and Cartwright, 1987; Sainz and Wolff, 1990; Freer *et al.*, 1997), but has been the subject of separate modelling exercises as well (Black and Reis, 1979; Gandar *et al.*, 1990; Bowman *et al.*, 1993) and will be discussed along with other issues throughout this chapter.

MODELLING AVAILABLE NUTRIENTS

Feed Intake

The feed intake level is among the most important nutritional factors controlling growth. However, the ability of models to predict voluntary feed intake accurately in a range of situations is limited. A satisfactory prediction requires an understanding of the control of intake. The current theories and factors influencing feed intake have been reviewed at a recent symposium (Allen, 1996; Fisher, 1996; Forbes, 1996; Illius and Jessop, 1996; Ketelaars and Tolkamp, 1996), and is discussed extensively by Poppi *et al.* in Chapter 3 of this volume. Briefly, physical and metabolic constraints have been suggested to determine feed intake. Physical constraints relate to the amount of material that can be accommodated in the gastrointestinal tract. Metabolic regulation involves factors including the requirement for nutrients at a certain production level and the balance of absorbed nutrients. Occasionally, behavioural influences (factors in the feeding environment such as social interactions) have been suggested also to affect feed intake. The intake over a 1-day period is equal to the sum of the amounts eaten in individual meals. Consequently, the relationships between daily intake and factors predominating in ending meals have been considered. There is, however, no consensus on how meal control relates to daily intake (Gill and Romney, 1994); it has also become clear that various factors influencing feed intake have to be interpreted in an integrated manner, rather than regarding the factors as mutually exclusive. For example, Poppi *et al.* (1994) developed a model to examine physical limits (rate of intake, faecal output, and rumen turnover) and metabolic limits (genetic limit to protein deposition, heat dissipation and ATP degradation via substrate cycling). Simulations using this model indicated that intake is often limited to a similar level by more than one factor, suggesting that different factors dominate intake regulation at different times of the day.

A number of models have been proposed to predict voluntary feed intake in ruminants, most of which are empirical. They relate feed intake to factors including body weight, body condition score, breed, chemical feed composition, energy content and physical form of the feed (Ingvartsen, 1994). Because of their empirical nature, they can provide potentially satisfactory predictions for the range of situations for which they have been developed. The few mechanistic models for predicting feed intake represent the rumen fermentation processes and particle dynamics, and energy and protein metabolism after absorption. An example of this type of modelling is presented by Sauvant *et al.* (1996). They combined a digestion submodel with a decision submodel for sheep. The digestion submodel describes flows of nutrients in the rumen and formation of volatile fatty acids (VFAs), based on dietary characteristics. The feeding decision submodel distinguishes between eating, ruminating and resting, and receives input from the digestion submodel. The choice among these activities is based on relative values of eating motivation and satiety status. Motivation and satiety are functions of rumen load, day–night rhythm, palatability and energy balance. Hence, this approach predicts intake from a selection of factors possibly involved. Although models based on physical limitations of feed intake can be quite successful, in particular when low-quality diets are fed (Fisher *et al.*, 1987), no models exist which can predict feed intake accurately in a wide range of dietary situations. Not surprisingly, feed intake is an input to many ruminant growth models, rather than being predicted. If one wishes to predict nutrient uptake in growth models, then simple regression equations applicable to the animal and diet of interest can be applied.

Rumen Fermentation

The fermentation processes in the rumen have a significant effect on the profile of nutrients available for absorption. In general, VFAs produced by fermentation of ingested feed is the most important source of energy, and microbial protein synthesized in the rumen is the most important source of amino acids for the host animal. However, rumen fermentation is rarely represented explicitly in models of ruminant growth; the absorbed nutrients are often inputs to the model (Gill *et al.*, 1984; France *et al.*, 1987; Di Marco *et al.*, 1989).

There has been considerable research on rumen fermentation processes, and several rumen models have been developed and described (Dijkstra and France, 1996; Dijkstra and Bannink, Chapter 13 this volume). Obviously, sound rumen models are required as part of mechanistic models to predict feed intake (discussed previously). Equally, the profile of nutrients available for absorption as predicted by rumen models can form the input for growth models. For lactating dairy cattle, Baldwin *et al.* (1987a) combined a rumen model and a post-absorptive metabolism model to predict the effect of diet on milk production and milk composition. The output of the rumen model constituted the major input for the metabolism model. Similarly, the combination of rumen and metabolism models for growing cattle may prove to be a powerful tool to evaluate feeding strategies and predict growth and growth composition.

Intestinal Digestion

Available nutrients arise primarily from digestion and subsequent absorption in the small intestinal tract, except for the VFAs which are absorbed mainly through the rumen wall. To date, little attention has been paid in whole-animal models to representing the mechanisms of intestinal digestion, and fixed digestibility coefficients usually have been applied. The intestinal digestion of microbial protein is fairly constant at 85% (Storm *et al.*, 1983) but that of feed components can vary widely. For example, true feed protein digestibility of a range of roughages and concentrates varied from 50 to 99% (Van Straalen and Tamminga, 1990). Fermentation of substrates in the caecum and colon is often not represented in whole-animal models. A few models represent this as a fixed proportion of material flowing out from the duodenum (Baldwin *et al.*, 1987b) or proportional to fermentation in the rumen (Dijkstra *et al.*, 1996). Of particular interest are the large amounts of endogenous protein secreted into the lumen of the intestinal tract. Endogenous protein secretion is influenced by the flow of dry matter (DM) in the intestine (in particular that of fibre) and the presence of antinutritional factors (reviewed by Tamminga *et al.*, 1995), but its representation in models has received very limited attention.

Recently, Bastianelli *et al.* (1996) presented a more mechanistic approach to simulate digestion and absorption in pigs. This approach to represent digestion in the intestines is useful in conceptualizing and discussing the simulation of nutrients available for absorption, even though the ruminant digestive tract may differ considerably from that of a pig. Briefly, four anatomical compartments were distinguished, i.e. the stomach, two parts of the small intestine and the large intestine. Digestion processes and nutrient flows into and out of each compartment were represented. In this model, the amount of feed nutrients absorbed from the small intestine depends on the rates of digestion, passage and absorption; the model also included a simple representation of endogenous secretions and hind-gut fermentation. Endogenous secretions were related to DM flow in a compartment, but the effects of the presence of antinutritional factors and of DM composition were not taken into account. Microbial growth in the hind-gut was determined mainly by carbohydrate availability, assuming a fixed efficiency of microbial growth, a fixed composition of microbial material, and a constant ratio of utilization of N substrates (ammonia and amino acids).

MODELLING POST-ABSORPTIVE NUTRIENT PARTITIONING

When constructing a nutrient-partitioning model, the choice of which body constituents to represent is among the first to be made and deserves careful consideration. Once the driving variables have been identified, one has to decide what biological processes to represent in the model and how the model should partition the absorbed nutrients into the defined body components. It is common practice to treat growth and maintenance, both in protein and energy metabolism, as separate entities. After subtracting the maintenance requirements from absorbed nutrients, the remaining energy or protein is deposited with a certain efficiency. These concepts have proven to be very useful both in research and in the field of application. For various reasons, however, this approach can be questioned. Apart from methodological problems (for a discussion, see ARC, 1980), biological processes that require energy for the purposes of maintenance and production are similar (e.g. protein turnover and ion pumping across membranes), and maintenance requirements depend on production level (Baldwin *et al.*, 1991). Therefore, by apportioning the needs for growth and maintenance separately, one is likely to limit progress in understanding the underlying mechanisms. In this section, attention is paid to (i) the choice of body components to represent in a growth simulation model; (ii) the importance of using individual nutrients as driving variables rather than total energy; (iii) the difference between the current representation of maintenance and production processes in models of growth, followed by an alternative approach: representation of biological processes, including its limitations; and (iv) a discussion of mechanisms regulating nutrient partitioning.

Choice of Body Components

Predicting the rate of gain of weight and the composition of the gain is one of the aims of every growth model. Whereas some models use body weight gain as an input variable to predict its composition (Korver *et al.*, 1988; Keele *et al.*, 1992), in most models the rate of body weight gain is defined as the sum of the gain of body components. The empty body can be considered as the sum of either chemical components, functional tissues (anatomical components) or a mixture of both. When representing (bio)chemical pathways in a model (e.g. protein and fat turnover), choosing chemical components (protein, fat, ash and water) has the obvious advantage that they are the direct products of metabolic reactions. In addition, many empirical models consider chemical body components because they are calibrated on whole-body chemical analyses (Loewer *et al.*, 1983; Fox and Black, 1984). Individual tissues, however, may differ considerably in metabolic rate and nutrient consumption. Therefore, in some models, chemical compartments (usually the body protein pool) are split up to represent functional tissues (protein in muscle, bone, hide and viscera: Gerrits *et al.*, 1997a; visceral and whole body protein: Di Marco *et al.*, 1989). Gill *et al.* (1989a,b) distinguished protein metabolism in ten tissue beds; this has the additional advantage of requiring less assumptions when fitting to most *in vivo* data at the organ or tissue level (Hanigan *et al.*, 1997). Representation of functional tissues (muscle, bone and adipose tissue) is rare, which is unfortunate because there is no direct relationship between the size of the chemical body protein and fat pools, and quantities of meat and adipose tissue. In this respect, wool is an exception, and is considered as a separate tissue in many models of sheep growth (Graham *et al.*, 1976; Arnold *et al.*, 1977; Gill *et al.*, 1984; Blackburn and Cartwright, 1987; Sainz and Wolff, 1990; Freer *et al.*, 1997). For the improvement of representation of biological mechanisms, for fitting to *in vivo* data, and for application in practice, moving towards the representation of chemical constituents within functional tissues would be an important step forward.

Energy Versus Nutrients

In many regression models, the difference between energy intake and the energy spent on maintenance is related to energy retention (Lofgreen and Garrett, 1968) or body weight gain (Fox and Black, 1984). The simplicity is the obvious advantage as well as the disadvantage of this approach. As discussed previously, there is a need to characterize body weight gain or energy retention in terms of chemical or anatomical constituents. From the feed input side, expressing intake in energy terms ignores its nutrient composition. Non-protein energy cannot be converted into protein; this has been the reason for many growth models including protein intake as a separate entity. In addition, the composition of absorbed non-protein energy (carbohydrates, fats and VFAs) will be of increasing importance in future growth models. The model of beef growth of Oltjen *et al.* (1986) may illustrate this; the model predicts body fat gain from the energy available after subtracting maintenance energy requirements and the energy required for protein gain from the energy intake, assuming a constant efficiency of energy utilization for growth. It does not always simulate acceptable rates of body fat gain, for which, according to Baldwin (1995), various factors may be responsible. Apart from errors in estimations of maintenance energy requirements or protein accretion rates, different metabolites are utilized for fat synthesis with a different efficiency, i.e. fatty acids (93–96%), glucose (80–85%) and acetate (65–75%). Considering the above, output from rumen simulation models, as discussed previously, can be of great value in improving models simulating growth of ruminants. In addition, in pre-ruminant calves, the model of Gerrits *et al.* (1997a) predicted increased rates of gain of protein and live weight at the expense of fat gain when dietary fat was exchanged for carbohydrates (Fig. 15.1). This illustrates the importance of distinguishing between energy sources in growth models.

The Maintenance and Production Approach Versus Modelling of Biological Processes

Basal energy expenditure

In fast growing animals, 30–50% of energy intake is spent on maintenance of body functions. Maintenance energy requirements are traditionally estimated as the *x*-intercept in the regression of energy retention on metabolizable energy intake. They can be considered to be the sum of body service functions. The major functions are listed in Table 15.1 and comprise 35–50% of basal energy expenditure. These are, however, dependent on each other. It may therefore be better to focus on energy expenditures associated with cell maintenance functions.

Obtaining satisfactory predictions at the whole-body level requires growth models to account completely for whole-body energy expenditure. This has led to the approach of defining maintenance energy requirements as a single energy-consuming process, which is currently used in most models. Energy

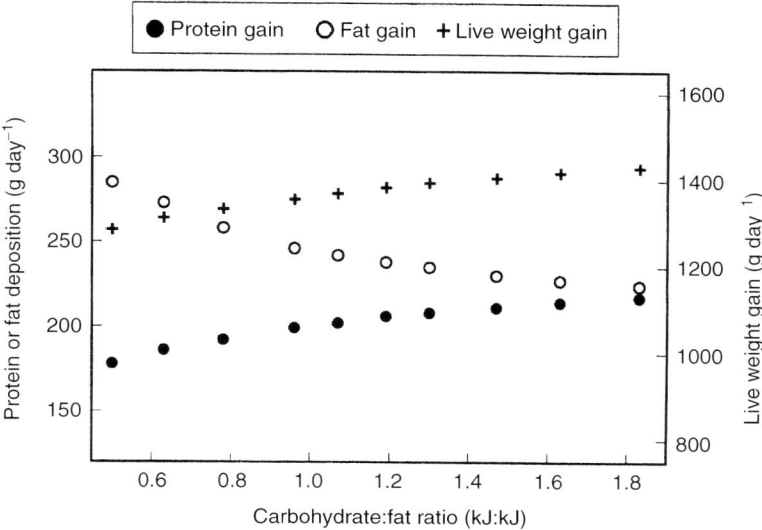

Fig. 15.1. Sensitivity of model predictions of the rate of gain of protein, fat and live weight of pre-ruminant calves to changes in the ratio between energy intake from carbohydrates and fat at a constant energy intake (from Gerrits *et al.*, 1997b).

Table 15.1. Energy expenditure in several major maintenance functions[a].

Function	Basal energy expenditure (%)
Service functions	
Kidney work	6–7
Heart work	9–11
Respiration	6–7
Nervous functions	10–15
Liver functions	5–10
Total	36–50
Cell maintenance	
Protein re-synthesis	10–20
Lipid re-synthesis	2–4
Ion transport	30–40
Total	40–60

[a]From Baldwin *et al.* (1991).

requirements for maintenance are then included as a constant, scaled by body weight raised to the power 0.75 or 0.67. Some models also include other variables such as growth rate and digestible energy intake in their regression equation (Graham *et al.*, 1976; Gill *et al.*, 1984). Apart from the strong correlation between energy intake and growth rate, this is a matter of semantics which can be avoided if maintenance energy requirements are defined rigorously as the energy needed to maintain energy balance (ARC,

1980). Refinement of this system is sought in relating maintenance energy expenditure to tissue size (Di Marco and Baldwin, 1989; Sainz and Wolff, 1990; Gerrits *et al.*, 1997a), accounting for different metabolic rates of tissues.

Breed differences in maintenance energy requirements

Differences in maintenance energy requirements are often attributed to altered body composition. The general view seems to be that the leanest breeds have the highest maintenance energy requirements (see Korver *et al.*, 1988; Webster, 1989), although there seems no consensus on this point. Ortigues *et al.* (1993) found maintenance energy requirements of lean and fat cows (Charolais) to be 516 and 536 kJ $LW^{-0.75}$, respectively, and attributed this (not significant) difference to the nutritional history of the cows and to the distribution of body protein over tissues differing in metabolic activity. In line with the latter, Webster (1989) suggested that the differences between breeds would be too large to be explained completely by differences in body fat content. He suggested that differences in the proportion of meta-bolically very active organs would also play a role. Considering this, the approach used by Baldwin *et al.* (1987a) in their model for lactating cows seems appropriate. Their approach, based on estimates of basal energy expenditure of lean body mass, fat and viscera, was also applied in the model of pre-ruminant calves of Gerrits *et al.* (1997a), showing good agreement with statistical estimates of whcle-body maintenance energy expenditure (Gerrits *et al.*, 1996).

Towards the representation of biological processes in energy metabolism

Considering the quantitative importance of maintenance energy, it is surprising that, in growth modelling, so little attention is paid to a mechanistic approach to allocate this expenditure to defined, biological processes. Gill *et al.* (1989b) made a first step in this direction by specifying ten tissue beds, and represent-ing protein turnover, the energy costs of peptide bond formation and proteo-lysis and associated costs cf ion (Na^+ and K^+) pumping in lambs, growing at two different rates. They found protein turnover and ion pumping to account for about 19 and 18–23% of whole-body energy expenditure, respectively.

Modelling biological processes instead of applying experimentally derived energetic efficiencies for tissue accretion (ARC, 1980; Koong *et al.*, 1982) causes problems (see Reeds, 1991). Present knowledge is insufficient to account completely for whole-body energy expenditure in well-defined processes. Obtaining satisfactory predictions at the whole-body level requires growth models to have a solid basis in calorimetry studies (either direct or indirect) and the gap between the sum of the defined processes and the whole-body energy expenditure has to be covered. This problem was encoun-tered in several research models representing biological processes to varying degrees. In these models, theoretical costs of protein synthesis and turnover rate were accounted for, but not the associated costs including, for example, ion pumping. This necessitated drains on the energy metabolite pools to cover the discrepancy between the energy costs, accounted for in the model and whole-body experimental data. Therefore, an additional energy-consuming process was included in the models of Di Marco and Baldwin (1989) ('balancing the ATP-zero pool'), France *et al.* (1987) ('substrate cycling'), Gerrits *et al.* (1997a) ('additional energy costs of growth'), Gill *et al.* (1984) ('ATP-degrader') and Sainz and Wolff (1990) ('balancing the ATP-zero pool').

In addition, when biological processes are represented together with a lump-sum for maintenance energy requirements there is a potential danger of double counting which has to be corrected for in case the processes involved are also accounted for in the maintenance energy requirements. Pettigrew *et al.* (1992) and Gerrits *et al.* (1997a), for example, reduced the maintenance energy requirements by 20–25% for these reasons.

Protein requirements for maintenance

According to classical protein metabolism concepts, nitrogen loss from the body can be factored into three distinct fractions: (i) endogenous urinary losses; (ii) metabolic faecal losses; and (iii) losses of skin and hair (Owens, 1987). Nevertheless, in protein metabolism, this concept of maintenance and production is not as widely accepted as in energy metabolism and is increasingly questioned (Millward *et al.*, 1990). Quantitatively, the contribution of the maintenance fractions to the total protein requirements of the growing animal is much smaller than that to the energy requirements. Nevertheless, the endogenous protein secreted into the lumen of the intestinal tract may well be of quantitative importance, as discussed previously, and is influenced by, for example, the flow of DM in the intestine (in particular that of fibre) and the presence of antinutritional factors.

Towards the representation of biological processes in protein metabolism

When representing biological processes in growth models, some representation of the rate of turnover of body proteins is necessary. Reeds and Mersmann (1991) stated that in quantitative terms, many more amino acids enter and leave body protein than are ever catabolized. Consequently, the impact of small proportional changes in either protein synthesis or degradation on the amino acid balance can be considerable. Apart from the energy requirements associated with protein turnover (discussed previously), there are several valid reasons for representing protein turnover as a biological process in growth models.

It is often assumed that the presence of catabolic enzymes in body tissues (especially liver) gives rise to inevitable catabolic amino acid losses (see, e.g., Liu *et al.*, 1995). Consequently, potential re-utilization of amino acids, degraded from body tissues, is lower than 100% and possibly depends on the amount of that amino acid passing the site of oxidation. Increasing the rate of protein turnover consequently should increase amino acid oxidation, which is a concept included in the model of Gerrits *et al.* (1997a). In ruminants, however, there is no quantitative information available on this relationship for individual amino acids. It has been known for quite some time that tissues other than liver are capable of oxidizing amino acids, e.g. branched chain amino acids in muscle tissue (Benevenga *et al.*, 1993). Recent research indicated that in ruminants, tissues of the gastrointestinal tract play an important role in the degradation of many amino acids, including essential ones (MacRae *et al.*, 1997). This argues for inclusion of protein turnover in growth models, specifically in different tissue beds.

In pig growth models, it is common practice to account for limiting amino acids (Moughan and Verstegen, 1988). Consequently, these models represent amino acid degradation due to dietary amino acid imbalance. In ruminants, this aspect has received very little attention, probably because it is difficult to predict the rumen output of individual amino acids from the mix of microbial, endogenous and dietary (by-pass) proteins. The Cornell CNCPS model predicts absorbed amino acids from microbial and dietary by-pass proteins (O'Connor *et al.*, 1993), but modelling of post-absorptive metabolism is confined to the whole animal, applying constant efficiencies of utilization after subtracting amino acid requirements for maintenance, as specified earlier. The approach of Gerrits *et al.* (1997a), specifically representing turnover of protein in hide, bone, visceral and muscle tissue and including the amino acid composition of these proteins, allows the possibility of calculating fluxes of individual amino acid, rather than total nitrogen. The amino acids needed for deposition can thus be calculated and, together with assumptions regarding the (inevitable) catabolic losses of amino acids, it is possible to predict the requirements for individual amino acids. The approach was developed for the pre-ruminant calf, in which the prediction of absorbed amino acids is relatively easy and is illustrated in Fig. 15.2. Apart from amino acid imbalance, other reasons for amino acid degradation, e.g. for ATP generation or gluconeo-genesis, should be represented in mechanistic growth models for growing ruminants (Di Marco *et al.*, 1989).

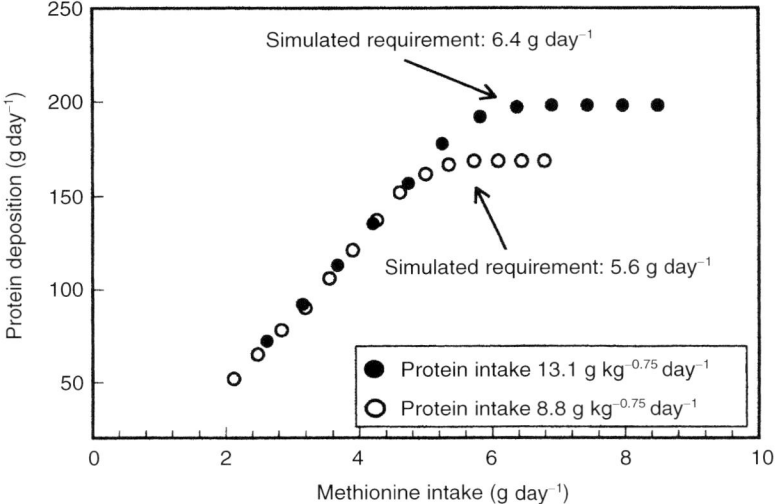

Fig. 15.2.   Simulation of the effect of a step-wise reduction of methionine intake on the average protein gain of pre-ruminant calves in the live weight range of 80–160 kg, at two protein intake levels. Protein-free energy intake was 820 kJ kg$^{-0.75}$ day^{-1} (from Gerrits *et al.*, 1997a).

REPRESENTATION OF MECHANISMS, REGULATION OF NUTRIENT PARTITIONING

Methods of Nutrient Partitioning in Growth Models

There are a number of theories concerning the regulation of nutrient partitioning which have been adopted in animal growth models. Several ruminant growth models partition available nutrients by assuming a genetic potential for growth and composition (Sanders and Cartwright, 1979; Loewer *et al.*, 1983; Oltjen *et al.*, 1986; Arnold and Bennett, 1991a,b). These models include a growth curve to represent this genetic potential, but the form of these curves differed between models. In the model of Sanders and Cartwright (1979), the curve related potential body weight gain to the proportion of mature weight in a linear way until about 1 year of age, and subsequently followed a Brody-type curve. In the model of Loewer *et al.* (1983), Gaussian growth curves of weight versus physiological age were parameterized for protein, fat, water and minerals. The model of Oltjen *et al.* (1986) applied growth curves for DNA accretion, which determined the maximum rate of protein synthesis. In these models, animals strive to reach a certain body composition, which can be restricted by nutrient supply. Excess nutrients are assumed to be stored as body fat. Priorities in energy partitioning for these models were first maintenance, second protein accretion and then fat accretion. The concept of using a growth curve to represent the potential mass of body components at any given time has the advantage of accommodating a mechanism for compensatory growth: after a period of nutrient restriction, the animal can deposit body tissues while striving to reach its potential. Representation of available nutrients and metabolic reactions, however, is limited in these models.

In non-ruminants, a concept, proposed by Whittemore and Fawcett (1976), has had a considerable impact on practical models of pig growth. They represented their rather empirical concept in a number of conceptual equations. It implied a preference for the animal to use ingested protein for tissue deposition, provided that the fat:protein ratio in the gain exceeds some minimum, fixed value; excess protein is oxidized and used for energy purposes. Similar equations were proposed for growing lambs (Black and Griffiths, 1975). Experimental work by Gerrits *et al.* (1996), however, showed this approach to be invalid for pre-ruminant calves. Pre-ruminant calves do not seem to strive to reach a minimal ratio of lipid to protein in the gain. The response of protein deposition rate to increased protein intakes was remarkably low: about 30% of the extra, ingested (milk) protein was deposited in body tissues. From these experiments, the reason for this low priority of calves for partitioning dietary protein into body protein remained unclear. It was suggested that calves might oxidize amino acids depending on the concentration of energy substrates available (fatty acids, amino acids, glucose).

A more metabolic approach, applied in many of the models already mentioned (Gill *et al.*, 1984, 1989; France *et al.*, 1987; Di Marco *et al.*, 1989; Gerrits *et al.*, 1997a), is based on the assumptions that the distribution of nutrients between body compartments is controlled primarily by substrate concentrations, and that these biological processes often follow the principles of saturable enzyme kinetics (Baldwin, 1995). These models simulate the partitioning of nutrients through intermediary metabolism into growth. In these models, the whole animal is represented by metabolite pools and body storage pools. The metabolite pools (usually small and relatively constant in size) serve to partition absorbed nutrients into body storage pools, which increase in size with time, representing the growth of the animal. The distribution of metabolites over various metabolic pathways is regulated by Michaelis–Menten-type equations, in which substrate concentrations stimulate, and concentration of end-products inhibit, the rate of a particular transaction (France and Thornley, 1984). Whereas the affinity and inhibition constants are often set by rule of thumb to calibrate model predictions on experimental data, the maximum reaction velocities depend primarily on the availability of enzymes required for the transaction and is a function of the size of the reaction site and potential drive for the particular process. Reaction velocities are therefore often scaled by the size of the tissue in which the transaction is expected to take place (France *et al.*, 1987; Gerrits *et al.*, 1997a). Generally, this makes these models more sensitive to changes in maximum velocities than to similar changes in other model parameters. The scaling factors used for different transactions are therefore a powerful means to account for changes in nutrient partitioning with increasing age or body weight.

Representation of Hormone Action

Several hormones have key roles in the partitioning of nutrients over body components. Plasma hormone concentrations can be used in metabolic models as a parameter affecting the utilization of a substrate for a specific transaction (Sainz, 1989). Representation of detailed hormone function, i.e. including hormone interactions, receptor numbers, affinities, etc., would make these models very complex and would have to be balanced against the objective of the whole-animal model. Hormone function is, however, rarely represented in models of animal growth, probably due to the complexity of endocrinological mechanisms. Moreover, in models which are nutrient driven, it is difficult to separate the regulating effects of hormones from their response to nutrient intake. It can be argued that representation of hormone function in metabolic models is useful only if the hormonal regulation of nutrient partitioning is clearly independent of nutrient intake. Examples of such situations may include nutrient partitioning during lactation or gestation, to represent differences between breeds and sexes; to represent effects of age; or to represent nutrient partitioning between wool and other tissues. Alternatively, of course, inclusion of hormone action can be a specific objective of the modelling exercise. The inclusion of hormone pools can also give rise to computational problems. Compared with metabolites, hormones are present at very low concentrations, and multiplication by metabolic volumes to give pool sizes still results in small pools relative to their rate of turnover. In modelling terms, solutions of models containing these so-called stiff equations may be unstable (France *et al.*, 1992).

Partitioning of Nutrients Between Wool and Other Tissues

The biology of wool growth has been well documented (Black and Nagorcka, 1993). The wool growth potential of a sheep seems to be determined largely by its genetic background. Nutrient restriction at critical periods of follicle development may limit the expression of these genetic potentials (Black and Reis, 1979) while restriction at other times just affects synthetic capacity. It seems that the partitioning of nutrients over wool and other tissues is less affected by nutrient intake than is the partitioning over other components such as meat and adipose tissue. For the representation of wool development in simulation models, several approaches are used. Arnold *et al.* (1977) and Graham *et al.* (1976) applied a fixed maximum wool growth rate potentially restricted by either energy or nitrogen intake. Sainz and Wolff (1990) followed a more mechanistic approach whereby the maximum rate of wool protein synthesis depends upon the amount of DNA in the body component in which skin and wool follicles are present and the availability of amino acids in the amino acid pool. In this model, energy restriction indirectly reduces wool growth rate through an increase in catabolic hormone concentration. In the model of Freer *et al.* (1997), the response of wool growth rate to changes in nutrient intake is calculated after accounting for a lag time effect. Wool growth rate is restricted either by intestinal protein availability or, at a low availability of metabolizable energy, by metabolizable energy intake, both after deduction of requirements for fetal growth and lactation. Furthermore, in this model, wool growth depends on potential fleece weight and day length, both parameters being dependent on the specified breed. Black and Reis (1979) and Bowman *et al.* (1993) focused specifically on simulation of wool growth rate. Bowman *et al.* (1993) extended an earlier model for predicting wool growth rate and wool quality parameters such as fibre diameter, staple length, staple strength and contamination with vegetable matter. They considered dry matter intake, relative to a defined maximum, as one of the important driving variables, and included a factor accounting for delay in the response of wool production to changes in intake level. They included the effect of photoperiod as a sine function, and corrected wool growth during pregnancy and lactation after deducting the energy used for pregnancy and lactation. A lower priority for wool growth in periods following strong nutritional limitation was accounted for. Furthermore, they developed an index to account for effects of fibre maturity on wool growth rate at any given stage of development, with effects being most pronounced directly after birth. The model of Black and Reis (1979) was restricted to predict the partitioning of sulphur amino acids (SAAs) from ingestion into skin, wool and other tissues. Sulphur-containing amino acids were assumed to limit wool growth. Their model was developed to predict the effects of absorption of SAAs, the genetic capacity of the animal to grow wool and of blood flow on the amino acid flux into wool. The genetic capacity was represented by a maximum potential clean wool growth rate, determined by specified follicle characteristics. Also, the maximum rate of uptake of SAAs into wool depended on the (partly genetically determined) proportion of three types of wool protein present, distinguished by their SAA content. The distribution of SAAs over wool and the remaining tissues was assumed to depend on the size of the SAA flux in skin and remaining tissues (determined by blood flow and SAA concentration) and the (genetically determined) maximal uptake into a particular wool protein class. Subsequently, the effects of changing assumptions on wool growth rate were tested. This model thus included a mechanism for competition of tissues for a particular nutrient; when increasing the demand of other tissues (for example during pregnancy or lactation), the model predicted decreased wool production rates, close to experimental observations.

Although the effects of nutrition on wool growth are more indirect compared with those on body weight gain and body composition, all models mentioned somehow account to some extent for effects of nutrition on wool growth rate. It is surprising, however, that the impact of intestinally available indispensable amino acids (in particular the SAAs) on wool growth rate has only been a key issue in the model of Black and Reis (1979).

Differences in Nutrient Partitioning Between Breeds

Various researchers have demonstrated differences in nutrient partitioning between sexes and breeds. When compared at the same weight, female cattle are generally fatter than castrates, which, in turn, are fatter than entire males (Robelin, 1986; Kirchgessner *et al.*, 1993). Also, when compared at a similar weight, beef breeds which are heavier at mature weight generally contain less fat in their empty bodies than do breeds of smaller mature weight (Campbell, 1988). The distribution of body fat between depots and the ratio between carcass muscle and bone tissue have also been reported to vary according to genotype (El Hakim *et al.*, 1986; Robelin, 1986). Unfortunately, research on differences between different genotypes of cattle is often rather descriptive in nature (see, e.g., El Hakim *et al.*, 1986; Robelin, 1986; Kirchgessner *et al.*, 1993). Reported differences in nutrient partitioning between genotypes, however, are large, and this, therefore, deserves attention in the development of animal growth models.

Leaner growth of certain breeds can be simulated by manipulating any of the parameters involved in protein deposition, in turn also affecting the fat deposition rate. The rate of protein turnover is likely to be important, while the higher preference of leaner breeds to use dietary protein for deposition may be related to a lower affinity of amino acids for catabolism. Furthermore, sex and breed may affect the regulation of fatty acid and fat metabolism but, unfortunately, little information is available on this subject. The mechanisms responsible, however, are likely to include hormone action. Therefore, once the model parameters to be modified by sex and breed are chosen, they can be altered by, for example, introducing parameters representing hormone function. An often used, more statistical, approach is to scale model parameters for each breed according to Taylor's rule (Taylor, 1980). Briefly, this rule is based on the observation that the rates of various functions in animals between species are proportional to their mature weight, raised to the power of 0.73. This approach has been adopted by Oltjen *et al.* (1986), for example, who scaled the maximum rates of DNA and protein synthesis, protein degradation and maintenance energy requirements according to this rule.

CONCLUSIONS

Several models of growth in ruminants have been developed. These models have been quite effective in predicting growth and in identifying areas of uncertainty with respect to current knowledge of growth processes. To date, most ruminant growth models focus on post-absorptive metabolism of nutrients. The integration of growth models and models which predict the nutrient availability resulting from feed intake, rumen fermentation and intestinal degradation processes will be an important step forward. In view of the differences in metabolic activity of individual organs and tissues, more effort is required to represent the functional tissues, rather than exclusively chemical components in the body. Finally, there is a lack of quantitative information on the variation in substrate affinity and its endocrine control resulting in a different nutrient partitioning between breeds and sexes.

ACKNOWLEDGEMENTS

We thank Dr Norm Adams for critically reviewing this chapter.

REFERENCES

Allen, M.S. (1996) Physical constraints on voluntary intake of forages by ruminants. *Journal of Animal Science* 74, 3063–3075.

Agricultural Research Council (1980) The nutrient requirements of ruminant livestock. Technical review by an agricultural research council working party. *The Nutrient Requirements of Ruminant Livestock*. Commonwealth Agricultural Bureaux, Slough, UK.

Arnold, R.N. and Bennett, G.L. (1991a) Evaluation of four simulation models of cattle growth and body composition, Part I. Comparison and characterization of the models. *Agricultural Systems* 35, 401–432.

Arnold, R.N. and Bennett, G.L. (1991b) Evaluation of four simulation models of cattle growth and body composition, Part II. Simulation and comparison with experimental growth data. *Agricultural Systems* 36, 17–41.

Arnold, G.W., Campbell, N.A. and Galbraith, K.A. (1977) Mathematical relationships and computer routines for a model of food intake, liveweight change and wool production in grazing sheep. *Agricultural Systems* 2, 209–226.

Baldwin, R.L. (1995) *Modeling Ruminant Digestion and Metabolism.* Chapman & Hall, London.

Baldwin, R.L., France, J., Beever, D.E., Gill, M. and Thornley, J.H.M. (1987a) Metabolism of the lactating cow. III. Properties of mechanistic models suitable for evaluation of energetic relationships and factors involved in the partition of nutrients. *Journal of Dairy Research* 54, 133–145.

Baldwin, R.L., Thornley, J.H.M. and Beever, D.E. (1987b) Metabolism of the lactating cow. II. Digestive elements of a mechanistic model. *Journal of Dairy Research* 54, 107–131.

Baldwin, R.L., Calvert, C.C. and Oberbauer, A.M. (1991) Growth control in the future. In: Pearson, A.M. and Dutson, T.R. (eds), *Growth Regulation in Farm Animals.* Elsevier, London, pp. 589–617.

Bastianelli, D., Sauvant, D. and Rérat, A. (1996) Mathematical modeling of digestion and nutrient absorption in pigs. *Journal of Animal Science* 74, 1873–1887.

Benevenga, N.J., Gahl, M. and Blemings, K.P. (1993) Role of protein synthesis in amino acid catabolism. *Journal of Nutrition* 123, 332–336.

Black, J.L. and Reis, P.J. (1979) Speculation on the control of nutrient partition between wool growth and other body functions. In: Black, J.L. and Reis, P.J. (eds), *Physiological and Environmental Limitations to Wool Growth.* University of New England Publishing Unit. Armidale, Australia, pp. 269–293.

Blackburn, H.D. and Cartwright, T.C. (1987) Description and validation of the Texas A & M sheep simulation model. *Journal of Animal Science* 65, 373–386.

Black, J.L. and Griffiths, D.A. (1975) Effects of live weight and energy intake on nitrogen balance and total N requirement of lambs. *British Journal of Nutrition* 33, 399–413.

Black, J.L. and Nagorcka, B.N. (1993) Wool growth. In: Forbes, J.M. and France, J. (eds), *Quantitative Aspects of Ruminant Digestion and Metabolism.* CAB International, Wallingford, UK, pp. 453–477.

Bowman, P.J., Cottle, D.J., White, D.H. and Bywater, A.C. (1993) Simulation of wool growth rate and fleece characteristics of Merino sheep in Southern Australia. Part 1, model description. *Agricultural Systems* 43, 287–299.

Campbell, R.G. (1988) Nutritional constraints to lean tissue accretion in farm animals. *Nutrition Research Reviews* 1, 233–253.

Dijkstra, J. and France, J. (1996) A comparative evaluation of models of whole rumen function. *Annales de Zootechnie* 45 (Suppl. 1), 175–192.

Dijkstra, J., France, J., Neal, H.D.St.C., Assis, A.G., Aroeira, L.J.M. and Campos, O.F. (1996) Simulation of digestion in cattle fed sugar cane, model development. *Journal of Agricultural Science, Cambridge* 127, 231–246.

Di Marco, O.N., Baldwin, R.L. and Calvert, C.C. (1989) Simulation of DNA, protein and fat accretion in growing steers. *Agricultural Systems* 29, 21–34.

El-Hakim, A., Eichinger, H. and Pirchner, F. (1986) Growth and carcass traits of bulls and veal calves of continental cattle breeds. 2. Carcass composition. *Animal Production* 43, 235–243.

Fisher, D.S., Burns, J.C. and Pond, K.R. (1987) Modeling *ad libitum* dry matter intake by ruminants as regulated by distension and chemostatic feedback. *Journal of Theoretical Biology* 126, 407–418.

Fisher, D.S. (1996) Modelling ruminant feed intake with protein, chemostatic and distention feedbacks. *Journal of Animal Science* 74, 3076–3081.

Forbes, J.M. (1996) Integration of regulatory signals controlling forage intake in ruminants. *Journal of Animal Science* 74, 3029–3035.

Fox, D.G. and Black, J.R. (1984) A system for predicting body composition and performance in growing cattle. *Journal of Animal Science* 58, 725–739.

France, J. and Thornley, J.H.M. (1984) *Mathematical Models in Agriculture.* Butterworths, London.

France, J., Gill, M., Thornley, J.H.M. and England, P. (1987) A model of nutrient utilization and body composition in beef cattle. *Animal Production* 44, 371–385.

France, J., Thornley, J.H.M., Baldwin, R.L and Crist, K.A. (1992) On solving stiff equations with reference to simulating ruminant metabolism. *Journal of Theoretical Biology* 156, 525–539.

Freer, M., Moore, A.D. and Donnelly, J.R. (1997) Grazplan, decision support systems for Australian grazing enterprises II. The animal biology model for feed intake, production and reproduction and the Grazfeed DSS. *Agricultural Systems* 54, 77–126.

Gandar, P.W., Kelly, K.E., Harris, P.M. and Dellow, D.W. (1990) Modelling growth in wool follicles. In: Robson, A.B. and Poppi, D.P. (eds), *Proceedings of the Third International Workshop on Modelling Digestion and Metabolism in Farm Animals.* September 4–6, 1989, Lincoln University, New Zealand, Lincoln University Press, pp. 189–205.

Gerrits, W.J.J., Tolman, G.H., Schrama, J.W., Tamminga, S., Bosch, M.W. and Verstegen, M.W.A. (1996) Effect of protein and protein-free energy intake on protein and fat deposition rates in preruminant calves of 80 to 240 kg live weight. *Journal of Animal Science* 74, 2129–2139.

Gerrits, W.J.J., Dijkstra, J. and France, J. (1997a) Description of a model integrating protein and energy metabolism in preruminant calves. *Journal of Nutrition* 127, 1229–1242.

Gerrits, W.J.J., France, J., Dijkstra, J., Bosch, M.W., Tolman, G.H. and Tamminga, S. (1997b) Evaluation of a model integrating protein and energy metabolism in pre-ruminant calves. *Journal of Nutrition* 127, 1243–1252.

Gill, M. and Romney, D. (1994) The relationship between the control of meal size and the control of daily intake in ruminants. *Livestock Production Science* 39, 13–18.

Gill, M., Thornley, J.H.M., Black, J.L., Oldham, J.D. and Beever, D.E. (1984) Simulation of the metabolism of absorbed energy-yielding nutrients in young sheep. *British Journal of Nutrition* 52, 621–649.

Gill, M., France, J., Summers, M., McBride, W. and Milligan, L.P. (1989a) Mathematical integration of protein metabolism in growing lambs. *Journal of Nutrition* 119, 1269–1286.

Gill, M., France, J., Summers, M., McBride, W. and Milligan, L.P. (1989b) Simulation of the energy costs associated with protein turnover and Na^+, K^+-transport in growing lambs. *Journal of Nutrition* 119, 1287–1299.

Graham, N.M., Black, J.L., Faichney, G.J. and Arnold, G.W. (1976) Simulation of growth and production in sheep – model 1. A computer program to estimate energy and nitrogen utilization, body composition and empty liveweight change, day by day for sheep of any age. *Agricultural Systems* 1, 113–138.

Hanigan, M.D., Dijkstra, J., Gerrits, W.J.J. and France, J. (1997) Modelling postabsorptive protein and amino acid metabolism in the ruminant. *Proceedings of the Nutrition Society* 56, 631–643.

Illius, A.W. and Jessop, N.S. (1996) Metabolic constraints on voluntary intake in ruminants. *Journal of Animal Science* 74, 3052–3062.

Ingvartsen, K.L. (1994) Models of voluntary feed intake in cattle. *Livestock Production Science* 39, 19–38.

Keele, J.W., Williams, C.B. and Bennett, G.L. (1992) A computer model to predict the effects of level of nutrition on composition of empty body gain in beef cattle, 1. Theory and development. *Journal of Animal Science* 70, 841–857.

Ketelaars, J.J.M.H. and Tolkamp, B.J. (1996) Oxygen efficiency and the control of energy flow in animals and humans. *Journal of Animal Science* 74, 3036–3051.

Kirchgessner, M., Schwarz, F.J., Otto, R., Reimann, W. and Heindl, U. (1993) Energie- und Nährstoffgehalte im Schlacht- und Ganzkörper wachsender Jungbullen, Ochsen und Färsen der Rasse Deutsches Fleckvieh bei unterschiedlicher Fütterungsintensität. *Journal of Animal Physiology and Animal Nutrition* 70, 266–277.

Koong, L.J., Falter, K.H. and Lucas, H.L. (1982) A mathematical model for the joint metabolism of nitrogen and energy in cattle. *Agricultural Systems* 9, 301–324.

Korver, S., Tess, M.W. and Johnson, T. (1988) A model of growth and growth composition for beef bulls of different breeds. *Agricultural Systems* 27, 279–294.

Liu, S.M., Lobley, G.E., MacLoad, N.A., Kyle, D.J., Chen, X.B. and Ørskov, E.R. (1995) Effects of long-term protein excess or deficiency on whole body protein turnover in sheep nourished by intragastric infusion of nutrients. *British Journal of Nutrition* 73, 829–839

Loewer, O.J., Smith, E.M., Taul, K.L., Turner, L.W. and Gay, N. (1983) A body composition model for predicting beef animal growth. *Agricultural Systems* 10, 245–256.

Lofgreen, G.P. and Garrett. W.N. (1968) A system for expressing the net energy requirements and feed values for growing and finishing beef cattle. *Journal of Animal Science* 27, 793–806.

MacRae, J.C., Bruce, L.A., Brown, D.S. and Calder A.G. (1997) Amino acid use by the gastrointestinal tract of sheep given lucerne forage. *Americal Journal of Physiology* 273, G1200–G1207.

Millward, D.J., Price, G.M., Pacy, P.J.H. and Halliday, D. (1990) Maintenance protein requirements, the need for conceptual re-evaluation. *Proceedings of the Nutrition Society* 49, 473–487.

Moughan, P.J. and Verstegen, M.W.A. (1988) The modelling of growth in the pig. *Netherlands Journal of Agricultural Science* 36, 145–166.

O'Connor, J.D., Sniffen, C.J., Fox, D.G. and Chalupa, W. (1993) A net carbohydrate and protein system for evaluating cattle diets, IV, predicting amino acid adequacy. *Journal of Animal Science* 71, 1298–1311.

Oltjen J.W. (1989) Modelling growth and metabolism in cattle. In: Robson, A.B. and Poppi, D.P. (eds), *Proceedings of the Third International Workshop on Modelling Digestion and Metabolism in Farm Animals*. September 4–6, 1989, Lincoln University, New Zealand, Lincoln University Press, pp. 255–262.

Oltjen, J.W., Bywater, A.C., Baldwin, R.L. and Garrett, W.N. (1986) Development of a dynamic model of beef cattle growth and composition. *Journal of Animal Science* 62, 86–97.

Ortigues, I., Petit, M. and Agabriel, J. (1993) Influence of body condition on maintenance energy requirements of Charolais cows. *Animal Production* 57, 47–53.

Owens, F.N. (1987) Maintenance protein requirements. In: Jarrige, R. and Alderman, G. (eds), *Feed Evaluation and Protein Requirement Systems for Ruminants*. Office for Official Publications of the European Community, Luxembourg, pp. 187–212.

Pettigrew, J.E., Gill, M., France, J. and Close, W.H. (1992) A mathematical integration of energy and amino acid metabolism of lactating sows. *Journal of Animal Science* 70, 3742–3761.

Poppi, D.P., Gill, M. and France, J. (1994) Integration of theories of intake regulation in growing ruminants. *Journal of Theoretical Biology* 167, 129–145.

Reeds, P.J. (1991) The energy cost of protein deposition. In: Wenk, C. and Boessinger, M. (eds), *Proceedings of the 12th Symposium on Energy Metabolism.* European Association of Animal Production Publication No. 58, Kartause Ittingen, Switzerland, pp. 473–479.

Reeds, P.J. and Mersmann, H.J. (1991) Protein and energy requirements of animals treated with β-adrenergic agonists, a discussion. *Journal of Animal Science* 69, 1532–1550.

Robelin, J. (1979) Influence de la vitesse de croissance sur la composition du gain de poids des bovins, variations selon la race et le sexe. *Annales de Zootechnie* 28, 209–218.

Robelin, J. (1986) Growth of adipose tissues in cattle; partitioning between depots, chemical composition and cellularity. A review. *Livestock Production Science* 14, 349–364.

Sainz, R.D. (1989) Models of metabolic regulation. In: Robson, A.B. and Poppi, D.P. (eds), *Proceedings of the Third International Workshop on Modelling Digestion and Metabolism in Farm Animals.* September 4–6, 1989, Lincoln University, New Zealand. Lincoln University Press, pp. 277–291.

Sainz, R.D. and Wolff J.E. (1990) Development of a dynamic, mechanistic model of lamb metabolism and growth. *Animal Production* 51, 535–549.

Sanders, J.O. and Cartwright, T.C. (1979) A general cattle production systems model. Part 2. Procedures used for simulating animal performance. *Agricultural Systems* 4, 289–309.

Sauvant, D., Baumont, R. and Faverdin, P. (1996) Development of a mechanistic model of intake and chewing activities of sheep. *Journal of Animal Science* 74, 2785–2802.

Schmidely, Ph. (1996) Growth in ruminants, a comparison of some mechanistic models. *Annales de Zootechie* 45 (Suppl. 1), 193–214.

Storm, E., Brown, D.S. and Ørskov, E.R. (1983) The nutritive value of rumen micro-organisms in the ruminant. 3. The digestion of microbial amino and nucleic acids in, and losses of endogenous nitrogen from, the small intestine of sheep. *British Journal of Nutrition* 50, 479–485.

Tamminga, S., Schulze, H., Van Bruchem, J. and Huisman, J. (1995) The nutritional significance of endogenous N-losses along the gastrointestinal tract of farm animals. *Archives of Animal Nutrition* 48, 9–22.

Taylor, St.C.S. (1980) Genetic size scaling rules in animal growth. *Animal Production* 30, 161–165.

Van Straalen, W.M. and Tamminga, S. (1990) Protein degradability in ruminant diets. In: Wiseman, J. and Cole, D.J.A. (eds), *Feedstuff Evaluation.* Butterworths, London, pp. 55–72.

Webster, A.J.F. (1989) Bioenergetics, bioengineering and growth. *Animal Production* 48, 249–269.

Whittemore, C.T. and Fawcett, R.H. (1976) Theoretical aspects of a flexible model to simulate protein and lipid growth in pigs. *Animal Production* 22, 87–96.

16 Modelling Growth and Lactation in Pigs

J.L. Black

John L. Black Consulting, Locked Bag 21, Warrimoo, Australia

INTRODUCTION

The growth and production characteristics of commercially raised pigs are affected by a wide range of factors relating to the genetic characteristics of individual animals, the chemical composition and physical form of the diets, the feeding strategies, the climatic environment and building characteristics, the penning and stocking arrangements as well as the prevalence and severity of disease. For reproducing sows, conceptus growth, milk yield and subsequent reproductive performance also depend on the composition of the diets and feeding regimes during gestation, the parity, number and size of piglets being fed by the sow and the stage of lactation. There is a great deal of information on how most of these factors influence growth and productivity of the pig, but the complexity of the interactions makes it virtually impossible to assess with accuracy the consequences of management changes on nutrient requirements, growth rate and body composition of the animals and on enterprise profitability. However, by transformation of the concepts and knowledge into mathematical equations which are then integrated using computer simulation modelling techniques, the vast store of information can be applied directly and rapidly to improving the management, animal welfare and profitability of individual pig production enterprises.

Several models of pig growth have been developed (Whittemore and Fawcett, 1974, 1976; Stombaugh and Oko, 1980; Tess *et al.*, 1983; Whittemore, 1983; Black *et al.*, 1986; Emmans, 1986; Moughan *et al.*, 1987; Burlacu *et al.*, 1988; Pomar *et al.*, 1991; Technisch Model Varkensvoeding (TMV), 1991; Bridges *et al.*, 1992; Pettigrew *et al.*, 1992a,b; de Lange *et al.*, 1998; National Research Council (NRC), 1998), and some have included aspects of reproductive performance (Black *et al.*, 1986; Pettigrew *et al.*, 1992a,b; NRC, 1998). Several models, for example the new NRC model, are static and do not predict changes in performance over time, but most pig models are dynamic. The majority of the existing models predict the utilization of amino acids and energy only, but the TMV model also predicts calcium and phosphorus utilization. Similarly, feed intake is predicted in a minority of the models (Black *et al.*, 1986; Emmans, 1986; Bridges *et al.*, 1992), but omission severely limits application of these models as feed intake is a major determinant of animal productivity. A few of the models (Black *et al.*, 1986; TMV, 1991; de Lange *et al.*, 1998; NRC, 1998) are being used as decision support systems to facilitate the management of commercial pig enterprises. The impact of computer models on enterprise profitability can be substantial (de Lange and Schreurs, 1995; Edwards, 1997; Smits, 1997; Black, 1998; de Lange *et al.*, 1998; Willis, 1998).

Models of most value as decision support systems for industry should be dynamic and constructed at a level that will allow accurate prediction of the efficiency of utilization of individual nutrient classes, changes in the quantity of the major components (protein, fat, water and ash) in the body and the contents of the digestive tract, as this may affect carcass yield (Black, 1998). In addition, for a lactating pig, milk yield and milk composition should be predicted. Described in this chapter are the principles behind the concepts used in models for pig growth and for predicting conceptus growth and milk yield in sows. Algorithms are not presented but these can be found in the referenced papers.

PRINCIPLES OF DYNAMIC MODELLING FOR PIG PRODUCTION

The principles behind dynamic modelling of nutrient utilization for simulating pig growth and production were outlined by Black and de Lange (1995). There are seven main steps in simulating changes in growth, body composition and reproductive performance of pigs, as follows: (i) body composition at the start of the simulation period; (ii) feed intake and the intake of individual nutrients; (iii) availability for metabolism of the consumed nutrients; (iv) nutrients, both energy and amino acids, required for maintenance of the animal's integrity; (v) nutrients used for growth and reproduction in relation to the genotypically controlled or exogenously modified growth potential, the physiological state of the pig, the climatic conditions and stresses associated with social inter-actions and disease; (vi) efficiency of nutrient use for different body functions, and (vii) final outcome of nutrient utilization in terms of net accretion of body tissues, gut contents, conceptus and fetal tissue, and milk yield. Each of these steps have been incorporated in the majority of pig models, using an iteration interval of 1 day, except for prediction of the effect of climate, where an iteration interval of 1 h has been used. The partition of nutrients between competing body functions has been achieved in these models by applying pre-determined priorities to individual physiological functions. The prediction of nutrient partition using Michaelis–Menten kinetics, where the iteration inter-val is in the order of minutes, has been attempted only by Pettigrew *et al.* (1992a,b).

BODY COMPOSITION AT THE START OF A SIMULATION

To predict the body composition of the animal at the end of a simulated period of time, the chemical composition of the body at the start of the simulation must be specified. The composition of the body affects various aspects of nutrient partitioning in animals. The energy requirement for maintenance is related directly to body protein mass (Graham *et al.*, 1974) and particularly to the mass of visceral organs (Ferrell, 1988). There is evidence also that the potential daily rate of protein and fat deposition is influenced by nutrient-imposed differences in body composition. Campbell and Dunkin (1983) observed that pigs given protein-inadequate diets were fatter and had faster rates of protein deposition following correction of the dietary deficiency than those fed adequate diets throughout the growing period.

In current pig simulation models, the initial weight of either body protein or body fat is either defined as a model input or calculated from other model inputs. De Lange (1995) defined the initial weight of body protein as a model input and then calculated initial body fat content from a minimum ratio of body fat to body protein which was assumed to be genotype dependent. Simple allometric and linear relationships were used to predict the initial weights of water and ash, respectively, from body protein. Ferguson and Gous (1993a) used a similar approach to that of de Lange (1995) except that a Gompertz function was employed to calculate body protein weight at the start of the simulation from estimates of body protein weight at birth and at maturity. Allometric equations were then used to predict body water and ash from body protein, and body fat was estimated from a ratio of body fat to protein at maturity. Alternatively, Black *et al.* (1986) used an equation developed by Searle and Griffiths (1976) which predicted body fat weight in relation to empty body weight and genotype for animals reared under near ideal conditions. The equation describes two linear relationships between body fat weight and empty body weight, with the slope of the relationships changing gradually from the pre-fattening to the fattening phase of growth. An equation of similar form was used to predict body protein weight in relation to the fat-free empty body. Parameters for these equations were genotype dependent and were derived from experiments in which pigs were grown from weaning to near maturity (Siebrits *et al.*, 1986; Whittemore *et al.*, 1988; Ferguson and Gous, 1993b). These equations are not entirely appropriate for situations where animals have been fed well below *ad libitum* prior to commencement of the simulation. The relationships between body protein, water and ash are known to be influenced slightly by the stage of animal maturity, genotype and previous nutritional history (Black, 1983; Stanks *et al.*, 1988). However, small inaccuracies in estimates of initial body composition are relatively unimportant in a dynamic model that simulates animal performance over a period in which body weight increases several fold.

PREDICTION OF FEED AND NUTRIENT INTAKE

Prediction of nutrient intake requires either a prediction of voluntary feed intake or knowledge of the specific feeding regime, as well as information on the composition of the diet eaten. Most builders of pig models have regarded the prediction of voluntary feed intake as too difficult (de Lange *et al.*, 1998). However, it has been predicted in the model of Black *et al.* (1986). Although, ultimately, voluntary intake is controlled by set points of centres in the hypothalamus (Revell and Williams, 1993), Black *et al.* (1986) assumed that each animal has a potential voluntary intake that is determined by its requirement for nutrients and particularly energy under ideal, thermoneutral, disease-free and stress-free conditions. The actual intake of an animal may be either increased or decreased from this potential, depending upon the dietary, climatic, social and disease environment that it experiences. A full description of the approach is given by Black (1995).

Potential Energy Demand

The potential energy demand of a non-reproducing pig that is assumed to drive voluntary feed intake is comprised of the sum of the energy required for maintenance and for tissue deposition. Maintenance is not constant in relation to metabolic live weight, but varies with several components that are described below. The potential rate of energy deposition can be assessed when pigs are grown from birth to maturity and given free access to a high-quality diet where feed intake is not limited by gut capacity, disease, the climatic or social environment. Ferguson and Gous (1993a) suggest that such a situation can be obtained by giving individually housed pigs a choice of a high- and a low-protein diet and a gradation of temperatures across a room. Several estimates of the potential rate of energy deposition have been made for pigs of different genotypes under assumed ideal conditions (Black *et al.*, 1986; Siebrits *et al.*, 1986; Whittemore *et al.*, 1988; Ferguson and Gous, 1993b).

Treatment of pigs with growth hormone is known to increase the rate of protein deposition and reduce the rate of fat deposition, with a resultant overall reduction in the rate of energy deposition (Campbell and Taverner, 1988). The voluntary feed intake of pigs treated with growth hormone is reduced significantly. Thus, pigs treated with growth hormone can be regarded as behaving like a different genotype. Provided the change in potential rate of energy deposition and maintenance energy requirements can be determined, voluntary feed intake of these animals can be predicted readily (Black, 1988).

The voluntary feed intake of the lactating sow is not determined simply by the energy demand for maintenance and lactation (Revell and Williams, 1993). Voluntary intake appears to be influenced largely by the body composition of the sow at the start of lactation, with feed intake being less in sows with greater than about 300 g of fat kg^{-1} empty body weight (Mullan and Williams, 1989, 1990). This concept has been incorporated into a simulation model of the reproducing sow (J.L. Black, unpublished) by calculating the maximum fractional rate of fat catabolism based on the results from Mullan and Williams (1989) and then assuming that this energy is available to the sow before calculating the feed intake needed for meeting the maintenance and lactation energy needs. Figure 16.1 provides a comparison of predicted and observed voluntary feed intake values for several experiments (King and Dunkin, 1986; Mullan and Williams, 1990; Mullan *et al.*, 1992; Revell *et al.*, 1998) and shows that the concepts predict much of the variation seen between treatments.

Essential Nutrient Deficiencies

There is evidence that both the potential rate of energy deposition and voluntary feed intake are reduced when pigs are given diets moderately deficient in either protein or essential amino acids (Likuski *et al.*, 1961; Anderson, 1979; Rogerson and Campbell, 1982; Henry *et al.*, 1992). Observations by R.P. Chapple (unpublished) indicate that the intake of metabolizable energy in growing pigs is maintained at potential rates until the protein deficiency is such that protein deposition falls to around 0.80 of its maximum rate; during this initial period, feed intake is maintained and fat deposition rate increases slightly. Once the protein deficiency is sufficient for the rate of protein deposition to be less than about 0.80 of the maximum rate, total energy deposition, fat deposition and voluntary feed intake fall linearly until energy intake is about 0.20 of its potential, when the pig is fed a protein-free

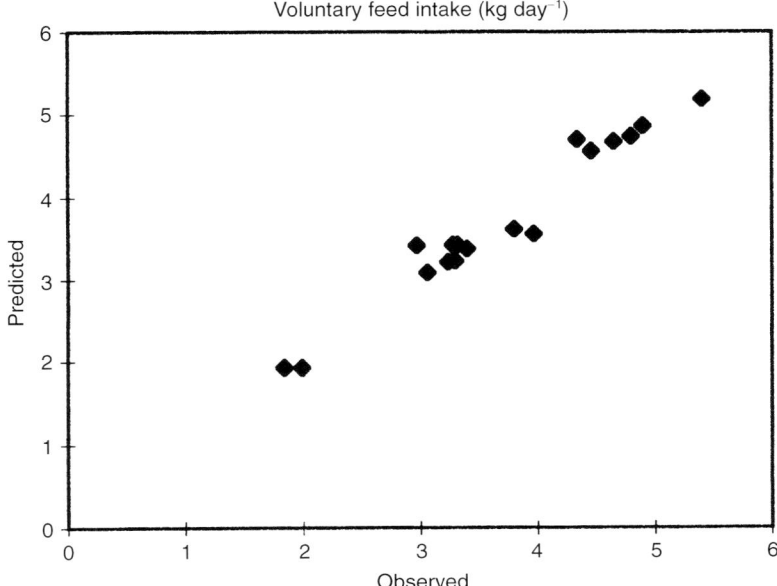

Fig. 16.1. A comparison of predicted and observed voluntary feed intake during lactation of sows fed a range of diets differing in protein and energy content during gestation and lactation.

diet. There is some evidence that the decline in feed intake in pigs fed protein-deficient diets is influenced by the biological value of the dietary protein, with a greater reduction in feed intake occurring when the amino acid pattern of the diet is poor (R.P. Chapple, unpublished).

Gut Capacity

There is considerable evidence indicating that the capacity of the digestive tract can influence the voluntary intake of pigs, particularly those of lighter weight. Black *et al.* (1986) reviewed the effects of reducing the digestible energy (DE) content of diets by adding inert substances on the voluntary feed intake of pigs ranging in weight from 10 to 107 kg. Pigs weighing less than about 20 kg appear unable to increase feed intake to compensate for a reduction in energy density of the diet below about 16 MJ kg^{-1}, whereas those weighing up to 50 kg appear unable to compensate when energy density of the diet falls below 14 MJ kg^{-1}. However, the intake of pigs over 70 kg continues to increase as energy density of the diet falls to 10 MJ kg^{-1}. The actual digestible energy content of the diet at which pigs are unable to increase feed intake further to compensate for a fall in energy density depends on the nature of the fibre, the bulk density of the diet (Zoiopoulos *et al.*, 1982), its fat content (Revell and Williams, 1993), the viscosity and water-holding capacity of the digesta (Eastwood *et al.*, 1983; Kyriakakis and Emmans, 1995) and its rate of flow throught the gut. An empirical estimate of the upper limit to feed intake set by gut capacity was established by Black *et al.* (1986) for pigs ranging in live weight from 10 to 200 kg. However, this representation cannot be regarded as satisfactory because the accuracy of predictions from simulation models can be extremely sensitive to the gut capacity-imposed limit to feed intake. Further research is required to define more precisely the effects of bulk density of the diet, its water-holding capacity, the nature and rate of digestion of fibre and other non-starch polysaccharides, and the effect of feeding regimes and of climatic conditions on the capacity of the digestive tract and the rate of flow of digesta through the tract.

Stocking Density and Number of Pigs in a Group

The effects both of stocking density measured as space allocation per pig for groups exceeding five animals, and of the number of pigs per pen when stocking density is low have been reviewed by Chapple (1993) and Morgan *et al.* (1998). Stocking density is expressed commonly in relation to floor area allocation (m^2) per live weight raised to a power (W, kg$^{0.67}$) because it accounts for pigs varying widely in weight. An area of 0.048 m^2 W$^{-0.67}$ is required for all pigs in the pen to lie recumbent, whereas an area of 0.034 m^2 W$^{-0.67}$ is required for all pigs to lie on their sternums. Feed intake appears to be depressed once the floor allocation falls below about 0.035–0.039 m^2 W$^{-0.67}$ and is depressed by about 20% when the allocation is only 0.020 m^2 W$^{-0.67}$.

In addition to the effect of high stock density, the intake of pigs given ample space appears to be reduced once there is social interaction with other animals. Gonyou *et al.* (1992) found that increasing the number of pigs per pen from one to five depressed feed intake by 8% despite there being adequate space for each pig. Similar results have been reported by de Haer and de Vries (1993) when individually housed pigs were compared with pens of eight pigs. Further increases in the number of pigs per pen above about ten does not appear to have any further effect on feed intake. Recent research suggests, however, that there may be an interaction between the effect of increasing the number of pigs per pen and the cleanliness of the environment on feed intake. Lee *et al.* (1997) examined the effect on feed intake of increasing the number of pigs in a pen from one to ten in a clean and a dirty commercial environment. The dirty environment was created by not cleaning the pen either before or during the experiment, whereas pens in the clean environment were cleaned before and daily during the experiment and the air was sprayed several times daily to reduce dust concentration. Feed intake of the group-penned pigs was significantly reduced compared with the single housed animals in the clean environment. However, in the dirty environment, the intake of the single and group-penned pigs was the same and not significantly different from the group-penned pigs in the clean environment.

Cold and Hot Climatic Conditions

When ambient temperature falls below the lower critical temperature of a pig, its energy expenditure increases and voluntary feed intake tends to increase, often after a delay, to compensate for the additional heat loss. Conversely, at temperatures above the evaporative critical temperature of the animal, which is the temperature at which respiration rate increases, feed intake declines. The magnitude of the rise in feed intake at low temperatures and the fall at high temperatures is influenced by diet composition, stocking arrangements and the weight of pigs. Close (1989) collated information on the effect of ambient temperature on the intake of metabolizable energy for pigs ranging in live weight from 18 to 90 kg and showed that feed intake fell for all animals at high ambient temperatures, whereas the increase in intake at low temperatures was greater in the heavier than in the lighter pigs.

The capacity of the digestive tract and the stocking arrangements appear capable of limiting the ability of pigs in cold situations to increase feed intake sufficiently to compensate for heat lost to the environment. Evidence was given above that the capacity of the digestive tract can limit feed intake, particularly for low body weight pigs, and this may prevent intake rising sufficiently to maintain energy balance, as was seen in the experiments with light weight pigs reported by Close (1989). Pigs in groups that are exposed to cold conditions spend more time huddling than in feeding activities compared with pigs in a thermally comfortable environment (Parker *et al.*, 1980). These observations indicate that feed intake of pigs exposed to cold conditions could be influenced by the number of pigs in the pen and be less for pigs housed in a group than for individually penned pigs. Nienaber *et al.* (1990, 1991) subjected pigs housed either separately or in groups of four to ambient temperatures 12°C below the estimated lower critical temperature and observed a 30% increase in feed intake for the cold-exposed single pig compared with only an 11% increase for the cold, group-housed pigs. This difference in feed intake could relate to the rate of heat loss from a pig when huddled in a group compared with the heat loss when it stands to eat. Presumably, the pig prefers to remain warm in the group and eats only when the hunger drive is high and then for shorter periods of time compared with pigs housed individually.

The decline in feed intake for pigs exposed to high temperatures has been shown in several studies to be negatively associated with an increase in body temperature (Fuller, 1965; Stahly and Cromwell, 1979; Schoenherr *et al.*, 1989; Giles, 1992; Lorschy, 1994). Giles (1992) observed from several experiments, that feed intake falls by approximately 0.4 for every 1°C increase in body temperature above about 39.2°C, and that animals cease eating when body temperature reaches about 41.3°C.

Disease

Almost all diseases cause a reduction in feed intake. Digestive diseases reduce the digestibility of the diet, result in scouring and, during the acute phase, cause a substantial reduction in feed intake. Respiratory diseases have their effect through increased fever initially, and subsequently through lung damage that reduces the animal's capacity for activity associated with feeding (Black *et al.*, 1998). During acute disease, stimulation of the immune system and release of the cytokines, interleukin-1 (IL-1), IL-6 and tumour necrosis factor (TNF) are likely to be the major reasons for the initial decline in feed intake in pigs with an acute inflammatory disease. IL-1, IL-6 and TNF are directly anorexigenic in mammals (Klasing and Johnstone, 1991), with IL-1 causing a greater response than TNF. The continued reduction of feed intake in pigs with chronic, severe lung damage is thought to be due primarily to a reduction in arterial oxygen saturation caused by a ventilation–perfusion inequality. The daily feed intake for a pig with pleuropneumonia was found by Bray (1996) to be only 0.24 of the pre-inoculation value compared with 0.97 for a control pig. The infected pig had a reduced time standing, smaller meals and a lower venous oxygen saturation than the saline-inoculated pig. These observations show that, relative to the control pig, the pleuropneumonia-infected pig removed more oxygen from the blood during standing and eating. Bray (1996) argued that the duration of the meal was shortened due to an inability of the pig to meet the oxygen demand for standing during long periods of time. Support for this suggestion comes from observations in humans with chronic, obstructive, pulmonary disease where muscle intracellular substrate concentrations were found at rest to be similar to those of unaffected people, but impaired muscular oxidative capacity was seen in the forearm of diseased patients following increasing amounts of muscular activity (Wuyam *et al.*, 1992).

There are no published simulation models that predict the effect of disease on the voluntary feed intake of pigs. However, more information of the type given above will allow prediction of intake from an understanding of the underlying mechanisms in relation to an estimation of the likely severity of the disease.

AVAILABILITY OF CONSUMED NUTRIENTS

The availability of nutrients can be defined as the proportion of a nutrient in the diet that is digested and absorbed in a form suitable for utilization. Digestibility is a measure of the disappearance of a nutrient as it passes through the digestive tract of an animal, and it is expressed as a percentage of the amount eaten. Energy digestibility is estimated commonly in pigs from its disappearance between the feed eaten and the faeces, whereas for amino acids, disappearance between the feed and the terminal ileum is used to estimate digestibility because of major changes to the amino acid composition of digesta in the hind-gut caused by microbial activity. Amino acids that disappear in the hind-gut are not absorbed by the pig, but are deaminated by microorganisms. Although simple sugars and starches are readily digested by intestinal enzymes, many soluble non-starch polysaccharides including oligo-saccharides, β-glucans, xylans and arabinoxylans are only fermented in the hind-gut. Similarly, dietary cellulose is degraded only by microorganisms in the large intestine, and the extent of digestion is influenced greatly by the degree of lignification and the amount of silicates associated with the fibre. A considerable amount of energy is lost during the microbial digestion of dietary substrates through heat of fermentation and the formation of methane. The amount of energy lost varies depending on the actual substrates degraded, but will generally be around 6% as heat of fermentation and between 6 and 18% incorporated into methane. Thus, whole-tract energy digestibility is not an accurate estimate of energy availability although it is used in most models.

The digestibility of nutrients is affected by the residence time of digesta in the gut, and it decreases as intake increases, particularly towards *ad libitum* (Low 1990). Digestibility also decreases when pigs are exposed to cold conditions because of an increase in the flow rate of digesta through the tract (Phillips *et al.*, 1982). Energy digestibility increases as a pig matures. Bakker (1996) observed a 2.25% higher energy digestibility when the same diet was fed to pigs weighing 90 kg compared with pigs weighing 60 kg. Similarly, van Barneveld (1997) showed that the digestibility of crude fibre in lupin seeds was 0.95 in sows but only 0.42 in pigs weighing about 50 kg. Most estimates of nutrient digestibility are made when animals are between about 30 and 50 kg and given intakes well below *ad libitum*. The digestibility in pigs of energy from cereal grains is influenced greatly by particle size, with the digestible energy content of maize increasing from 14.7 to 16.1 MJ kg^{-1} as particle size was decreased from 1200 to 400 μm (Wondra *et al.*, 1995). Saturated fatty acids are digested less efficiently than unsaturated fatty acids, and long chain fatty acids less efficiently than short chain fatty acids (Freeman *et al.*, 1968). The absorption of saturated fatty acids is increased by the presence of unsaturated fatty acids and monoglycerides. Stahly (1984) showed that when the ratio of unsaturated to saturated fatty acids was greater than 1.5, the digestibility of total fat was in the range of 0.85–0.92. However, when the ratio was between 1.0 and 1.3, the digestibility of total fat was as low as from 0.35 to 0.75. The digestibility of dietary fat can be affected also by the presence of insoluble dietary fibre. Bakker (1996) found that the digestibility of an animal fat added to a maize-based diet was 0.91, whereas replacement of maize with either cellulose or soybean hulls to 0.30 of the diet reduced fat digestibility to 0.83 and 0.87, respectively. None of the current pig models predict the effects of all these interactions on energy digestibility, and the values used in models often overestimate the actual digestibility, particularly for animals fed at rates near *ad libitum*. However, models of digesta transit through the digestive tract, enzyme hydrolysis of dietary ingredients and the kinetics of nutrient absorption (Usry *et al.*, 1991; Bastianelli and Sauvant, 1998) provide a sound basis for predicting nutrient availability when integrated into whole-animal models.

The main factors determining digestibility of dietary protein are the amino acid composition and tertiary structure, the degree of denaturation during processing and storage and the presence of antinutritional factors such as trypsin inhibitors, lectins and tannins that interfere with the action of proteases. Antinutritional factors are more prominent in legume seeds than in cereal grains. Phytates can also bind some amino acids and influence their availability to the animal (Barnett *et al.*, 1993). It is also important to distinguish between apparent and true digestibility of amino acids. Apparent digestibility is estimated by subtracting ileal amino acid flow from amino acid intake. Amino acids in the ileum which are of dietary origin cannot be distinguished form those of endogenous origin that are derived from enzymes, sloughed cells and mucoproteins present in the gut. Estimates of true or real digestibility attempt to correct for the endogenous amino acids. Many different methods have been used to measure endogenous amino acid flow at the terminal ileum, and results vary widely (Moughan *et al.*, 1998). Estimates of true amino acid digestibility can be as much as 15 percentage units higher than those of apparent digestibility for amino acids such as cysteine and proline because of their high content in soft keratins and mucoproteins, respectively. Ideally, pig simulation models should predict the true availability of amino acids and assume the endogenous gut losses to be part of the animal requirements for amino acids.

Some nutrients that are digested and absorbed may be in a form unsuitable for metabolism and are therefore unavailable to the animal. Lysine appears to be particularly susceptible to being absorbed in forms that are unavailable for metabolism because of the ease with which sugars react with its free ε-amino group forming early Maillard compounds that are absorbed but excreted quantitatively in the urine. Similarly, many amino acids form D-isomers during processing. Some of these cannot be reconverted to the L-form within the animal for use in protein synthesis and are excreted from the body. Factors contributing to the unavailability of absorbed amino acids have been discussed by Batterham (1992).

The availability of nutrients in most animal growth models is assumed to be constant. Predictions from these models are highly sensitive to variation in nutrient availability when that nutrient is limiting performance. Hence, these values need to be set with care.

NUTRIENTS REQUIRED FOR MAINTENANCE

Nutrients are required to sustain life, and several processes must be represented either explicitly or implicitly within the models if predictions are to be accurate over a wide range of situations. The important body processes maintaining the integrity of an animal require both energy, particularly in the form of ATP, and amino acids.

Maintenance Energy Requirements

Major energy-demanding processes for maintenance of an animal include those associated with blood flow, respiration, muscle tone, ion balance and tissue turnover, those associated with activity and the ingestion of feed, those associated with the control of body temperature and those resulting from the presence of disease. A recent review of the literature shows that estimates of the metabolizable energy requirements for maintenance in pigs, when expressed as body weight (W kg^{-1}) raised to the 0.75 power, range from 385 to 670 kJ W$^{-0.75}$, with a mean of 445 kJ W$^{-0.75}$ (NRC, 1998). Clearly, the maintenance energy requirement of pigs is not constant, and the causes of the variation must be considered explicitly if pig simulation models are to predict growth and production accurately.

Fasting Heat Production

The energy requirements of the first group of processes listed above have been measured as the heat production of a fasting animal when lying recumbent. Fasting heat production is related to the body composition of the animal and to its relative rate of growth, which reflects differences in the relative mass of visceral organs and in the rate of tissue turnover. The magnitude of the variation in fasting heat production was illustrated by Koong *et al.* (1983) who fed pigs from either obese or lean genotypes over different growth trajectories from 27 to 41 kg live weight. One group from each genotype (HL) was fed to gain 19 kg over the first period of 35 days and then to lose 5 kg over the next 35 days. A second group (MM) was fed to gain 7 kg during both the 35-day periods, and a third group (LH) was fed to lose 5 kg in the first period and gain 19 kg in the second 35 days. The results given in Table 16.1 show that fasting heat production ranged from 329 to 491 kJ W$^{-0.75}$

Table 16.1.   Effect of different nutritional regimes on weight of the gastrointestinal (GI) tract and liver and on fasting heat production of obese and lean pigs.

	Obese			Lean		
	HL	MM	LH	HL	MM	LH
Final weight (kg)	40.8	41.6	39.6	40.2	42.0	41.6
GI tract (g)	1204	1489	1664	1444	1682	1827
Liver (g)	407	549	623	454	550	616
Fasting heat production ($kJ\ W^{-0.75}\ day^{-1}$)	345	384	442	329	445	491

From Koong *et al.* (1983).

day^{-1} and was strongly correlated with the total weight of the gastrointestinal tract and liver.

Turner and Taylor (1983) propose that there is an absolute basal maintenance which occurs when an animal is at zero energy balance in equilibrium with the diet and level of feed intake. The weight of gut, liver and other visceral organs are also in equilibrium with the situation. As feed intake is raised to cause growth, the gut and liver particularly increase in weight to a greater degree than other tissues. Ferrell (1988) presented results to show that the total viscera and nervous tissue of sheep represent only 0.11 of body mass but account for 0.58 of oxygen consumption, whereas the remaining muscle, fat, skin and bone represent 0.89 of body mass and only 0.42 of energy expenditure. Muscle also has a higher rate of oxygen consumption per unit mass than body fat. Thus, it would be ideal for a model of pig growth to represent viscera, muscle and the remainder of the body as separate state variables. In addition, protein turnover contributes from 0.15 to 0.25 of basal energy expenditure and increases markedly with increased growth and productivity (Gill and Oldham, 1993). Knap and Schrama (1996) have developed a simulation procedure that models protein turnover in six body protein pools (muscle, connective tissue, liver, blood plasma, gastrointestinal and 'other' proteins) in the pig.

Activity

Normal activity associated with feeding and group dynamics can account for 20% or more of the maintenance energy requirements of pigs. Noblet *et al.* (1993) found that activity accounted for 15% of total heat production of adult non-breeding sows in individual metabolism cages. The values ranged from 8.1 to 20.3 for individual animals. Activity is known to contribute to a greater proportion of heat production in young (Halter *et al.*, 1980) and group-housed animals (Verstegen *et al.*, 1987) than in individually housed animals used by Noblet *et al.* (1993).

The energy cost of standing has been measured at approximately 30 kJ $W^{-0.75}$ 100 min^{-1} for pigs ranging in weight from 30 to 200 kg (Noblet *et al.*, 1993). This value is about five times more than required by sheep and 12 times more than the rat. The energy cost of eating a diet containing 15.6 MJ kg^{-1} of digestible energy was measured at 100 kJ kg^{-1} diet (Noblet *et al.*, 1993). The energy cost of eating is closely related to the nature of the diet consumed (Halter *et al.*, 1980). Ideally, pig growth models should include predictions of activity associated with feeding, the energy cost of eating and other non-feeding, 'play' activity.

Maintenance of Body Temperature

The energy required for maintenance of a pig increases substantially when ambient temperature falls below the animal's lower critical temperature due to the extra heat produced to maintain body temperature. The increase in energy expenditure is related to several factors including the number of pigs in the pen, air speed, floor temperature and the type and proportion of the pig's skin that is wet. The increase in maintenance energy requirement associated with heat loss to the environment can be calculated readily using the relationships developed by Bruce and Clark (1979) and Black *et al.* (1986). The amount of energy expended in the cold increases with decreasing ambient temperature to a maximum known as summit metabolism which can approach ten times fasting heat production (Curtis, 1983). However, summit metabolism can be maintained for only 1–2 h before body temperature starts to decline precipitously and death occurs (Black *et al.*, 1998). The energy expenditure of a pig also increases when the ambient temperature is above the animal's evaporative critical temperature, due to an increase in rate of respiration. Although the energy cost of increased respiration rate is considerable (Ingram and Legge, 1969), relative to other energy demands an increase in respiration is likely to add only slightly to the maintenance energy needs of the pig.

Disease

Disease increases the energy requirements for maintenance of pigs relative to non-diseased animals eating similar amounts of feed. Bray (1996) showed that the oxygen consumption of resting pigs infected with *Actinobacillus pleuropneumonia* increased substantially despite a fall in voluntary feed intake. Depending on the severity of the disease as indicated by the extent of lung damage, maintenance requirements were increased by as much as 0.7 and 0.5, respectively, at 2 and 6 days after infection (Black *et al.*, 1998). Similar observations have been made with humans suffering from sepsis who have increased oxygen consumption (Clevenger, 1993) and an increase of 0.5 in heat production. The administration of an endotoxin to sheep, which resulted in a 1.5°C increase in body temperature, was shown to be associated with an increase of 0.33 in heat production (Baracos *et al.*, 1987). No pig models currently predict explicitly the effects of disease on energy maintenance requirements. More information is required on the effects of disease on energy expenditure, but it is important to make an empirical adjustment to the maintenance requirement of diseased pigs in models that are to have widespread commercial application.

Pregnancy and Lactation

Although the conceptus is fast growing tissue, comparisons between pregnant and non-pregnant gilts indicate that maintenance energy requirement is not significantly affected by pregnancy when expressed per kg $W^{0.75}$ (Close *et al.*, 1985). There is however, some evidence suggesting that there is an increase in heat production of between 5 and 10% for lactating sows resulting from the heat production associated with the synthesis of milk (NRC, 1998).

Maintenance Amino Acid Requirements

Amino acids required to maintain the integrity of a pig include those that are inevitably catabolized, those lost from the body through the digestive tract and those lost in sloughed skin and in hair growth.

Inevitable catabolism
Traditionally, nitrogen loss through the inevitable catabolism of amino acids has been assumed to equal either the urinary nitrogen excretion of animals fed protein-free diets or the value obtained by extrapolation to zero of regression equations relating urinary nitrogen loss to nitrogen intake (Black *et al.*, 1986). Estimates of endogenous urinary nitrogen loss range from 126 to 207 mg $W^{-0.75}$ (Armstrong and Mitchell, 1955; Tullis *et al.*, 1986) with a mean of 155 mg $W^{-0.75}$ (J.L. Black, unpublished). The composition of amino acids contributing to this lost nitrogen is known to vary substantially from that of whole-animal protein, with certain amino acids such as lysine and the other branched chain amino acids being highly conserved, while methionine is conserved to a far lesser extent (Hegsted and Chang, 1965; Baker *et al.*, 1966a,b; Said and Hegsted, 1970). The need for methionine, particularly, is influenced by its function as a methyl donor and is affected by the presence in the diet of other methyl donors such as choline and betaine. A variety of amino acid patterns have been assumed for inevitable catabolism, including the ratio in whole-body protein.

Endogenous gut losses
There is strong evidence that endogenous ileal amino acid loss is related closely to feed dry matter intake (DMI) and varies only slightly with body weight. Butts *et al.* (1993) increased feed intake of pigs weighing 32 kg in eight increments from 0.9 to 2.17 kg day^{-1} and found that although ileal nitrogen loss increased from 2.77 to 5.90 g day^{-1}, it changed little when expressed as

g kg⁻¹ of DMI. Hodgkinson *et al.* (1997) found recently that endogenous ileal nitrogen loss fell by approximately 0.02 g kg⁻¹ of DMI for pigs over the weight range from 40 to 85 kg when fed at 10% of metabolic live weight. In addition to DMI, endogenous ileal nitrogen is affected by the fibre and protein contents of the diet and by antinutritional factors such as trypsin inhibitors and lectins (Schulze, 1994). Schulze (1994) measured the effect of cellulose and other fibre sources on endogenous ileal nitrogen loss and found a relatively constant effect of dietary fibre irrespective of its source, of approximately 6 g kg⁻¹ of DMI kg⁻¹ of neutral detergent fibre (NDF) intake. Hodgkinson *et al.* (1997) observed that endogenous ileal nitrogen loss increased by approximately 4 g kg⁻¹ of DMI as the protein content of a purified diet was increased from 5 to 20%. Shulze (1994) showed that 5 g kg⁻¹ of trypsin inhibitor in the diet increased endogenous ileal nitrogen loss by about 4 g kg⁻¹ of DMI, whereas 1 g of lectin kg⁻¹ diet increased it by 1.18 g kg⁻¹ of DMI. The results from these experiments provide information which can be used to predict the effects of various dietary factors on endogenous nitrogen losses from the gut, and these concepts should be included in accurate pig simulation models.

Proteins contributing to the gut endogenous nitrogen loss include enzymes of pancreatic and intestinal epithelial cell origin, sloughed cells and secreted mucins. The sloughed cells are rich in cysteine contained in soft keratins, whereas the mucin proteins are rich in threonine and proline. The amino acid pattern of endogenous losses from the gut will depend on the relative contribution of these individual protein sources. However, little information is currently available on sources of endogenous protein and, traditionally, the mean composition, of amino acids flowing from the ileum has been used for predicting amino acid requirements resulting from endogenous gut secretions.

Integument

Amino acids are lost daily from pigs through the shedding of skin and hair. The proteins are predominantly keratins and contain a high proportion of cysteine. Moughan (1989) estimated the daily nitrogen loss via the integument of pigs to be 15 mg $W^{-0.75}$, and used the amino acid composition given by Schulz and Oslage (1976) to determine the amino acid requirements for this process.

NUTRIENTS USED FOR GROWTH AND REPRODUCTION

Although maintenance and growth are a continuum in the process of nutrient metabolism, the two processes have been separated traditionally by nutritional scientists. Growth of the empty body is determined by the rate of deposition of protein, fat, water and ash. To describe fully the growth characteristics of a pig, it is necessary to know the net rate of tissue accretion over the full range of feed intakes from starvation to *ad libitum* in a non-limiting environment and over the range of body weights from birth to maturity. Differences between pigs must be described in relation to their strain and sex, and nutrient intake, and the effects of climate and disease must also be determined. It is particularly important to understand how the rate of protein deposition changes as energy intake increases.

Protein Deposition in Relation to Energy Intake

The effect of energy intake on the rate of protein deposition in entire male pigs weighing 75 kg is illustrated in Fig. 16.2 (Dunkin *et al.*, 1986); this general relationship applies to all animals. The rate of protein deposition increases linearly as the energy intake from a nutritionally balanced diet increases until the maximum deposition rate achievable for the particular 'genotype' of pig and its body weight is reached, and then it remains constant. The slope of the initial linear portion of the relationship reflects the relative rates of protein and fat deposition and varies between entire male and female pigs (Campbell *et al.*, 1985) and between strains of pig (Campbell and Taverner, 1985). The slope of the relationship falls also as pigs increase in weight, and the form of the change has been described by Black *et al.* (1986). With most young pigs weighing less than about 40 kg and heavier pigs from strains selected for extremely fast rates of protein growth, the voluntary energy intake of the pig is insufficient for the potential rate of protein deposition to be achieved and for the relationship to plateau. Treatment of pigs with exogenous hormones can

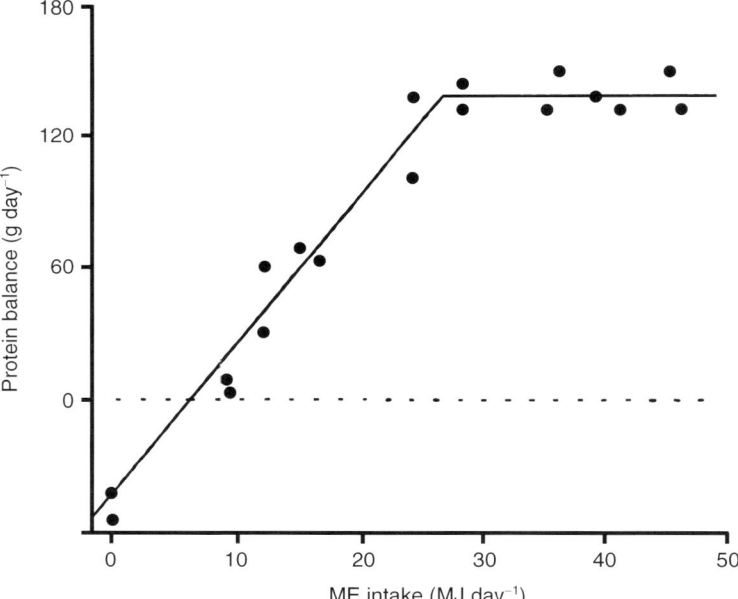

Fig. 16.2. Relationship between metabolizable energy (ME) intake (MJ day⁻¹) and the rate of protein deposition in entire male pigs weighing 75 kg. From Dunkin *et al.* (1986).

increase dramatically the slope of the relationship between protein deposition and energy intake. The daily administration of growth hormone has been shown to increase growth rate, improve the efficiency of feed use and produce leaner carcasses (Campbell *et al.*, 1988,1989). The rate of protein deposition was increased by approximately 35%, fat deposition fell by almost 40% and voluntary energy intake was reduced by 10%. The normally observed differences in performance characteristics between entire male, female and castrated pigs are virtually eliminated by treatment with exogenous growth hormone, but differences between strains of pig remain (Campbell *et al.*, 1989). Both the slope of the relationship and the potential rate of protein deposition at the plateau must be described over the full range of body weights before the growth characteristics of an animal are known in sufficient detail to predict its potential growth rate accurately.

Potential Rate of Protein Deposition

The plateau of the relationship in Fig. 16.2 can be regarded as the potential of the pig to deposit body protein. This potential rate of protein deposition can be determined at any body weight by feeding animals sufficient energy to ensure that the maximum rate has been achieved. This maximum will generally not occur in young pigs unless they are fed extremely large amounts of a highly digestible diet, as was achieved by Hodge (1974) when pigs up to 20 kg were offered reconstituted cow's milk *ad libitum*. Nevertheless, pigs have been fed highly digestible conventional diets under ideal conditions and the rate of protein deposition assumed to be close to maximal. The shape of the resulting relationships between potential rate of protein deposition and body weight varies widely between strain and sex of pig (Black *et al.*, 1986; Siebrits *et al.*, 1986; Whittemore *et al.*, 1988; Ferguson and Gous, 1993b). In some obese genotypes, the maximum rate of protein deposition is reached at live weights around 30–40 kg, with the rate declining rapidly as the pig matures, whereas in other genotypes selected for fast rates of lean growth, the maximum is not achieved until the pig weighs about 100 kg, and rates close to the maximum are maintained over a wide weight range.

Several mathematical procedures have been used to predict the potential rate of protein deposition as the pig matures. The Gompertz equation was used by Ferguson and Gous (1993a). Although the parameters for the equation are easy to establish experimentally, the equation is inflexible because there is only one set of parameters and one curve that can fit a specific maximum rate of deposition and a specific maximum body protein content. The Gompertz equation does not allow the maximum rate of protein deposition to occur at any body weight other than $1/e$ of final maximum protein weight, nor does it provide for either a steeper or flatter rise to and fall from the maximum rate of protein deposition when plotted against body weight (Black *et al.*, 1995). The Gompertz equation is a special case of a family of curves from the Richards equation which predicts a sigmoidal growth pattern where the inflection point can occur at any fraction of mature weight. Both the Richards equation and that used in the AUSPIG model allow the maximum rate of protein deposition to occur at any body weight, and the AUSPIG equation allows the 'steepness' of the relationship between rate of body protein deposition and body protein content to vary (Black *et al.*, 1995). Although the Richards and AUSPIG equations have wide flexibility, it is difficult experimentally to set the parameters for these equations. Estimates must be made for the daily rates of protein deposition, body protein content and empty body weight as the pigs grow in an ideal environment from near birth to maturity. Despite these difficulties, the potential protein deposition curve with body weight has been described for a range of genotypes (Davies *et al.*, 1993).

Effect of Ambient Temperature

The climatic conditions to which a pig is exposed can influence the rates of protein and fat deposition. Under conditions of constant feed intake, the rate of protein deposition is affected little by moderate depressions in temperature below the lower critical temperature, and the additional energy required to maintain body temperature comes at the expense of fat deposition (Verstegen *et al.*, 1973; Le Dividich and Noblet, 1982). Consequently, under conditions of restricted feeding, pigs exposed to moderate cold become leaner. Alternatively, when individually housed pigs exposed to cold are given free access to feed and can increase consumption sufficiently to compensate for the additional heat loss, there appear to be no effects of temperature on the rates of either fat or protein deposition (Verstegen *et al.*, 1985; Rinaldo and Le Dividich, 1991). However, there is evidence (Panaretto, 1968) that, under more extreme conditions of cold when metabolic rate approaches half summit, the concentrations of plasma cortisol increase and the rate of protein deposition declines substantially, and animals may become fatter that those of the same weight reared under thermoneutral conditions.

When pigs exposed to hot temperatures are given the same feed intake as pigs under thermoneutral conditions, the rates of either fat or protein deposition and body composition are not affected by temperature (Rinaldo and Le Dividich, 1991). However, when feed is freely available, the intake of hot pigs is less than for pigs in a comfortable environment, and the rate of fat deposition declines more than the rate of protein deposition, with the pigs becoming leaner (Verstegen *et al.*, 1973; Close *et al.*, 1978). This response to a reduction in feed intake is seen typically in pigs housed under ideal environmental conditions (Campbell, 1988) and there is no evidence for a direct effect of temperature on the body composition of pigs other than through a change in energy intake. This appears to be true even under conditions of extreme heat because L.R. Giles (unpublished) observed no change in the rate of protein deposition or in plasma cortisol concentrations in pigs when body temperatures were elevated to over 40°C for 48 h.

Effect of Social Stress and Disease

Placing pigs in groups compared with housing in single pens has been shown to reduce feed intake, increase plasma cortisol concentrations (Lee *et al.*, 1997) and increase body fat content (Chapple, 1993). Similarly, both acute and chronic disease decreases feed intake and stimulates the release of cytokines and cortisol (Klasing *et al.*, 1991). Cytokines, particularly IL-1 and TNF, synergize the effects of cortisol to stimulate proteolysis and reduce protein accretion rates in diseased pigs (Zamir *et al.*, 1992). Thus both social stress and disease alter the utilization of nutrients by decreasing the rate of protein deposition relative to fat deposition.

Conceptus Growth and Milk Production

The potential rates of energy and protein deposition in the conceptus are affected by both the stage of gestation and the number of fetuses (Noblet *et al.*, 1985). As pregnancy proceeds, the rate of increase in weight of fluids and placenta decreases relative to the fetuses, and the dry matter content of the gravid uterus increases substantially. However, there is little change in the relative energy and nitrogen content of the conceptus dry matter (Noblet *et al.*, 1985). The growth of the conceptus has been described satisfactorily by the Gompertz equation adjusted for litter size (Noblet *et al.*, 1985; Black *et al.*, 1986). The potential rates of energy and protein deposition have been calculated from the predicted conceptus growth rate and the estimated changes in the energy and protein content of the maturing gravid uterus (Black *et al.*, 1986). The amino acid composition of the conceptus has been assumed to be similar to that of whole-body protein. Birth weight of piglets increases with energy intake of the sow as it is raised to around 25 MJ of metabolizable energy per day, with further increases resulting in an increase in the rate of maternal tissue deposition. This suggests that conceptus growth is maximized when energy intake is around 0.5–0.6 of the potential intake of the gestating sow. Black *et al.* (1986) used a curvilinear equation to predict conceptus growth rate when the intake of metabolizable energy by the sow was less than 0.5 of its potential.

Milk yield is affected by stage of lactation, litter size, genotype and parity of the sow (Elsley, 1970). Milk yield increases to peak at around day 30 of lactation. The energy content of milk is relatively constant throughout lactation, whereas the protein content falls during the first 3 weeks and then rises steeply (Elsley, 1970). Potential milk yield has been predicted in relation to the day of lactation (Black *et al.*, 1986) using the equation of Wood (1969) adjusted for litter size and parity. The potential requirements for energy and amino acids for lactation have been obtained by multiplying the potential milk yield by the energy and protein content of milk (Black *et al.*, 1986). The amino acid content of milk have been given by Duee and Jung (1973) and NRC (1998). Lactating sows have an enormous capacity to mobilize body fat and protein and maintain potential milk yield early in lactation despite low intake of either protein or energy (Etienne *et al.*, 1984; Mullan and Williams, 1989). Mullan and Williams (1989, 1990) also showed that when feed intake is restricted during lactation, sows can loose up to 45 kg live weight over a 4-week lactation, with daily rates of fat and protein mobilization being as high as 835 and 165 g day^{-1}, respectively. The results from this experiment have been used to calculate potential fractional rates of fat and protein mobilization during lactation. Initially it was assumed by J.L. Black (unpublished) that these fractional tissue mobilization rates would be achieved under all circumstances and the substrates released would be available to supplement the energy and amino acids from the diet for meeting the needs of milk production. However, simulation of several experiments, in which the response in piglet growth to dietary protein content was investigated, showed that the rate of protein mobilization must be inversely related to the degree of the dietary amino acid deficiency. The experiments showed that the amount of body protein mobilized declined as the protein content of the diet increased, despite there still being inadequate protein for maximum milk production. Consequently, the fractional rate of protein mobilization within the AUSPIG model was made to be a function of the degree of amino acid deficiency to meet the demands of maximum milk production. The likely maximum fractional rates of fat and protein mobilization can be calculated from the experiment of Mullan and Williams (1989) and used to predict to impact of energy and amino acid intake on milk yield as lactation proceeds.

Amino Acid Deficiency

The total requirement for each amino acid for each body function can be calculated from the level of production permitted by the intake of metabolizable energy. In the model of Black *et al.* (1986), the requirement for each essential amino acid plus the combinations of cysteine and methionine and of tyrosine and phenylalanine, and of total nitrogen then are compared with the actual amino acids available for metabolism. If the availability of any amino acid is less than the total required, priority is given to the components of maintenance, inevitable catabolism, endogenous gut losses and skin and hair loss. In non-reproducing animals, body protein deposition is decreased to the permitted level by the supply of the most limiting amino acid. In reproducing animals, priority after maintenance is given to either conceptus growth or milk production after accounting for the amount of body protein that must be catabolized to supply the limiting amino acids. A maximum fractional rate of body protein catabolism is set for each physiological state, and conceptus growth and milk yield are decreased if this rate must be exceeded to allow the level of production permitted by the actual energy intake.

EFFICIENCY OF NUTRIENT USE

The efficiencies with which each essential amino acid and total available nitrogen are used can be calculated from the quantities available relative to the total demand by the animal. The marginal efficiency of total nitrogen use or the biological value of the dietary nitrogen can be estimated from these values. The order of limiting amino acids can be determined, and the amount of each amino acid needed to overcome any deficiency or excess amino acids that will be catabolized can also be calculated. The nitrogen in catabolized amino acids and other dietary non-protein nitrogen sources is converted to urea and the urea excreted at an energy cost of 21.8 kJ g^{-1} of N (Blaxter and Martin, 1962). The efficiency of energy use depends on the chemical composition of the tissues synthesized and the actual nutrients providing the energy for this synthesis and for the ATP needed for maintaining the integrity of the animal. It is possible from knowledge of the stoichiometry of aggregated biochemical reactions to calculate the average efficiency of use of the total energy of different nutrients for different body functions (Table 16.2). There are particularly large differences in the efficiency with which the major nutrient classes are used for the synthesis of triglycerides. The efficiency of energy use for protein synthesis has been assumed to be 0.54 by ARC (1980) and 0.53 by Moughan and Verstegen (1988). However, Knap and Scharma (1996) considered the energy cost of turnover of existing protein as well as the cost of synthesis on protein deposited. Estimates of the efficiency of utilization of metabolizable energy for conceptus growth range from about 0.21 to 0.72 (Close *et al.*, 1985). Noblet *et al.* (1990) suggest that a value of 0.50 should be used based on the experiment of Noblet and Etienne (1987), but Black *et al.* (1986) adopted a value of 0.60. The estimated efficiency of utilization of metabolizable energy for milk production covers a much narrower range of 0.65–0.79, with a suggested value of 0.72 (Noblet *et al.*, 1990). Metabolizable energy is used with an efficiency of 100% when providing heat to maintain the body temperature of an animal below its lower critical temperature.

Table 16.2. Estimates of the biochemical efficiency with which different nutrient classes are used for different metabolic purposes.

	Energy yield (%)			
	Microbial fermentation			
Nutrient class	Heat	Methane	ATP	Lipid
Fatty acids	—	—	66	90
Glucose	—	—	68	74
Amino acids	—	—	58	53
Digested fibre	6	10	50	62

FINAL OUTCOME OF NUTRIENT UTILIZATION

The final outcome of nutrient utilization requires the prediction of rate of gain in body protein, fat, water and ash, and changes in the weight of gut contents, the weight of the conceptus and live weight. The daily gain in body protein is calculated from net nitrogen retention in the body, and the proportion of protein in the fat-free lean is often used to estimate fat-free gain (Black *et al.*, 1986). Frequently, the gain in body fat is calculated from body energy balance after allowing for the energy deposited in body protein by assuming the energy content of protein and fat to be 23 and 39.3 MJ kg^{-1}, respectively. Some models also predict estimates of animal performance that are of commercial importance to pig production such as feed:gain ratio, back-fat thickness, carcass yield and the percentage of lean meat (Black *et al.*, 1986).

CONCLUSIONS

The major steps required for the development of computer models that simulate nutrient utilization, growth and reproduction of animals are now well understood. Detailed information is available for pigs on the most important factors determining the requirements of energy and amino acids for maintenance, body growth, conceptus growth and milk production. The effects of strain and sex of pig, diet composition, climate and stocking arrangements have been included in several pig simulation models which predict animal performance accurately under a wide range of circumstances. Although disease is known to have a major effect on the performance of many commercially raised pigs, no models yet predict explicitly the impact of disease on growth and reproduction. Considerable information is now becoming available on the physiological and metabolic effects of various diseases in pigs, and the next major challenge is to include these concepts into a mechanistic representation of disease in pig simulation models.

REFERENCES

Agricultural Research Council (1980) *The Nutrient Requirements of Pigs*. Commonwealth Agricultural Bureaux, Slough, UK.

Anderson, G.H. (1979) Control of protein and energy intake: role of plasma amino acids and brain neurotransmitters. *Canadian Journal of Physiology and Pharmacology* 57, 1043–1057.

Armstrong, D.Z.G. and Mitchell, H.H. (1955) Protein nutrition and the utilization of dietary protein at different levels of intake by growing swine. *Journal of Animal Science* 14, 49–68.

Baker, D.H., Becker, D.E., Norton, H.W., Jensen, A.H. and Harmon, B.G. (1966a) Quantitative evaluation of the threonine, isoleucine, valine and phenylalanine needs of adult swine for maintenance. *Journal of Nutrition* 88, 391–396.

Baker, D.H., Becker, D.E., Norton, H.W., Jensen, A.H. and Harmon, B.G. (1966b) Quantitative evaluation of the tryptophan, methionine and lysine needs of adult swine for maintenance. *Journal of Nutrition* 89, 441–447.

Bakker, G.C.M. (1996) Interaction between carbohydrates and fat in pigs: impact on energy evaluation in pigs. PhD thesis, University of Wageningen, The Netherlands.

Baracos, V.E., Whitmore, W.T. and Gale, R. (1987) The metabolic cost of fever. *Candian Journal of Physiology and Pharmacology* 65, 1248.

Barnett, B.J., Clarke, W.A. and Batterham, E.S. (1993) Has phytase a proteolytic effect in diets for weaner pigs? In: Batterham, E.S. (ed.), *Manipulating Pig Production IV.* Australasian Pig Science Association, Werribee, Victoria, Australia, p. 227.

Bastianelli, D. and Sauvant, D. (1998). Digestion, absorption and excretion. In: Kyriazakis, I. (ed.), *A Quantitative Biology of the Pig.* CAB International, Wallingford, UK, pp. 249–273.

Batterham, E.S. (1992) Availability and utilisation of amino acids for growing pigs. *Nutrition Research Reviews* 5, 1–18.

Black, J.L. (1983) Growth and development of lambs. In: Haresign, W. (ed.), *Sheep Production.* Butterworths, London, pp. 21–58.

Black, J.L. (1988) Animal growth and its regulation. *Journal of Animal Science* 66 (Suppl. 3), 1–22.

Black, J.L. (1995) Modelling energy metabolism in the pig – critical evaluation of a simple model. In: Moughan, P.J., Verstegen M.W.A. and Visser-Reyneveld, M.I. (eds), *Modelling Growth in the Pig.* Wageningen Pers, The Netherlands, pp. 87–102.

Black, J.L. (1998) Application of computer simulation models in intensive animal production. In: *Proceedings of the Australian Poultry Science Symposium. Volume 10.* University of Sydney, Sydney, Australia, pp. 18–25.

Black, J.L. and de Lange, C.F.M. (1995) Introduction to the principles of nutrient partitioning for growth. In: Moughan, P.J., Verstegen M.W.A. and Visser-Reyneveld, M.I. (eds), *Modelling Growth in the Pig.* Wageningen Pers, The Netherlands, pp. 33–45.

Black, J.L., Campbell, R.G., Williams, I.H., James, K.J. and Davies, G.T. (1986) Simulation of energy and amino acid utilisation in the pig. *Research and Development in Agriculture* 3, 121–145.

Black, J.L., Davies, G.T., Bray, H.J., Giles, L.R. and Chapple, R.P. (1995) Modelling the effects of genotype, environment and health on nutrient utilisation. In: Danfaer, A. and Lescoat, P. (eds), *Modelling Nutrient Utilisation in Farm Animals.* Research Centre Foulum, Denmark, pp. 85–105.

Black, J.L., Bray, H.J. and Giles, L.R. (1998) The thermal and infectious environment. In: Kyriazakis, I. (ed.) *Quantitative Biology of the Pig.* CAB International, Wallingford, UK, pp. 71–97.

Blaxter, K.L. and Martin, A.K. (1962) Utilisation of protein as a source of energy in fattening sheep. *British Journal of Nutrition* 16, 397–407.

Bray, H.J. (1996) The physiological response of growing pigs to pleuropneumonia. PhD thesis, University of Sydney.

Bridges, T.C., Turner, L.W., Usry, J.L. and Nienber, J.A. (1992) Modelling the physiological growth of swine. *Transactions of the American Society of Agricultural Engineers* 46, 285–303.

Bruce, J.M. and Clark, J.J. (1979) Models of heat production and critical temperature for growing pigs. *Animal Production* 28, 353–369.

Burlacu, G., Burlacu, R., Columbeanu, I. and Alexandru, G. (1988) Contributions to the study of mathematical modelling of energy and protein balance simulation in non-ruminant animals. *Bulletin of the Academy of Agriculture and Forestry Sciences* 17, 185–200.

Butts, C.A., Moughan, P.J., Smith, W.C., Renyolds, G.W. and Garrick, D.J. (1993) The effect of food dry matter intake on endogenous ileal amino acid excretion determined under peptide alimentation in the 50 kg liveweight pig. *Journal of the Science of Food and Agriculture* 62, 235–243.

Campbell, R.G. (1988) Nutritional constraint to lean tissue accretion in farm animals. *Nutrition Research Reviews* 1, 233–253.

Campbell, R.G. and Dunkin, A.C. (1983) The influence of protein nutrition in early life on growth and development of the pig. 1. Effect on growth performance and body composition. *British Journal of Nutrition* 50, 605–617.

Campbell, R.G., Steele, N. C., Caperna, T.J., McMurtry, J.P., Solomon, M.B. and Mitchell, A.D. (1988) Interrelationships between energy intake and exogenous growth hormone administraion on performance, body composition, and protein and energy metabolism of growing pigs weighing 25 to 55 kg live weight. *Journal of Animal Science* 66, 1653–1655.

Campbell, R.G., Steele, N. C., Caperna, T.J., McMurtry, J.P., Solomon, M.B. and Mitchell, A.D. (1989) Interrelationships between sex and exogenous growth hormone administration on performance, body composition and fat and protein accretion of growing pigs. *Journal of Animal Science* 67, 177–186.

Campbell, R.G. and Taverner, M.R. (1988) Genotype and sex effects on the responsiveness of growing pigs to exogenous porcine growth hormone (PGH) administration. *Journal of Animal Science* 66, 257.

Campbell, R.G. and Taverner, M.R. (1985) Effect of strain and sex on protein and energy metabolism in growing pigs. In: Moe, R.W., Tyrell, H.F. and Reynolds, P.J. (eds), *Energy Metabolism of Farm Animals*. European Association of Animal Production Publication No. 32, pp. 78–81.

Campbell, R.G., Taverner, M.R. and Curic, D.M. (1985) Effects of sex and energy intake between 48 and 90 kg live weight on protein deposition in growing pigs. *Animal Production* 40, 497–503.

Chapple, R.P. (1993) Effect of stocking arrangement on pig performance. In: Batterham, E.S. (ed.), *Manipulating Pig Production IV*. Australasian Pig Science Association. Attwood, Victoria, pp. 87–97.

Clevenger, F.W. (1993) Nutritional support in the patient with the systemic inflammatory response syndrome. *The American Journal of Surgery* 165, 68S–74S.

Close, W.H. (1989) The influence of the thermal environment on the voluntary food intake of pigs. In: Forbes, J.M., Varley, M.A. and Lawrence, T.L.J. (eds), *The Voluntary Food Intake of Pigs*. British Society of Animal Production, Edinburgh, UK, pp. 87–96.

Close, W.H., Mount, L.E. and Brown, D. (1978) The effects of plane of nutrition and environmental temperature on the energy metabolism of the growing pig. II. Growth rate, including protein and fat deposition. *British Journal of Nutrition* 40, 423–431.

Close, W.H., Noblet, J. and Heavens, R.P. (1985) Studies on the energy metabolism of the pregnant sow. 2. The partition and utilization of metabolisable energy intake in pregnant and non-pregnant animals. *British Journal of Nutrition* 53, 267–279.

Curtis, S.E. (1983) *Environmental Management in Animal Agriculture*. Iowa State University Press, Ames, Iowa.

Davies, G.T., James, K.J. and Black, J.L. (1993) *AUSPIG – A Decision Support System for Pig Farm Management: User Guide and Reference, Version 3.00*. CSIRO, Division of Animal Production, Sydney.

de Lange, C.F.M. (1995) Framework for a simplified model to demonstrate principles of nutrient partitioning for growth in the pig. In: Moughan, P.J., Verstegen M.W.A. and Visser-Reyneveld, M.I. (eds), *Modelling Growth in the Pig*. Wageningen Pers, The Netherlands, pp. 71–85.

de Lange, C.F.M. and Schreurs, H.W.E. (1995) Principles of model application. In: Moughan, P.J., Verstegen M.W.A. and Visser-Reyneveld, M.I. (eds), *Modelling Growth in the Pig*. Wageningen Pers, The Netherlands, pp. 187–208.

de Lange, C.F.M., Marty, B.J., Szkotnicki, B. and Birkett, S. (1998) Pig growth modelling as an advisory tool: the Canadian experience. In: Pearson G. (ed.) *Using Technology to Optimise Nutrition*. Extension Publication, Monogastric Research Centre, Massey University, Palmerston North, New Zealand, pp. 1–15.

de Haer, L.C.M. and de Vries, A.G. (1993) Feed intake patterns of and feed digestibility in growing pigs housed individually or in groups. *Livestock Production Science* 33, 277–292.

Duee, P.H. and Jung, J. (1973) Composition en acides amines du lait de truie. *Annales de Zootechnie* 22, 243–247.

Dunkin, A.C., Black, J.L. and James, K.J. (1986) Relation between energy intake and nitrogen retention in entire male pigs weighing 75 kg. *British Journal of Nutrition* 55, 201–207.

Eastwood, M.A., Robertson, J.A., Brydon, W.G. and MacDonald, D. (1983) Measurement of water holding properties of fibre and their faecal bulking capacity in man. *British Journal of Nutrition* 50, 539–547.

Edwards, A.E. (1997) Weaner–grower nutrition review. In: Bryden, D.I. (ed.) *Pig Production. Proceedings* 285. Post Graduate Committee in Veterinary Science, University of Sydney, Sydney, pp. 145–169.

Elsley, F.W.H. (1970) Nutrition and lactation in the sow. In: Falconer, I.R. (ed.) *Lactation*. Butterworths, London, pp. 393–411.

Emmans, G.C. (1986) A model of the food intake, growth and body composition of pigs fed *ad libitum*. *Animal Production* 42, 471.

Etienne, M., Noblet, J. and Desmoulin, B. (1984) Mobilisation des reserves corporelles chez la truie primipare en lactation. *Reproduction, Nutrition, Developpement* 25, 341–344.

Ferguson, N.S. and Gous, R.M. (1993a) Evaluation of pig genotypes. 1. Theoretical aspects of measuring genetic parameters. *Animal Production* 56, 233–243.

Ferguson, N.S. and Gous, R.M. (1993b) Evaluation of pig genotypes. 2. Testing experimental procedure. *Animal Production* 56, 245–249.

Ferrell, C.L. (1988) Contribution of visceral organs to animal energy expenditures. *Journal of Animal Science* 66 (Suppl. 3), 23–34.

Freeman, C.P., Holme, D.W. and Annison, E.F. (1968) The determination of the true digestibilities of interesterified fats in young pigs. *British Journal of Nutrition* 22, 651–660.

Fuller, M.F. (1965) The effect of environmental temperature on the nitrogen metabolism and growth of the young pig. *British Journal of Nutrition* 19, 531–546.

Giles. L.R. (1992) Energy expenditure of growing pigs at high ambient temperatures. PhD thesis, University of Sydney.

Gill, M. and Oldham, J.D. (1993) Growth. In: Forbes, J.M. and France, J. (eds), *Quantitative Aspects of Ruminant Digestion and Metabolism*. CAB International, Wallingford, UK, p. 383–403.

Gonyou, W.H., Chapple, R.P. and Frank, G.R. (1992) Productivity, time budgets and social aspects of eating in pigs penned in groups of five or individually. *Applied Animal Behaviour Science* 34, 291–301.

Graham, N.McC., Searle, T.W. and Griffiths, D.A. (1974) Basal metabolic rate in young sheep. *Australian Journal of Agricultural Research* 25, 957–971.

Halter, H.M., Wenk, C. and Schurch, A. (1980) Effect of feeding level and feed composition on energy utilisation, physical activity and growth performance of piglets. In: Mount, L.E. (ed.), *Energy Metabolism*. European Association of Animal Production Report No. 26. Butterworths, London, pp. 195–198.

Hegsted, D.M. and Chang, Y. (1965) Protein utilisation in growing rats at different levels of intake. *Journal of Nutrition* 87, 19–31.

Henry, V., Seve, B., Colleaux, P., Ganier, P., Saligaut, C. and Jego, P. (1992) Interactive effects of dietary levels of tryptophan and protein on voluntary feed intake and growth performance in pigs, in relation to plasma free amino acids and serotonin. *Journal of Animal Science* 70, 1873–1887.

Hodge, R.W. (1974) Efficiency of feed conversion and body composition of the preruminant lamb and the young pig. *British Journal of Nutrition* 32, 113–126.

Hodgkinson, S.M., Moughan, P.J. and Reynolds, G.W. (1997) Effect of live weight on endogenous ileal nitrogen and amino acid excretion in the growing pig. In: Cranwell, P.D. (ed.), *Manipulating Pig Production VI*. Australasian Pig Science Association, Werribee, Victoria, Australia, p. 235.

Ingram, D.L. and Legge, K.F. (1969) The effect of environmental temperature on respiratory ventilation in the pig. *Respiration Physiology* 8, 1–12.

King, R.H. and Dunkin, A.C. (1986) The effect of nutrition on the reproductive performance of first-litter sows. 4. The relative effect of energy and protein intakes during lactation on the performance of sows and their piglets. *Animal Production* 42, 319–325.

Klasing, K.C. and Johnstone, B.J. (1991) Monokines in growth and development. *Poultry Science* 70, 1781–1789.

Klasing, K.C., Johnstone, B.J. and Benson, B.N. (1991) Implications of an immune response on growth and nutrient requirements of chicks. In: Haresign, W. and Cole, D.J.A. (eds), *Recent Advances in Animal Nutrition*. Nottingham University Press, Nottingham, UK, pp. 135–146.

Knap, P.W. and Schrama, J.W. (1996) Simulation of growth in pigs: approximation of protein turn-over parameters. *Animal Science* 63, 533–547.

Koong, L.J., Nienaber, J.A. and Mersmann, H.J. (1983) Effects of plane of nutrition on organ size and fasting heat production of genetically obese and lean pigs. *Journal of Nutrition* 113, 1616–1631.

Kyriazakis, I. and Emmans, G.C. (1995) The voluntary fed intake of pigs given feeds based on wheat bran, dried citrus pulp and grass meal, in relation to measurements of feed bulk. *British Journal of Nutrition* 73, 191–207.

Le Dividich, J. and Noblet, J. (1982) Growth rate and protein and fat gain in early weaned piglets housed below thermoneutrality. *Livestock Production Science* 9, 731–742.

Lee, C., Golden, S.E., Harrison, D.T., Giles, L.R., Bryden, W.L., Downing, J.A. and Wynn, P.C. (1997) Effect of group size and environment on weaner pig performance and plasma cortisol concentration. In: Cranwell, P.D. (ed.), *Manipulating Pig Production VI*. Australasian Pig Science Association, Werribee, Victoria, Australia, p. 301.

Likuski, H.J.A., Bowland, J.P. and Berg, R.T. (1961) Energy digestibility and nitrogen retention by pigs and rats fed diets containing non-nutritive diluents and varying protein level. *Canadian Journal of Animal Science* 41, 89–101.

Lorschy, M.L. (1994) The physiological regulation of heat exchange in the lactating sow exposed to high ambient temperatures. PhD thesis, University of Sydney.

Low, A.G. (1990) Nutritional regulation of gastric secretion, digestion and emptying. *Nutrition Research Reviews* 3, 229–252.

Morgan, C.A., Nielsen, B.L., Lawrence, A.B. and Mendl, M.T. (1998) Describing the social environment and its effects. In: Kyriazakis, I. (ed.), *Quantitative Biology of the Pig*. CAB International, Wallingford, UK, pp. 99–125.

Moughan, P.J. (1989) Simulation of the daily partitioning of lysine in the 50 kg liveweight pig – a factorial approach to estimating requirements for growth and maintenance. *Research and Development in Agriculture* 6, 1–14.

Moughan, P.J. and Verstegen, M.W.A. (1988) The modelling of growth in the pig. *Netherlands Journal of Agricultural Research* 36, 145–166.

Moughan, P.J., Smith, W.C. and Pearson, G. (1987) Description and validation of a model simulating growth in the pig (20–90 kg liveweight). *New Zealand Journal of Agricultural Research* 27, 501–507.

Moughan, P.J., Souffrant, W.B. and Hodgkinson, S.M. (1998) Physiological approaches to determining gut endogenous amino acid flows in the mammal. *Archives of Animal Nutrition* 51, 237–252.

Mullan B.P., Brown, W. and Kerr, M. (1992) The response of the lactating sow to ambient temperature. *Proceedings of the Nutrition Society of Australia* 17, 215.

Mullan B.P. and Williams. I.H. (1989) The effect of body reserves at farrowing on the reproductive performance of first-litter sows. *Animal Production* 48, 449–457.

Mullan B.P. and Williams. I.H. (1990) The chemical composition of sows during their first lactation. *Animal Production* 51, 375–387.

National Research Council (1998) *Nutrient Requirements of Swine*, 10th Revised edn. National Academic Press, Washington, DC.

Nienaber, J.A., McDonald, T.P., Hahn, G.L. and Chen, Y.R. (1990) Eating dynamics of the growing–finishing swine. *Transactions of the American Society of Agricultural Engineers* 33, 2011–2018.

Nienaber, J.A., McDonald, T.P., Hahn, G.L. and Chen, Y.R. (1991) Group feeding behaviour of swine. *Transactions of the American Society of Agricultural Engineers* 34, 289–294.

Noblet, J. and Etienne, M. (1987) Metabolic utilisation of energy and maintenance requirements in pregnant sows. *Livestock Production Science* 16, 243–257.

Noblet, J., Close, W.H., Heavens, R.P. and Brown, D. (1985) Studies on the energy metabolism of the pregnant sow. 1. Uterus and mammary tissue development. *British Journal of Nutrition* 53, 251–265.

Noblet, J., Dourmad, J.Y. and Etienne, M. (1990) Energy utilisation in pregnant and lactating sows: modeling of energy requirements. *Journal of Animal Science* 68, 562–572.

Noblet, J., Shi, X.S. and Dubois, S. (1993) Energy cost of standing activity in sows. *Livestock Production Science* 34, 127–136.

Panaretto, B.A. (1968) Some metabolic effects of cold stress on undernourished non-pregnant ewes. *Australian Journal of Agricultural Science* 19, 273–282.

Parker, R.O., Williams, P.E.V., Aherne, F.X. and Young, B.A. (1980) Serum concentration changes in protein, glucose, urea, thyroxine and triiodothyronine and thermostability of neonatal pigs farrowed at 25 and 10°C. *Canadian Journal of Animal Science* 60, 503–511.

Pettigrew, J.E., Gill, M., France, J. and Close, W.H. (1992a) A mathematical integration energy and amino acid metabolism of lactating sows. *Journal of Animal Science.* 70, 3742–3761.

Pettigrew, J.E., Gill, M., France, J. and Close, W.H. (1992b) Evaluation of a mathematical model of sow metabolism. *Journal of Animal Science.* 70, 3762–3773.

Phillips, P.A., Young, B.A. and McQuitty, J.B. (1982) Liveweight, protein deposition and digestibility in growing pigs exposed to low temperature. *Canadian Journal of Animal Science* 62, 95–108.

Pomar, C., Harris, D.L. and Minvielle, F. (1991) Computer simulation model of swine production systems I. Modelling the growth of young pigs. *Journal of Animal Science* 69, 1468–1488.

Revell, D.K. and Williams, I.H. (1993) Physiological control and manipulation of voluntary feed intake. In: Batterham, E.S. (ed.), *Manipulating Pig Production IV.* Australasian Pig Science Association, Attwood, Victoria, pp. 55–80.

Revell, D.K., Williams, I.H., Mullan, B.P., Ranford, J.L. and Smits, R.J. (1998) Body composition at farrowing and nutrition during lactation affects the performance of primiparous sows. I. Voluntary feed intake, weight loss and plasma metabolites. *Journal of Animal Science* 76, 1729–1737.

Rinaldo, D. and Le Dividich, J. (1991) Assessment of optimal temperature for performance and chemical body composition of growing pigs. *Livestock Production Science* 29, 61–75.

Rogerson, J.C. and Campbell, R.G. (1982) The response of early-weaned piglets to various levels of lysine in diets of moderate energy content. *Animal Production* 35, 335–339.

Said, A.K. and Hegsted, D.M. (1970) Response of adult rats to low dietary levels of essential amino acids. *Journal of Nutrition* 100, 1363–1375.

Schoenherr, W.D., Stahly, T.S. and Cromwell, G.L. (1989) The effects of dietary fat and fibre addition on yield and composition of milk from sows housed in a warm or hot environment. *Journal of Animal Science* 67, 482–495.

Schulz, E and Oslage, H.J. (1976) Ansatz von amino auren in Organen, Geweben und Gesamttierkorpern von wachsenden Schweinen. *27 Jahrestagung der Europaischen Veneinigung fur Tierzucht.* Zurich.

Schulze, H. (1994) Endogenous ileal nitrogen losses in pigs. PhD thesis, Wageningen, The Netherlands.

Searle, T.W. and Griffiths, D.A. (1976) The body composition of growing sheep during milk feeding, and the effect on composition of weaning at various body weights. *Journal of Agricultural Science* 86, 483–493.

Siebrits, F.K., Kemm, E.H., Ras, M.N. and Barnes, P.M. (1986) Protein deposition in pigs as influenced by sex, type and live mass. I. The pattern and composition of protein deposition. *South African Journal of Animal Science* 16, 23–27.

Smits, R. (1997) How AUSPIG is being applied on farms. Its place and its limitations. In: Bryden, D.I. (ed.), *Pig Production. Proceedings* 285. Post Graduate Committee in Veterinary Science, University of Sydney, Sydney, pp. 173–180.

Stahly, T.S. (1984) Use of fats in diets for growing pigs. In: Wiseman, J. (ed.), *Fats in Animal Nutrition.* Butterworths, London, pp. 313–331.

Stahly, T.S. and Cromwell, G.L. (1979) Effect of environmental temperature and dietary fat supplement on the performance and carcass characteristics of growing and finishing swine. *Journal of Animal Science* 49, 1478–1488.

Stanks, M.H., Cooke, B.C., Fairbairn, C.B., Fowler, N.G., Kirby,.P.S., McCracken, K.J., Morgan, C.A., Palmer, F.G. and Peers, F.G. (1988) Nutrient allowances for growing pigs. *Research and Development in Agriculture* 5, 71–88.

Stombaugh, D.P. and Oko, A. (1980) Simulation of nutritional–environmental interactions in growing swine. In: Mount, L.E. (ed.), *Energy Metabolism.* Butterworths, London, pp. 209–215.

Tess, M.W., Bennett, G.L. and Dickerson, G.E. (1983) Simulation of genetic changes in life cycle efficiency of pork production. *Journal of Animal Science* 56, 336–353.

Technisch model Varkensvoeding (1991) *Proefstation voor de Varkenshouderij.* Infomatie model. Rosmalen, The Netherlands.

Tullis, B.J., Whittemore, C.T. and Phillips, P. (1986) Compensatory nitrogen retention in growing pigs following a period of deprivation. *British Journal of Nutrition* 56, 259–267.

Turner, H.G. and Taylor, C.S. (1983) Dynamic factors in models of energy utilisation with particular reference to maintenance requirement of cattle. *World Review of Nutrition and Dietetics* 42, 135–190.

Usry, J.L., Turner, L.W., Stahly, T.S., Bridges, T.C. and Gates, R.S. (1991) GI tract simulation model of the growing pig. *Transactions of the American Society of Agricultural Engineers* 34, 1879–1890.

van Barneveld, R.J. (1997) *Understanding the Nutritional Value of Lupins for Pigs.* Final Report, Pig Research and Development Corporation, Canberra.

Verstegen, M.W.A., Close, W.H., Start, I.B. and Mount, L.E. (1973) The effect of environmental temperature and plane of nutrition on heat loss, energy retention and deposition of protein and fat in groups of growing pigs. *British Journal of Nutrition* 30, 21–35.

Verstegen, M.W.A., Brandsma, H.A. and Mateman, G. (1985) Effect of ambient temperature and feeding level on slaughter quality in fattening pigs. *Netherlands Journal of Agricultural Science* 33, 1–15.

Verstegen, M.W.A., Verhagen, J.M.F. and Den Hartog, L.A. (1987) Energy requirements of pigs during pregnancy: a review. *Livestock Production Science* 16, 75–89.

Whittemore, C.T. (1983) Development of recommended energy and protein allowances for growing pigs. *Agricultural Systems* 11, 159–186.

Whittemore, C.T. and Fawcett, R.H. (1974) Model responses of the growing pig to the dietary intake of energy and protein. *Animal Production* 19, 221–231.

Whittemore, C.T. and Fawcett, R.H. (1976) Theoretical aspects of a flexible model to simulate protein and lipid growth in pigs. *Animal Production* 22, 87–96.

Whittemore, C.T., Tullis, J.B. and Emmans, G.C. (1988) Protein growth in pigs. *Animal Production* 46, 437–445.

Willis, S. (1998) AUSPIG in action – a Queensland experience. In: Kratzmann, S. and Fearon, P. (eds), *Pan Pacific Pork Expo Seminar Day Proceedings.* Australasian Pig Institute, Queensland, pp. 41–47.

Wondra, K.J., Hancock, J.D., Kennedy, G.A., Behnke, K.C. and Wondra, K.R. (1995) Effects of reducing particle size of corn in lactation diets on energy and nitrogen metabolism in second-parity sows. *Journal of Animal Science* 73, 427–432.

Wood, P.D.P. (1969) Factors affecting the shape of the lactation curve in cattle. *Animal Production* 11, 307–316.

Wuyam, B., Payen, J.F. and Levy, P., Bessaidne, H., Reutenauer, H., Le Bas, J.F. and Benabid, A.L. (1992) Metabolism and aerobic capacity of skeletal muscle in chronic respiratory failure related to chronic obstructive pulmonary disease. *European Respiratory Journal* 5, 157–162.

Zamir, O., Hasselgren, P.O., Kunkel, S.L., Frederick, J., Higashiguchi, T. and Fischer, J.E. (1992) Evidence that tumor necrosis factor participates in regulation of muscle proteolysis during sepsis. *Archives of Surgery* 127, 170.

Zoilopoulos, P.E., English, P.R. and Topps, J.H. (1982) High-fibre diets for *ad libitum* feeding of sows during lactation. *Animal Production* 35, 25–33.

17

Modelling the Utilization of Dietary Energy and Amino Acids by Poultry

M.G. MacLeod

Roslin Institute (Edinburgh), Roslin, Midlothian, UK

INTRODUCTION

The perception that it is relatively inexpensive to carry out experiments with poultry has had both positive and negative effects on the development of a modelling approach. The positive effects are that there is a large body of data to underpin modelling and also that validation should be less costly than with other species. The negative effect is that there may be a temptation to carry out a new experiment for every new set of circumstances. Digestibility and metabolizability measurements should probably be exempted from this charge, since bio-assay is sufficiently rapid and precise to remain the method of choice over predictive methods (Sibbald, 1976; Farrell, 1978; McNab and Blair, 1988; Bourdillon *et al.*, 1990). However, adherence to an empirical approach in poultry nutrition has meant that relatively few scientists have made an active contribution to the development of predictive models. Although the most challenging long-term prospects may lie in mechanistic modelling, the practical use of empirical models of responses to nutrients (Curnow, 1973; Fisher *et al.*, 1973) continues to be important.

This chapter briefly describes some empirical and mechanistic models, as well as models which have some characteristics of both. In keeping with the title of the book, it concentrates on models of the utilization of nutrients for growth and egg production, rather than models of the patterns of growth and egg production themselves. A brief review of poultry growth models was published by Zoons *et al.* (1991). France *et al.* (1996) made an instructive comparison of commonly used growth equations with a novel, flexible, function which can be used to describe a range of hyperbolic curves. A number of predictive models of the requirements of laying hens for feed or energy have been published, most of which have been empirical (Byerly, 1979; Muramatsu *et al.*, 1989, 1994; Chwalibog, 1992; Pesti *et al.*, 1992; Sakomura *et al.*, 1993a,b). The problem of scaling maintenance requirements in relation to the body weight of growing animals is discussed in papers by Emmans (1987, 1995) and Emmans and Fisher (1986), drawing on the ideas of Taylor (1970) and Taylor and Young (1968).

A tangible sign of the acceptance of empirical models of responses to nutrients is that such models are now incorporated in proprietary computerized feedback loop systems which control food provision (and even food composition) in response to the flock's 'goodness of fit' to a target growth curve. MacLeod and Waibel (1997) describe an experimental example of such a system.

ENERGY EVALUATION

There is a tendency for discussion about feedstuff evaluation to concentrate on dietary energy values. This chapter will follow the pattern of giving most space to energy, because that is where there seems to be most scope for refinement. However, responses to individual chemical entities, especially amino acids, will also be considered. There are biological reasons for treating energy as a special case:

1.   Dietary energy has a 'summarising' role in the control of food intake, the intake of individual nutrients is strongly influenced by the nutrient:energy ratio.
2.   There may be an argument for describing a diet in terms of chemical composition but this should not obscure the fact that some of the bird's control mechanisms may, in effect, perceive the substrates as contributors to energy supply rather than identifying them as specific chemicals. It is therefore ideal to be able to describe a feedstuff or ingredient in terms of its biological energy value and its chemical composition.
3.   Because birds are homeotherms, there are interactions between dietary energy and the thermal environment. The interaction is clearest at the level of maintenance energy requirement, but ambient temperature may also affect net efficiency of energy utilization (Geraert et al., 1988).

A survey of feed evaluation in Europe (de Boer and Bickel, 1988) listed the wide range of energy evaluation systems in use. The section on poultry was conspicuous for its comparatively high degree of consensus, with almost all countries using a metabolizable energy (ME) system (ME = gross energy – (faecal + urinary energy)). ME is a reliable index of what is available to the bird for maintenance and production, but not a predictor of how efficiently the bird then uses what is available. Its low variability (Hill and Anderson, 1958) is a consequence of ignoring many of the bird's metabolic responses to its food. ME is assumed to be linearly additive, which is practically convenient but not exhaustively tested (MacLeod et al., 1996).

Advances in energy evaluation for poultry are likely to come from taking into account the differing biochemical efficiencies with which the chemical constituents of the diet are utilized (Millward et al., 1976). This is most likely to be achieved by a modelling-based approach, of which there have already been attempts at varying levels of mechanism and aggregation (Nehring and Haenlein, 1973; de Groote, 1974a; Emmans, 1994; MacLeod, 1994). Although demonstrably capable of distinguishing between the energetic effects of feeding different classes of chemical substrate (Tasaki and Kushima, 1979), or between grossly different compound feeds (MacLeod, 1990, 1992), calorimetric experiments need copious replication in order to detect the effects of more subtle changes in diet composition. These relatively small effects, which may still be of biological and commercial significance, are more likely to be detected by a valid predictive model than by any but an extremely large experiment.

Empirical Predictive Modelling of Net Energy

Net energy (NE) is that part of the dietary energy which is available to the animal for maintenance and production (i.e. NE = ME − heat increment, where heat increment is the increase in heat production which occurs when food is ingested). A 'productive energy' system (in most respects analogous to NE) was used commercially from about 1946 to about 1960, following the work of Fraps and Carlyle (1939) and Fraps (1946). This system was based on comparative slaughter measurements of energy retention and estimates of maintenance energy requirement calculated from body weight. Productive energy was discarded in favour of ME because of its lack of precision (ranges of up to ± 20% for a single feedstuff). Much of this variability may have resulted from the measurement technique, in which inter-individual variation was combined with errors inherent in the comparative slaughter method and in the method of calculating maintenance requirement from bird weight.

The Rostock net energy (NEF) model

Work at Rostock (Nehring, 1967; Schiemann, 1967; Schiemann *et al.*, 1971; Nehring and Haenlein, 1973) demonstrated that excessive variability is not an inevitable characteristic of NE systems. Extensive measurements with several species showed that NE (for fat deposition) can be predicted from digestible fat, protein and carbohydrate contents of feedstuffs with a coefficient of variation of about ± 5%. The Rostock NEF system (net energy for fat deposition) was based on a large number of calorimetric trials on the main agricultural species. The multiple regression models derived from these trials predict NEF from measurements of digestible crude protein (P), digestible crude fat (F) and digestible crude fibre + digestible N-free extract (largely carbohydrate, C). The model for the domestic fowl (derived from adult cockerels) is:

$$\text{NEF (kJ g}^{-1}) = 10.8P + 33.4F + 13.4C \ (\pm \text{ cv } 5.2\%). \tag{17.1}$$

The corresponding equation for ME is:

$$\text{ME (kJ g}^{-1}) = 17.8P + 39.8F + 17.7C. \tag{17.2}$$

The coefficients relating NEF to ME (and therefore equivalent to k_f, the net efficiency of energy utilization for fattening) are 0.60, 0.84 and 0.78 for protein, fat and carbohydrate, respectively. The NEF model was based specifically on the utilization of energy for fat deposition, a decision cogently argued by Nehring (1967). The NEF system appeared to satisfy many of the criteria for a practical NE system. Its non-use in most countries may have stemmed from its reliance on digestibility coefficients, which would have been identified as a particular problem in the case of poultry.

A modification of the Rostock NEF model for poultry

De Groote's (1974a) proposed NE method for poultry by-passed the uncertainty associated with digestibility coefficients by deriving NE values from existing ME data. This was done for each feedstuff by multiplying its ME value by the relative proportions (by weight) of crude protein, crude fat and starch plus sugar, each of which was in turn multiplied by an experimentally estimated utilization coefficient. The coefficients were 0.60, 0.90 and 0.75 for protein, fat and carbohydrate, respectively. The calculation was therefore:

$$NE = ME (0.60P + 0.90F + 0.75C)/(P + F + C) \qquad (17.3)$$

where P, F and C are g kg^{-1} of protein, fat and carbohydrate. In striving for simplicity, de Groote's system neglected differences in the digestibility of the protein, fat and carbohydrate fractions both within and between feedstuffs. Perhaps more importantly, by using proportions of feedstuff by weight, it appeared to neglect the fact that fat has about twice the ME value of protein and carbohydrate and that the utilization coefficients (0.60, 0.90 and 0.75) referred to ME and not to mass. De Groote (1975) rationalized the low ME value assigned to fat by quoting the low digestibility coefficient (~ 0.5) cited by the Rostock group. The use of this low value does seem to detract from the de Groote method; it appears also not to be consistent with the ME values given for individual animal and plant fats in Table 6 of his paper (de Groote, 1974a). The NE values in the latter table can be recalculated on the basis of proportions of ME rather than weight by the following equation:

$$NE_1 = ME (0.60 \times 17.8P + 0.90 \times 39.8F + 0.75 \times 17.7C)/$$
$$(17.8P + 39.8F + 17.7C) \qquad (17.4)$$

where 17.8, 39.8 and 17.7 are ME coefficients from the Rostock model. Table 17.1 shows that the original NE values in de Groote (1974a), calculated on a weight basis, have a negative error, which increases with fat content. Table 17.2 allows comparison of the relative energy values of ingredients calculated as ME, NE or NE_1. In each case, wheat is given a value of 100. The most important feature of the comparison is the magnitude of the difference between relative replacement values expressed as ME and as NE or NE_1. The disagreement varies according to the concentrations of fat and protein and the ratios of these concentrations to one another. Few independent tests of the de Groote NE model have appeared in the literature. In survey data from chickens 0–28 days old, Fisher and Wilson (1974) calculated that, of the 39% variation in growth rate unexplained by variation in dietary ME concentration, 18% could be accounted for by recalculating intake on NE by the de Groote method. In the case of food conversion efficiency, the improvement was even

Table 17.1. The NE values calculated by the de Groote model on the basis of proportions of ingredient as protein, carbohydrate and fat by weight (NE, Equation 17.3) or on the basis of proportional energetic contribution to ME (NE$_1$, Equation 17.4). The net efficiency associated with NE$_1$ is tabulated as k_1.

Ingredient	ME	NE	NE$_1$	NE − NE$_1$ diff (%)	k_1
Wheat	12.9	9.4	9.5	−1.0	0.74
Maize	14.4	10.6	10.8	−1.3	0.75
Rye	12.0	8.7	8.8	−0.8	0.74
Barley	11.4	8.4	8.4	−0.0	0.74
Oats	11.0	8.0	8.2	−2.2	0.75
Sorghum	13.9	10.2	10.3	−1.1	0.74
Wheat shorts	8.5	6.0	6.2	−2.7	0.73
Wheat flour middlings	11.1	7.9	8.1	−2.0	0.73
Wheat bran	5.4	3.9	4.0	−3.1	0.73
Wheat white middlings	12.4	9.0	9.1	−1.0	0.73
Tapioca	12.4	9.3	9.3	−0.3	0.75
Rice germ meal	11.5	8.6	9.0	−4.5	0.78
Dried whey	8.0	5.8	5.8	−0.4	0.73
Molasses	8.2	6.0	6.0	−0.0	0.73
Sugar	15.5	11.5	11.6	−0.8	0.75
Peanut meal	11.1	7.1	7.1	−0.7	0.64
Sesame meal	8.3	5.4	5.5	−2.0	0.66
Fish meal	13.0	8.4	9.0	−6.5	0.69
Soybean meal (440 g kg^{-1} of CP)	9.4	6.0	6.1	−0.8	0.65
Soybean meal (500 g kg^{-1} of CP)	10.3	6.5	6.7	−1.7	0.64
Sunflower meal	7.2	4.6	4.6	−1.4	0.65
Herring meal	13.4	8.5	9.0	−5.3	0.67
Blood meal	12.0	7.2	7.3	−0.7	0.61
Meat-and-bone scraps	8.5	5.5	5.9	−6.5	0.69
D,L-Methionine	14.7	8.8	8.8	0.0	0.60
L-Lysine-HCl	11.0	6.6	6.6	0.0	0.60
Dried skim milk	10.5	7.1	7.2	−0.5	0.68
Brewer's yeast	8.7	5.6	5.7	−1.6	0.65
Alfalfa (160 g kg^{-1} of CP)	4.8	3.2	3.3	−2.9	0.70
Alfalfa (180 g kg^{-1} of CP)	5.7	3.8	4.0	−3.7	0.70
Alfalfa (200 g kg^{-1} of CP)	6.3	4.2	4.4	−4.3	0.70
Maize gluten meal (420 g kg^{-1} of CP)	11.5	7.5	7.6	−1.8	0.66
Maize gluten meal (620 g kg^{-1} of CP)	14.4	9.3	9.5	−2.6	0.66
Hydrocarbon yeast (BP)	10.6	6.5	6.6	−1.6	0.62
Cottonseed meal	7.6	4.8	4.8	−1.0	0.63
Lard	34.0	30.6	30.6	0.0	0.90
Tallow	31.0	27.9	27.9	0.0	0.90
Soybean oil	37.7	33.9	33.9	0.0	0.90

Table 17.2. Relative substitution values of raw materials on the basis of ME and on the basis of NE calculated by Equations 17.3 and 17.4.

Ingredient	ME	NE	NE_1	NE_1/ME
Wheat	100.0	100.0	100.0	1.00
Maize	111.4	113.0	113.4	1.02
Rye	92.9	93.0	92.8	0.99
Barley	88.6	88.9	88.8	1.00
Oats	85.1	85.3	86.3	1.01
Sorghum	107.5	108.3	108.4	1.01
Wheat shorts	65.6	64.2	65.3	1.00
Wheat flour middlings	85.7	84.1	85.0	0.99
Wheat bran	42.2	41.2	42.0	1.00
Wheat white middlings	96.4	95.6	95.7	0.99
Tapioca	96.4	98.8	98.2	1.02
Rice germ meal	89.0	91.1	94.3	1.06
Dried whey	62.0	61.5	61.2	0.99
Molasses	63.6	63.9	63.3	0.99
Sugar	120.1	122.6	122.2	1.02
Peanut meal	85.7	75.0	74.8	0.87
Sesame meal	64.6	56.9	57.5	0.89
Fish meal	100.6	89.5	94.4	0.94
Soybean meal (440 g kg^{-1} of CP)	72.7	64.0	63.9	0.88
Soybean meal (500 g kg^{-1} of CP)	80.2	69.5	70.0	0.87
Sunflower meal	55.5	48.7	48.9	0.88
Herring meal	103.5	90.3	94.2	0.91
Blood meal	92.9	76.8	76.6	0.82
Meat-and-bone scraps	66.1	58.9	62.1	0.94
D,L-Methionine	113.6	93.4	92.5	0.81
L-Lysine-HCl	85.2	70.1	69.4	0.81
Dried skim milk	81.2	75.7	75.3	0.93
Brewer's yeast	67.2	59.2	59.5	0.89
Alfalfa (160 g kg^{-1} of CP)	36.9	34.2	34.9	0.95
Alfalfa (180 g kg^{-1} of CP)	43.8	40.5	41.6	0.95
Alfalfa (200 g kg^{-1} of CP)	48.7	38.8	46.2	0.95
Maize gluten meal (420 g kg^{-1} of CP)	89.3	79.6	80.3	0.90
Maize gluten meal (620 g kg^{-1} of CP)	111.4	98.6	100.2	0.90
Hydrocarbon yeast (BP)	82.5	68.7	69.1	0.84
Cottonseed meal	59.1	50.6	50.7	0.86
Lard	263.6	324.7	321.9	1.22
Tallow	240.3	296.3	293.4	1.22
Soybean oil	292.2	360.3	356.8	1.22

NE is based on proportions of protein, fat and carbohydrate by weight, NE_1 on proportions by ME value. Wheat is given the relative value of 100 as a standard for comparison. The economic importance of using different models of feeding value may lie in the effects on the substitution values of diet ingredients.

greater; unexplained variation was reduced from 27 to 8%. In the finisher stage, ME and NE scales gave similar correlations with production responses. De Groote's own (1974b) comparison of ME and NE systems claimed improved performance and higher returns over food costs in diets formulated on the basis of NE. As Farrell (1979) pointed out, however, the NE diets almost always contained more fat than the ME-formulated diets; preliminary tests of de Groote's system by Farrell and co-workers gave variable results for the economic advantages of the NE system. It can be argued, of course, that the first level of validation of an NE system should be at the biological level, in terms of predictive power; a diet formulated on the basis of NE need not automatically be less costly. The long-term economic benefits would come from more precise formulation of diets. It is surprising, therefore, that more attention has not been given to the de Groote model, this would at least have obviated *ad hoc* adjustments applied to ME by practical nutritionists.

Prediction of Energy Value by Metabolic Simulation

Nehring (1967) alluded to the use of ATP synthesis and breakdown as measures of the energy value of the diet or its chemical constituents. This principle was described in greater detail by later authors (Scheele *et al.*, 1973; Schulz, 1975, 1978; Livesey, 1984, 1985). Although much of this work was not targeted at avian species, similarities between taxonomic groups are more important than the differences at this level of biological organization.

The model of Livesey (1984) calculates energy yield (as ATP) from carbohydrates, fats and proteins, which have been absorbed across the gut wall and are available for cellular catabolism. It treats all substances purely as energy sources and therefore corresponds with an ME form of evaluation. In fact, by assuming oxidation of amino acids, it simulates the correction of ME to zero nitrogen retention, as is usually done in practice. ATP yields calculated stoichiometrically for individual amino acids, fatty acids, glycerol and glucose were related to gross energy contents of proteins and fats calculated from bond energies to give values for the gross chemical energy corresponding to a molecule of cytoplasmic ATP. Errors in this relationship are potentially large, depending on the stoichiometry selected for mitochondrial oxidative phosphorylation, on the degree of uncoupling of oxidation and phosphorylation, and on the proportion of amino acids oxidized via gluconeogenesis. Taking account of these sources of error reduces the error range in ATP yield to about ±10%. Most of the variation in ATP yield from protein was explicable in terms of real differences in heat of combustion depending on amino acid composition. Relatively little of the variation was due to differences in efficiency of ATP generation (compare Schulz, 1975). The residual uncertainty in food energy equivalents of cytoplasmic ATP could probably be reduced by better information about the energy costs of absorption from the gut and translocation across other membranes. A further paper by Livesey (1985) examined the effects of oxidation–phosphorylation uncoupling on the relationships among the energy equivalents of carbohydrate, fat and protein. For a food evaluation system, in which the prime objective is to obtain reliable relative replacement values for ingredients, these scientifically important finer details may not be of critical importance.

A Simulation Model to Predict Dietary Net Energy Yield for Poultry

Previous models suggested for an NE system have been highly aggregated in chemical terms (Nehring and Haenlein, 1973; de Groote, 1974). Such systems may have the advantage of simplicity but, with better description of diet composition, we can use more information to predict response. MacLeod (1994), therefore, set out to develop a mechanistically based model of nutrient metabolism as a basis for an NE type of food evaluation. The design specifications were that: (i) variations in efficiency of utilization (heat increment) should be taken into account; (ii) the system should, wherever possible, be based on the chemical composition of ingredients; (iii) it should allow for different rates and compositions of product synthesis; (iv) it should avoid incorporating more assumptions than necessary, but should allow for refinement as more information accumulates; and (v) it should provide, in the first instance, a reliable standard for substitution of ingredients for one another, i.e. a standard that is correct in *relative* terms. This relative standard should, however, lay the foundation for *absolute* correctness.

A simplified flow diagram of the model is shown in Fig. 17.1. The simulation is structured as a number of independent programme units, which can be refined individually without interacting unpredictably with other parts of the programme. The model incorporates several empirical relationships predicting whole-animal responses (e.g. food intake, maintenance requirement). Since this is primarily a food evaluation model, the main reason for predicting intake is to ensure that energy intake is within reasonable limits. All the model would be able to do with a large excess of energy above requirements and above its capacity for fat deposition would be to simulate 'burning it off' as heat (regulatory diet-induced thermogenesis). This would be biologically unrealistic because the domestic fowl does not usually exhibit a capacity for regulatory diet-induced thermogenesis.

Whenever possible, the simulation uses experimentally determined values of the digestibilities of chemical entities within the ingredient (Heartland Lysine Inc., 1990; Longstaff and McNab, 1991; Shafey and McDonald, 1991; Zuprizal *et al.*, 1993).

The stoichiometric foundation of the simulation was derived largely from Schulz (1978). Because of the differences between mammalian (urea-excreting, ureotelic) and avian (uric acid-excreting, uricotelic) amino acid metabolism, however, different stoichiometric coefficients were used for amino acid breakdown. The energy cost of uric acid synthesis is accounted for and amino acid compositions of proteins in body, feathers and egg were compiled from various sources (Lunven *et al.*, 1973; Hakansson *et al.*, 1978b; Blair *et al.*, 1981; Nitsan *et al.*, 1981; Hurwitz *et al.*, 1983b).

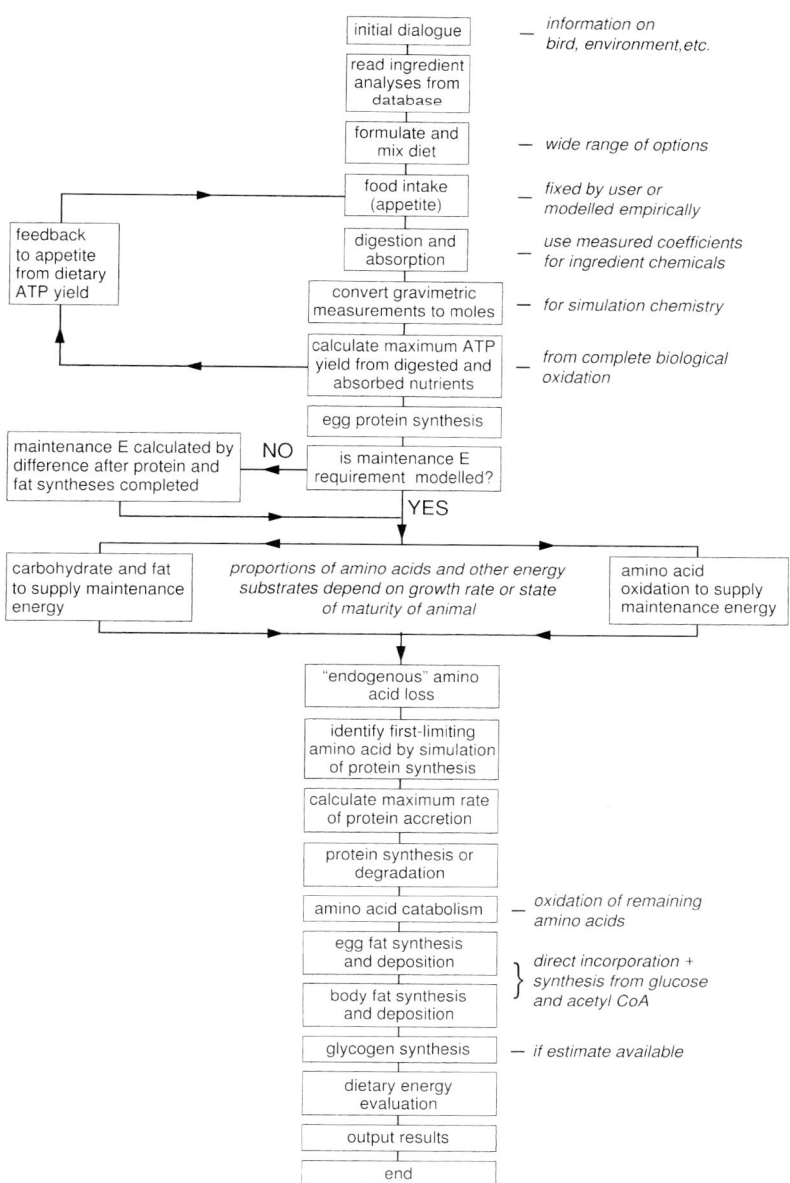

Fig. 17.1. Flowchart of the simulation model of MacLeod (1994) showing the key programme units and their hierarchical arrangement. Throughout the programme, energy utilization and deposition are recorded as the breakdown and synthesis of ATP (not shown in the flowchart, to avoid repetitive detail).

The original information on the existing ingredients was derived largely from analyses at the Roslin Institute (e.g. Blair *et al.*, 1981; McNab and Scougall, 1982), but new ingredients can be added and existing analyses edited by the user. If an amino acid analysis for a new ingredient is not available, the simulation can make estimates from a measurement of crude protein concentration and a statement of the most closely related feedstuff for which a full analysis exists (e.g. another legume species, another grain species). The model predicts how consumed nutrients are partitioned between different biological processes (e.g. body growth, egg production, body maintenance) and how efficiently the nutrients are used, in energy terms. The model has been designed particularly with food evaluation in view, so the prediction of utilization efficiency is given priority in constructing the model. For instance, hierarchies of biological processes (e.g. maintenance, followed by egg synthesis, followed by body growth) are usually imposed, rather than allowing simultaneous competition for resources on the basis of different affinity constants. The exception is to set up a competitive interaction between protein accretion and oxidation in the metabolic fate of amino acids; the comparative affinities of these two processes are summarized by a growth potential term determined by the age and genotype of the bird being simulated. The hierarchies used in the simulation are hypotheses based on the interpretation of published experiments; it is implicit in their use that functionally, if not intellectually, the animal also has priorities in the use of nutrients. Such priorities have developed through natural (or artificial) genetic selection. The physiological control (endocrine, neural, etc.) of these priorities is sometimes known, but is outside the scope of this model.

One of the effects of building amino acid oxidation into maintenance, even when there is sufficient carbohydrate and fat to satisfy energy requirements, is to simulate a maintenance requirement for protein, removing the need to specify an endogenous loss. It also simulates, or may even explain, a generalized protein requirement over and above that for specific amino acids, which could be ascribed to 'inevitable amino acid catabolism' (Moughan, 1994). The simulation can be used to generate energy values for individual feedstuffs, although it is intended to simulate the use of realistic mixtures. Using single-ingredient values additively may lose some of the advantages of simulation but is still likely to give a better prediction of animal response than ME values used additively. Validation experiments and sensitivity analysis of the simulation are briefly described by MacLeod (1994, 1998).

In MacLeod (1998), k_g (net efficiency of energy utilization for growth) and NE (net energy g^{-1} of food) were predicted while food intake, maintenance energy requirement and rate of protein accretion were varied in turn, with the two other parameters held constant (Table 17.3). It is clear that, as in experiment, the net availability of energy is strongly related to the proportion of energy supplied as protein and, particularly, the proportion used for protein synthesis. The maximum rate of protein accretion permitted by the simulation was similar to the rate recorded in MacLeod (1990) on a similar food intake. The relatively low sensitivity of the simulation model to the selected key parameters, combined with a higher sensitivity to diet composition, suggest that the simulation may be quite robust in its application to feed evaluation.

Table 17.3. Sensitivity of the simulation model of MacLeod (1994) to food intake, maintenance energy requirement and rate of protein accretion.

	Food intake (g)	TME intake (kJ day^{-1})	Maintenance (kJ day^{-1})	Protein retention varied (g day^{-1}) proportion of dry matter gain		k_g	NE (kJ g^{-1})
Food intake varied	79	987	525	5.88	0.49	0.732	9.14
	92	1149	525	7.39	0.55	0.735	9.18
	105	1311	525	8.74	0.44	0.737	9.20
	118	1474	525	8.74	0.37	0.743	9.28
	131	1636	525	8.74	0.33	0.747	9.33
Maintenance varied	105	1311	394	8.74	0.38	0.741	9.26
	105	1311	460	8.74	0.41	0.739	9.23
	105	1311	525	8.74	0.44	0.737	9.20
	105	1311	591	8.62	0.46	0.734	9.18
	105	1311	656	8.34	0.49	0.733	9.15
Protein retention	105	1311	525	6.68	0.35	0.745	9.31
	105	1311	525	7.71	0.39	0.741	9.26
	105	1311	525	8.74	0.44	0.737	9.20
	105	1311	525	8.89	0.44	0.736	9.20
	105	1311	525	8.89	0.44	0.736	9.20

Effective Energy

Emmans (1984, 1994) has published a well-argued method for estimating the effective energy of a diet or ingredient, in which ME is adjusted for the heat increment of feeding by applying linear coefficients to five measurable components of the interaction between the animal and its diet. The coefficients were derived empirically but are broadly interpretable in relation to biological mechanisms. Effective energy is equal to ME less predicted heat increment. The concept, therefore, falls within the category of NE systems, but the term 'effective energy' avoids some of the semantic and logical problems associated with using NE to describe diets, when it more correctly describes the animal's response. Since effective energy can be applied across species, it will, without doubt, be described elsewhere in this book, but I will try to draw attention to aspects particularly relevant to poultry.

ME is chosen as the starting point for the effective energy calculation. ME_c (subscript c for conventional) is defined as the gross energy (GE) of the diet less energy losses as faecal energy (FE), urinary energy (UE) and combustible gases (MTHE):

$$ME_c \text{ (kJ day}^{-1}) = GE - (FE + UE + MTHE) \tag{17.5}$$

The production of combustible gases (largely methane) by poultry is negligible, which allows simplification of the equation to the form used in poultry feed evaluation. Correction of ME to zero nitrogen retention (NR), to give ME_n, is also permissible, corresponding with standard procedure in the case of poultry.

$$ME_n(kJ\ day^{-1}) = ME_c - a(6.25\ NR)$$
$$= (h_p - a) \times PR + h_l \times LR + H \tag{17.6}$$

where PR and LR are the rates of retention of protein and lipid (g day^{-1}), h_p and h_l are the heats of combustion of protein (23.8 kJ g^{-1}) and lipid (39.6 kJ g^{-1}), and H is heat production (kJ day^{-1}). Although poultry excrete uric acid rather than urea, the trans-species nitrogen correction (a) of 5.63 kJ g^{-1} of protein retained (35.2 kJ g^{-1} of N) agrees closely with the 36.5 kJ g^{-1} of N conventionally used for poultry. The ME obtained from the diet must either be retained in the animal's body or lost as heat. If we know or can predict the performance of the bird in terms of protein and fat retention (carbohydrate deposition being negligible in the long term), prediction of ME requirement needs only a prediction of heat production. Most of Emmans (1994) is concerned with this. Heat production is described as having two components, i.e. fasting heat production (FHP) and heat increment of feeding (HIF). FHP is given by:

$$FHP = -(h_p - a) \times PR - h_l \times LR \tag{17.7}$$

where PR and LR are protein and fat retentions (which are negative in the fasted bird). Maintenance heat production (MH) is calculated with the simplifying assumption that the fasted bird is catabolizing only lipid:

$$MH = FHP - w_u \times FUN \tag{17.8}$$

where w_u (kJ g^{-1} of N) is the heat production associated with the synthesis and excretion of urinary N, and FUN is urinary nitrogen loss during fasting (g day^{-1}). Heat increment for maintenance (HIM, kJ day^{-1}), ignoring methane production in the case of poultry, is given by the equation:

$$HIM = w_d \times FOM + w_u \times UN \tag{17.9}$$

where w_d (kJ g^{-1}) is heat production associated with the production of faecal organic matter, FOM. Maintenance ME requirement (MEM, kJ day^{-1}) is given by

$$MEM = MH + HIM \tag{17.10}$$

Under conditions of positive protein and fat retention (and ignoring methane production),

$$HIF = w_d \times FOM + w_u \times UN + w_p \times PR + w_l \times LR \tag{17.11}$$

(where w_p and w_l are the heat productions associated with protein and lipid deposition, respectively, and HIF is the heat increment of feeding) and ME requirement (kJ day^{-1}) is given by:

$$ME = ER + MH + HIF \tag{17.12}$$

where ER is energy retention. For poultry, the coefficients for the different components of heat increment were estimated from the feeding and comparative slaughter experiments of Hakansson *et al.* (1978a,b). Effective energy requirement (EERQ, kJ day^{-1}), taking into account the energy contents of protein and lipid, the energy costs of depositing them and the energy cost of nitrogenous excretion, was shown to be given by:

$$EERQ = MH + 50 \times PR + 56 \times LR \tag{17.13}$$

The effective energy of an ingredient (EE) can be expressed as:

$$EE = ME_N - w_d \times FOM - 0.16\ w_u \times DCP + 12z \times DCL \tag{17.14}$$

where DCL is digestible crude lipid (g g^{-1}) and z is the proportion of retained lipid which comes directly from feed lipid. Although the concept of 'effective energy' is applicable across species and genetic lines, the actual values for raw materials do differ between genotypes, and the *de novo* measurement of the effective energy of individual ingredients can be time-consuming (Farrell *et al.*, 1998).

AMINO ACID REQUIREMENTS

Response Models

The amino acid response models of Curnow (1973) and Fisher *et al.* (1973) have been extremely influential. Curnow (1973) showed how the smooth population (flock egg output) response to intake of an amino acid could be explained on the basis of individuals having an abrupt threshold response (at maintenance intake) followed by a rectilinear response until an abrupt plateau response is attained. The smooth population curve then results from biological (assumed to be normal) variation in the maintenance intake and the intake for maximum response. Curnow (1973) derived an exact equation for the response line. Fisher *et al.* (1973) described a computer simulation method, using Monte Carlo random sampling techniques (for a brief description of these, see France and Thornley, 1984) to simulate a flock of birds. In economic terms, the optimum intake of an individual amino acid would be where the gradient of the response curve was equal to the ratio of the cost of an additional unit of amino acid to the financial return from an additional unit of egg production. The gradient of the curve at any point can be calculated by differentiation. Above the optimum intake, the cost of extra amino acid provision would exceed the return from extra egg production. Although designed for egg production, the 'Reading model' (Curnow, 1973; Fisher *et al.*, 1973) has also been applied successfully to growing broilers (Boorman and Burgess, 1986). Curnow and Torenbeek (1996) explained how use of the model for an individual amino acid, assuming that other amino acids are not limiting production, can result in overestimation of optimum intakes. They propose a model and optimization procedure which allows simultaneous estimation of the optimal intakes of several amino acids. A strength of the models just mentioned is that they deal explicitly with the fact that poultry scientists are usually either content or constrained to deal with the response of a flock rather than those of individuals.

More Mechanistic Approaches to Predicting Amino Acid Requirements

An alternative approach to determining amino acid requirements is to estimate total requirement from the growth of different components of the body or eggs, using models that quantify how much of each amino acid is needed for each component. Several examples were published by Hurwitz and Bornstein (1973, 1977) and Hurwitz *et al.* (1978, 1983a,b). Hurwitz and Bornstein (1973, 1977) proposed two models of protein and amino acid requirements of the laying hen. Both models (A and B, Equations 17.15 and 17.16 below) employed the same, previously measured, values for maintenance requirement and rate of protein accretion, but went on to make some mechanistically based hypotheses related to the synthesis of egg protein. In both models, it was assumed that yolk protein was synthesized on a steady-state basis, with its constituent amino acids coming directly from food. However, models A and B differed in assumptions related to albumen and shell membrane proteins. Model A assumed that all the albumen and membrane proteins were synthesized at the time of albumen formation. Model B assumed that only the ovoglycoproteins and membranes were synthesized at this time and that the ovalbumins were synthesized continuously and stored until required for egg formation. Hurwitz and Bornstein hypothesized that proteins synthesized during the short period of egg white deposition must necessitate the breakdown of body protein. Models A and B are shown below; both give requirements in total dietary amino acid, allowing for a digestibility of 0.85:

Model A $\qquad A = 1.85Wk_m + 0.21Gk_t + EP(63k_y + 158k_t)$ $\qquad$ (17.15)

Model B $\qquad A = 1.85Wk_m + 0.21Gk_t + EP(63k_y + 59k_o + 52k_t)$ $\qquad$ (17.16)

where A is the requirement for an individual amino acid (g day^{-1}); W is body weight (kg); G is weight gain (g day^{-1}); E is egg weight (g); P is egg production (eggs bird^{-1} day^{-1}); and k_m, k_t, k_y, *and* k_o are the proportions of the individual amino acid in the total of all amino acids in maintenance, body tissues, yolk and ovomucoid respectively.

The ideas of Hurwitz and his colleagues formed the foundation for refinements by Smith (1978a,b), who also reviewed the physiological background to the models of Hurwitz and Bornstein (1973, 1978). These models, especially version B, were quite successful in predicting requirements under a range of conditions (Hurwitz and Bornstein, 1977), although calculations by Fisher (1980) and experimental work by Hiramoto *et al.* (1990) suggest that variation in protein synthesis rate due to egg protein deposition is small compared with the total rate of protein synthesis in the bird. Hurwitz *et al.* (1978) also described some success in the estimation of amino acid requirements for growing chicks from maintenance requirement, body weight gain, the proportion of feather protein in body gain and the amino acid composition of carcass and feathers. A broadly similar approach was used for turkeys and gave calculated amino acid requirements which were in good agreement with experimentally determined values (Hurwitz *et al.*, 1983a,b).

Martin *et al.* (1994) described a method in which the requirements for four functions were calculated, i.e. body protein gain, body protein maintenance, feather protein gain and feather protein maintenance. Accretion of body protein and feather protein were described accurately by Gompertz functions. Allometric relationships were then used to estimate other components of growth, which were summed to give the growth rate of the whole body.

Using appropriate amino acid compositions for body and feather protein, the requirements for essential amino acids (AAR, g day^{-1}) were calculated by the following equation:

$$AAR = [a \, dBP/dt + b \, dFP/dt]/0.8 + [aMP_B + bMP_F] \qquad (17.17)$$

where a is the amino acid content of body protein (g kg^{-1}), dBP/dt is the rate of body protein accretion (kg day^{-1}), b is the amino acid content of feather protein (g kg^{-1}), dFP/dt is the rate of feather protein accretion (kg day^{-1}), MP_B is maintenance protein for the body (kg day^{-1}) and MP_F is maintenance protein for feathers (kg day^{-1}). The constant 0.8 is an estimate of the efficiency of amino acid utilization. A prediction of food intake is necessary to allow amino acid requirements to be expressed as a proportion of the diet. Food intake (DFI, 'desired food intake', predicted as in Emmans and Fisher, 1986) and amino acid requirements as concentrations in the diet (DAA, g kg^{-1} diet) were estimated as:

$$DAA = AAR/DFI \qquad (17.18)$$

where AAR is amino acid requirement (g bird^{-1} day^{-1}) and DFI is 'desired food intake' (kg day^{-1}).

There have been few attempts at mechanistic modelling of amino acid metabolism in poultry. Saunderson (1988) used a set of simple models to examine the possible role of amino acid metabolism in the differences between genetic lines of broilers selected for fatness (FL) or leanness (LL). No significant differences had been shown in any individual metabolic pathway examined in these lines, but modelling demonstrated that a number of small changes could combine to produce values for fat and protein deposition which approximated to the observed differences in body composition.

CONCLUSIONS

Although comparatively few poultry nutritionists have been directly involved in the process of predictive modelling, a number of robust and useful models have been developed. However, not enough have been published in recent years to displace some of the earlier feed evaluation models from this chapter.

Some of those in current practical use are subject to commercial confidentiality and could not be described.

Hesitancy in the commercial uptake of NE models is worth mentioning. Because of the potential for biochemical interactions, the degree of linear additivity of NE values of individual ingredients is often mentioned as a barrier to their use in feed formulation. However, if ME values are linearly additive, while NE values are not, it can only be because ME does not detect the interactions which are seen as complicating an NE system. Emmans (1994) appears to have chosen his words carefully in stating that 'As effective energy values…are additive *to the extent that ME values are additive*, they can be used to formulate diets using linear programming'. The same might well be said for other NE systems: even if the additivity is not perfect, prediction of animal response might still be more accurate than with a system (such as ME) which takes little account of post-absorptive metabolism. A possible criticism is that NE is a property not of the food but of the birds' response to the food. This is clearly true, but must be true of any meaningful feed evaluation system.

Modelling will play an increasingly important part in poultry nutrition as a way of organizing the large body of existing knowledge. The direct interfacing of nutritional models with commercial processes such as feed formulation, feeding and climatic control is a particularly interesting prospect.

REFERENCES

Blair, J.C., Harber, C.D., McNab, J.M., Mitchell, G.G. and Scougall, R.K. (1981) *Analytical Data of Poultry Feedstuffs. 1. General and Amino Acid Analyses, 1977–1980*. Occasional Publication No 1, ARC Poultry Research Centre, Roslin, Midlothian, UK.

Boorman, K.N. and Burgess, A.D. (1986) Responses to amino acids. In: Fisher, C. and Boorman, K.N. (eds), *Nutrient Requirements of Poultry and Nutritional Research*. Poultry Science Symposium 19. Butterworths, London, pp. 99–123.

Bourdillon, A., Carre, B., Conan, L., Duperray, J., Huyghebaert, G., Leclercq, B., Lessire, M., McNab, J. and Wiseman, J. (1990) European reference method for the *in vivo* determination of metabolizable energy with adult cockerels: reproducibility, effect of food intake and comparison with individual laboratory methods. *British Poultry Science* 31, 557–565.

Byerly, T.C. (1979) Prediction of the food intake of laying hens. In: Boorman, K.N. and Freeman, B.M. (eds), *Food Intake Regulation in Poultry. Poultry Science Symposium 14*. Longmans, Edinburgh, UK, pp. 327–363.

Chwalibog, A. (1992) Factorial estimation of energy requirement for egg production. *Poultry Science* 71, 509–515.

Curnow, R.N. (1973) A smooth population response curve based on an abrupt threshold and plateau model for individuals. *Biometrics* 29, 1–10.

Curnow, R.N. and Torenbeek, R.V. (1996) Optimal diets for egg production. *British Poultry Science* 37, 373–382.

de Boer, F. and Bickel, H. (1988) Livestock feed resources and feed evaluation in Europe. Present situation and future prospects. *Livestock Production Science* 19, 1–410.

de Groote, G. (1974a) Utilisation of metabolizable energy. In: Morris, T.R. and Freeman, B.M. (eds), *Energy Requirements of Poultry. Poultry Science Symposium 9.* Longmans, Edinburgh, UK, pp. 113–134.

de Groote, G. (1974b) A comparison of a new net energy system with the metabolizable energy system in broiler diet formulation, performance and profitability. *British Poultry Science* 15, 75–95.

de Groote, G. (1975) Net energy systems for chickens. In: *Proceedings of the 1975 Georgia Nutrition Conference.* Atlanta, Georgia, pp. 9–30.

Emmans, G.C. (1984) An additive and linear energy scale. *Animal Production* 38, 538.

Emmans, G.C. (1987) Growth, body composition and feed intake. *World's Poultry Science Journal* 43, 208–227.

Emmans, G.C. (1994) Effective energy: a concept of energy utilization applied across species. *British Journal of Nutrition* 171, 801–821.

Emmans, G.C. (1995) Problems in modelling the growth of poultry. *World's Poultry Science Journal* 51, 77–89.

Emmans, G.C. and Fisher, C. (1986) Problems in nutritional theory. In: Fisher, C. and Boorman, K.N. (eds), *Nutrient Requirements of Poultry and Nutritional Research. Poultry Science Symposium 19.* Butterworths, London, pp. 59–64.

Farrell, D.J. (1978) Rapid determination of metabolizable energy of foods using cockerels. *British Poultry Science* 19, 303–308.

Farrell, D.J. (1979) Energy systems for pigs and poultry: a review. *Journal of the Australian Institute of Agricultural Science* 1979, 21–34.

Farrell, D.J., Smulders, A., Mannion, P.F., Smith, M. and Priest, J. (1998) The effective energy of six poultry diets measured in young and adult birds. In: McCracken, K.J., Unsworth, E.F. and Wylie, A.R.G. (eds), *Energy Metabolism of Farm Animals.* CAB International, Wallingford, Oxon, UK, pp. 371–374.

Fisher, C. (1980) Protein deposition in poultry. In: Buttery, P.J. and Lindsay, D.B. (eds), *Protein Deposition in Animals.* Butterworths, London, pp. 251–270.

Fisher, C. and Wilson, B.J. (1974) Response to dietary energy concentration by growing chickens. In: Morris, T.R. and Freeman, B.M. (eds), *Energy Requirements of Poultry. Poultry Science Symposium 9.* Longmans Ltd, Edinburgh, UK, pp. 151–184.

Fisher, C., Morris, T.R. and Jennings, R.C. (1973) A model for the description and prediction of the response of laying hens to amino acid intake. *British Poultry Science* 14, 469–484.

France, J. and Thornley, J.H.M. (1984) *Mathematical Models in Agriculture.* Butterworths, London.

France, J., Dijkstra, J., Thornley, J.H.M. and Dhanoa, M.S. (1996) A simple but flexible growth function. *Growth, Development and Ageing* 60, 71–83.

Fraps, G.S. (1946) Composition and productive energy of poultry feeds and rations. *Texas Agricultural Station Bulletin* No. 678.

Fraps, G.S. and Carlyle, E.C. (1939) The utilization of the energy of feed by growing chickens. *Texas Agricultural Experiment Station Bulletin* No. 571.

Geraert, P.A., MacLeod, M.G. and Leclercq, B. (1988) Energy metabolism in genetically fat and lean chickens: diet- and cold-induced thermogenesis. *Journal of Nutrition* 118, 1232–1239.

Hakansson, J., Eriksson, S. and Svensson, S.A. (1978a) *The Influence of Feed Energy Level on Feed Consumption, Growth and Development of Different Organs of Chicks.* Report No 57. Swedish University of Agricultural Sciences. Sveriges Lantbruksuniversitet, Uppsala, Sweden.

Hakansson, J., Eriksson, S. and Svensson, S.A. (1978b) *The Influence of Feed Energy Level on Chemical Composition of Tissues and on the Energy and Protein Utilisation of Broiler Chicks*. Report No. 59. Swedish University of Agricultural Sciences. Sveriges Lantbruksuniversitet, Uppsala, Sweden.

Heartland Lysine, Inc. (1990) *True Digestibility of Essential Amino Acids for Poultry*. Heartland Lysine, Inc., Chicago, Illinois.

Hill, F.W. and Anderson, D.L. (1958) Comparison of metabolizable energy and productive energy determination with growing chicks. *Journal of Nutrition* 64, 587–603.

Hiramoto, K., Muramatsu, T. and Okumura, J. (1990) Protein synthesis in tissues and in the whole body of laying hens during egg formation. *Poultry Science* 56, 969–978.

Hurwitz, S. and Bornstein, S. (1973) The protein and amino acid requirements of laying hens: suggested models for calculation. *Poultry Science* 52, 1124–1134.

Hurwitz, S. and Bornstein, S. (1977) The protein and amino acid requirements of laying hens: experimental evaluation of models of calculation. 1. Application of two models under various conditions. *Poultry Science* 56, 969–978.

Hurwitz, S., Sklan, D. and Bartov, I. (1978) New formal approaches to the determination of energy and amino acid requirements of chicks. *Poultry Science* 57, 197–205.

Hurwitz, S., Frisch, Y., Bar, A., Eisner, U., Bengal, I. and Pines, M. (1983a) The amino acid requirements of growing turkeys. 1. Model construction and parameter estimation. *Poultry Science* 62, 2208–2217.

Hurwitz, S., Plavnik, I., Bengal, I., Talpaz, H. and Bartov, I. (1983b) The amino acid requirements of growing turkeys. 2. Experimental validation of model-calculated requirements for sulfur amino acids and lysine. *Poultry Science* 62, 2387–2393.

Livesey, G. (1984) The energy equivalents of ATP and the energy values of food proteins and fats. *British Journal of Nutrition* 51, 15–28.

Livesey, G. (1985) Mitochondrial uncoupling and the isodynamic equivalents of protein, fat and carbohydrate at the level of biochemical energy provision. *British Journal of Nutrition* 53, 381–389.

Longstaff, M. and McNab, J.M. (1991) The inhibitory effects of hull polysaccharides and tannins of field beans *Vicia faba* L. on the digestion of amino acids, starch and lipid and on digestive enzyme activities in young chicks. *British Journal of Nutrition* 65, 199–216.

Lunven, P., Le Clement de St Marcq, C., Carnovale, E. and Fratoni, A. (1973) Amino acid composition of the hen's egg. *British Journal of Nutrition* 30, 189–194.

MacLeod, M.G. (1990) Energy and nitrogen intake, expenditure and retention at 20° in growing fowl given diets with a wide range of energy and protein contents. *British Journal of Nutrition* 64, 625–637.

MacLeod, M.G. (1992) Energy and nitrogen intake, expenditure and retention at 32° in growing fowl given diets with a wide range of energy and protein contents. *British Journal of Nutrition* 67, 195–206.

MacLeod, M.G. (1994) Dietary energy utilisation: prediction by computer simulation of energy metabolism. In: Aguilera, J. (ed.), *Proceedings of the 13th Symposium on Energy Metabolism of Farm Animals*. September 18–24, 1994, Mojacar, Spain, pp. 237–240.

MacLeod, M.G. (1997) Dietary energy utilisation: sensitivity of a net energy model to varying estimates of food intake, maintenance requirement and protein deposition. In: McCracken, K.J., Unsworth, E.F. and Wylie, A.R.G. (eds), *The Energy Metabolism of Farm Animals*. CAB International, Wallingford, Oxon, UK, pp. 323–326.

MacLeod, M.G. and Waibel, P.E. (1997) Computerised control of growth rate and measurement of metabolic rate in the domestic fowl and the turkey. *Growth, Development and Ageing* 61, 101–106.

MacLeod, M.G., McNab, J.M., Bernard, K. and Wilson S. (1996) A possible synergistic interaction in the utilisation of blood meal and hydrolysed feather meal. *British Poultry Science* 37, S64.

Martin, P.A., Bradford, G.D. and Gous, R.M. (1994) A formal method of determining the dietary amino acid requirements of laying-type pullets during their growing period. *British Poultry Science* 35, 709–724.

McNab, J.M. and Blair, J.C. (1988) Modified assay for true and apparent metabolizable energy based on tube feeding. *British Poultry Science* 29, 697–707.

McNab, J.M. and Scougall, R.K. (1982) The tryptophan content of some feedstuffs for poultry. *Journal of the Science of Food and Agriculture* 33, 715–721.

Millward, D.J., Garlick, P.J. and Reeds, P.J. (1976) The energy cost of growth. *Proceedings of the Nutrition Society* 35, 339–349.

Moughan, P.J. (1994) Modelling amino acid absorption and metabolism in the growing pig. In: D'Mello, J.P.F. (ed.), *Amino Acids in Farm Animal Nutrition*. CAB International, Wallingford, UK, pp. 133–154.

Muramatsu, T., Isariyodom, S., Umeda, I. and Okumura, J. (1989) Computer-simulated growth prediction of replacement pullets with special reference to seasonal changes in feed intake. *Poultry Science* 68, 771–780.

Muramatsu, T., Kawamura, H., Tanaka, Y. and Okumura, J. (1994) Prediction of egg production and growth performance by using a computer simulation model for laying hens. *Japanese Poultry Science* 31, 313–326.

Nehring, K. (1967) Investigation on the scientific basis for the use of net energy for fattening as a measure of feed value. In: Blaxter, K.L., Kielanowski, J. and Thorbek, G. (eds), *Energy Metabolism of Farm Animals, Proceedings of the 4th Symposium*, Warsaw, pp. 5–20.

Nehring, K. and Haenlein, G.F.W. (1973) Feed evaluation and ration calculation based on net energy. *Journal of Animal Science* 36, 949–964.

Nitsan, Z., Dvorin, A. and Nir, I. (1981) Composition and amino acid content of carcass, skin and feathers of the growing gosling. *British Poultry Science* 22, 79–84.

Pesti, G.M., Dorfman, J.H. and Gonzalez-A., M.J. (1992) Model for predicting egg output and metabolizable energy intake of laying pullets. *British Poultry Science* 33, 543–552.

Sakomura, N.K., Rostagno, H.S., Soares, P.R. and Sanchez, G. (1993a) Determination of equations to predict the nutritional requirement of energy in broiler breeders and laying hens (in Portuguese). *Revista da Sociedade Brasileira de Zootecnia* 22, 723–731.

Sakomura, N.K., Rostagno, H.S., Couto, H.P. and Macari, M. (1993b) Energy requirements for white leghorn layers by using prediction equations (in Portuguese). *Revista da Sociedade Brasileira de Zootecnia* 22, 732–744.

Saunderson, C.L. (1988) Amino acid and protein metabolism in genetically lean and fat lines of chickens. In: Leclercq, B. and Whitehead, C.C. (eds), *Leanness in Domestic Birds. Genetic, Metabolic and Hormonal Aspects*. Butterworths, London, pp. 363–374.

Scheele, C.W., Janssen, W.M.M.A. and Verstegen, M.W.A. (1973) Feed energy that can be stored in high energy phosphate bonds, as a determinant for predicting the energy value of poultry feeds. In: Vermorel, M. (ed.), *Proceedings of the 7th European Association of Animal Production Symposium on the Energy*

Metabolism of Farm Animals. Vichy, France. Guy de Bussac, Clermont-Ferrand, pp. 277–280.

Schiemann, R. (1967) The scientific demands made of a system for evaluating feeds on energy sources and progress made towards their realisation. In: Blaxter, K.L., Kielanowski, J. and Thorbek, G. (eds), *Energy Metabolism of Farm Animals, Proceedings of the 4th Symposium*, Warsaw, pp. 31–40.

Schiemann, R., Nehring, K., Hoffmann, L., Jentsch, W. and Chudy, A. (1971) *Energetische Futterbewertung und Energienormen.* Deutscher Landwirtschaft Verlag, Berlin.

Schulz, A.R. (1975) Computer-based method for calculation of the available energy of proteins. *Journal of Nutrition* 105, 200–207.

Schulz, A.R. (1978) Simulation of energy metabolism in the simple-stomached animal. *British Journal of Nutrition* 39, 235–254.

Shafey, T.M. and McDonald, M.W. (1991) The effects of dietary concentrations of minerals, source of protein, amino acids and antibiotics on the growth of and digestibility of amino acids by broiler chickens. *British Poultry Science* 32, 535–544.

Sibbald, I.R. (1976) A bioassay for true metabolizable energy in feedstuffs. *Poultry Science* 55, 303–308.

Smith, W.K. (1978a) The amino acid requirements of laying hens: models for calculation. 1. Physiological background. *World's Poultry Science Journal* 34, 81–96.

Smith, W.K. (1978b) The amino acid requirements of laying hens: models for calculation. 2. Practical application. *World's Poultry Science Journal* 34, 129–136.

Tasaki, I. and Kushima, M. (1979) Heat production when single nutrients are given to fasted cockerels. In: Mount, L.E. (ed.), *Proceedings of the 8th European Association of Animal Production Symposium on the Energy Metabolism of Farm Animals.* Butterworths, London, pp. 253–256.

Taylor, St.C.S. (1970) Models of maintenance requirement in livestock. In: Jones, J.W.G. (ed.), *The Use of Models in Agricultural and Biological Research.* Grassland Research Institute, Hurley, UK, pp. 107–117.

Taylor, St.C. S. and Young, G.B. (1968) Equilibrium weight in relation to food intake and genotype in twin cattle. *Animal Production* 30, 161–165.

Zoons, J., Buyse, J. and Decuypere, E. (1991) Mathematical models in broiler raising. *World's Poultry Science Journal* 47, 243–255.

Zuprizal, Larbier, M., Chagneau, A.M. and Geraert, P.A. (1993) Influence of ambient temperature on true digestibility of protein and amino acids of rapeseed and soybean meals in broilers. *Poultry Science* 72, 289–295.

18 Modelling Growth in Fish

Y. Cui and S. Xie

Institute of Hydrobiology, The Chinese Academy of Sciences, Wuhan, Hubei, China

INTRODUCTION

Growth models, especially bioenergetics models, have been widely used by animal nutritionists to formulate feeding systems for domestic mammals and birds (Baldwin and Sainz, 1995). However, only some initial attempts have been made to use growth models to formulate feeding systems for fish (Cho, 1992). As a result, feeding charts for cultured fish are often formulated based on experience (Austreng *et al.*, 1987; Cho, 1992). As rates of food consumption and growth in fish are highly variable and are affected by a number of factors, including water temperature, fish weight and food quality (Brett and Groves, 1979), it is important to develop predictive growth models that can be used to formulate feeding systems for fish. In ecological studies of fish, modelling growth has a long history (Ricker, 1979). These models are aimed to predict fish growth or food consumption under natural conditions and have been widely used in fishery management (Ricker, 1979; Hansen *et al.*, 1993). One group of growth models describes changes in fish size over time, e.g. the von Bertalanffy model which describes the relationship between body length and the age of fish in a given population (von Bertalanffy, 1957). Such models have no applications to fish culture. The other group of models predicts growth rate from a series of predictor variables. In these models, the predictor variables considered are usually ration level, water temperature and fish size, which are the major factors affecting the growth rate and energy budget of fish under natural (but not aquaculture) conditions (Brett and Groves, 1979). There are two types of predictive models: the empirical models that are developed using the technique of regression analysis of the relationship between growth rate and predictor variables, and the bioenergetics models that are based on the principles of bioenergetics.

©CAB *International* 2000. *Feeding Systems and Feed Evaluation Models*
(eds M.K. Theodorou and J. France)

The extensive experiences of fish ecologists in growth modelling will be useful for fish nutritionists in developing models that are useful for the formulation of feeding systems for fish. The purpose of this chapter is to describe the growth models, especially the bioenergetics models, that are developed and widely used by fish ecologists, and to discuss their potential applications in formulating feeding systems for fish. The chapter is organized in three parts. First, there is a brief introduction to the empirical growth models; then, a detailed description is provided on the bioenergetics models; finally, the discussion considers the potential applications of these models in formulating feeding systems for fish.

EMPIRICAL GROWTH MODELS

The empirical growth models are developed by relating the growth rate of fish to predictor variables using the technique of step-wise regression. Examples of empirical models include the models for the three-spined stickleback *Gasterosteus aculeatus* L. (Allen and Wootton, 1982), the black rockfish *Sebastes melanops* (Boehlert and Yoklavich, 1983) and the European minnow *Phoxinus phoxinus* (L.) (Cui and Wootton, 1988).

Cui and Hung (1995) used this approach to develop a prototype feeding–growth table (a table listing optimum ration levels and the corresponding growth rates under various conditions) for white sturgeon, *Acipenser transmontanus* Richardson, a new aquaculture species for which no commonly used feeding tables have been developed. Using data from a series of experiments on the growth–ration relationship at different water temperatures, empirical models for growth rate and optimum ration level were developed:

$$G = -4.43 + 0.566T - 0.0151T^2 + 0.00849 \ln(L_R + 0.1)T^2$$

$$- 0.00122 \ln(L_R + 0.1)T^2 \ln W \tag{18.1}$$

and

$$\ln L_{Ropt} = -2.88 - 0.25 \ln W + 0.400T - 0.0077T^2 \tag{18.2}$$

where G is growth rate (% day^{-1}), L_{Ropt} is optimum ration level (% body weight day^{-1}), T is water temperature (°C), L_R is ration level (% body weight day^{-1}) and W is body weight of fish (g).

The growth model provided accurate predictions for the growth of white sturgeon in independent experiments (Cui and Hung, 1995). The prototype feeding–growth model was formulated by combining the growth model and the model for optimum ration level (Table 18.1).

A different type of empirical growth model was proposed by Cho (1992) for cultured salmonids based on the concept of thermal unit growth coefficient (C_{TG}):

$$C_{TG} = (W_t^{1/3} - W_0^{1/3})/\Sigma(TD) \tag{18.3}$$

Table 18.1. A prototype feeding–growth table listing the suggested ration level and expected growth rate for various sizes of white sturgeon at different temperatures.

Temp. (°C)	Weight (g)						
	50	100	200	400	600	800	1000
	RL/GR	RL/GR	RL/GR	RL/GR	RL/GR	RL/GR	RL/GR
10	0.53/–0.45	0.44/–0.45	0.37/–0.43	0.31/–0.38	0.26/–0.35	0.26/–0.31	0.25/–0.29
12	0.84/0.15	0.70/0.09	0.59/0.08	0.50/0.10	0.45/0.13	0.42/0.16	0.40/0.18
14	1.25/0.75	1.05/0.61	0.88/0.53	0.74/0.49	0.67/0.50	0.62/0.51	0.59/0.53
16	1.74/1.34	1.47/1.09	1.23/0.91	1.04/0.80	0.94/0.77	0.87/0.76	0.82/0.76
18	2.29/1.92	1.93/1.52	1.62/1.22	1.36/1.01	1.23/0.93	1.14/0.89	1.08/0.87
20	2.84/2.45	2.38/1.89	2.00/1.45	1.68/1.12	1.52/0.98	1.41/0.91	1.34/0.86
22	3.29/2.91	2.77/2.18	2.33/1.58	1.95/1.12	1.77/0.92	1.64/0.80	1.55/0.73
24	3.60/3.26	3.02/2.34	2.54/1.59	2.13/1.00	1.93/0.74	1.79/0.58	1.70/0.48
26	3.69/3.43	3.10/2.34	2.61/1.44	2.19/0.74	1.98/0.42	1.84/0.23	1.74/0.10

RL, suggested ration level (% body weight day^{-1}); GR, expected growth rate (% body weight day^{-1}). RL and GR are predicted from empirical models developed by Cui and Hung (1995).

where W_t is the final and W_0 the initial body weight of fish, T is water temperature (°C) and D is duration of each growth period in days. The estimated final body weight can be calculated as:

$$W_t = [W_0^{1/3} + \Sigma(C_{TG}TD)]^3 \tag{18.4}$$

The model was reported to follow closely the actual growth curves of rainbow trout, lake trout, brown trout, chinook salmon and Atlantic salmon in the Fish Nutrition Laboratory of the University of Guelph, Ontario, Canada. Whether this model has general applications to other species needs further investigations. The model assumes that the growth of fish is proportional to the temperatures summed over days, and this assumption remains untested. Also, the model does not take into account the effect of ration level and can only be used in situations where the fish are fed to satiation.

Empirical models are usually accurate in describing the data set from which the models are developed (Allen and Wootton, 1982; Boehlert and Yoklavich, 1983; Cui and Wootton, 1988; Cui and Hung, 1995). Such models, however, are difficult to interpret biologically, and extrapolation of the model outside the experimental conditions that produce the data may lead to unrealistic predictions. Also, the empirical models are unable to predict food consumption and waste production. The bioenergetics models are expected to be more general and to be able to predict all the components of the energy budget.

BIOENERGETICS MODELS FOR FISH

In this section, a brief review of basic concepts of fish bioenergetics is provided, and the basic concepts and procedures of bioenergetics modelling are described. Details of such models are demonstrated by describing three examples of bioenergetics models.

Basic Concepts of Fish Bioenergetics

Several reviews are available on fish bioenergetics (Brett and Groves, 1979; Jobling, 1994), and only a brief introduction to the basic concepts is given here. Different terms and symbols in fish bioenergetics are used by ecologists and nutritionists. Because most bioenergetics models are developed by fish ecologists, the ecological terms and symbols will be used in this chapter, and a description of the nutritional terms can be found in Beamish and Trippel (1990).

The basic energy budget for fish can be written as:

$$G = C - F - U - R \tag{18.5}$$

where G is energy deposited in fish tissues (growth), C is energy consumed as food (food consumption), F is energy lost as faeces (faecal production), U is energy lost in nitrogenous excretory products (mainly ammonia and urea) (nitrogenous excretion) and R is energy lost as heat (metabolism) (Brett and Groves, 1979). R is divided into three components:

$$R = R_s + SDA + R_a \tag{18.6}$$

where R_s is standard metabolism, SDA is specific dynamic action and R_a is activity metabolism. Standard metabolism, an approximation of basal metabolism used in animal energetics, is defined as the metabolic rate of resting, unfed fish. Two problems affect the accurate measurement of the standard metabolism. First, fish are rarely at complete rest. This may be overcome by measuring the metabolic rate of fish forced to swim at various speeds and extrapolating the speed to zero (Beamish *et al.*, 1989). The second problem is that the 'unfed' status is difficult to standardize. The metabolic rate of fish may continue to decline for weeks after starvation (Jobling, 1980). Because of these problems, various approximations of standard metabolism, such as routine metabolism, lower routine metabolism, resting metabolism and fasting metabolism, are used in the literature (Jobling, 1994). Because of the lack of standardized measuring methods, the above terms can be regarded as synonyms. SDA, usually measured as the post-prandial increase in metabolism, is defined as the metabolic cost of food utilization. Sources of SDA may include the digestion, absorption and storage of nutrients, deamination of amino acids, urea synthesis and synthesis and turnover of tissue components (especially protein) (Jobling, 1981, 1983b). Activity metabolism is the metabolic cost associated with swimming activity. SDA and activity metabolism are often difficult to separate technically, because the activity of fish may also increase after feeding. The term feeding metabolism has been proposed to represent the sum of SDA and activity metabolism (Ursin, 1967; Cui *et al.*, 1994).

Growth

Growth rate of fish is often expressed in terms of specific growth rate (SGR, % day^{-1}): SGR = $100(\ln W_t - \ln W_0)/t$, where W_t is the final and W_0 the initial body weight of fish and t is the duration of the growth period in days. Ration level is an important factor affecting the growth of fish. The typical SGR–ration relationship has been suggested to be a decelerating curve (Fig. 18.1). SGR initially increases rapidly with increased ration and, when ration level is high, the increase in SGR with ration slows down (Brett and Groves, 1979; Jobling, 1994). The conversion efficiency (or growth efficiency), defined as the ratio of body weight gain (or energy gain) to food consumption, increases initially with increased ration to a peak, and then decreases with further increases in ration. The ration at which conversion efficiency is maximal is called the optimum ration. Recent studies, however, showed that linear SGR–ration relationships are not uncommon (Cui *et al.*, 1994, 1996a,b). Jobling (1995) even suggested that the curvilinear SGR–ration relationship may be an artefact caused by the inaccurate quantification of uneaten food at high rations. However, the curvilinear relationship was reported in several studies using live animals as food, for which the uneaten food can be quantified accurately (Allen and Wootton, 1982; Cui and Wootton, 1988; Cortes and Gruber, 1994) (Fig. 18.2).

At the maximum ration, SGR of fish increases with water temperature to a peak at the optimum temperature for growth, and decreases with further

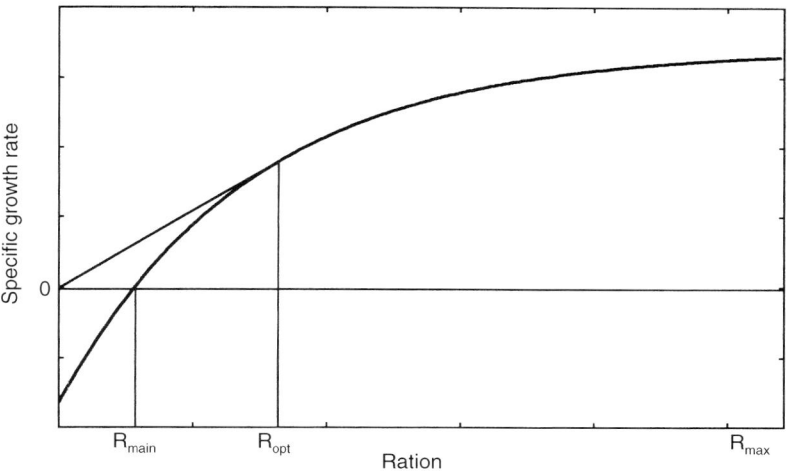

Fig. 18.1. Typical growth–ration relationship for fish. R_{main}, maintenance ration; R_{opt}, optimum ration; R_{max}, maximum ration.

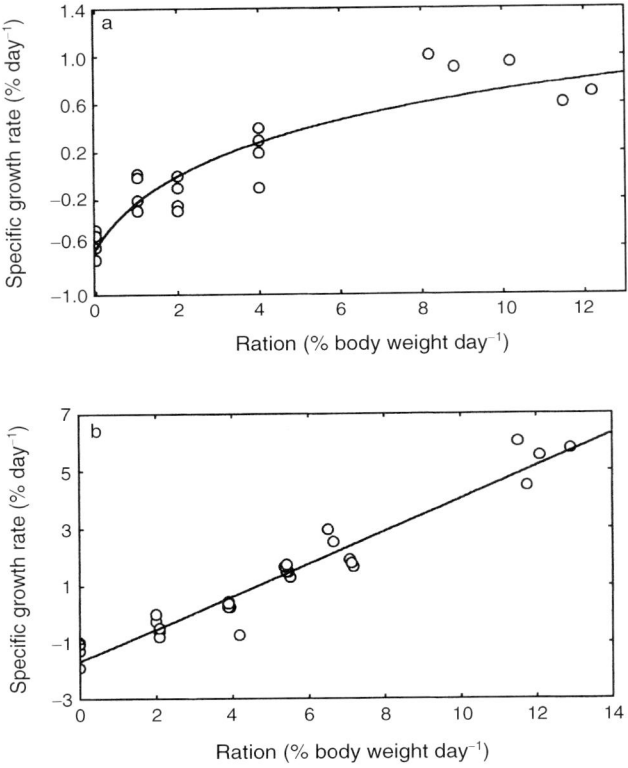

Fig. 18.2. Two types of growth–ration relationships in fish. (a) 1–5 g European minnow at 15°C (Cui and Wootton, 1988); (b) 2.4 g white sturgeon at 19°C (Cui *et al.*, 1996b). Live oligochaete worms were used as food in both studies.

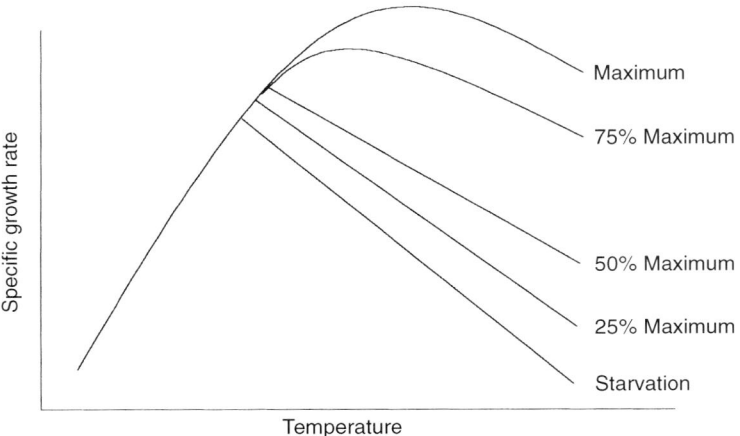

Fig. 18.3. Typical effects of temperature on the growth rate of fish fed at different ration levels. Intermediate ration levels are expressed as percentages of maximum rations.

increases in temperature. At a fixed, restricted ration, SGR decreases with increased temperature because of increased cost of standard metabolism at higher temperatures (Fig. 18.3) (Brett and Groves, 1979; Allen and Wootton, 1982; Cui and Wootton, 1988). At the maximum ration, SGR usually decreases with increased body weight (Jobling, 1983a; Cui *et al.*, 1996a,b). At a fixed, restricted ration, SGR is expected to increase with increased body weight, since the weight-specific standard metabolism decreases with body weight. This pattern has been shown in a study on the white sturgeon, *A. transmontanus* (Fig. 18.4) (Cui *et al.*, 1996b).

Food Consumption

The maximum rate of food consumption is affected mainly by water temperature and body weight of fish. The maximum consumption increases with increased water temperature to a peak at the optimum temperature, and then decreases with further increases in water temperature (Brett and Groves, 1979). A power function has been used to describe the relationship between maximum consumption (C_{max}) and body weight (W) in fish:

$$\ln C_{max} = a + b \ln W \tag{18.7}$$

The weight exponent (b) is typically less than 1, suggesting that weight-specific consumption decreases with body weight.

Faecal Production

Faecal production is usually expressed in terms of digestibility (or absorption efficiency), which is defined as: $100(C - F)/C$, where C is food consumption and F is faecal production. Digestibility is largely dependent on the type of

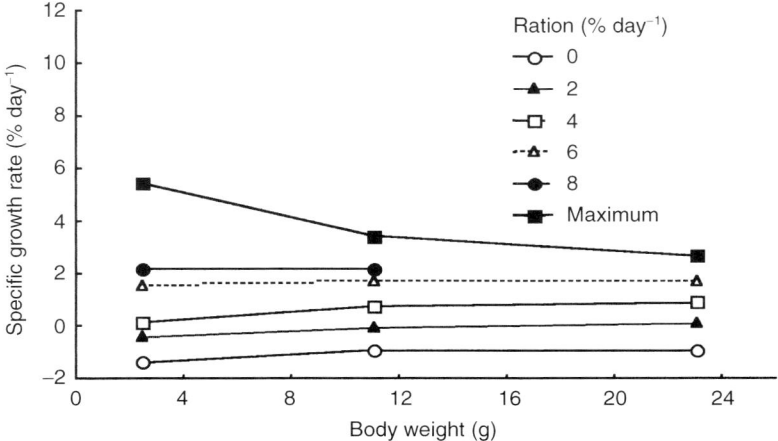

Fig. 18.4. Effects of body weight on the specific growth rate of white sturgeon fed at different ration levels (Cui *et al.*, 1996b).

food, and is typically over 85% for animal foods and less than 60% for plant foods (Pandian and Vivekanandan, 1985). Whether digestibility changes with ration level is still controversial (Cui *et al.*, 1994).

Nitrogenous Excretion

Nitrogenous excretion can be divided into endogenous and exogenous excretion. The former comprises the nitrogenous excretory products resulting from breakdown of tissue proteins and is affected mainly by water temperature and fish size. The latter results from direct deamination of amino acids ingested and absorbed from the food and is mainly determined by the quantity and quality of food (Jobling, 1994).

Metabolism

Standard metabolism is a function of water temperature and body weight of fish (Brett and Groves, 1979; Jobling, 1994). An increase in water temperature usually results in an increase in standard metabolism. The relationship between standard metabolism (R_s) and body weight (W) is typically a power function: $\ln R_s = a + b\ln W$. Most estimates of the weight exponent (b) lie within the range of 0.65–0.9 (Jobling, 1994), suggesting that the weight-specific rate of standard metabolism usually decreases with increased body weight.

Most of the estimates of SDA lie within the range of 10–20% of food energy (Jobling, 1981). The SDA coefficient (the proportion of food energy spent in SDA) is expected to be affected by food composition and several other factors (Beamish and Trippel, 1990).

Many studies have been made on the relationship between metabolic rate and swimming speed of fish under forced swimming conditions (Soofiani and Priede, 1985; Dabrowski, 1986), but few studies have reported the actual cost of activity metabolism in fish under natural or cultured conditions. Some estimates were made on the activity cost under natural conditions using heart rate or muscle electrogram telemetry (Demens *et al.*, 1996; Thorarensen *et al.*, 1996), but the accuracy of such estimates remains uncertain.

BASIC MODELLING PROCEDURES

The first bioenergetics model for fish was proposed by Kitchell *et al.* (1974). Models using similar principles, however, were proposed earlier by other researchers (Ursin, 1967; Kerr, 1971). Models based on that proposed by Kitchell *et al.* (1974) were soon applied to many fish species and used extensively in fish ecology and fishery management (see models compiled by

Hewett and Johnson (1992) and reviews in Brandt and Hartman (1993)). Hewett and Johnson (1992) developed computer software that incorporates the existing bioenergetics models for various fish species and can be adapted to other species once the parameters are established. The model is based on the energy budget equation:

$$G = C - F - U - R_s - SDA - R_a \tag{18.8}$$

Submodels are developed to describe each of the components on the right-hand side of the equation. These submodels are functions of a series of predictor variables, usually ration level, water temperature and body weight. Such functions and their parameters are either obtained empirically or adapted from other species or general knowledge using simplifying assumptions. Values for C, F, U, R_s, SDA and R_a can thus be calculated for the conditions specified, and growth calculated as the difference between C and other components. Total body energy content on day t (E_t) can be calculated as:

$$E_t = E_{t-1} + G \tag{18.9}$$

where G is energy growth in day t–1. E_t is converted into body weight on day t (W_t):

$$W_t = E_t/E_w \tag{18.10}$$

where E_w is the energy content per unit weight of fish. E_w was regarded as a constant in many early bioenergetics models (Kitchell *et al.*, 1977; Kitchell and Breck, 1980; Bevelhimer *et al.*, 1985), but it would be more appropriate to treat E_w as a function of predictor variables, as several factors, including ration level and body weight, are known to affect the body composition of fish (Cui and Wootton, 1989; Cui *et al.*, 1996a).

Using a computer program, growth over a period of time can be predicted. It is also possible to predict food consumption from growth data by performing simulations using a number of ration levels ranging from starvation to maximum. The ration level that produces the predicted final weight closest to the observed final weight is the predicted ration level. Figure 18.5 shows the flow diagram of the bioenergetics model for the European minnow (Cui and Wootton, 1989).

The Model for Yellow Perch and Walleye

The bioenergetics models for yellow perch (*Perca flavescens*) and walleye (*Stizostedion vitreum vitreum*), developed by Kitchell *et al.* (1977), are probably the most influential, and served as a standard for later bioenergetics modelling in fish. The two models are similar in structure and the same values are used for many parameters.

A complex function is used to describe the maximum food consumption (C_{max}, % body weight day^{-1}):

$$C_{max} = C_{mm}r_c \tag{18.11}$$

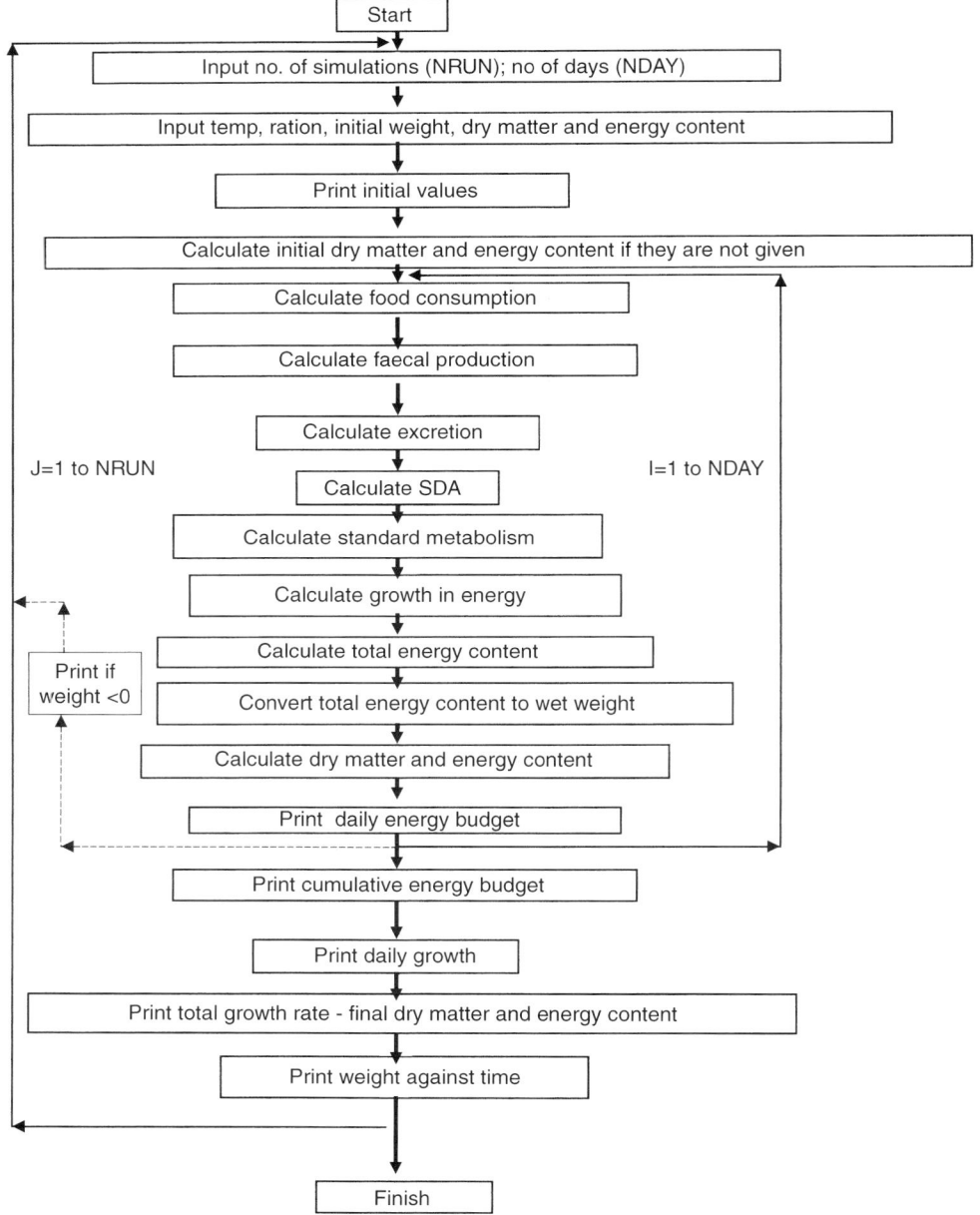

Fig. 18.5. Flow diagram for the bioenergetics model for the European minnow. Redrawn from Cui and Wootton (1989).

C_{mm} is the maximum consumption at the optimum temperature for food consumption and is a function of body weight (B, g): $C_{mm} = a_1 B^{b1}$. r_c is a temperature (T, °C)-dependent function:

$$r_c = (V^X)(e^{X(1-V)}) \tag{18.12}$$

$$V = (T_m - T)/(T_m - T_o) \tag{18.13}$$

$$X = (W^2(1 + (1 + 40/Y)^{1/2})^2)/400 \tag{18.14}$$

$$W = (\ln Q)(T_m - T_o) \tag{18.15}$$

$$Y = (\ln Q)(T_m - T_o + 2) \tag{18.16}$$

where Q approximates the Q_{10} value (proportional increase in consumption as temperature increases by 10°C) for maximum food consumption below the optimum temperature. This function can describe the response of C_{max} to water temperature over the entire temperature range. The value of r_c reaches a maximum of 1.0 at the optimum temperature, T_o, and declines rapidly to 0 at the maximum temperature, T_m.

Faecal production (F) is regarded as a function of food consumption (C), temperature (T) and ration level (P):

$$F = C\alpha_1 T^{\beta_1} e^{\gamma_1 P} \tag{18.17}$$

where P is expressed as a proportion of C_{max}.

A similar function is used to describe nitrogenous excretion (U):

$$U = C\alpha_2 T^{\beta_2} e^{\gamma_2 P}. \tag{18.18}$$

A function similar to that used for C_{max} is used to model standard metabolism:

$$R_s = R_{sm} r_R \tag{18.19}$$

$$r_R = (V^X)(e^{X(1-V)}) \tag{18.20}$$

where R_{sm} is the maximum rate of standard metabolism at the optimum temperature for respiration (T_o), and V and X are functions of T_o, T_m and Q.

SDA is regarded as a constant fraction of food consumption, and R_a is regarded as a fixed multiple of R_s. The energy content of fish tissues is assumed to be a constant, and values for most of the parameters were derived from published data on other species, i.e. the European perch (*Perca fluviatilis*), the bluegill sunfish (*Lepomis macrochirus*), the largemouth bass (*Micropterus salmoides*) and the brown trout (*Salmo trutta*).

The model was used to simulate growth of yellow perch and walleye in two lakes. However, it was not validated against experimental data, and it is difficult to judge how accurate the model predictions are.

The Minnow Model

Cui and Wootton (1989) developed a bioenergetics model for the European minnow, *Phoxinus phoxinus* (L.). The equations and parameters in almost all the submodels, except those for SDA, were derived, using the technique of regression analysis, from the authors' own empirical data on the bioenergetics of the minnow.

Maximum rate of food consumption (C_{max}, kJ day^{-1}) is a function of body weight (W, g) and water temperature (T, °C) (P_i denotes parameters):

$$C_{max} = P_3 W^{P_4} T^{P_5} \tag{18.21}$$

This model is unable to predict the decline in C_{max} when water temperature is above the optimum temperature, but temperatures above the optimum temperature may rarely occur in the natural habitats of the fish (Wootton, 1990).

Faecal production (F, kJ day^{-1}) is a function of food consumption (C, kJ day^{-1}) and water temperature:

$$F = P_6 C^{P_7} e^{P_8 T} \tag{18.22}$$

Nitrogenous excretion by feeding fish (U_f, kJ day^{-1}) is a function of food consumption and water temperature:

$$U_f = P_9 + P_{10}C + P_{11}T \tag{18.23}$$

Nitrogenous excretion by starving fish (U_s, kJ day^{-1}) is a function of body weight and water temperature:

$$U_s = P_{12} W^{P_{13}} \exp(P_{14}T + P_{15}T^2) \tag{18.24}$$

SDA (kJ day^{-1}) is assumed to be a constant fraction of food consumption:

$$SDA = P_{16}C \tag{18.25}$$

The value for P_{16} is assumed to be 0.15, the average SDA coefficient published for other fish species.

Standard metabolism (R_s, kJ day^{-1}) is a function of body weight and water temperature:

$$R_s = P_{17} W^{P_{18}} e^{P_{19}T} \tag{18.26}$$

Activity metabolism (R_a, kJ day^{-1}) is a function of food consumption, body weight and water temperature:

$$R_a = P_{20} + P_{21}C + P_{22}W + P_{23}T^2 + P_{24}WT \tag{18.27}$$

Both dry matter content (D, %) and energy content (E, kJ g^{-1} dry matter) of fish tissues are functions of ration level (L_R, % body weight day^{-1}), body weight and water temperature:

$$D = \exp(P_{25} + P_{26}\ln(L_R + 1) + P_{27}\ln(L_R + 1)\ln T + P_{28}\ln W \ln T) \tag{18.28}$$

$$E = \exp(P_{29} + P_{30}\ln W + P_{31}\ln T + P_{32}\ln(L_R + 1)\ln W + P_{33}\ln W \ln T + P_{34}\ln(L_R + 1)\ln W \ln T) \tag{18.29}$$

Sensitivity analysis showed that model predictions are sensitive to errors in the parameters for the dry matter and energy contents of fish tissues and metabolic rate. The model predicted a linear growth–ration relationship (Fig. 18.6), which is different from the curvilinear relationship observed for this species (Cui and Wootton, 1988). Validation of the model against independent laboratory data at three ration levels (starvation, 3% and maximum) showed that the model overestimated growth at higher rations (Fig. 18.7).

To identify the submodels that are responsible for the failure of the model, observed values of various components of the energy budget in the validation experiment were substituted into the model, and the degree of improvement in the model predictions was used as an indication of the relative importance of error in the submodel in causing the failure of the bioenergetics model. The

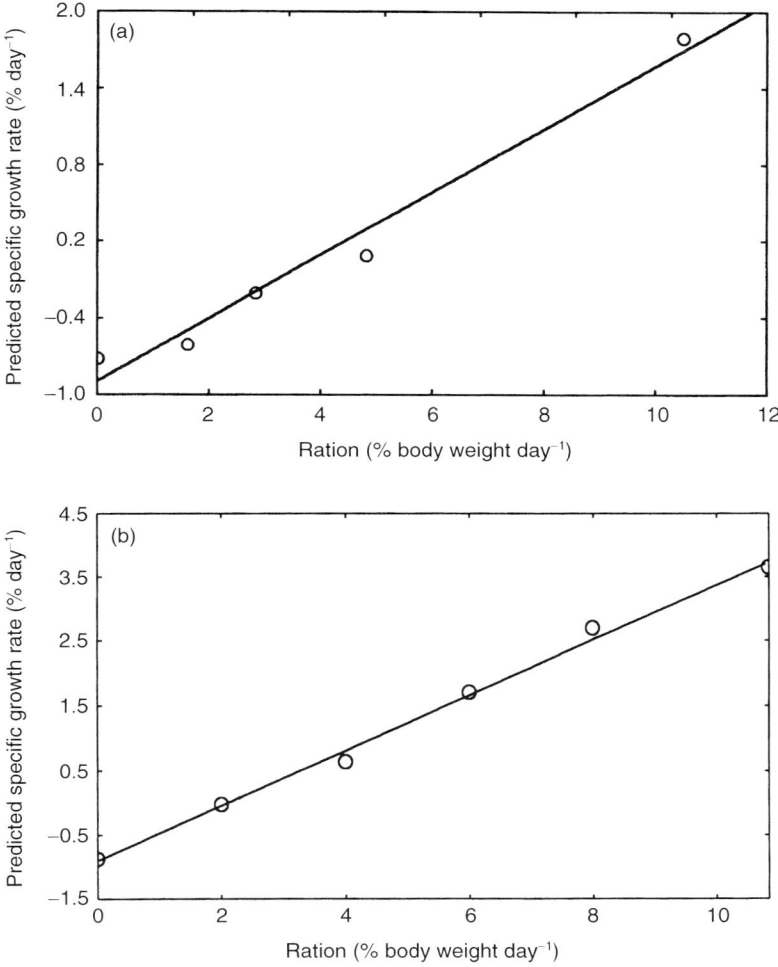

Fig. 18.6. Predicted growth–ration relationship using bioenergetics models for two fish species. Points represent predicted values and lines are fitted by regression to show the tendencies. (a) 1–5 g European minnow at 15°C (Cui and Wootton, 1989); (b) 12 g grass carp at 30°C (Y. Cui, unpublished data).

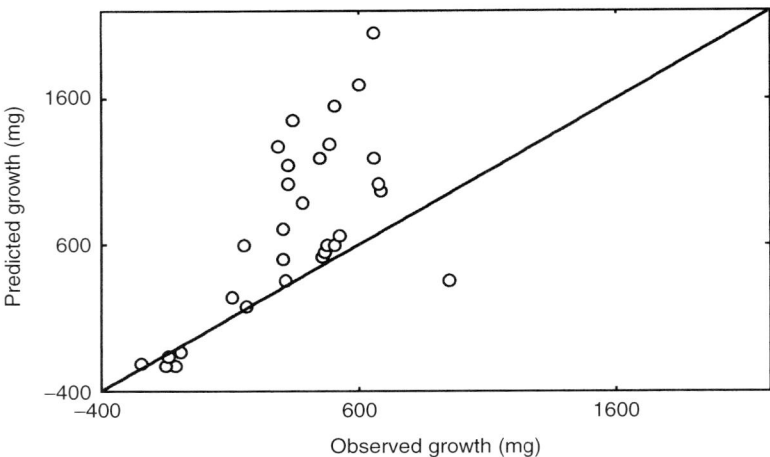

Fig. 18.7. Relationship between predicted and observed wet weight growth over 21 days in an experiment on the European minnow fed at three ration levels. Predictions were made using a bioenergetics model. The straight line passes through the origin and has a slope of 1. Points above this line represent overestimates and those below this line represent underestimates (Cui and Wootton, 1989).

results suggested that inaccurate parameters or structures for the submodels for the energy content of fish tissues (Equations 18.8 and 18.9) and metabolism (Equations 18.5, 18.6 and 18.7) are the major causes for the failure of the model.

The Grass Carp Model

The grass carp (*Ctenopharyngodon idella*) is a herbivorous fish feeding mainly on aquatic macrophytes. Based on the experiments on the energy budget of the grass carp conducted in our laboratory (Cui *et al.*, 1993, 1994, 1995, 1996a), we developed a bioenergetics model for this species (Y. Cui, unpublished data).

Maximum ration level (L_{Rmax}, % body weight day^{-1}) is a power function of water temperature (T, °C) (Cui *et al.*, 1995):

$$L_{Rmax} = 0.04T^{2.32}. \tag{18.30}$$

The effect of body size on the maximum ration level was not considered because Cui *et al.* (1996a) did not find significant effects of size on maximum ration level in grass carp over the range 12.8–95.2 g.

Faecal production (F, kJ day^{-1}) is a fraction of food energy (C, kJ day^{-1}):

$$F = fC. \tag{18.31}$$

The parameter *f* is dependent on the diet. A default value of 0.5 is assumed for macrophytes if no data on digestibility is available.

Nitrogenous excretion by starving fish (U_s, kJ day^{-1}) is a function of body weight (W, g) and water temperature (T, °C) (Cui *et al.*, 1993):

$$U_s = 0.00109W^{0.71}e^{0.077T}. \tag{18.32}$$

Nitrogenous excretion by feeding fish (U_f, kJ day^{-1}) is a fraction of food energy (Cui *et al.*, 1994):

$$U_f = 0.049C. \tag{18.33}$$

Metabolism is divided into fasting and feeding metabolism. Fasting metabolism (R_{fa}, kJ day^{-1}) is a function of body weight and water temperature (Cui *et al.*, 1993):

$$R_{fa} = 0.00926W^{0.753}e^{0.077T}. \tag{18.34}$$

Cui *et al.* (1994) showed that the proportion of metabolizable energy (C − F − U) spent in feeding metabolism is independent of ration level, and the average value is 67.2%. Thus feeding metabolism (R_{fe}, kJ day^{-1}) is modelled as:

$$R_{fe} = 0.672(C - F - U). \tag{18.35}$$

The energy content of fish tissues (E, kJ g^{-1} wet weight) is a function of ration level (L_R, % body weight day^{-1}) (Cui *et al.*, 1994):

$$E = 2.767 + 0.0111L_R. \tag{18.36}$$

The model predicted a linear relationship between SGR and ration size (Fig. 18.6b). This is in agreement with the observed growth–ration relationship for grass carp (Cui *et al.*, 1994).

Data from three experiments were used for the validation of the model, called experiments V1, V2 and V3, respectively. V1 was a 21-day experiment conducted at 30°C in which energy budgets were determined for grass carp (initial weight: 12 g) fed duckweed (*Spirodela polyrhiza*) at five ration levels ranging from starvation to maximum (Cui *et al.*, 1994). Three groups of fish were tested at each ration level. V2 was a 21-day experiment conducted at 27.3°C in which energy budgets were determined for grass carp (initial weight: 3.8 g) fed maximum rations of duckweed (a mixture of *Lemna minor* and *S. polyrhiza*) (Cui *et al.*, 1992). Fourteen fish were tested individually. V3 was a 15-day experiment conducted at 30°C in which energy budgets were determined for three size groups of grass carp fed maximum rations of lettuce leaves (*Lactuca sativa* var. *asparagina*) (Cui *et al.*, 1996a). Five tanks of fish were tested in each size group. The initial weights of the three size groups were 12.8, 37.2 and 95.2 g, respectively.

Initial weight, initial body energy content, water temperature, ration level and digestibility of food were input into the model to predict final body weight. The model produced good predictions for experiments V1 and V2; for experiment V3, the model produced good predictions for the smallest size group (12.8 g) but overestimated the growth of the larger size groups (Fig. 18.8).

To identify the model components that were responsible for the discrepancies between the predicted and observed growth in the 37.2 and

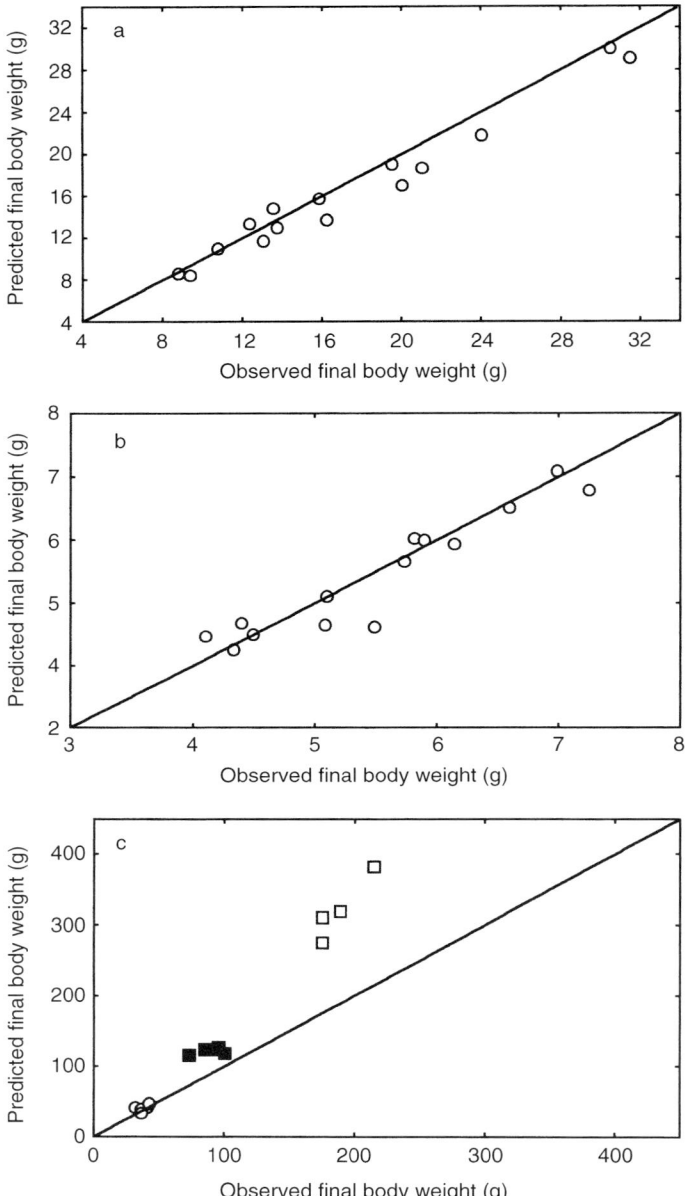

Fig. 18.8.   Relationship between the model-predicted and observed final weight for grass carp in three validation experiments. Predictions were made using a bioenergetics model. The straight line passes through the origin and has a slope of 1. Points above this line represent overestimates and those below this line represents underestimates. (a) Experiment V1. (b) Experiment V2. (c) Experiment V3, size group: ○: 12.8 g; ■: 37.2 g; □: 95.2 g (Y. Cui, unpublished data).

Table 18.2. Changes in the model predictions of mean final weight of fish in two size groups of grass carp in an experiment (validation experiment V3) when observed values were used for certain components of the bioenergetics model for the grass carp (Y. Cui, unpublished data).

	Predicted final weight (g)	
	37.2 g group	95.2 g group
Observed weight	90.2	185.8
Original model	124.1	310.9
Excretion and metabolism = observations[a]	115.7	305.6
Energy content = observations[b]	89.7	187.1

[a]Modified model in which observed values were used for excretion and metabolism.
[b]Modified model in which observed values were used for the final energy content of fish body.

95.2 g size groups in experiment V3, the observed values of the energy budget components and energy content of the fish were substituted in the model. When observed values were used for excretory loss and metabolism, there was little improvement in the model predictions. However, when observed final energy content of fish was substituted in the model, there was a great improvement in model predictions for both the 37.2 and 95.2 g size groups, and the predicted final weights were very close to the observed values (Table 18.2). The analysis suggested that failure to predict the growth of 37.2 and 95.2 g fish was caused mainly by inaccurate predictions of the final energy content of the fish. The model assumes that body energy content of fish is independent of body weight, whereas in the experiments energy content increased substantially with increases in body weight (Cui *et al.*, 1996a).

DISCUSSION

Although the bioenergetics models have been widely used in fish ecology and fishery management (Brandt and Hartman, 1993; Hansen *et al.*, 1993), it is surprising that few models have been validated rigorously. Ney (1993) listed published validations of bioenergetics models using field data. Four of the six validations showed poor fits between model predictions and field estimates. The validity of using field data for validation is itself questionable. Field estimates of food consumption, usually based on circadian patterns of gut contents and gastric evacuation (Elliott and Persson, 1978), are subject to uncertain errors. Validations should be made using accurate laboratory data. If the model fails to predict fish growth or consumption under laboratory conditions, one should not expect it to work under field or aquaculture conditions. The minnow and grass carp models, described in this chapter, derived most of the parameters from experiments on the species modelled, but still failed to provide accurate predictions of laboratory growth.

The realism of bioenergetics models is also questionable. Because one of the major purposes of bioenergetics modelling is to predict growth from food consumption or to predict food consumption from growth, it is important that the model should be able to predict a realistic growth–ration relationship. Most authors of bioenergetics models did not provide such a prediction. The growth–ration relationship in the European minnow is a decelerating curve (Cui and Wootton, 1988), but the minnow model predicted a linear relationship (Cui and Wootton, 1989). In most bioenergetics models for fish, components related to ration level include faecal production, nitrogenous excretion and SDA. These components are usually modelled as constant fractions of food consumption. Therefore, it is expected that most of these models would predict a linear growth–ration relationship.

As suggested by the validation of the bioenergetics models for the European minnow and grass carp, the failure of the models to predict the growth accurately are caused mainly by inaccurate predictions of metabolism and energy content of fish tissues. Modelling SDA as a constant fraction of food energy may not be justified. The SDA coefficient will change with the

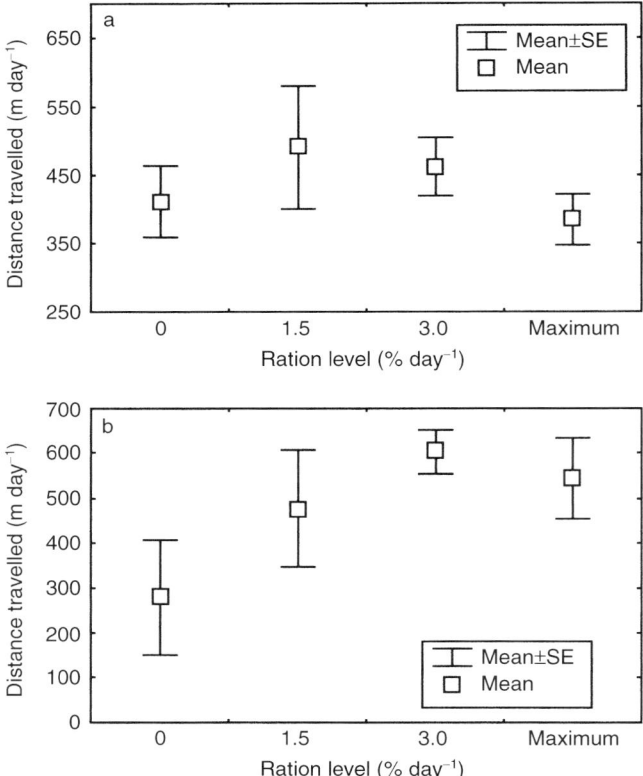

Fig. 18.9. The effect of ration level on the activity level (total distance travelled per day) for two fish species during long-term growth. (a) Nile tilapia; (b) gibel carp (X. Qian and Y. Cui, unpublished data).

composition of food, especially the protein content (Cho *et al.*, 1982; Ross *et al.*, 1992). The SDA coefficient has been reported to change with ration level, water temperature and body weight (see review by Beamish and Trippel (1990)). In most bioenergetics models, activity metabolism is arbitrarily assumed to be a multiple of standard metabolism (Hewett and Johnson, 1992). Very little is known about the factors that determine activity metabolism. The major difficulty with estimating activity metabolism lies in the accurate quantification of the activity level for fish. In our laboratory, we continuously measured, using a computer-controlled video-monitoring system, the activity level of Nile tilapia (*Oreochromis niloticus*) and gibel carp (*Carassius auratus gibelio*) fed various rations in a 24-day growth trial (X. Qian and Y. Cui, unpublished data). In both species, the activity level was highest at an intermediate ration level, but the differences among ration groups were not statistically significant (Fig. 18.9). In most studies of metabolism in fish, the metabolic rate was measured by respirometry over a short (typically 24 h) period. It is not clear whether the short-term measurements are representative of the level during long-term growth. Long-term respirometry, in which metabolic rate is monitored continuously during a growth trial, should provide more realistic measurements of metabolism (Brafield, 1985; Ross *et al.*, 1992).

Many factors are known to affect the energy content of fish tissues, including food composition (Hung *et al.*, 1987), ration level (Cui *et al.*, 1994), water temperature (Elliott, 1976) and fish size (Cui *et al.*, 1996a). In most bioenergetics models for fish, the energy content of fish tissues was assumed to be either a constant or a function of body weight (Hewett and Johnson, 1992). No models have been developed to provide accurate predictions of energy content of fish tissues from a combination of predictor variables.

In conclusion, current bioenergetics models developed by fish ecologists are not accurate enough for formulating optimum ration levels and energy requirements of fish under aquaculture conditions. More knowledge is needed on the quantitative effects of predictor factors on the SDA, activity metabolism and energy content of fish tissues. The effects of important factors that affect fish growth and metabolism under aquaculture conditions, such as food composition, dissolved oxygen, ammonia, pH and culture density, need more investigations. For the moment, the empirical growth models may be more practical, but the application is limited to the formulations of feeding and growth tables.

Some initial attempts have been made in cultured fishes to develop nutrient-based growth models that are used for domestic mammals and birds (Machiels and Henken, 1986; Conceicao, 1997). These models may provide a promising alternative to current bioenergetics modelling. The framework of the models is similar to that used in other bioenergetics models, but the models take into account the metabolism of dietary protein and lipid, and are supposed to be able to predict the effect of food composition on growth. Simplifying assumptions were made on the total metabolic rate. The models predict fish body composition based on the proportions of dietary amino acids, fatty acids and carbohydrates that are used for heat production and tissue synthesis, but these proportions are calculated from untested assumptions. A more in depth understanding of nutritional biochemistry and nutrient metabolism in fish is needed to develop models along this line.

ACKNOWLEDGEMENTS

We would like to thank Professor Silas Hung for commenting on an early draft of the manuscript.

REFERENCES

Allen, J.R.M. and Wootton, R.J. (1982) The effect of ration and temperature on the growth of the three-spined stickleback, _Gasterosteus aculeatus_ L. _Journal of Fish Biology_ 20, 409–422.

Austreng, E., Storebakken, T. and Asgard, T. (1987) Growth rate estimates for cultured Atlantic salmon and rainbow trout. _Aquaculture_ 60, 157–160.

Baldwin, R.L. and Sainz, R.D. (1995) Energy partitioning and modeling in animal nutrition. _Annual Review of Nutrition_ 15, 191–211.

Beamish, F.W.H. and Trippel, E.A. (1990) Heat increment: a static or dynamic dimension in bioenergetics models? _Transactions of the American Fisheries Society_ 119, 649–661.

Beamish, F.W.H., Howlett, J.C. and Medland, T.E. (1989) Impact of diet on metabolism and swimming performance in juvenile lake trout, _Salvelinus namaycush_. _Canadian Journal of Fisheries and Aquatic Sciences_ 46, 384–388.

Bevelhimer, M.S., Stein, R.A. and Carline, R.F. (1985) Assessing significance of physiological differences among three esocids with a bioenergetics model. _Canadian Journal of Fisheries and Aquatic Sciences_ 42, 57–69.

Boehlert, G.W. and Yoklavich, M.M. (1983) Effects of temperature, ration, and fish size on growth of juvenile black rockfish. _Environmental Biology of Fishes_ 8, 17–28.

Brafield, A.E. (1985) Laboratory studies of energy budgets. In: Tytler, P. and Calow, P. (eds), _Fish Energetics, New Perspectives_. Croom Helm, London, pp. 257–281.

Brandt S.B. and Hartman K.J. (1993) Innovative approaches with bioenergetics models: future applications to fish ecology and management. _Transactions of the American Fisheries Society_ 122, 731–735.

Brett, J.R. and Groves, T.D.D. (1979) Physiological energetics. In: Hoar, W.S., Randall D.J. and Brett, J.R. (eds), _Fish Physiology_ Vol. 8. Academic Press, London, pp. 279–352.

Cho, C.Y. (1992) Feeding systems for rainbow trout and other salmonids with reference to current estimates of energy and protein requirements. _Aquaculture_ 100, 107–123.

Cho, C.Y., Slinger, S.J. and Bayley, H.S. (1982) Bioenergetics of salmonid fishes: energy intake, expenditure and productivity. _Comparative Biochemistry and Physiology_ 73B, 25–41.

Conceicao, L.E.C. (1997) Growth in early life stages of fishes: an explanatory model. PhD thesis, Wageningen Agricultural University, Wageningen, The Netherlands.

Cortes, E. and Gruber, S.H. (1994) Effect of ration size on growth and gross conversion efficiency of young lemon sharks, _Negaprion brevirostris_. _Journal of Fish Biology_ 44, 331–341.

Cui, Y. and Hung, S.S.O. (1995) A proto-type feeding-growth table for white sturgeon. *Journal of Applied Aquaculture* 5, 25–34.

Cui, Y. and Wootton, R.J. (1988) Bioenergetics of growth of a cyprinid, *Phoxinus phoxinus* (L.): the effect of ration and temperature on growth rate and efficiency. *Journal of Fish Biology* 33, 763–773.

Cui, Y. and Wootton, R.J. (1989) Bioenergetics of growth of a cyprinid, *Phoxinus phoxinus* (L.): development and testing of a growth model. *Journal of Fish Biology* 34, 47–64.

Cui, Y., Liu, X., Wang, S. and Chen, S. (1992) Growth and energy budget of young grass carp, *Ctenopharyngodon idella* Val., fed plant and animal diets. *Journal of Fish Biology* 42, 231–238.

Cui, Y., Wang, S. and Chen, S. (1993) Rates of metabolism and nitrogen excretion in starving grass carp in relation to body weight. *Acta Hydrobiologica Sinica* 17, 375–376.

Cui, Y., Chen, S. and Wang, S. (1994) Effect of ration size on the growth and energy budget of the grass carp, *Ctenophayrngodon idella* Val. *Aquaculture* 123, 95–107.

Cui, Y., Chen, S. and Wang, S. (1995) Effect of temperature on the energy budget of the grass carp, *Ctenopharyngodon idella* Val. *Oceanologia et Limnologia Sinica* 26, 169–174.

Cui, Y., Chen, S. and Wang, S. (1996a) Effect of body size on the growth and energy budget of young grass carp (*Ctenopharyngodon idella* Val.). *Acta Hydrobiologica Sinica* 20 (Suppl.), 172–177.

Cui, Y., Hung, S.S.O. and Zhu, X. (1996b) Effect of ration and body size on the energy budget of white sturgeon. *Journal of Fish Biology* 49, 863–876.

Dabrowski, K.R. (1986) Active metabolism in larval and juvenile fish: ontogenetic changes, effect of water temperature and fasting. *Fish Physiology and Biochemistry* 1, 125–144.

Demens, E., McKinley, R.S., Weatherley, A.H. and McQueen, D.J. (1996) Activity patterns of largemouth and smallmouth bass determined with electromyogram biotelemetry. *Transactions of the American Fisheries Society* 125, 434–439.

Elliott, J.M. (1976) Body composition of brown trout (*Salmo trutta* L.) in relation to temperature and ration size. *Journal of Animal Ecology* 45, 273–289.

Elliott, J.M. and Persson, L. (1978) The estimation of daily rates of food consumption for fish. *Journal of Animal Ecology* 47, 977–991.

Hansen M.J., Boisclair, D., Brandt, S.B., Hewett, S.W., Kitchell, J.F., Lucaus, M.C. and Ney, J.J. (1993) Applications of bioenergetics models to fish ecology and management: where do we go from here? *Transactions of the American Fisheries Society* 122, 1019–1030.

Hewett, S.W. and Johnson, B.L. (1992) *Fish Bioenergetics Model 2*. University of Wisconsin, Sea Grant Institute, WIS-SG-92-250, Madison.

Hung S.S.O., Moore, B.J., Bordner, C.E. and Conte F.S. (1987) Growth of juvenile white sturgeon (*Acipenser transmontanus*) fed different purified diets. *Journal of Nutrition* 117, 328–334.

Jobling, M. (1980) Effects of starvation on proximate chemical composition and energy utilisation of plaice, *Pleuronectes platessa* L. *Journal of Fish Biology* 17, 325–334.

Jobling, M. (1981) The influence of feeding on the metabolic rate of fish: a short review. *Journal of Fish Biology* 18, 385–400.

Jobling, M. (1983a) Growth studies with fish – overcoming the problem of size variation. *Journal of Fish Biology* 22, 153–157.

Jobling, M. (1983b) Towards an explanation of specific dynamic action (SDA). *Journal of Fish Biology* 23, 549–555.

Jobling, M. (1994) Bioenergetics: feed intake and energy partitioning. In: Rankin, J.C. and Jensen, F.B. (eds), *Fish Ecophysiology*. Chapman & Hall, London, pp. 1–44.

Jobling, M. (1995) Feeding of charr in relation to aquaculture. *Nordic Journal of Freshwater Researches* 71, 102–112.

Kerr, S.R. (1971) A simulation model of lake trout growth. *Journal of Fisheries Research Board of Canada* 28, 815–819.

Kitchell, J.F. and Breck, J.E. (1980) Bioenergetics model and foraging hypothesis for sea lamprey (*Petromyzon marinus*). *Canadian Journal of Fisheries and Aquatic Sciences* 37, 2159–2168.

Kitchell, J.F. Koonce, J.F., O'Neill, R.V, Shugart, H.H., Magnuson, J.J. and Booth, R.S. (1974) Model of fish biomass dynamics. *Transactions of the American Fisheries Society* 103, 786–798.

Kitchell, J.F., Stewart, D.J. and Weininger, D. (1977) Applications of a bioenergetics model to yellow perch (*Perca flavescens*) and walleye (*Stizostedion vitreum vitreum*). *Journal of Fisheries Research Board of Canada* 34, 1922–1935.

Machiels, M.A.M. and Henken, A.M. (1986) A dynamic simulation model for growth of the African catfish, *Clarias gariepinus* (Burchell 1822). I. Effect of feeding level on growth and energy metabolism. *Aquaculture* 56, 29–52.

Ney, J.J. (1993) Bioenergetics modeling today: growing pain on the cutting edge. *Transactions of the American Fisheries Society* 122, 736–748.

Pandian, T.J. and Vivekanandan, E. (1985) Energetics of feeding and digestion. In: Tytler, P. and Calow, P. (eds), *Fish Energetics, New Perspectives*. Croom Helm, London, pp. 99–124.

Ricker W.E. (1979) Growth rates and models. In: Hoar, W.S., Randall D.J. and Brett, J.R. (eds), *Fish Physiology* Vol. 8. Academic Press, London, pp. 678–743.

Ross, L.G., McKinney, R.W., Cardwell, S.K., Fullarton, J.G., Roberts, S.E.J. and Ross, B. (1992) The effects of dietary protein content, lipid content and ration size on oxygen consumption and specific dynamic action in *Oreochromis niloticus* L. *Comparative Biochemistry and Physiology* 103A, 537–578.

Soofiani, N.M. and Priede, I.G. (1985) Aerobic metabolism scope and swimming performance in juvenile cod, *Gadus morhua* L. *Journal of Fish Biology* 26, 127–138.

Thorarensen, H., Gallaugher, E. and Farrela, A.P. (1996) The limitations of heart rate as a predictor of metabolic rate in fish. *Journal of Fish Biology* 49, 226–236.

Ursin, E. (1967) A mathematical model of some aspects of fish growth, respiration and mortality. *Journal of Fisheries Research Board of Canada* 24, 2355–2453.

Von Bertalanffy, L. (1957) Quantitative law in metabolism and growth. *Quarterly Review of Biology* 32, 217–231.

19 The Nutrition of Companion Animals

A.C. Longland[1], M.K. Theodorou[1] and I.H. Burger[2]

[1]*Institute of Grassland and Environmental Research, Plas Gogerddan, Aberystwyth, Ceredigion, UK; [2]Waltham Centre for Pet Nutrition, Melton Mowbray, Leicestershire, UK*

INTRODUCTION

Companion animals are those species which have a special association with man and are dependent on him to some degree (Greenhalgh and Corbin, 1998), the most popular in the UK being dogs, cats, fish, birds, horses, certain rodents and rabbits. Frequently the relationship between man and the companion animal is sentimental, although for dogs and horses the relationship may be more practical, their being used for racing, hunting, herding, aiding the disabled or for military and police work. The aim of feeding companion animals, irrespective of their role, is to promote a long and healthy life and, to achieve this goal, many are fed at or near their maintenance levels throughout adulthood. Feeding to maintain rather than increase body weight and to accommodate changes in metabolism during ageing are pertinent considerations for the companion animal, but are largely irrelevant in farmed species. Therefore, conceptually, the nutrition of companion animals is more akin to that of humans than that of the other species described in this book, which are seldom allowed to achieve their natural life span but are fed to elicit the most effective production of meat, milk, eggs or wool.

The 1996 statistics for companion animals in the UK suggest that approximately one-quarter and one-fifth of UK households owned dogs and cats, respectively, representing a total of 7.5 million of both species, and the number is rising (PFMA, 1997). The size of the companion animal food market is thus considerable and is estimated to be worth \$9.2 billion in Europe alone (FEDIAF, 1996).

With the exception of horses, which are discussed in Chapter 11, this chapter will review the nutrition of dogs, cats, birds, fish, rabbits and rats through their different life stages of growth, maintenance, gestation and lactation. For dogs, the nutrition of working and senior animals is also discussed. Clinical nutrition, to combat or aid recovery from specific conditions, will only be touched upon where it is normally associated with the ageing process, but is otherwise outside the scope of this text.

NUTRIENT REQUIREMENTS

The term nutrient requirement refers to the quantity of an essential nutrient needed by the animal for a given physiological state (Morris and Rogers, 1994). Although single values for a particular life stage may be presented, these can only be guidelines, as the nutrient requirements of any given individual within a population will vary in relation to: (i) its physiological state; (ii) the metabolic function that is to be sustained or optimized; and (iii) the genetic potential of the animal. Values quoted by the National Research Council (NRC) usually reflect minimum requirements and are often determined with highly purified ingredients; they give no margin of safety for varying bio-availability of nutrients in relatively unprocessed diets and take no account of variation of ingredients, interactions between them or indeed interlaboratory variation in analysis of samples. Therefore, the recommended mimima presented here should be regarded with caution, particularly when formulating diets from raw materials. Furthermore, some nutrients are toxic above a certain level and, in practical terms, the amounts of such nutrients in a diet must lie between deficiency on the one hand and toxicity on the other. Wherever possible, the nutrient requirements for companion animals presented here are based on the results of experimental trials on the target species. Where such information is not available, recommendations are based on diets that have been shown to support the target species throughout the different life stages.

ENERGY

Energy is required for growth, maintenance, reproduction and lactation. Energy is provided by carbohydrate, fat and protein, which typically have gross energy (GE) values of 17.4, 39.3 and 23.3 kJ g^{-1}, respectively. The difference between GE intake and that in the faeces is the digestible energy (DE) of a diet, metabolizable energy (ME) being the DE less the energy in urine and combustible gases. Due to the complexities involved in measuring the combustible gases, reported ME values usually reflect DE minus urinary energy. As ME values are reasonable estimates of the energy available to the animal, ME values are quoted for preference in this chapter.

Fat

In addition to being a concentrated energy source, fat provides essential fatty acids (EFAs) and is a carrier for the fat-soluble vitamins. The EFAs are the (*n*-6) series, consisting of linoleic, γ-linolenic and arachidonic acids. Requirements for the (*n*-6) EFAs are usually expressed in terms of linoleic acid as this can usually be converted into linolenic and arachidonic acids via desaturation or elongation, respectively (and because linoleate and arachidonate are normally minor constituents of most fats). The α-linolenate (*n*-3) family are also EFAs for some species (Holman *et al.*, 1982; McCartney, 1996). Signs of EFA deficiency include scurfy coats, hair loss, anaemia, fatty liver and reduced fertility (Wills, 1996a). In addition to the nutritional characteristics of fat, the organoleptic properties of diets can be enhanced by its inclusion (NRC, 1986).

Carbohydrates

All animals require glucose to supply energy to the tissues and as a substrate for the production of glycoproteins, etc. Glucose is provided either directly, through degradation of carbohydrates or via the metabolism of gluconeogenic substrates such as glycerol or amino acids. Carbohydrates not digested by the animals' enzymes are potentially available for fermentation by the gut microflora, the resulting volatile fatty acids (VFAs) being used as an energy source.

Protein and Amino Acids

Plants and microorganisms can synthesize proteins from nitrogenous compounds, but animals cannot synthesize the amino group and thus require a supply of dietary amino acids. Some amino acids can, via transamination, be elaborated from other amino acids, but others, the indispensable amino acids (IAAs), cannot be produced in this manner in sufficient quantities by the tissues to support normal function. For most species, there are ten IAAs; arginine, histidine, isoleucine, leucine, lysine, methionine, phenylalanine, threonine, tryptophan and valine. Arginine has been shown to be involved in the release of several metabolic hormones, including insulin, glucagon, growth hormone and prolactin, and has also been implicated in the immune response (Barbul *et al.*, 1981). The branched chain amino acids, leucine, isoleucine and valine, are required for protein synthesis, protein turnover and energy metabolism (Adibi, 1980). Methionine and cysteine are the sulphur amino acids and are limiting amino acids in many protein sources (Allison *et al.*, 1974). Various factors can affect the sulphur amino acid requirement, for example in diets deficient in methyl donors such as choline the requirement for methionine is higher. Phenylalanine and tyrosine are the aromatic amino acids and may substitute for one another to some extent.

The IAA profiles required at each of the life stages will differ because the utilization of amino acids for protein turnover is different from that for net tissue accretion. Consequently, the amount of protein required depends on the source: smaller amounts of highly digestible protein with amino acid profiles which closely complement the animals' requirement need to be fed than those which are less digestible and match the animals' demands less precisely.

Minerals

The ash fraction of a diet consists of minerals; the macrominerals that are required at relatively high concentrations and the microminerals of which only trace amounts are needed. Minerals are essential for maintaining, at various levels, acid–base balances, tissue structure and osmotic pressure, in addition to being key components in many enzyme systems. There are many inter-relationships between minerals, and feeding excessive levels of some minerals may prove injurious.

Macrominerals

Calcium and phosphorus have a close metabolic association, and the ratio of these is of importance for many species. Their utilization is dependent upon an adequate supply of vitamin D. Calcium deficiency is usually associated with osteoporosis, detachment of the teeth, bone resorption, rickets in young animals, tetany and reproductive incompetence. Calcium is also involved in blood clotting. Phosphorus deficiency can result in rickets, pica, poor growth and osteomalacia. A high phosphorus:calcium ratio can lead to symptoms of calcium deficiency. Potassium is required for neural transmission, fluid balance, muscle metabolism and immune function, and its deficiency results in poor growth, muscular paralysis and lesions to the kidney and heart. Sodium and chloride are the major electrolytes of extracellular fluids, deficiencies leading to fatigue, poor water balance, retarded growth and hair loss. Proper muscle and nervous tissue function depend on a correct magnesium and calcium balance, and magnesium ions are essential for enzymes involved with energy metabolism. Magnesium deficiency can lead to anorexia, poor growth, ataxia and convulsions, and is enhanced by elevated levels of dietary calcium and phosphorus (Bunce *et al.*, 1962a,b). Iron is an essential component of haemoglobin, myglobin and haem enzymes, deficiency being characterized by anaemia. Iron is more available from animal rather than plant sources, and soya protein is thought to reduce absorption of iron, zinc and manganese (Burger, 1995). Iron also interacts with calcium metabolism. Copper is a constituent of many enzyme systems and is also closely linked to iron metabolism; copper deficiencies can impair haemoglobin synthesis. Excess levels of both iron and copper can be toxic. Zinc is required for the activity of enzymes involved in the metabolism of nucleic acids, carbohydrates, proteins and fats. Zinc is also needed for maintaining healthy keratinized tissues and germinal epithelia, and deficiencies are characterized by poor growth, anorexia and testicular atrophy (Burger, 1995). Iodine is needed for thyroid function, deficiencies leading to hypothyroidism and goitre, compromised utilization of calcium and phosphorus and reproductive failure. Manganese is important in bone formation and reproduction. There are strong interactions between selenium and vitamin E (NRC, 1978a), selenium being a component of glutathione peroxidase which protects cells against lipid peroxides. Selenium deficiencies result in muscular weakness, anorexia and, ultimately, coma and death.

Microminerals

Chromium is needed for insulin function; fluoride for tooth and bone development and possibly reproductive function; nickel for membrane function; molybdenum is a cofactor for many enzymes; silica is required for skeletal development and connective tissue formation; vanadium for reproduction, growth and fat metabolism.

Vitamins

Vitamins are organic compounds essential to metabolic regulation. They may be fat soluble (A, D, E and K) or water soluble (the B vitamins and vitamin C). The former are capable of being stored in the body, whereas the water-soluble vitamins are not stored and must be ingested regularly. Vitamin A is necessary for visual pigments, maintenance of cellular structure and for healthy skin, mucous membranes, tooth and bone development. Deficiencies cause night blindness, infertility, seborrhoeic coat conditions and xerophthalmia. Toxicity results in liver damage and ankylosis of the joints. Vitamin C is required for protein synthesis and a number of other intracellular processes but, with the exception of the guinea pig, can be synthesized by most companion animal species, precluding the need for its dietary provision. Vitamin D requirements are associated with dietary levels of calcium and phosphorus and it stimulates calcium absorption from the gut. In some species, vitamin D may be synthesized in cutaneous lipid after exposure to the sun; deficiencies result in rickets in growing animals and osteomalacia in adults. Excess vitamin D can result in hypercalcaemia, calcification of lungs, kidneys and gut, deformation of teeth and jaw, and death. Vitamin E, together with selenium, protects cellular membranes against oxidative damage: in diets rich in easily oxidized polyunsaturated fatty acids (PUFAs) the requirement for this vitamin is increased. Vitamin E deficiency can cause skeletal muscular dystrophy, reduced reproductive performance and a compromised immune system. Vitamin K may be synthesized by the gut bacteria and is involved in blood clotting; if bacterial synthesis is reduced, haemorrhaging may occur. Thiamin (B_1) is involved in carbohydrate metabolism, symptoms of deficiencies include anorexia, neurological abnormalities and heart failure. Riboflavin (B_2) is required in oxidative enzyme systems and for cellular growth, deficiencies leading to lesions of the eye and skin and testicular hypoplasia. Pyridoxine (B_6) is required for protein metabolism; deficiencies are exacerbated with high-protein diets and cause anorexia, anaemia and kidney damage, especially in cats. Cyanocobalamin (B_{12}) is involved in fat and carbohydrate metabolism and is needed for myelinization of nerves. Deficiencies lead to pernicious anaemia and neurological defects. Pantothenic acid is a constituent of coenzyme A and is involved in nutrient metabolism; deficiencies are rare but if they do occur are manifested by reduced growth, fatty liver, gastrointestinal disorders and convulsions. Biotin is synthesized by gut bacteria and is involved in lipid and amino acid metabolism. Raw egg contains avidin which will complex with biotin rendering it unavailable. Biotin deficiencies lead to poorly keratinized structures. Folate can be synthesized by the gut bacteria and is required for DNA synthesis and maturation of red blood cells in bone marrow. Symptoms of deficiency include anaemia and leucopenia. Choline is a constituent of phospholipids and is a precursor of the neurotransmitter, acetylcholine.

DOGS

Energy

Dogs can live for up to 20 years and exhibit a very large intraspecific variation in mature body size, ranging from a 1 kg Chihuahua to a 115 kg St Bernard, necessitating the need to express energy in terms of metabolic body weight (kg).

Maintenance

A number of values for the maintenance requirements for adult dogs have been reported, reflecting the wide range of breeds (and, therefore, different body shapes and coat types) kept under a variety of environmental conditions and the exercise regimes used in the measurements. Values of 423, 490 and 523 kJ of ME $W^{-0.75}$ day^{-1} were determined for Labrador–Retrievers, Beagles and Siberian Huskies, respectively, housed in indoor kennels (maintained at ~ 20°C) with outdoor runs (at −6 to 24°C) (Finke, 1991). Manner (1991), however, using eight breeds of dog, suggested values of 387 and 450 kJ of ME $W^{-0.75}$ day^{-1}, respectively, for resting and active dogs maintained at 20°C. These values are somewhat lower than the 678 kJW$^{-0.64}$ day^{-1} for dogs at rest and 643 kJW$^{-0.73}$ day^{-1} for active dogs (covering 10.5 km day^{-1}) obtained by Burger and Johnston (1991) using whole-body calorimetry. On the other hand, Kienzle and Rainbird (1991) observed that apart from Great Danes and Newfoundlands, which were above and below the average respectively, the maintenance energy requirement for most breeds was close to 495 kJ of ME $W^{-0.75}$ day^{-1}. The above values for the maintenance energy requirement of dogs are not comprehensive and they are lower than the 552 kJ of ME $W^{-0.75}$ day^{-1} given by the NRC (1974). For practical purposes, these values should be taken as initial guidelines for diet formulation: the amounts fed can then be adjusted according to the responses of individual animals.

Growth

Post-weaning, puppies grow rapidly and generally attain about half of their adult weight at 6 months. Thereafter, different breeds of dog will grow at different relative rates, larger breeds taking longer than smaller breeds to reach their mature weight. Puppies initially require twice as much energy as an adult on a body weight basis, but this difference reduces with increased puppy size. Thus by the time puppies are 40 and 80% of their mature weight, they need 1.6- and 1.2-fold of the adult requirement, respectively (Wills and Morris, 1996). However, research on Great Danes suggests that for this breed at least, puppies at weaning and at 5 months old (40% of mature weight) require

3.6- and 2.8-fold, respectively, of the adult maintenance energy requirement (Meyer and Zentec, 1991). High levels of energy intake can cause rapid growth at the expense of proper skeletal development, and several of the large and giant breeds are disproportionately susceptible to skeletal disorders such as hypertrophic osteodystrophy and osteochondrosis.

Gestation and lactation

The majority of fetal growth of puppies occurs in the last third of gestation and, despite early development of mammary and uterine tissues, the energy requirements of the gravid bitch do not increase markedly before this time (Wills, 1996b). During the final trimester of pregnancy, the energy requirements of the bitch should be satisfied by increasing her ration by 15% per week so that at parturition she is consuming 60% more than at pre-conception (Legrand-Defretin and Munday, 1995). At peak lactation, depending on the age and size of the litter, bitches may need up to four times their maintenance energy requirement. Thus to enable lactating bitches to ingest sufficient food to remain in energy balance, they should be offered several small, nutritionally balanced, nutrient-dense meals per day.

Lipid

The apparent digestibility of various plant and animal fats by dogs has been shown to vary from 0.8 to 0.95 (James and McCay, 1950; Orr, 1965). The current NRC (1985) recommendation for dietary fat for dogs at maintenance is 50 g kg^{-1} (to include 10 g of linoleic acid kg^{-1}); in diets containing 16.7 MJ of ME kg^{-1}, these values are very similar to the recommendations of Wills (1996b) of 3.2 g of fat MJ ME^{-1} (to include 0.66 g of linoleic acid MJ ME^{-1}) for all life stages. Puppies from 2 to 6 months have been shown to grow well on diets which supply 13–76% of the energy from fat and 20–25% of the energy from high-quality protein (Rosmos *et al.*, 1976). This demonstrates that if the IAA requirements are met, puppies grow well on diets with a wide range of fat contents. For giant breed puppies, Lepine (1998) has indicated that diets containing 140 g of fat kg^{-1} of diet are suitable.

Studies on the appropriate levels of fat for diets for breeding bitches are few, but it appears that as long as the balance of nutrients to energy are met, satisfactory reproductive performance will occur. Ontko and Phillips (1958) fed diets containing 80 and 160 g of fat kg^{-1} to breeding bitches, resulting in satisfactory reproductive performance.

Carbohydrates

Carbohydrates are a relatively cheap energy source, and cereal starches, particularly those which have been dextrinized, are normally well digested by dogs, coefficients in excess of 0.84 being reported (Rosmos *et al.*, 1981). Dogs do not appear to have a specific requirement for carbohydrate provided that there are sufficient dietary gluconeogenic amino acids and glycerol (Blaza *et al.*, 1989); however, adequate levels may be hard to attain at times of high demand such as gestation and lactation, resulting in bitches becoming hypoglycaemic and ketotic before parturition, accompanied by high mortality of pups (Rosmos *et al.*, 1981). As in rats, fetal abnormalities and resorption may also occur in bitches fed carbohydrate-free diets at conception (Taylor *et al.*, 1983). It is therefore recommended that bitches receive some available carbohydrate in their diets, particularly if protein levels are close to the nutritional requirement.

Although dietary fibre is not normally regarded as essential for dogs, some inclusion in the diet may be beneficial, though excessive amounts may lead to a reduction in availability of other nutrients (NRC, 1985). However, there is recent evidence that feeding fermentable dietary fibre aids in maintaining glucose homeostasis in diabetic dogs (McBurney *et al.*, 1998) and fructo-oligosaccharides (FOSs) may have beneficial effects in managing a number of conditions, including corneal lipidosis (Diez *et al.*, 1998).

Protein and Amino Acids

Maintenance levels of protein for adult dogs have been given as 9.6 g of protein MJ ME^{-1} (Wills, 1996a), corresponding to 160 g of protein kg^{-1} in a diet containing 16.7 MJ of ME kg^{-1}. These values are somewhat higher than the range of 65–120 g of casein kg^{-1} quoted by Kade *et al.* (1948), thus accommodating the use of protein sources of lower biological value than casein.

Puppies require protein for tissue synthesis during growth as well as for maintenance. Protein quality is important, depending upon both the IAA profile of the protein source and their availability. Work by Nierinck *et al.* (1991) illustrates this point for although the amino acid profiles of three protein sources of animal origin (lung, tripe and mince) and soya meal are similar, crude protein digestibility coefficients averaged 0.92 for the animal proteins and 0.76 for the soya meal. The diets in this study were formulated such that the intakes and digestibility of the four protein sources by Beagles should have resulted in all of the amino acids being available to them at above the NRC (1985) recommended minima. However, protein deficiencies may develop if vegetarian diets are formulated to lower protein levels without knowledge of protein availability. Indeed, in a field study of diets fed to vegetarian dogs, Kienzle and Engelhard (1998) found that protein supply to adults was marginal and some puppies showed signs of protein deficiency. Minimum protein levels of 110–220 g of protein kg^{-1} for growing dog diets have been quoted (Johnson, 1993), and the NRC (1985) concluded that 120 g of protein kg^{-1} diet from a high-quality source such as lactalbumin should meet growth requirements, although 150 g of protein kg^{-1} may be marginal for giant breeds. Wills (1996b) regarded 13.2 g of protein MJ ME^{-1} (equivalent to

220 g of protein kg^{-1} in diets containing 16.7 MJ ME kg^{-1}) as being suitable for growth of most breeds, although Lepine (1998) suggests that 260 g of protein kg^{-1} is more appropriate for giant breeds. As for other species, excess protein in the diet is either metabolized to produce energy or is stored as fat. However, when high-fat diets are fed, the protein requirement should be increased to maintain a balanced protein to energy ratio: a diet containing 200 g of fat kg^{-1} required protein levels of 250 g kg^{-1} (Ontko *et al.*, 1957).

There is a paucity of information on the protein requirements of breeding bitches, but lactating bitches fed a carbohydrate-free diet required 300 g of protein kg^{-1} compared with 160 g kg^{-1} in diets containing carbohydrate (Kienzle and Meyer, 1989). Wills (1996b) suggested that 13.2 g of protein MJ ME^{-1} (equivalent to 220 g of protein kg^{-1} in a 16.7 MJ ME kg^{-1} diet) was sufficient for reproduction. Protein sources fed to lactating bitches should be of high biological value for maximum utilization, and minimum bulk.

Milner (1979a,b) established that the ten IAAs outlined above were required for optimum growth and nitrogen balance in growing Beagles. Current NRC recommendations for IAAs in diets for dogs are summarized in Table 19.1.

Minerals

As there are limited data on the mineral requirements of dogs, the interactions between dietary minerals, their bio-availability and the differing requirements between breeds makes recommendations at best a compromise. The mineral requirements of dogs presented in Table 19.1 are based on the NRC (1985) recommendations and those of Burger (1995). The values for calcium and phosphorus, however, may not meet the needs of some of the large or giant breeds, and Lepine and Reinhart (1998) recommend that the diet of large or giant breed dogs should contain 8 g of calcium, 6.7 g of phosphorus, 260 g of high-quality protein and 140 g of fat kg^{-1} of diet to allow a moderate growth rate and enable a managed increase in body mass commensurate with healthy skeletal development.

Vitamins

The vitamin requirements of dogs at all life stages are given in Table 19.1. Young, growing dogs require twice as much vitamin E as adults, and Harris and Embree (1963) have suggested a vitamin E:PUFA ratio (g kg^{-1}) of 0.6:1 as a minimum to protect against PUFA peroxidation. Dogs cannot synthesize vitamin D to any significant extent, and it must therefore be supplied in the diet, although specific supplementation is probably unnecessary as most diets inherently contain sufficient vitamin D to meet the needs of dogs.

Table 19.1. Indispensable amino acid, mineral and vitamin requirements of dogs.

	Growth[a]		Maintenance[b]		Reproduction[b]	
	kg^{-1} diet	MJ ME^{-1}	kg^{-1} diet	MJ ME^{-1}	kg^{-1} diet	MJ ME^{-1}
Amino acids						
Arginine (g)	5.04	0.33	1.08[a]	0.07	NS	NS
Histidine (g)	1.84	0.12	1.09[a]	0.07	NS	NS
Isoleucine (g)	3.54	0.23	2.31[a]	0.15	NS	NS
Leucine (g)	5.85	0.38	4.16[a]	0.27	NS	NS
Lysine (g)	5.04	0.33	2.47[a]	0.16	NS	NS
Total sulphur amino acids (g)	3.85	0.25	1.54[a]	0.10	NS	NS
Total aromatic amino acids (g)	7.24	0.47	4.31[a]	0.28	NS	NS
Threonine (g)	4.62	0.30	2.16[a]	0.14	NS	NS
Tryptophan (g)	1.54	0.10	0.62[a]	0.04	NS	NS
Valine (g)	3.85	0.25	2.92[a]	0.19	NS	NS[b]
Minerals						
Calcium (g)	6.00	0.39[b]	6.00	0.39	10.16	0.66
Phosphorus (g)	4.4	0.29	4.62	0.30	8.16	0.53
Potassium (g)	4.4	0.29	4.62	0.30	4.62	0.30
Sodium (g)	0.6	0.04	0.62	0.04	0.77	0.05
Chloride (g)	0.9	0.06	0.85	0.055	0.924	0.06
Magnesium (mg)	400	26.0	352.7	22.9	352.70	22.9
Iron (mg)	32	2.10	36.81	2.39	73.92	4.80
Copper (mg)	2.9	0.19	4.62	0.30	6.47	0.42
Manganese (mg)	5.1	0.33	4.62	0.30	4.62	0.30
Zinc (mg)	35.6	2.30	46.05	2.99	46.00	2.99
Iodine (mg)	0.6	0.40	0.62	0.04	1.34	0.09
Selenium (mg)	0.11	0.006	0.11	0.006	0.11	0.006
Vitamins						
Vitamin A (IU)	3773[b]	245[b]	3773	245	4605	299
Vitamin D (IU)	404	26	400	26	462	30
Vitamin E (mg)	22[b]	3[b]	27.72	1.8	46.20	3
Vitamin K (mg)	NR	NR	NR	NR	NR	NR
Thiamin (mg)	1.0	0.06	0.93	0.06	0.92	0.06
Riboflavin (mg)	2.46	0.16	2.31	0.15	2.31	0.15
Pantothenic acid (mg)	10	0.65	10.16	0.66	10.16	0.66
Niacin (mg)	11.0	0.72	11.09	0.72	11.09	0.72
Pyridoxine (mg)	1.1	0.07	1.09	0.07	1.08	0.07
Folic acid (µg)	200	13	200	13	200	13
Vitamin B$_{12}$ (µg)	25.00	1.62	24.95	1.62	25.00	1.62
Choline (mg)	1247	81	1155	75	1155	75
Biotin	NR	NR	NR	NR	NR	NR

[a] Data from NRC (1985) based on diets containing 15.4 MJ ME kg^{-1}; [b] data from Burger (1995).

NS = not stated; NR = no requirement when natural ingredients are fed as sufficient should be produced by the intestinal microflora.

Senior Animals

The population of elderly animals is increasing and, in a survey of the UK dog population in 1991, 35% were over 7 years old (Davies, 1996). As a consequence, considerable research is being focused on the nutrition of elderly dogs with a view to alleviating some of the avoidable effects of ageing.

Defining the elderly dog is best undertaken in terms of breed, as small to medium sized dogs are known to live longer than the large and giant breeds. The age at which small and medium breeds are considered 'geriatric' averages at 11 years, whereas for large and giant breeds the onset of old age is some 3–4 years earlier (Goldston, 1995). The maintenance energy requirements of dogs decrease with age, such that that of Labrador–Retrievers decreased from 593 kJ of DE $W^{-0.75}$ for 2-year-olds to 464 kJ of DE $W^{-0.75}$ for those older than seven (Kienzle and Rainbird, 1991). This reduction was thought to be due to a decline in (i) fat free body mass, (ii) basal metabolic rate and (iii) the activity of elderly dogs. However, the intakes of senior dogs are often reduced and thus it is not always necessary to lower the energy density of the diet in order to prevent obesity (Kienzle and Rainbird, 1991). The digestibility of nutrients does not appear to be altered in ageing dogs (Sheffey *et al.*, 1985) although it is possible that their protein requirement increases with age (Wannemacher and McCoy, 1966). However, a balance needs to be struck between feeding elderly dogs higher levels of protein and inducing renal stress and Mundt (1991) recommended 4.5 g of highly digestible protein $W^{-0.75}$. Furthermore, only highly digestible carbohydrates should be fed to elderly dogs in order to minimize post-ileal formation of bacterial metabolites (Meyer *et al.*, 1989). Utilization of fatty acids may change with age, and that of EFAs has been reviewed by Turek and Hayek (1998). It has been shown that a reduction in desaturation of fatty acids may occur with ageing (Biagi *et al.*, 1991): supplementation of fatty acid products of desaturases (e.g. γ-linolenic acid) may help by-pass such effects, and increased supplementation of diets with *n*-3 fatty acids may also be of benefit to the elderly dog (Hayek, 1998).

Comparatively little is known about the mineral and vitamin requirements of geriatric dogs. However, there may be a requirement for higher levels of vitamin E (Mundt, 1991), and reduced serum levels of copper and zinc in ageing dogs may suggest that they have a higher requirement for these (Keen *et al.*, 1981). Elderly dogs also have a reduced ability to incorporate skeletal calcium (Weigel and Alexander, 1981), suggesting an extra dietary calcium requirement. Indeed, Mundt (1991) suggests that the levels of vitamins and minerals required to support growing dogs are appropriate for senior dogs.

Exercise

The two most extreme forms of exercise for dogs are embodied in racing Greyhounds which are required to perform intense activity in a sprint for 30 s or so per race, and sled dogs (Huskies) which may be required to perform in endurance competitions in which they may run 100 miles per day for

10–14 days (Hinchcliff *et al.*, 1998). Other working dogs such as foxhounds and herding dogs perform somewhere in between these two extremes. Although the exhaustion of glycogen reserves is the chief limiting factor for any continued effort, other factors such as providing a suitable electrolyte/water regime to reduce dehydration and to buffer acid accumulation must also be catered for when formulating diets for racing animals. Greyhounds may require 10–20% more energy above their maintenance requirements, and Grandjean (1996) has suggested that Greyhound diets should contain 400–450 g of carbohydrate, 200–250 g of fat and 350 g of protein kg^{-1} of diet, with short and or medium chain fatty acids contributing up to 25% of the total dietary lipid. Unlike human athletes, carbohydrate loading in dogs can lead to exertional rhabdomyolysis, and therefore should not be employed to enhance the performance of racing dogs (Kronfeld, 1973). However, high-fat diets combined with appropriate training appear to result in glycogen sparing, reduced lactic acid accumulation and thus decreased muscle fatigue (Kronfeld *et al.*, 1994). Hultman *et al.* (1971) and Mutch and Bannister (1983) demonstrated that overall carbohydrate oxidation accounts for 10–15% of the energy supply to muscles in Huskies, whereas free plasma fatty acids account for 70–90%. Therefore, as a general rule, as the length of a race increases and the intensity of performance decreases, the content of dietary lipid should be elevated at the expense of carbohydrate (Grandjean, 1996). Recommendations for racing huskies are 300 g of highly digestible carbohydrate and 350 g of fat kg^{-1} of diet which, as for Greyhounds, should contain 25% short and medium chain fatty acids; the EFAs should be supplied at twice the maintenance value, the ratio between *n*-6 and *n*-3 fatty acids being maintained at 5.5:1 (Grandjean, 1996). For endurance races, protein should be supplied at about 21.5 g of MJ ME^{-1}. For other working dogs, dietary fat contents should rise from 140 to 220 g kg^{-1} as work becomes more prolonged, while protein should rise from 300 to 400 g kg^{-1} of diet to provide a high protein:energy ratio (Grandjean, 1996). Calcium, phosphorus and trace element levels should be twice, and those of magnesium three times the maintenance levels in diets for prolonged endurance race dogs, with intermediate levels for racing Greyhounds or sprint sled dogs (Grandjean, 1996). Although dogs do not sweat in the same way as humans and horses, sled dogs produce large volumes of urine, resulting in substantial sodium loss, which, if not supplied in the diet, may cause hyponatraemia (Hinchcliff *et al.*, 1998). In order to ensure adequate amounts of vitamins for racing dogs, dietary levels of vitamins A, D and K can be doubled and those of vitamins B, C and E tripled (Grandjean, 1996).

The energy costs of thermoregulation in dogs are high, and thus a drop in ambient temperature from 15 to 8°C results in a 25% increase in the ME requirement (Blaza, 1982). Dogs working in hot, humid conditions may require 50–100% more energy than similar dogs in less stressful circumstances (Costill, 1981). Furthermore, there is an increased need for water in working dogs. Compared with a 20 kg house dog which loses about 1400 ml of water per day via urine, faeces and respiration, 20 kg sprint and distance sled dogs will lose about two- and fourfold this amount, respectively (Reynolds *et al.*, 1998), and thus suitable hydration is essential for dogs undergoing strenuous work for many hours. There is evidence to suggest that inducing a mild state of hyperhydration via dilute glycerol solutions may result in more enduring rehydration than more traditional glucose–electrolyte solutions (Reynolds *et al.*, 1998).

CATS

Domestic cats (*Felis domesticus*), unlike dogs, are obligate carnivores in that although they can obtain substantial amounts of nutrients from foods of plant origin, they require certain fatty acids and proteins that can only be obtained from animal sources.

Energy

The mature body weight of the domestic cat ranges from 2 to 6 kg and, because there is little variation in body size compared with dogs, many authors express energy requirements in terms of body weight as opposed to metabolic body weight. Kendall *et al.* (1983) found no benefit in terms of precision of using mass exponents of body weight (in kg) of $W^{0.75}$ or $W^{0.67}$ compared with unity.

The maintenance energy requirements of large (Earle and Smith, 1991) or inactive (Kendall *et al.*, 1983) cats range from 209 to 293 kJ of ME W^{-1} day^{-1}, whereas slightly higher values of 335–380 kJ of ME W^{-1} day^{-1} have been reported for active cats with access to runs (Miller and Allison, 1958). Neutering cats reduces their energy requirements, such that the resting metabolic rates of gonadectomized male and female cats were 28 and 33% less, respectively, than their sexually intact counterparts (Root *et al.*, 1996). Thus energy intakes of neutered cats should be regulated to prevent their becoming obese.

Kittens grow rapidly in the first 6 months, during which time they attain three-quarters of their mature size, weight gains after this time being associated with developmental rather than skeletal changes (Legrand-Defretin and Munday, 1995). From 6 to 8 weeks of age, males weigh more than their female littermates, and this pattern continues throughout life. Kittens normally start to eat solid food at about 4 weeks old, and intakes initially are in the region of 10–40 kJ W^{-1} day^{-1}. By 6 weeks, they may consume between 250 and 350 kJ W^{-1} day^{-1}, and by 8–10 weeks they may ingest in excess of 800 kJ W^{-1} day^{-1} (Munday and Earle, 1991). Thereafter, ME intakes decline per unit of body weight, although they remain fairly high during the rapid growth phase in the first 6 months, ME requirements at 25–30 weeks are approximately 420 kJ of ME W^{-1} day^{-1}, but by the time cats are 1 year old, their requirements stabilize to around 250 kJ of ME W^{-1} day^{-1} for confined cats and 355 kJ of ME W^{-1} day^{-1} for those which are more active (Miller and Allison, 1958).

During gestation and lactation, nutrients must be supplied to meet the needs of the mother, to support fetal growth and allow for effective milk production. It is therefore recommended that the diet during this period is nutrient dense, highly palatable and digestible. The normal gestation period of the cat is 64 days and, unlike most mammals, as soon as a queen has conceived she will increase her food intake, such that her body weight increases steadily, virtually from the first day of gestation. In a comprehensive review of the subject, Loveridge and Rivers (1989) showed that over the entire pregnancy, the average weight gain was 39% of the pre-conception weight of the queen. Following parturition, only 40% of this extra weight was lost, the remainder being lost during lactation. These authors concluded that this unusual weight gain profile during gestation in cats was due to net uterine tissue accretion during early pregnancy, which was presumed to act as an energy store to be mobilized during lactation. To accommodate the steady increase in maternal body weight during gestation, queens require a rising level of nutrition, and Loveridge, (1986) showed that the voluntary intakes of 64 queens rose steadily from 1.13 to 2 MJ of ME per cat per day (equivalent to 389 to 456 MJ of ME W^{-1} day^{-1}). Despite being offered *ad libitum* access to nutrient-rich, palatable diets during lactation, queens often lose body weight during this time. Loveridge (1986) reported that the mean weight of queens decreased from 4.2 kg post-partum to 3.4 kg after 6 weeks lactation. The energy requirements of lactating queens depends on the size and age of the litter, as kittens may gain more than 100 g per week. Thus, a queen with one kitten in the first 2 weeks of lactation required 48% of the energy of her counterpart with six kittens, corresponding values being 45, 41, 33 and 28% in weeks 3, 4, 5 and 6 of lactation, bearing in mind that from week 4 some 5% of the energy from solid food would have been ingested by the kittens, rising to 30% during weeks 6 and 7 (Legrand-Defretin and Munday, 1995; NRC, 1986).

Lipid

Cats have a high tolerance to fat, diets containing 640 g of fat kg^{-1}were fed successfully to cats without causing ketonuria, increased fat excretion or cardiovascular changes (Humphreys and Scott, 1962). Diets containing 250–300 g of fat kg^{-1} are often fed, and high-fat diets tend to be palatable (Kendall, 1984). Kane *et al.* (1981) found that cats preferred diets containing 250 g of fat kg^{-1} compared with those containing 100 or 500 g of fat kg^{-1}. The fat contents of commercially available cat foods range from 75 to 300 g of fat kg^{-1} of diet (Kelly, 1996).

Cats are deficient in certain desaturase enzymes and are unable to convert linoleic acid to arachidonate, which must therefore be supplied in the diet; arachidonate is only found in animal sources (MacDonald *et al.*, 1983, 1984a,b,c). Without a source of arachidonate, queens cannot produce viable young (MacDonald *et al.*, 1984c). The NRC (1986) regard 5 and 0.2 g of linoleic and arachidonic acids per kg of diet, respectively, as the minimum levels of inclusion for cats. However, Wills (1996a) suggests that 1 and 0.02 g of linoleic and arachidonic acids, respectively, MJ ME^{-1} (equivalent to 20.9 and 0.42 g in a diet containing 20.9 MJ of ME kg^{-1}) should be supplied in diets for cats at all life stages.

Carbohydrates

Cats do not necessarily require carbohydrates, if sufficient fat and protein are provided from which the metabolic requirement for glucose can be met. However, commercial cat diets frequently contain more than 400 g of carbohydrate kg^{-1} which they may utilize well: digestibility coefficients for glucose and starch being 0.99 and 0.96, respectively (Trudell and Morris, 1975). However, consistent with their carnivorous nature, cats are unable to tolerate very high carbohydrate diets (Wills, 1996a). Dietary fibre is fermented to varying degrees by cats, and some dietary inclusion of fibre may be useful to encourage peristalsis, improve faecal consistency and to lower the energy density of diets for inactive cats.

Protein and Amino Acids

Although minimal requirements of the ten essential amino acids have been determined in growing cats, there are few corresponding values for maintenance and reproduction. Compared with less strict carnivores, cats have a limited ability to regulate transaminase and urea cycle enzyme activity to accommodate changes in dietary protein levels, resulting in comparatively high nitrogen losses (Rogers *et al.*, 1977). As a consequence, both kittens and adult cats have higher protein requirements than most other companion mammals (Rogers and Morris, 1982).

The protein content of diets for cats at maintenance recommended by Kelly and Wills (1996) and Burger (1995) is 15 g of protein MJ ME^{-1}, which equates to 250 g of protein kg^{-1} in diets containing 16.7 MJ of ME kg^{-1}. The recommendations for the protein contents of diets of both growing kittens and breeding females are the same; the NRC (1986) suggests 240 g of protein kg^{-1} diet, whereas Kelly and Wills (1996) and Burger (1995) recommend 17 g of protein MJ ME^{-1} (equivalent to 284 g of protein kg^{-1} in diets containing 16.7 MJ of ME kg^{-1}). The IAA requirements of growing cats are summarized in Table 19.2. It is noteworthy that arginine is an IAA for cats evidenced by the onset of hyperammonaemia within 3 h of feeding a single arginine-free meal (Morris and Rogers, 1978a,b).

Taurine
Taurine is an amino sulphonic acid which occurs in the free form in all animal tissues. Most mammals can synthesize adequate levels of taurine from methionine and cysteine. Additionally, the bile salts of most mammals contain a mixture of taurine and glycine conjugates and, when in a state of taurine deficiency, preferentially synthesize glycine conjugates. However, in the cat, bile salts are conjugated exclusively with taurine (Rabin *et al.*, 1976), and thus

Table 19.2. Indispensable amino acid, mineral and vitamin requirements of cats.

	Growth[a]		Maintenance[c]		Reproduction[c]	
	kg^{-1} diet	MJ ME^{-1}	kg^{-1} diet	MJ ME^{-1}	kg diet^{-1}	MJ ME^{-1}
Amino acids						
Arginine (g)	10.03	0.48	10[b]	0.48[b]	NS	NS
Histidine (g)	2.93	0.14	NS	NS	NS	NS
Isoleucine (g)	5.02	0.24	NS	NS	NS	NS
Leucine (g)	11.91	0.57	NS	NS	NS	NS
Lysine (g)	7.94	0.38	3.3[b]	0.16[b]	NS	NS
Total sulphur amino acids (g)	7.52	0.36	3.2[b]	0.16[b]	NS	NS
Total aromatic amino acid (g)	8.57	0.41	NS	NS	NS	NS
Threonine (g)	7.11	0.34	NS	NS	NS	NS
Tryptophan (g)	1.46	0.07	NS	NS	NS	NS
Valine (g)	6.06	0.29	NS	NS	NS	NS
Minerals						
Calcium (g)	8.00	0.38	8.40[d]	0.40[d]	12.50	0.60
Phosphorus (g)	6.06	0.29	6.30	0.30	10.00	0.48
Potassium (g)	3.97	0.19	6.30	0.30	6.30	0.30
Sodium (g)	0.48	0.023	0.50	0.023	1.0[d], 2.5[c]	0.05[d], 0.12[c]
Chloride (g)	19.01	0.91	NS	NS	NS	NS
Magnesium (mg)	397	19	376	18	627	30
Iron (mg)	79.42	3.80	84[c], 125[d]	4[c], 6[d]	125	6
Copper (mg)	5.02	0.24	5[c], 6.3[d]	0.24[c], 0.3[d]	6.30	0.30
Manganese (mg)	5.02	0.24	5[c], 6.3[d]	0.24[c], 0.6[d]	12.50	0.60
Zinc (mg)	50.16	2.40	50.00	2.40	50.16	2.40
Iodine (mg)	0.355	0.017	0.4[c], 1.25[d]	0.02[c],0.06[d]	1.25	0.06
Selenium (mg)	0.10	0.005	0.10	0.006	0.10	0.006
Vitamins						
Vitamin A (IU)	3333	1600	3333	160	6000	329
Vitamin D (IU)	500	24	500[d], 1000[c]	12[d], 24[c]	1254	60
Vitamin E (mg)[e]	30	1.43	17	0.8	30[d], 100[c]	3[d], 5[c]
Vitamin K (μg)	100	5.00	125	6	125	6
Thiamin (mg)	5	0.24	5	0.24	6	0.3
Riboflavin (mg)	4	0.20	4	0.20	6	0.3
Pantothenic acid (mg)	5	0.24	5	0.24	12.5	0.6
Niacin (mg)	40	2.00	40	2.0	56	2.7
Pyridoxine (mg)	4	0.20	4	0.20	5	0.24
Folic acid (μg)	800	38.00	800	38	1250	60
Vitamin B_{12} (μg)	20	1.00	21	1.0	25	1.2
Choline (mg)	2400	114.00	2508	120	2508	120
Biotin (μg)	70[c], 125[d]	3.34[c], 6[d]	90[c], 125[d]	4.2[c], 6[d]	90[c], 125[d]	4.2[c], 6[d]
Taurine (mg) (canned)	3114[c]	150[c]	3114	150	3114	150
Taurine (mg) (dry food)	1254[c]	60[c]	1254	60	1254	60
Taurine mg (all diets)	400	19	400[1]	19[1]	400[1]	19[1]

[a]Data from NRC (1986) based on diets containing 20.9 MJ of ME kg^{-1}; [b]Data from Burger and Smith (1990); [c]Data from Burger (1995); [d]Data from Kelly and Wills (1996); [e]For diets containing high levels of polyunsaturated fatty acids, vitamin E levels should be quadrupled for all life stages.

production of bile salts represents a net loss of taurine to the cat (Sturman *et al.*, 1978). Cats also exhibit poor hepatic synthesis of taurine from methionine and this, together with losses via bile salts, accounts for the susceptibility of felines to taurine-deficient diets which is manifested in feline central retinal degeneration where affected cats can become blind. Dietary levels of taurine recommended by NRC (1986) are 400 mg kg^{-1} diet for growth and maintenance and 500 mg kg^{-1} diet for reproduction. However, the taurine requirements reported in Burger (1995) are much higher at 60 and 149 mg MJ ME^{-1} for commercial dry and canned foods respectively (equivalent to 1000 and 2500 mg kg^{-1} diet containing 16.67 MJ of ME kg^{-1}diet).

Minerals

Calcium and phosphorus deficiencies result in hyperparathyroidism in the cat (Bennet, 1976) with a generalized decrease in bone density due to resorption of bone calcium to restore serum levels. Dietary magnesium levels of 400 mg kg^{-1} diet are recommended for growth (NRC, 1986) but excessive magnesium intakes are associated with the formation of urinary calculi (Kallfellz *et al.*, 1980). The mineral requirements of cats are given in Table 19.2.

Vitamins

Cats require pre-formed vitamin A for, unlike many other mammals, they are unable to convert β-carotene into vitamin A (Ahmad, 1931), deficiency leading to conjunctivitis and photophobia (Scott *et al.*, 1964). Cats cannot synthesize sufficient vitamin D on exposure to sunlight and again require a dietary supply: both vitamins A and D are present in animal tissues and, as cats are obligate carnivores, they obtain sufficient amounts of these from their prey. Growing kittens require greater dietary levels of niacin than many other mammals, as felines are unable to synthesize adequate levels of niacin from tryptophan. Excessive levels of vitamins, especially A and D, can be toxic; feeding cats liver or certain canned fish diets for long periods can lead to A and D hypervitaminosis, respectively. Many cats are fed fish-based diets which contain high levels of PUFAs which can auto-oxidize to produce potentially harmful peroxides and, therefore, appropriate intakes of vitamin E are essential for cats fed such diets. Of the water-soluble vitamins, thiamin is most likely to be lacking, and certain species of fish contain thiaminase (Smith and Prout, 1944) and thus should not be fed raw. Furthermore, as thiamin is very sensitive to heating (up to 90% may be lost during canning), it is important that sufficient thiamin is added to processed diets to ensure adequate availability. Thiamin deficiency leads to neurological defects and death (Everet, 1944). The vitamin requirements of cats are given in Table 19.2.

Senior cats

Cats are generally considered to be geriatric from 12 years old (Goldston, 1995). The incidence of disease is greater with elderly cats, chronic renal failure and hyperthyroidism being fairly common disorders (Peterson and Graves, 1992) leading to low intakes and weight loss. Feeding highly digestible, palatable diets is recommended as senior cats have a reduced digestive capacity. Neoplasia is also prevalent in the geriatric cat population, and feeding vitamin A and E may help to inhibit this and certain other degenerative diseases (Legrand-Defretin and Munday, 1995).

BIRDS

The most popular species of companion bird in Northern Europe are the budgerigar (*Melopsittacus undulatus*) and the canary (*Serinus canarius*), the former making up approximately 65% and the latter 15–20% of the total population, the balance largely consisting of finches, parrots, parakeets, pigeons and quails (Nott and Taylor, 1995). In the wild, these birds eat seeds from a wide variety of plant species (Nott, 1992). South American and African parrots eat fruit, shoots (Nott and Taylor, 1995) and insects (Long, 1984) as well as seeds, whilst the lories and lorikeets are nectar feeders (Nott and Taylor, 1995). The naturally varied diets of such birds in the wild ensures that their nutrient requirements are met but, if young captive birds are offered diets of limited diversity, they may become loath to accept new foods, leading to nutritional deficiencies (Lawton, 1988). The information below largely pertains to the psttiacines and canaries and is not comprehensive. As there is a paucity of data on the nutrient requirements of companion birds, other authors frequently present data extrapolated from the NRC (1984) to which the reader is referred for additional information.

The avian metabolic rate is higher than that of mammals, with body temperatures averaging 41–42°C (Nott and Taylor, 1995), and birds excrete surplus nitrogen in the form of uric acid, a process requiring 3.25 times more energy than the excretion of an equivalent amount of urea. Consequently, as birds require more energy pro-rata than mammals, they must process, digest and absorb food with greater efficiency. The nutrient requirements of birds are affected by the physiological state of the animal, and factors such as age, breeding and moulting may result in increased nutrient demands.

Energy

Various estimates have been made of the maintenance energy requirements for different types of companion birds. Aschoff and Pohl (1970) quoted a value of 535 kJ $W^{-0.715}$ day^{-1} for finches, and Earle and Clarke (1991) reported the maintenance energy requirement of budgerigars to be 1.59 MJ W^{-1} day^{-1}, a value some threefold that of a domestic chicken (NRC, 1984). Young birds grow rapidly, budgerigar chicks increasing their birth weight more than 12-fold in the first 10 days of life. In order to achieve this growth rate and simultaneously developing plumage, young birds devote a large proportion of energy intake to growth (Brooke and Birkhead, 1991). Breeding substantiality raises energy demands; over an average 78-day breeding period from pairing to fledging of 28 budgerigar families (hens average body weight 50 g, cocks average body weight 56 g and three chicks), mean intakes were 242 kJ per family, with a peak of daily energy demand at day 60 of approximately 500 kJ per family (Earle and Clarke, 1991).

Protein

Protein has an important role in birds in terms of beak, feather and claw production (Earle and Clarke, 1991). The protein requirement of the adult budgerigar has been estimated as 100 g kg^{-1} of the diet (Drepper *et al.*, 1988), and this may increase two- or threefold during moulting or breeding, especially when crop milk is produced. There are scant data on the amino acid requirements of companion birds but, in a study on budgerigars, values of 20, 35 and 35 g kg^{-1} of crude protein were given for lysine, arginine and methionine plus cysteine (the sulphur amino acids being required for feather formation), respectively (Drepper *et al.*, 1988). Seeds commonly fed to companion birds, such as red and white millet, groats and canary grass seed, typically contain from 110 to 160 g of protein kg^{-1}, with apparent protein metabolizability in budgerigars of 0.72–0.90 (Earle and Clarke, 1991). It has been estimated that 200 g of protein and a minimum of 8 g of lysine kg^{-1} of diet is required for optimal growth of cockateils (*Nymphicus hollandicus*) (Roudybush and Grau, 1991), although Ullrey *et al.* (1991) recommend 240 g of protein kg^{-1} for all life stages of psittacines. Thirteen grams of protein MJ ME^{-1} with 0.62 and 0.82 mg of lysine and sulphur amino acids, respectively, enabled rapid growth rates in canaries (Kamphues and Meyer, 1991). In addition to the ten IAAs required by adult birds, glycine is also regarded as indispensable for growing chicks as its hepatic synthesis is limited in these birds, yet it is essential for collagen, feather and uric acid production; 10 g of glycine kg^{-1} diet is recommended (Featherstone, 1976). Ullrey *et al.* (1991) reported that growth and reproduction of psittacines was supported by an extruded cereal-based diet, the amino acid content of which is shown in Table 19.3.

Lipid

Fat is a useful source of energy for birds, and Ullrey *et al.* (1991) reported that 20 g of fat kg^{-1} supported satisfactory performance in psittacines; it has been suggested that linoleic acid should constitute 10 g kg^{-1} of the diet (NRC, 1984). Companion birds are often prone to obesity, and restricted fat intakes are therefore recomended. The seeds usually fed to budgerigars contain 35–70 g fat and 17–24 g of linoleic acid kg^{-1} of diet (Earle and Clarke, 1991), which suggests that such birds are unlikely to suffer EFA deficiency.

Table 19.3. Indispensable amino acid, mineral and vitamin requirements of companion birds, ornamental fish, rabbits and rats (kg^{-1} diet).

	Birds[a]	Fish[b]	Rabbits[c]		Rats[d]	
			Growth	Reproduction	Growth	Reproduction
Amino acids						
Arginine (g)	13	16	6.0	NS	4.3	4.3
Histidine (g)	NS	8	3.0	NS	2.8	2.8
Isoleucine (g)	11	9	6.0	NS	6.2	6.2
Leucine (g)	NS	13	11.1	NS	10.7	10.7
Lysine (g)	12	22	6.5	NS	9.2	9.2
Total sulphur amino acids (g)	9	12	6.0	NS	10.0	10.0
Total aromatic amino acids (g)	NS	25	11.1	NS	10.2	10.2
Threonine (g)	9.5	15	6.0	NS	6.2	6.2
Tryptophan (g)	2.4	3	2.0	NS	2.0	2.0
Glycine (g)	10	NS	NS	NS	NS	NS
Valine (g)	NS	14	7.0	NS	7.4	7.4
Minerals						
Calcium (g)	11	3–7	4.0	4.5 (7.5)	5.0	6.3
Phosphorus (g)	8	4–6	2.2	3.7 (5)	3.0	3.4
Potassium (g)	7	6–12	6.0	6.0	3.6	3.6
Sodium (g)	2	12–30[e]	2.0	2.0	0.5	0.5
Chloride (g)	2	NR	3.0	3.0	0.5	0.5
Magnesium (mg)	1500	400–700	350	350	500	600
Iron (mg)	150	200	50[e]	50[e]	35	75
Copper (mg)	20	3.0	3.0	3.0	5.0	8.0
Manganese (mg)	65	13	8.5	2.5	10.0	10
Zinc (mg)	120	80–200	50[e]	50[e]	12	25
Iodine (mg)	1	NS	0.2	0.2	0.15	0.15
Selenium (mg)	0.3	0.5–1.0	0.1	0.1	0.15	0.40
Vitamin						
Vitamin A (IU)	8000	10,000*	580[$]	>1160	2333	2333
Vitamin D (IU)	500	2400	1000[e]	1000[e]	1000	1000
Vitamin E (IU)	250	100	40	40	18	18
Vitamin K (mg)	4000	10,000	0.2	0.2	1000	1000
Thiamin (mg)	6	10	10[e]	10[e]	4.0	4.0
Riboflavin (mg)	6	10	10[e]	10[e]	10[e]	4.0
Pantothenic acid (mg)	20	50	20[e]	20[e]	10	10
Niacin (mg)	55	50	180	NS	15	15
Pyridoxine (mg)	6	10	39	NS	6.0	6.0
Folic acid (µg)	900	10	1000[e]	1000[e]	1000	1000
Vitamin B$_{12}$ (µg)	25	20	10[e]	10[e]	50	50
Choline (mg)	1700	2000	1200	1500[e]	750	750
Biotin (µg)	300	1000*	200[e]	200[e]	200	200

[a]Data from Ullrey *et al.* (1991) of the nutrient composition of an extruded diet fed to birds (psittacines) which supported good performance at all life stages, as data on exact requirements for each nutrient are unavailable. For information on nutrient requirements of poultry, readers are advised to consult the NRC (1984).

[b]Data for indispensable amino acids are for carp reported by Nose (1979). Data for mineral requirements are those quoted by Burger (1995), the values being ranges depending on species and life stage, and are not maxima and minima. Data for vitamin requirements of ornamental fish are estimated minima quoted by McCartney (1996), except where marked with an asterisk where data is from Burger (1995).

[c]Data are from NRC (1977), values for vitamins in parentheses denote requirements for lactation.

[$]Rabbits at all life stages also have a requirement for 30 mg carotene kg^{-1} of diet.

[d]Data are from NRC (1993); [e]data from Kelly and Wills (1996).

NS = not stated; NR = not required.

Minerals

The calcium requirement for birds has been calculated as 6–10 g kg^{-1} of diet depending on factors such as egg laying, etc. Most birdseeds are deficient in calcium, hence the need to provide appropriate supplements, but excess calcium can reduce iodine absorption (Taylor, 1954). Phosphorus in seed is often in the form of phytate and therefore of limited availability, values of 0.1–1 g kg^{-1} of diet as available phosphorus being quoted by Earle and Clarke (1991). Kamphues and Meyer (1991) suggested a phosphorus requirement of 0.32 and 0.57 g MJ ME^{-1} for slow and rapidly growing canaries, respectively. The mineral content of the extruded diet of Ullrey *et al.* (1991) which supported good performance of psittacines is given in Table 19.3.

Vitamins

Seeds lack β-carotene and do not contain vitamin A and, therefore, vitamin A deficiencies are relatively common in seed-fed birds (Pitts, 1983). Supplementing budgerigars with an extra 12 µg of vitamin A per day per bird was regarded as sufficient (Baker, 1990), with 750 µg being the maximum tolerated, as oversupplementation can prove lethal. Vitamin D can be synthesized from sterols in the cutaneous layer of budgerigars on exposure to UV light or sunlight, and a daily requirement of 25 mg per bird has been suggested (Baker, 1990). Vitamin D deficiency can lead to poor feathering, reduced egg production, thin shells and rickets (Garlisch and Wyatt, 1971). The vitamin content of the diet fed to psittacines by Ullrey *et al.* (1991) is shown in Table 19.3.

ORNAMENTAL FISH

The nutrition of ornamental fish is one of the least researched areas of companion animal nutrition; much of the information given in various articles on the subject is derived from data obtained for food fish (McCartney, 1996). As food fish are dealt with in Chapter 18, only some general points will be addressed here. The interest in ornamental fish keeping has increased dramatically worldwide over the past 20 years, such that it is estimated that there are more ornamental fish kept than all the other companion animals added together (Pannevis, 1995).

A number of fish species at different life stages with different nutritional requirements and feeding habits are often kept together in a community tank, yet they frequently are fed a single diet. Moreover, the diet must be able to satisfy the needs of fish which are surface, middle or bottom feeders and which may have different diurnal feeding patterns. The safest option is to feed diets with specifications that will satisfy the needs of the most nutritionally demanding fish and yet will not result in environmental pollution, either directly in the form of uneaten food or indirectly through excretion.

Energy

As fish are poikilotherms, their energy requirements change with environmental temperature, although it is generally accepted that the maintenance energy requirements of fish are usually 5–10% of that of birds and mammals (Smith, 1989). Other contributory factors to explain such comparatively low energy requirements are that fish excrete ammonia rather than urea and their movement through water is more energetically efficient than that of land animals (Pannevis, 1995). The maintenance energy requirement of goldfish increased approximately threefold when water temperature was raised form 20 to 24°C, and that of pond-kept Koi carp more than doubled when the water temperature increased from 10 to 20°C but, when water temperatures dropped below, 5°C these fish barely fed at all, living off body reserves for many months (McCartney, 1996). However, for a 5 g goldfish at 20°C, McCartney (1996) suggests that the maintenance energy requirement is 40 kJ of DE W^{-1}. Calculating from results reported by Pannevis and Earle (1994a) for the maintenance energy requirements of four popular ornamental fish at 26°C, the following values (J of DE g body weight^{-1} day^{-1}) were obtained: Neon Tetras (374), Leopard Danio (422), Kribensis (178) and Moonlight Gourami (272). To put this into context for the fish owner, in the case of average sized adult Neon Tetras and Moonlight Gouramis, these values equate to 0.6 and 4.9 flakes of a commercial fish food per day. Although for reasons of economy, carbohydrates are fed as energy sources to food fish, carbohydrates are relatively poorly utilized by many fish. Therefore, the majority of the dietary energy for ornamental fish is provided in the form of lipid and protein. Optimum ratios for digestible protein to digestible energy are believed to be in the region of 19–28 mg of protein kJ DE^{-1} (NRC, 1993a).

Protein

The requirement for protein by fish ranges from 250 to 500 g kg^{-1} and should be of high quality to prevent pollution; indeed many of the health problems encountered with ornamental fish are related to nitrate build up due to overfeeding and/or insufficient filtering (McCartney, 1996). The quantitative requirement by fish for IAAs is high (NRC, 1993a), and feeding a protein source that closely matches the needs of the species in question will result in maximal growth and minimal excretion. For these reasons, fish meal is regarded as the most appropriate protein source in ornamental fish diets.

Lipid

Fat is required as a major energy source by ornamental fish and it is recommended that diets contain between 100 and 200 g of fat kg^{-1} diet (Cowey and Sergeant, 1979). It is of note that fats with a low melting point should be avoided when feeding fish in cold environments, as solidification in the digestive tract could prove fatal. The EFAs required depend on both the species and the feeding habits of the fish. Freshwater species are generally able to elongate fatty acids or desaturate the shorter chain EFAs and therefore require either linolenic (*n*-3) or linoleic (*n*-6) acids, or both. Marine herbivores require a source of *n*-3 fatty acids, and linolenic acid is usually sufficient. The piscivores, however, require an EFA profile that reflects the composition of their prey, and need the longer, more unsaturated eicosapentanoic acid (EPA) (20:5) (*n*-3) or docosahexaenoic acid (DHA) (22:6) (*n*-3). McCartney (1996) recommends the inclusion of 10 g each of linoleic and linolenic acids per kg in freshwater fish diets and 20 g of linolenic acid per kg of diet for marine herbivores, this being replaced with DHA or a mixture of EPA and DHA for marine carnivores.

Minerals

Unlike terrestrial animals, fish can absorb some minerals from the environment via fins, gills or oral epithelia, or by drinking, thus reducing their dietary requirement for minerals. Calcium deficiency has seldom been detected in ornamental fish due to their being fed high levels of calcium-rich fish meal. Deficiencies of sodium, potassium and chlorine are also rare, as they are readily absorbed from the environment. Requirements of other minerals by ornamental fish are shown in Table 19.3. A problem encountered with present-ing food in water is leaching of water-soluble nutrients into the environment prior to their ingestion by the fish. For example, Pannevis and Earle (1994b) showed that 90% of vitamin B_{12}, 66% of vitamin C, 47% of pantothenic acid and up to 27% of choline, folate and pyridoxine was lost from the diet after only 30 s exposure. Furthermore, freshwater fish cannot absorb leached vitamins, and the dilution effect of a large body of water would render drink-ing by marine species ineffective in terms of leached vitamin recapture. As the average time for flaked feeds to reach the bottom of a 30 cm deep tank is 90 s (Pannevis, 1995), bottom feeders are potentially most at risk from hypovitaminosis when flaked diets are fed, due to both longer dietary expo-sure to water and competition from surface and middle water feeders. It is recommended that both flaked diets which float and fast sinking pelleted diets are fed in community tanks to allow fish of all feeding habits maximum benefit from their diets. When formulating diets for ornamental fish estimated leaching characteristics should be considered.

RABBITS

The domestic rabbit (*Oryctolagus cuniculus*) varies in mature size from 1 to more than 6 kg. Rabbits are herbivores, and practise coprophagy from the time they start to eat solid food, whereby, in addition to hard faecal pellets, they produce soft, dark pellets which are ingested by the rabbit on their emergence from the anus. Coprophagy, together with hind-gut fermentation, is thought to provide most of the B vitamin requirements of the rabbit and some bacterial protein. Coprophagy also allows further digestion of nutrients by virtue of their being passed through the digestive tract on successive occasions.

Energy

Rabbits derive their dietary energy largely from fresh and conserved forages and cereal grains. Rabbits are less efficient at utilizing dietary fibre compared with other herbivores, the average crude fibre digestibility coefficient in the rabbit being 0.14 compared with 0.22 for pigs, 0.33 for the guinea pig and 0.41 for the horse (Maynard and Loosli, 1969). The gut microflora of the rabbit differs from that of other herbivorous species in that it consists almost entirely of *Bacteriodes* species (Fuller and Moore, 1971), whereas *Escherichia coli* and *Lactobacillus* sp. are either absent, or present in very low numbers (Smith, 1965). However, it has been calculated from caecal VFA concentrations that 10–12% of the daily maintenance energy requirement of rabbits may be met from fermentation of fibre (Lebas, 1975a), and some fibre in the diet is necessary for maintaining normal digestive function, a minimum of 120 g of crude fibre kg^{-1} of diet being recommended (Lebas, 1975b). Rabbits digest both starch and sugars efficiently (NRC, 1977), and like other species will utilize protein for energy when other sources are scarce. The energy requirements of rabbits have been reported as: 418.4 kJ $W^{-0.75}$ day^{-1} for maintenance, 798–882 kJ $W^{-0.75}$ day^{-1} for growth, 567 and 840 kJ $W^{-0.75}$ day^{-1} for early and late gestation respectively, and 1260 kJ $W^{-0.75}$ day^{-1} for lactation (Tobin, 1996).

Lipid

Rabbits showed a preference for diets containing up to 100 g of oil kg^{-1} of diet over those containing 0, 50 or 200 g of oil kg^{-1} (Cheeke, 1974), and young rabbits grew faster on diets containing 100–250 g of oil kg^{-1} than those on diets containing 50 g of oil kg^{-1} (Thacker, 1956). However, the tendency for rabbits to become obese has led to a recommendation of 20–50 of oil kg^{-1} in diets for non-lactating animals (Tobin, 1996). Recent work by Pascual *et al.* (1999) showed that lactating does had higher milk yields with greater energy content, enabling higher litter weight gains and reduced mortality of nursing young, when fed diets containing 99 or 117 g of lipid kg^{-1} of diet compared to those fed a diet containing 26 g of lipid kg^{-1}.

Protein and Amino Acids

Rabbits require essential amino acids in the diet, as the bacterial protein obtained via coprophagy is inadequate in this regard. The NRC (1977) recommendations for amino acids for rabbits as a percentage of the diet are:

arginine, 0.6; lysine, 0.65; and methionine plus cysteine, 0.6, and these levels will both support rapid growth and provide a safety margin for those at maintenance. In contrast to the situation with fibre, rabbits are able to utilize forage protein more efficiently than a number of other monogastrics. Thus protein digestibility coefficients for alfalfa meal were 0.75 and 0.50 for rabbits and pigs, respectively (Slade and Hintz, 1969). Although coprophagy does not appear to provide many essential amino acids, it may help to maintain the nitrogen balance in mature animals fed low-quality protein sources. For example, adult rabbits fed gelatin which were allowed to engage in coprophagy were able to maintain a positive nitrogen balance but were unable to do so when coprophagy was prevented (Kennedy and Hirschberger, 1974). Although exact recommendations as to appropriate levels of dietary protein in diets for rabbits at different life stages depend on protein quality, the generally accepted levels of protein are 120, 160, 150 and 170 g of crude protein kg^{-1} of diet for maintenance, growth, pregnancy and lactation (NRC, 1977).

Minerals

Calcium and phosphorus levels of 4.5 and 3.7 g kg^{-1}, respectively, were regarded as adequate for growth and gestation (Chapin and Smith, 1967), the corresponding values for lactation being 7.5 and 5.0 (Lebas *et al.*, 1971). Although these values are slightly lower than those recommended by Tobin (1996), these amounts should not be exceeded greatly, for rabbits excrete excess dietary calcium in the urine and thus high-calcium diets can lead to the formation of uroliths. The mineral requirements of rabbits are summarized in Table 19.3.

Vitamins

Although adequate supplies of the vitamin B complex should be supplied via coprophagy, it is prudent to supply some dietary B vitamins, especially for animals housed on wire-mesh floors. Although it might also be expected that coprophagy would supply the vitamin K requirement, there have been reports of deficiencies affecting reproduction, and therefore there is a case for providing dietary vitamin K. Rabbits are highly sensitive to vitamin E deficiency resulting in muscular dystrophy and cardiac problems, therefore diets should always be supplemented with this vitamin, taking into account its relatively rapid decline in feeds (5–20% per month) during storage (Tobin, 1966). Rabbits also require β-carotene which, in addition to being a precursor of vitamin A, also appears to enhance reproductive performance, and levels of at least 30 mg of β-carotene kg^{-1}diet have been recommended (NRC, 1977). It is of note that although β-carotene is present in fresh green feed, up to 80% may be lost during haymaking. The vitamin requirements of rabbits are given in Table 19.3.

RATS

Domestic rats (*Rattus norvegicus*) are omnivorous monogastrics which in their natural habitat eat grains, seeds, small invertebrates and vertebrates. Most commercial diets for rats, however, are largely vegetarian, being based on milk products, vegetable oils, cereal grains and legumes such as peas, beans and alfalfa.

Energy

The maintenance energy requirement for rats quoted by NRC (1993b) is 470 kJ $W^{-0.75}$ day^{-1}, whereas that for growth is reported to be 606 kJ $W^{-0.75}$ day^{-1} (Tobin, 1996). Although protein appears to be more important for successful reproduction than energy (Menaker and Navia, 1973), the daily ME requirement for early gestation is approximately 600 kJ $W^{-0.75}$ day^{-1}, but may increase to 1110 kJ $W^{-0.75}$ day^{-1} during later stages of pregnancy (NRC, 1993b). The energy requirement of lactating females depends on the number of pups suckled but is usually in the order of two- to fourfold that of dry females (Nelson and Evans, 1961). In addition to large increases in feed intake, during lactation, female rats mobilize large stores of body fat that they accrue during gestation. Naismith *et al.* (1983) reported that such body fat supplied much of the energy required during lactation. Nevertheless, it has been estimated that the daily ME requirement of rats at peak lactation ranges from 1.3 MJ $W^{-0.75}$ day^{-1} (NRC, 1993b) to 1.8 MJ $W^{-0.75}$ day^{-1} (Tobin, 1996).

Lipid

Digestibility of lipids by rats varies with source, processing and level of dietary inclusion. Thus digestibility coefficients for lipid from diets containing either 50 g kg^{-1} of cocoa butter or corn oil by adult rats were 0.92 and 0.59, respectively, though in the latter case when the inclusion level was increased to 200 g kg^{-1} the corresponding coefficient was 0.71 (Apgar *et al.*, 1987). The digestibility coefficient of lipid from bland lard was 0.94, but only 0.63 for hydrogenated lard in adult female rats (Crockett and Deuel, 1947). Linolenic acid (*n*-6) is the major EFA required by rats, and work by Bourre *et al.* (1989, 1990) indicated that 1.2 and 0.2 g of linoleic and linolenic acids, respectively, per kg of diet were the minimum requirements of EFA by rats, but these levels are substantially lower than the 6.0 g of linoleic acid kg $^{-1}$ recommended by the NRC (1978a). Dietary levels of 50 g of lipid kg $^{-1}$ appear to be sufficient for all classes of rat, and many sources of lipid provide adequate amounts of the EFA at this level of inclusion (NRC, 1993b). A number of studies have shown that high levels (up to 500 g kg^{-1}) of lipid inclusion in the diet can have deleterious effects on health (Kollmorgen *et al.*, 1983), reproduction (Richardson *et al.*, 1964) and growth of nursing pups (Rolls and Rowe, 1982).

Carbohydrates

Rats require dietary carbohydrates and can utilize glucose, fructose, starch, maltose and a number of sources of dietary fibre. However, poor performance was observed in galactose-fed rats (Day and Pigman, 1957), and xylose was found to be toxic to adult rats (Booth *et al.*, 1953). Carbohydrate-free diets cannot support reproduction (Taylor *et al.*, 1983) or lactation in rats (Koski *et al.*, 1990).

Protein and Amino Acids

Protein contents of 50 and 70 g kg^{-1} of diet for high-quality and natural ingredient protein sources, respectively, have been reported as being suitable for maintenance of rats (Bricker and Mitchell, 1947; Dibak *et al.*, 1986). Early work suggested that the least amount of protein which had a balanced amino acid profile required to support maximum weight gain in young rats was 100–150 g kg^{-1} of a low fibre diet containing 17 MJ of ME kg^{-1} (Mitchell and Beadles, 1952). This is consistent with the NRC (1978a) recommendation of 120 g of protein kg^{-1} for growing rats when a highly digestible protein source with a balanced amino acid profile was used. However, when non-purified dietary ingredients are used, 180–250 g of crude protein kg^{-1} is required. Studies by Nelson and Evans (1958) on lactating females indicated that although diets containing 180 g of casein kg^{-1} supported maximum growth in suckling young, 240 g was required for the dam to maintain or gain body weight.

The current (NRC, 1993b) recommendations for IAAs to support growth and reproduction in rats are largely the same, with lower values for maintenance (Table 19.3.) In addition to these requirements, there appears to be a need for dietary asparagine during pregnancy as asparagine-free diets were associated with impaired neurological development of the young (Newburg and Fillios, 1979).

Minerals

Dietary concentrations of calcium and phosphorus of 5 and 3 g kg^{-1} of diet, respectively, are recommended by the NRC (1993b) for growth and maintenance of rats, a calcium:phosphorus molar ratio of at least 1.3:1 being required. The lactating rat has an increased demand for these minerals, and Brommage (1989) indicated that a lactating female could pass 200 mg of calcium and 140 mg of phosphorus into the milk in 1 day, and that this increased need was met via higher feed intakes and enhanced absorption of these minerals: NRC (1993b) suggests 6.3 g of calcium and 3.7 g of phosphorus kg^{-1} of diet as being suitable. Magnesium-deficient rats may have reduced litter sizes or malformed young (Hurley *et al.*, 1976a,b). Rats have a specific require-ment for iodine, their utilization of this mineral being high and being largely used for thyroid hormone function (Gross, 1962). Although Tobin (1996) suggests that 75 mg of manganese kg^{-1} of diet is appropriate for all classes of rat, the current NRC recommendation is 10 mg kg^{-1}, due to reports by Davis *et al.* (1971) that excess manganese has a deleterious affect on iron metabolism. Table 19.3 shows the mineral requirements of rats.

Vitamins

High levels of vitamin A and E appear to interfere with vitamin K metabolism such that the onset of vitamin K deficiency is more rapid in diets rich in vitamins A and E (Ralli and Dumm, 1953). Biotin-deficient rats have a depressed immune response, develop alopecia and may develop an abnormal gait (Rabin, 1983). Pantothenic acid deficiencies in rats result in death after 4–6 weeks exposure to the deficient diet, largely due to heamorrhagic necrosis of the adrenals, reduced antibody production (Ralli and Dumm, 1953) and thiamin deficiency. The vitamin requirements of rats are given in Table 19.3.

CONCLUSIONS

The diversity of species addressed in this chapter dictated that each was dealt with in less depth than in those chapters dedicated to a single species. The overriding feature to emerge from this chapter was the variability in nutrient requirements both within classes of companion animals and also within a species, particularly dogs. The data reported here are valid for the target animal within a prescribed set of experimental conditions, but clearly it would be a gargantuan and probably impossible task to determine the nutritive requirements of every breed at each life stage under every conceivable exercise regime and set of housing conditions. Therefore, the values presented here should be regarded as guidelines. The unusual metabolism of the cat, rendering it dependent on at least some animal-derived nutrients, is of interest, and the specific needs of the increasing population of geriatric dogs and cats are areas for further research. The requirement of companion birds from the time they are hatched for a variety of seed types suitably supplemented to meet their nutritional demands must be considered and, as more avian species become popular as companion birds, further investigations as to their specific needs will be necessary. The challenges of feeding communities of many species of ornamental fish with diametrically opposed needs continue to require detailed research into effective feeding strategies as well as nutrient requirements. Although much of the data on rabbits was obtained from animals kept under laboratory conditions, many companion rabbits live in outdoor hutches, and therefore allowances must be made for extra nutrient demands under variable conditions. However, several points have emerged which are common for all companion animals in that highly digestible diets with 'ideal protein' in the form of amino acids which closely conform to the animals requirements are desirable to reduce pollution and to prevent imbalances with associated wastage of poorly utilized nutrients. From the foregoing chapters, it is clear that companion animal nutrition is a relatively recent area of research in comparison with that of farmed livestock for which models to predict the responses of animals to different feeding systems abound. It is to be hoped that future research will yield similar models for companion animals.

REFERENCES

Adibi, S.A. (1980) Roles of branched chain amino acids in metabolic regulation. *Journal of Laboratory Clinical Medicine* 95, 475–477.

Ahmad, B. (1931) The fate of carotene after absorption in the animal. *Biochemical Journal* 25, 1195–1204.

Allison, J.B., Anderson, J.A. and Seely, R.D. (1947) Some effects of methionine on the utilisation of nitrogen in the adult dog. *Journal of Nutrition* 33, 361–370.

Apgar, J.L., Shively, C.A. and Tarka, S.M. Jr (1987) Digestibility of cocoa butter and corn oil and their influence on fatty acid distribution in rats. *Journal of Nutrition* 117, 660–665.

Aschoff, J. and Pohl, H. (1970) Rhythmic variation in energy metabolism. *Federation Proceedings* 29, 1541–1552.

Baker, J. (1990) Dangers in vitamin overdose. *Cage Aviary Birds* 31, 5–6.

Barbul, A., Sisto, D., Wasserkrug, H.L. and Efron, G. (1981) Arginine stimulates lymphocytic immune response in healthy human beings. *Surgery* 90, 244–257.

Bennet, D. (1976) Nutrition and bone diseases in the dog and cat. *Veterinary Record* 98, 313–315.

Biagi, P.L., Bordoni, A., Hrelia, S., Celadon, M. and Horrobin, D.F. (1991) Gamma-linoleic acid dietary supplementation can reverse the ageing influence on rat liver microsome delta-6-desaturase activity. *Biochimica et Biophysica Acta* 1083, 187–192.

Blaza, S.E. (1982) Energy requirements of dogs in cool conditions. *Canine Practice* 9, 10–15.

Blaza, S.E., Booles, D. and Burger, I.H. (1989) Is carbohydrate essential for pregnancy and lactation in dogs? *Nutrition of the Dog and Cat. Waltham Symposium No. 7.* Cambridge University Press, Cambridge, pp. 229–242.

Bourre, J.M., Francois, M., Youyoo, A., Dumont, O., Piciotti, M., Pascal, G. and Durand, G. (1989) The effects of dietary α-linolenic acid on the composition of nerve membranes, enzymatic activity, amplitude of electrophysiological parameters, resistance to poisons and performance of learning tasks in rats. *Journal of Nutrition* 119, 1880–1892.

Bourre, J. M., Piciotti, M., Dumont, O., Pascal, G. and Durand, G. (1990) Dietary linoleic acid and polyunsaturated fatty acids in rat brain and other organs: minimal require-ments of linoleic acid. *Lipids* 25, 465–472.

Booth, A.N., Wilson, R.H. and De, F. (1953) Effects of prolonged ingestion of xylose on rats. *Journal of Nutrition* 49, 347–355.

Bricker, M.L. and Mitchell, H.H. (1947) The protein requirements of the adult rat in terms of the protein contained in egg, milk, and soy flour. *Journal of Nutrition* 34, 491–505.

Brommage, R. (1989) Measurement of calcium and phosphorus fluxes during lactation in the rat. *Journal of Nutrition* 119, 428–438.

Brooke, M. and Birkhead, T. (1991) *The Cambridge Encyclopaedia of Ornithology.* Cambridge University Press, Cambridge, UK.

Bunce, G.E., Chiemchaisri, Y. and Phillips, P.H. (1962a) The mineral requirements of the dog 1V. Effect of certain dietary and physiological factors on the magnesium deficiency syndrome. *Journal of Nutrition* 76, 23–29.

Bunce, G.E., Jenkins, K.J. and Phillips, P.H. (1962b) The mineral requirements of the dog 111. The magnesium requirement. *Journal of Nutrition* 76, 17–22.

Burger, I.H. (1995) A basic guide to nutrient requirements. In: Burger, I.H. (ed.), *The Waltham Book of Companion Animal Nutrition.* Pergamon Press, Oxford, UK, pp. 5–24.

Burger, I.H. and Johnson, J.V. (1991) Dogs large and small: the allometry of energy requirements within a single species. *Journal of Nutrition* 121, S18–S21.

Chapin, R.E. and Smith, S.E. (1967) The calcium tolerance of growing and reproducing rabbits. *Cornell Veterinarian* 57, 480–491.

Cheeke, P.R. (1974) Feed preferences of adult male Dutch rabbits. *Laboratory Animal Science* 24, 601–604.

Costill, D.L. (1981) Fats and carbohydrates as determinants of athletic performance. In: Heskell, W., Skala, J. and Whittman, J. (eds), *Proceedings of the Symposium on the Nutritional. Determinants of Athletic Performance.* Bull Publishing, San Francisco, pp. 16–28.

Cowey, C.B. and Sergeant J.R (1979) Nutrition. In: Hoar, S.W. and Randall, D.J. (eds), *Fish Physiology,* Vol. VIII. Academic Press Inc., San Diego, pp. 1–61.

Crockett, M.E. and Deuel, H.J., Jr (1947) A comparison of the coefficient of digestibility and rate of absorption of several natural and artificial fats as influenced by melting point. *Journal of Nutrition* 33, 187–194.

Davies, M. (1996) An introduction to geriatric veterinary medicine. In: Davies, M. (ed.), *Canine and Feline Geriatrics.* Blackwell Science, Cambridge, UK, pp. 1–11.

Davis, C.D., Ney, D.M. and Gregor, J.L. (1990) Manganese, iron and lipid interactions in rats. *Journal of Nutrition* 120, 507–513.

Day, H.G. and Pigman, W. (1957) Carbohydrates in nutrition. In: Pigman, W. (ed.), *The Carbohydrates.* Academic Press, New York, pp. 779–806.

Dibak, O., Krajcovicova-Kudlackova, M., Grancicova, E. and Jankovicova, M. (1986) Body composition and physiological casein and wheat gluten protein requirements of 180-day old rats. *Physiologia Bohemoslovaca* 35, 71–80.

Diez, M., Grauwels, M., Henroteaux, M. and Istasse, L. (1998) Use of fructo-oligosaccharides (FOS) in a dog with corneal lipidosis. *Proceedings of the 2nd Annual Conference of the European Society of Veterinary and Comparative Medicine.* Vienna, Austria, p. 34.

Drepper, K., Menke, K.H., Schulz, G. and Wachter Vormann, W. (1988) Investigations on the protein and energy requirements of adult budgerigars (*Melopsittacus undulatus*) in cage husbandry. *Kleinterpraxis* 33, 57–62.

Earle, K.E. and Clarke, N.R. (1991) The nutrition of the budgerigar (*Melopsittacus undulatus*). *Journal of Nutrition* 121, S186–S192.

Earle K.E. and Smith, P.M. (1991) Digestible energy requirements of adult cats at maintenance. *Journal of Nutrition* 121, S45–S46.

Everett, G.M. (1944) Observations on the behaviour and neurophysiology of acute thiamine deficient cats. *American Journal of Physiology* 141, 439–442.

Featherstone, W.R. (1976) Glycine–serine interrelations in the chick. *Federation Proceedings* 35, 1910–1913.

FEDIAF (European Pet Food Federation) (1996) *The European Pet Food Industry*, Brussels, Belgium.

Finke, M.D. (1991) Evaluation of the energy requirement of adult kennel dogs. *Journal of Nutrition* 121, S22–S28.

Fuller, R. and Moore, J.H. (1971) The effect of rabbit intestinal microflora of diets which influence serum cholesterol levels. *Laboratory Animals* 5, 25–30.

Garlisch, J.D. and Wyatt, R.D. (1971) Effects of vitamin D_3 on calcium retention and eggshell calcification. *Poultry Science* 50, 950–956.

Goldston, R.T. (1995) Introduction and overview of geriatrics. In: Goldston, R.T. and Hoskins, J.D. (eds), *Geriatrics and Gerontology of the Dog and Cat* W.B Saunders Company, Philadelphia, pp. 1–9.

Grandjean, D. (1996) Nutrition of racing and working dogs In: Kelly, N. and Wills, J. (eds), *Manual of Companion Animal Nutrition and Feeding*. British Small Veterinary Association, Cheltenham, Gloucestershire, UK, pp. 63–92.

Greenhalgh, J.F.D. and Corbin, J.E. (1998) Role of companion animals for the quality of human life. *Proceedings of the 8th World Symposium on Animal Production*. Seoul National University, Seoul, Korea, pp. 449–458.

Gross, J. (1960) Iodine and bromine. In: Comar, C.L. and Bronner, F. (eds), *Mineral Metabolism*, Vol. 2, Part B. Academic Press, New York, pp. 221–285.

Harris, P.L. and Embree, N.D. (1963) Quantitative consideration of the effects of polyunsaturated fatty acid content of the diet upon the requirement for vitamin E. *American Journal of Clinical Nutrition* 13, 385–390.

Hayek, M.G. (1998) Age related changes in physiological function in the dog and cat: nutritional implications In: Rheinhart, G.A. and Carey, D.P. (eds), *Recent Advances in Canine and Feline Nutrition, Volume 2. Iams Nutrition Symposium Proceedings*. Orange Frazer Press, Ohio, pp. 363–380.

Hinchcliff, K.W., Rheinhart, G.A., Burr, J.R., Schreier, C.J. and Swenson, R.A. (1998) Effect of racing on water metabolism, serum sodium and potassium concentrations, renal hormones and urine composition of Alaskan sled dogs. In: Rheinhart, G.A. and Carey, D.P. (eds), *Recent Advances in Canine and Feline Nutrition, Volume 2. Iams Nutrition Symposium Proceedings*. Orange Frazer Press, Ohio, pp. 283–294.

Holman, R.T., Johnson, S.B. and Hatch, T.F. (1982) A case of human linolenic acid deficiency involving neurological abnormality. *American Journal of Clinical Nutrition* 36(6), 1254–1255.

Hultman, E., Bergstrom, I .and Roch-Norland, A.E. (1971) Glycogen storage in human skeletal muscle. In: Pernow, B. and Saltin, B. (eds), *Muscle Metabolism During Exercise*. Plenum Press, New York, pp. 273–280.

Humphreys, E.R. and Scott, P.P. (1962) The addition of herring and vegetable oils to the diet of cats *Proceedings of the Nutrition Society* 21, xviii.

Hurley, L.S., Cosens, G. and Therialut, L.L. (1976a) Teratogenic effects of magnesium deficiency in rats. *Journal of Nutrition* 106, 1254–1260.

Hurley, L.S., Cosens, G. and Therialut, L.L. (1976b) Magnesium, calcium and zinc levels of maternal and foetal tissues in magnesium-deficient rats. *Journal of Nutrition* 106, 1261–1264.

James, W.T. and McCay, W.C. (1950) A study of effects of intake, activity and digestive efficiency in different types of dog. *American Journal of Veterinary Research* 11, 412–417.

Johnson, V. (1993) Protein requirements of dogs. *Waltham International Focus* 3, 9–14.

Kade, C.F., Jr, Phillips, J.H. and Phillips, W.A. (1948) The determination of the minimum requirement of the adult dog for maintenance of energy balance. *Journal of Nutrition* 36, 109–114.

Kallfelz, F.A., Bresset, J.D. and Wallace, J.R. (1980) Urethral obstruction in randon source SPF male cats induced by dietary magnesium. *Feline Practice* 10, 250–270.

Kamphues, J. and Meyer, H. (1991) Basic data for factorial derivation of energy and nutrient requirements for growing canaries. *Journal of Nutrition* 121, S207–208.

Kane, E., Morris, J.G. and Rogers, Q.R. (1981) Acceptability and digestibility by adult cats of diets made with various sources and levels of fat. *Journal of Animal Science* 53, 1516–1523.

Keen, C.L., Lonnerdal, B. and Fisher, G.L. (1981) Seasonal variations and effects of age on serum copper and zinc values in the dog. *American Journal of Veterinary Research* 42, 347–350.

Kelly, N. (1996) Food types and evaluation. In: Kelly, N. and Wills, J. (eds), *Manual of Companion Animal Nutrition and Feeding*. British Small Veterinary Association, Cheltenham, Gloucestershire, UK, pp. 22–42.

Kelly, N. and Wills, J. (1996) Appendix I. In: Kelly, N. and Wills, J. (eds), *Manual of Companion Animal Nutrition and Feeding*. British Small Veterinary Association, Cheltenham, Gloucestershire, UK, pp. 253–257.

Kendall, P.T. (1984) The use of fat in dog and cat diets. In: Wiseman, J. (ed.), *Fats in Animal Nutrition*. Butterworth Press, Boston, 383 pp.

Kendall, P.T., Blaza, S.E. and Smith, P.M. (1983) Comparative digestible energy contents of adult Beagles and domestic cats for body weight maintenance. *Journal of Nutrition* 113, 1946–1955.

Kennedy, L.G. and Hershberger, T.V. (1974) Protein quality for the non-ruminant herbivore. *Journal of Animal Science* 39, 506–511.

Kienzle, E. and Englehard, R. (1998) What is going on in vegetarian carnivores? Results of a field study. *Proceedings of the 2nd Annual Conference of the European Society of Veterinary and Comparative Medicine*. Vienna, Austria, p. 12.

Kienzle, E. and Meyer, H. (1989) The effects of carbohydrate-free diets containing different levels of protein on reproduction in the bitch. In: Burger, I.H. and Rivers, J.W.P (eds), *Nutrition of the Dog and Cat*. Cambridge University Press, Cambridge, UK, pp. 243–258.

Kienzle, E. and Rainbird, A. (1991) Maintenance energy requirement of dogs – what is the correct value for calculation of metabolic body weight in dogs? *Journal of Nutrition* 121, S39–S40.

Kollmorgen, G.M., King, M.M., Kosanke, S.D. and Do,C. (1983) Influence of dietary fat and indomethacin on growth of transplantable mammary tumours in rats. *Cancer Research* 43, 4714–4719.

Koski, K.G., Hill, F.W. and Lönnerdal, B. (1990) Altered lactational performance in rats fed low carbohydrate diets and its effect on growth of neo-natal rat pups. *Journal of Nutrition* 120, 1028–1036.

Kronfield, D.S. (1973) Diet and performance of racing sled dogs. *Journal of the American Veterinary Medical Association* 162, 470–473.

Kronfield, D.S., Ferrante, P.L. and Grandjean, D. (1994) Optimal nutrition for athletic performance with emphasis on fat adaptation in dogs and horses. *Journal of Nutrition* 124, 2745–2753.

Lawton, M.P.C. (1988) Nutritional disease. In: Price, C.J. (ed.), *Manual of Parrots, Budgerigars and Other Psittacine Birds*. British Small Animal Veterinary Association, Cheltenham, UK, pp. 157–162.

Lebas, F. (1975a) Influence of the dietary energy content on the growth performance of the rabbit. *Annales de Zootechnie* 24, 281–288.

Lebas, F. (1975b) *The Meat Rabbit: Nutritional Requirements and Feeding Practices.* Itavi, Paris.

Lebas, F., Besancon, P. and Abouyoub, A. (1971) Mineral composition of rabbits milk: variation according to stage of lactation. *Annales de Zootechnie* 20, 487–495.

Legrand-Defretin, V. and Munday, H.S. (1995) Feeding dogs and cats for life. In: Burger, I.H. (ed.), *The Waltham Book of Companion Animal Nutrition.* Pergamon Press, Oxford, UK, pp. 57–68.

Lepine, A. (1998) Nutritional management of the large breed puppy. In: Rheinhart, G.A. and Carey, D.P. (eds), *Recent Advances in Canine and Feline Nutrition, Volume 2. Iams Nutrition Symposium Proceedings.* Orange Frazer Press, Ohio, pp. 53–62.

Lepine, A.J. and Rheinhart, G.A. (1998) The role of nutrition in skeletal development In: *Proceedings of the 2nd Annual Conference of the European Society of Veterinary and Comparative Nutrition.* Vienna, Austria, p. 21.

Long, J.L. (1984) The diets of three species of parrot in the south of Western Australia. *Australian Wildlife Research* 11, 357–372.

Loveridge, G.G. (1986) Body weight changes and energy intakes of cats during gestation and lactation. *Animal Technology* 37, 7–15.

Loveridge, G.G. and Rivers, J.P.W. (1989) Bodyweight changes and energy intakes of cats during pregnancy and lactation. In: Burger, I.H. and Rivers, J.P.W (eds), *Nutrition of the Dog and Cat.* Cambridge University Press, Cambridge, UK, pp. 113–132.

MacDonald, M.L., Rogers, Q.R. and Morris, J.G. (1983) Role of linoleate as an essential fatty acid for the cat independent of arachidonate synthesis. *Journal of Nutrition* 113, 1422–1433.

MacDonald, M.L., Anderson, B.C. Rogers, Q.R. and Buffington, C.A. (1984a) Essential fatty acid requirements of cats; pathology of essential fatty acid deficiency. *American Journal of Veterinary Research* 45, 1310–1316.

MacDonald, M.L., Rogers, Q.R. and Morris, J.G. (1984b) Nutrition of the domestic cat, a mammalian carnivore. *Annual Review of Nutrition* 4, 521–562.

MacDonald, M.L., Rogers, Q.R. and Morris, J.G. (1984c) Effects of linoleate and arachidonate deficiencies on reproduction and spermatogenesis of the cat. *Journal of Nutrition* 114, 719–726.

Manner, K. (1991) Energy requirements for maintenance of adult dogs. *Journal of Nutrition* 121, S37–S38.

Maynard, L.A. and Loosli, J.K. (1969) *Animal Nutrition.* McGraw-Hill, New York.

McBurney, M.I., Massimino, S.P., Field, C.J., Sunvold, G.D. and Hayek, M.G. (1998) Modulation of intestinal function and glucose homeostasis in dogs by ingestion of fermentable dietary fibres. In: Reinhart, G.A. and Carey, D.P. (eds), *Recent Advances in Canine and Feline Nutrition Vol 11. Proceedings of the Iams Nutrition Symposium.* Orange Frazer Press, Ohio, pp. 113–122.

McCartney, A. (1996) Ornamental fish nutrition and feeding. In: Kelly, N. and Wills, J. (eds), *Manual of Companion Animal Nutrition and Feeding.* British Small Animal Veterinary Association, Cheltenham, UK, pp. 244–252.

Menaker, L. and Navia, J.M. (1973) Appetite regulation in the rat under various physiological conditions: the role of dietary protein and calories. *Journal of Nutrition* 103, 347–352.

Meyer, H. and Zentec, J. (1991) Energy requirements of growing Great Danes. *Journal of Nutrition* 121, S35–S36.

Meyer, H., Arndt, J., Bechfield, T., Elbers, H. and Schienemann, H. (1989) Preceacal and post-ileal protein digestibility. In: Meyer, H. (ed.), *Contributions to Digestive Physiology in Dogs.* Verlag Paul Parey, Berlin, pp. 59–77.

Miller, S.A. and Allison, J.B. (1958) The dietary nitrogen requirements of the cat. *Journal of Nutrition* 64, 493–501.

Milner, J.A. (1979a) Assessment of the indispensable and dispensable amino acids for the immature dog. *Journal of Nutrition* 107, 1161–1167.

Milner, J.A. (1979b) Assessment of the essentiality of methionine, threonine, tryptophan, histidine and isoleucine in immature dogs. *Journal of Nutrition* 108, 1351–1357.

Mitchell, H.H. and Beadles, J.R. (1952) The determination of the protein requirement of the rat for maximum growth under conditions of restricted consumption of food. *Journal of Nutrition* 47, 133–145.

Morris, J.G., and Rogers, Q.R. (1994) Assessment of the nutritional adequacy of pet foods through the life cycle. *Journal of Nutrition* 124, 2520S–2534S.

Morris, J.C. and Rogers, Q.R. (1978a) Ammonia intoxication in the cat as a result of a dietary deficiency of arginine. *Science* 199, 431–432.

Morris, J.C. and Rogers, Q.R. (1978b) Arginine an essential amino-acid for the cat. *Journal of Nutrition* 108, 1944–1953.

Munday, H.S. and Earle, K.E. (1991) The energy requirements of the queen during lactation and kittens from birth to 12 weeks. *Journal of Nutrition* 121, S43–S44.

Mundt, H.C. (1991) Nutrition of old dogs. *Journal of Nutrition* 121, S541–S542.

Mutch, B.J.C and Bannister, E.W. (1983) Ammonia metabolism in exercise and fatigue, a review. *Medical Science of Sport and Exercise* 15, 9–84.

Naismith, D.J., Richardson, D.P. and Pritchard, A.E. (1983) The utilisation of protein and energy during lactation in the rat, with particular regard to the use of fat accumulated during pregnancy. *British Journal of Nutrition* 48, 433–441.

National Research Council (1974) *Nutrient Requirements of Dogs.* National Academy of Science, Washington, DC.

National Research Council (1977) *Nutrient Requirements of Rabbits.* National Academy of Science, Washington, DC.

National Research Council (1978a) *Nutrient Requirements of Laboratory Animals.* National Academy of Science, Washington, DC.

National Research Council (1978b) *Nutrient Requirements of Cats.* National Academy of Science, Washington, DC.

National Research Council (1984) *Nutrient Requirements of Poultry.* National Academy of Science, Washington, DC.

National Research Council (1985) *Nutrient Requirements of Dogs.* National Academy of Science, Washington, DC.

National Research Council (1986) *Nutrient Requirements of Cats.* National Academy of Science, Washington, DC.

National Research Council (1993a) *Nutrient Requirements of Cold-water Fish.* National Academy of Science, Washington, DC.

National Research Council (1993b) *Nutrient Requirements of Laboratory Animals.* National Academy of Science, Washington, DC.

Neirinck, K., Istasse, L., Gabriel, A., Van Eanaeme, C. and Bienfait, J.M. (1991) Amino acid composition and digestibility of four protein sources for dogs. *Journal of Nutrition* 121, S45–S46.

Nelson, M.M. and Evans, H.M. (1961) Dietary requirements for lactation in rats and other laboratory animals. In: Kon, S.K. and Cowie, T. (eds), *Milk, The Mammary Gland and its Secretion*, Vol 11. Academic Press, New York, pp. 137–194.

Nelson, M.M. and Evans, H.M. (1958) Sulfur amino acid requirement for lactation in the rat. *Proceedings of the Society of Experimental Biological Medicine* 99, 723–725.

Newburg, D.S. and Filios, L.C. (1979) A requirement for dietary asparagine in pregnant rats. *Journal of Nutrition* 109, 2190–2197.

Nott, H.M.R. (1992) The nutritional requirements of psittacines. *Waltham International Focus* 2, 2–7.

Nott, H.M.R. and Taylor, E.J. (1995) Nutrition of pet birds. In: Burger, I.H. (ed.), *The Waltham Book of Companion Animal Nutrition*. Pergamon Press, Oxford, UK, pp. 69–84.

Ontko, J.A., Wuthier, R.E. and Phillips, P.H. (1957) The effect of increased dietary fat upon the protein requirement of the growing dog. *Journal of Nutrition* 62, 163–169.

Orr, N.W.M. (1965) The food requirements of antarctic sledge dogs. In: Graham-Jones, O. (ed.), *Canine and Feline Nutritional Requirements*. Pergamon Press, Oxford, UK, pp. 101–112.

Pannevis, M.C. (1995) Nutrition of ornamental fish. In: Burger, I.H. (ed.), *The Waltham Book of Companion Animal Nutrition*. Pergamon Press, Oxford, UK, pp. 85–96.

Pannevis, M.C. and Earle, K.E. (1994a) Maintenance energy requirements of five popular species of ornamental fish. *Journal of Nutrition* 124, S2616–S2618.

Pannevis, M.C. and Earle, K.E. (1994b) Nutrition of ornamental fish: water soluble viatmin leaching and growth in *Paracheirodon innesi*. *Journal of Nutrition* 124, S2633–S2635.

Pascaul, J.J., Cervea,C., Blas, E. and Fernandez-Carmona, J. (1999) Effect of high fat diets on performance and milk composition of multiparous rabbit does. *Animal Science* 68, 151–162.

Peterson, M.E. and Graves, T.K. (1992) Diagnosisis and treatment of occult hyperparathyroidism in cats. In: Sokolowski, J.H. and Campfield, W.W. (eds), *Proceedings of the 15th Waltham /OSU Symposium for the Treatment of Small Animal Diseases: Endocrinology*. Kal Kan Foods Inc., Vernon, California, pp. 7–12.

Pet Food Manufacturers Association (1997) *PFMA Profile*. London.

Pitts, C. (1983) Hypovitaminosis A in psittacines. In: Kirk, R.W. (ed.), *Current Veterinary Therapy VIII*. W.B. Saunders Company, Philadelphia, pp. 622–625.

Rabin, B. (1983) Inhibition of experimentally induced autoimmunity in rats by biotin deficiency. *Journal of Nutrition* 113, 2316–2322.

Rabin, B., Nicolosi, R.J. and Hayes, K.C. (1976) Dietary influence on bile acid conjugation in the cat. *Journal of Nutrition* 106, 1241–1246.

Ralli, E.P. and Dumm, M.E. (1953) Relation of pantothenic acid to adrenal cortical function. *Vitamins and Hormones* 11, 133–154.

Reynolds, A.J. and Rheinhart, G.A. (1998) The role of fat in the formulation of performance rations: focus on fat sources. In: Rheinhart, G.A. and Carey, D.P. (eds), *Recent Advances in Canine and Feline Nutrition, Volume 2. Iams Nutrition Symposium Proceedings*. Orange Frazer Press, Ohio, pp. 277–282.

Reynolds, A.J., Sneddon,K., Rheinhart, G.A., Hinchcliff, K.W. and Swenson, R.A. (1998) Hydration strategies for exercising dogs. In: Rheinhart, G.A. and Carey, D.P. (eds), *Recent Advances in Canine and Feline Nutrition, Volume 2. Iams Nutrition Symposium Proceedings*. Orange Frazer Press, Ohio, pp. 249–248.

Richardson, L.R., Godwin, J., Wilkes, S. and Cannon, M. (1964) Reproductive performance of rats receiving various levels of dietary protein and fat. *Journal of Nutrition* 82, 257–262.

Rogers, Q.R. and Morris, J.G. (1982) Do cats really need more protein? *Journal of Small Animal Practise* 23, 521–532.

Rogers, Q.R., Morris, J.G. and Freedland, R.A. (1977) Lack of hepatic enzymic adaptaion to low and high levels of dietary protein in the adult cat. *Enzyme* 22, 348–350.

Rolls, B.J. and Rowe, E.A. (1982) Pregnancy and lactation in the obese rat: effects on maternal and pup weights. *Physiological Behaviour* 29, 393–400.

Root, M.V., Johnson, S.D. and Olson, P.N. (1996) Effect of prepubertal and post pubertal gonadectomy on heat production measured by indirect calorimetry in male and female domestic cats. *American Journal of Veterinary Research* 57, 371–374.

Rosmos, D.R., Belo, P.S., Bennick, M.R., Bergen, W.G., and Leveille, G.A (1976) Effects of dietary carbohydrate, fat and protein on growth, body composition, and blood metabolite levels in the dog. *Journal of Nutrition* 106, 1452–1456.

Rosmos, D.R., Palmer, H.J., Muiruri, K.L. and Bennick, M.R. (1981) Influence of a low carbohydrate diet on performance of pregnant and lactating dogs. *Journal of Nutrition* 111, 678–689.

Roudybush, T.E. and Grau, C.R. (1991) Cockatiel (*Nymphicus hollandicus*) nutrition. *Journal of Nutrition* 121, S206.

Scott, P.P., Greaves, J.P. and Scott, M.G. (1964) Nutritional blindness in the cat. *Experimental Eye Research* 3, 357–360.

Sheffey, B.E., Williams, A.J., Zimmer, J.F. and Ryan, G.D. (1985) Nutrition and metabolism of the geriatric dog. *Cornell Veterinarian* 75, 324–347.

Slade, L.M. and Hintz, H.F. (1969) Comparison of digestion in horses, ponies, rabbits and guinea pigs. *Journal of Animal Science* 28, 842–843.

Smith, D.C. and Prout, L.M. (1944) Development of thiamine deficiency in the cat on a diet of raw fish. *Proceedings of the Society of Experimental Biological Medicine* 56, 1–4.

Smith, H.W. (1965) Observations of the flora of the alimentary tract of animals and factors affecting its composition. *Journal of Pathology and Bacteriology* 89, 95–122.

Smith, P.R. (1989) Nutritional energetics. In: Halver, J.E. (ed.), *Fish Nutrition*, 2nd edn. Academic Press, New York, pp. 1–29.

Sturman, J.A., Rassin, D.K., Hayes, K.C. and Gaull, G.E. (1978) Taurine deficiency in the kitten: exchange and turnover of (35S) taurine in the brain, retina and other tissues. *Journal of Nutrition* 108, 1462–1476.

Taylor, S. (1954) Calcium as a goitrogen. *Journal of Clinical Endocrinology and Metabolism* 14, 1412–1422.

Taylor, S.A., Shrader, R.E., Koski, K.G. and Zeman, F.J. (1983) Maternal and embryonic response to a carbohydrate-free diets fed to rats. *Journal of Nutrition* 113, 253–267.

Thacker, E.J. (1956) The dietary fat level in the nutrition of the rabbit. *Journal of Nutrition* 58, 243–249.

Tobin, G. (1996) Small pets, food types, nutrient requirements and nutritional disorders. In: Kelly, N. and Wills, J. (eds), *Manual of Companion Animal Nutrition and Feeding*. British Small Animal Veterinary Association, Cheltenham, UK, pp. 208–225.

Trudell, J.I. and Morris, J.G. (1975) Carbohydrate digestion in the cat. *Journal of American Science* 41, 329.

Turek, J.J. and Hayek, M.G. (1998) Effect of omega-6:omega-3 fatty acid ratios on cytokine production in adult and geriatric dogs. In: Rheinhart, G.A. and Carey, D.P. (eds), *Recent Advances in Canine and Feline Nutrition, Volume 2. Iams Nutrition Symposium Proceedings*. Orange Frazer Press, Ohio, pp. 305–324.

Ullrey, D.E., Allen, M.E. and Baer, D.J. (1991) Formulated diets versus seed mixtures for psittacines. *Journal of Nutrition* 121, S193–S205.

Wannemacher, R.W. and McCoy, J.R. (1966) Determination of the optimal dietary protein requirement of young and old dogs. *Journal of Nutrition* 88, 66–74.

Weigel, J. and Alexander, J.W (1981) Ageing and the musculoskeletal system. *Veterinary Clinicians of North American Small Animal Practices* 11, 749–746.

Wills, J. (1996a) Adult maintenance. In: Kelly, N. and Wills, J. (eds), *Manual of Companion Animal Nutrition and Feeding*. British Small Animal Veterinary Association, Cheltenham, Gloucestershire, UK, pp. 44–46.

Wills, J. (1996b) Reproduction and lactation. In: Kelly, N. and Wills, J. (eds), *Manual of Companion Animal Nutrition and Feeding*. British Small Veterinary Association, Cheltenham, Gloucestershire, UK, pp. 47–51.

Wills, J.M. and Morris, J.G. (1996) Feeding puppies and kittens. In: Kelly, N. and Wills, J. (eds), *Manual of Companion Animal Nutrition and Feeding*. British Small Animal Veterinary Association, Cheltenham, UK, pp. 52–61.

Index